Frontiers in the Actin Cytoskeleton

Frontiers in the Actin Cytoskeleton

Special Issue Editor
Francisco Rivero

MDPI • Basel • Beijing • Wuhan • Barcelona • Belgrade

Special Issue Editor
Francisco Rivero
University of Hull
UK

Editorial Office
MDPI
St. Alban-Anlage 66
4052 Basel, Switzerland

This is a reprint of articles from the Special Issue published online in the open access journal *International Journal of Molecular Sciences* (ISSN 1422-0067) from 2019 to 2020 (available at: https://www.mdpi.com/journal/ijms/special_issues/actin-cytoskeleton).

For citation purposes, cite each article independently as indicated on the article page online and as indicated below:

LastName, A.A.; LastName, B.B.; LastName, C.C. Article Title. *Journal Name* **Year**, *Article Number*, Page Range.

ISBN 978-3-03936-565-4 (Hbk)
ISBN 978-3-03936-566-1 (PDF)

© 2020 by the authors. Articles in this book are Open Access and distributed under the Creative Commons Attribution (CC BY) license, which allows users to download, copy and build upon published articles, as long as the author and publisher are properly credited, which ensures maximum dissemination and a wider impact of our publications.

The book as a whole is distributed by MDPI under the terms and conditions of the Creative Commons license CC BY-NC-ND.

Contents

About the Special Issue Editor vii

Francisco Rivero
Editorial of Special Issue "Frontiers in the Actin Cytoskeleton"
Reprinted from: *Int. J. Mol. Sci.* **2020**, *21*, 3945, doi:10.3390/ijms21113945 1

Yi Liu, Keyvan Mollaeian, Muhammad Huzaifah Shamim and Juan Ren
Effect of F-actin and Microtubules on Cellular Mechanical Behavior Studied Using Atomic Force Microscope and an Image Recognition-Based Cytoskeleton Quantification Approach
Reprinted from: *Int. J. Mol. Sci.* **2020**, *21*, 392, doi:10.3390/ijms21020392 5

Kiyotaka Tokuraku, Masahiro Kuragano and Taro Q. P. Uyeda
Long-Range and Directional Allostery of Actin Filaments Plays Important Roles in Various Cellular Activities
Reprinted from: *Int. J. Mol. Sci.* **2020**, *21*, 3209, doi:10.3390/ijms21093209 19

Itallia Pacentine, Paroma Chatterjee and Peter G. Barr-Gillespie
Stereocilia Rootlets: Actin-Based Structures That Are Essential for Structural Stability of the Hair Bundle
Reprinted from: *Int. J. Mol. Sci.* **2020**, *21*, 324, doi:10.3390/ijms21010324 35

L. Shannon Holliday, Lorraine Perciliano de Faria and Wellington J. Rody Jr.
Actin and Actin-Associated Proteins in Extracellular Vesicles Shed by Osteoclasts
Reprinted from: *Int. J. Mol. Sci.* **2020**, *21*, 158, doi:10.3390/ijms21010158 48

Tohnyui Ndinyanka Fabrice, Thomas Fiedler, Vera Studer, Adrien Vinet, Francesco Brogna, Alexander Schmidt and Jean Pieters
Interactome and F-Actin Interaction Analysis of *Dictyostelium discoideum* Coronin A
Reprinted from: *Int. J. Mol. Sci.* **2020**, *21*, 1469, doi:10.3390/ijms21041469 67

David R. J. Riley, Jawad S. Khalil, Jean Pieters, Khalid M. Naseem and Francisco Rivero
Coronin 1 Is Required for Integrin β2 Translocation in Platelets
Reprinted from: *Int. J. Mol. Sci.* **2020**, *21*, 356, doi:10.3390/ijms21010356 87

Vikash Singh, Anthony C. Davidson, Peter J. Hume and Vassilis Koronakis
Arf6 Can Trigger Wave Regulatory Complex-Dependent Actin Assembly Independent of Arno
Reprinted from: *Int. J. Mol. Sci.* **2020**, *21*, 2457, doi:10.3390/ijms21072457 107

Eva Kollárová, Anežka Baquero Forero, Lenka Stillerová, Sylva Přerostová and Fatima Cvrčková
Arabidopsis Class II Formins AtFH13 and AtFH14 Can Form Heterodimers but Exhibit Distinct Patterns of Cellular Localization
Reprinted from: *Int. J. Mol. Sci.* **2020**, *21*, 348, doi:10.3390/ijms21010348 120

Vedud Purde, Florian Busch, Elena Kudryashova, Vicki H. Wysocki and Dmitri S. Kudryashov
Oligomerization Affects the Ability of Human Cyclase-Associated Proteins 1 and 2 to Promote Actin Severing by Cofilins
Reprinted from: *Int. J. Mol. Sci.* **2019**, *20*, 5647, doi:10.3390/ijms20225647 136

Almudena García-Ortiz and Juan Manuel Serrador
ERM Proteins at the Crossroad of Leukocyte Polarization, Migration and Intercellular Adhesion
Reprinted from: *Int. J. Mol. Sci.* **2020**, *21*, 1502, doi:10.3390/ijms21041502 158

Dureen Samandar Eweis and Julie Plastino
Roles of Actin in the Morphogenesis of the Early *Caenorhabditis elegans* Embryo
Reprinted from: *Int. J. Mol. Sci.* **2020**, *21*, 3652, doi:10.3390/ijms21103652 179

Francine Parker, Thomas G. Baboolal and Michelle Peckham
Actin Mutations and Their Role in Disease
Reprinted from: *Int. J. Mol. Sci.* **2020**, *21*, 3371, doi:10.3390/ijms21093371 189

Silvia Pelucchi, Ramona Stringhi and Elena Marcello
Dendritic Spines in Alzheimer's Disease: How the Actin Cytoskeleton Contributes to
Synaptic Failure
Reprinted from: *Int. J. Mol. Sci.* **2020**, *21*, 908, doi:10.3390/ijms21030908 205

**Ximena Báez-Matus, Cindel Figueroa-Cares, Arlek M. Gónzalez-Jamett,
Hugo Almarza-Salazar, Christian Arriagada, María Constanza Maldifassi,
María José Guerra, Vincent Mouly, Anne Bigot, Pablo Caviedes and Ana M. Cárdenas**
Defects in G-Actin Incorporation into Filaments in Myoblasts Derived from Dysferlinopathy
Patients Are Restored by Dysferlin C2 Domains
Reprinted from: *Int. J. Mol. Sci.* **2020**, *21*, 37, doi:10.3390/ijms21010037 228

Liam Caven and Rey A. Carabeo
Pathogenic Puppetry: Manipulation of the Host Actin Cytoskeleton by *Chlamydia trachomatis*
Reprinted from: *Int. J. Mol. Sci.* **2020**, *21*, 90, doi:10.3390/ijms21010090 244

About the Special Issue Editor

Francisco Rivero, a scientist at the Hull York Medical School, University of Hull, has focused on research on the roles of actin-binding proteins in a variety of model systems since 1994. He obtained a Ph.D. in Medicine in 1994 from the Complutense University of Madrid. He has occupied research posts at the Max Planck Institute for Biochemistry near Munich and at the Institute for Biochemistry of the University of Cologne. Currently, he is Section Editor-in-Chief of *Cells*, a member of the editorial board of *Scientific Reports* and *BMC Molecular and Cell Biology*, and a regular peer reviewer for *International Journal of Molecular Sciences*. He is author of more than 80 scientific publications in international journals and has several edited books; his H-index is 37. He is a member of the British Society for Cell Biology and the Platelet Society.

Editorial

Editorial of Special Issue "Frontiers in the Actin Cytoskeleton"

Francisco Rivero

Centre for Cardiovascular and Metabolic Disease, Hull York Medical School, Faculty of Health Sciences, University of Hull, Hull HU6 7RX, UK; Francisco.rivero@hyms.ac.uk; Tel.: +44-01482-466433

Received: 28 May 2020; Accepted: 28 May 2020; Published: 30 May 2020

The actin cytoskeleton is of fundamental importance for eukaryotic cell homeostasis. It contributes to developing and maintaining cell shape and tissue integrity and is crucial for cell migration, movement of organelles, vesicle trafficking, and the completion of cell division. Impressive advances have been made in recent years towards understanding the intricacies of the microfilament system's organization and function. This Special Issue of *IJMS* covers a broad range of cutting-edge aspects related to the actin cytoskeleton.

The mechanical properties of the cell are intimately linked to the highly dynamic and, at the same time, highly crosslinked cytoskeletal structures that occupy the cytoplasm. Liu et al. [1] use atomic force microscopy indentation coupled to image recognition-based cytoskeleton quantification to quantify the effect of F-actin and microtubule morphology, achieved by various levels of depolymerization, on cellular mechanical properties, and conclude that living cells are able to sense and adapt to the polymerization state of their cytoskeleton components. Dozens of actin-binding proteins (ABPs) orchestrate the dynamic remodeling of the actin cytoskeleton and integrate it with the cell signaling machinery. ABPs have unique localizations within the crowded cytoplasm, in which they diffuse within seconds. How does a specific ABP find where to bind when the cytoplasm offers endless possibilities? Tokuraku et al. [2] discuss the concept of allosteric regulation of ABP localization that arises from cooperative conformational changes propagating along actin filaments upon binding of ABPs like myosin, tropomyosin, cofilin, and others. This phenomenon is proposed to play an important role in the formation and regulation of actin structures like stress fibers, lamellipodia, and filopodia. It probably also applies to more elaborate structures like the inner ear hair cell stereocilia. Hair cells are the specialized neuroepithelial cells responsible for detecting sound and head movements. Each of these cells carries an apical bundle of stereocilia that upon deflection activate ion channels, causing depolarization, neurotransmitter release, and excitation of auditory or vestibular nerves. Pacentine et al. [3] review the current knowledge on the morphology, composition, and role of the rootlet, a specialized structure that anchors the actin core of the stereocilium to the cell body.

Actin and ABPs play roles not only in structures built in the cytoplasm but also in extracellular vesicles released by the cells and used for intercellular communication. Holliday et al. [4] review the actin and actin-associated proteome of extracellular vesicles released by osteoclasts. They report the presence of members of several of the major classes of ABPs, suggesting that they are important for the formation of extracellular vesicles and for their regulatory function on osteoblasts. The power of proteomics is also showcased in the study of Fabrice et al. [5]. Here, the authors describe the interactome of the *Dictyostelium discoideum* amoeba coronin A. Coronins are evolutionary conserved cytoskeleton remodeling proteins, but actin-independent roles, particularly in signaling, are emerging. The study found coronin A in complex with a number of ABPs, but also several uncharacterized proteins, metabolic enzymes, and a transcription factor that will provide fodder for future studies. Co-sedimentation experiments in this study also put into question the relevance of the interaction of coronin A with actin, suggesting that the phenotypes observed in coronin A-deficient amoebae are mainly the result of altered signaling. In mammalian cells, one of the interaction partners of coronin

1 is the cytoplasmic tail of integrin β2, a component of lymphocyte-associated antigen 1, the fourth most abundant integrin in platelets. Coronins 1, 2, and 3 are abundant in platelets, but their roles are poorly understood. Riley et al. [6] describe the characterization of mouse platelets deficient in coronin 1 and show that the protein is dispensable for most cellular processes, most likely due to functional overlap among coronins, but is required for translocation of integrin β2 to the platelet surface upon stimulation with thrombin.

Nucleation-promoting factors activate the Arp2/3 complex to trigger actin nucleation. Several of those factors have been extensively studied over the last few decades, including the Wave complex. This complex is itself activated by interaction with activated small GTPases like Rac1. Singh et al. [7] investigate Arf6, a member of the ADP ribosylation factor family involved in a wide array of cellular functions. Some members of the Arf family, like Arf5 and Arl1, cooperate with Rac1 to recruit the Wave complex. In their study, Singh et al. now show that another Arf family member, Arf6, is not only capable of activating the Wave complex indirectly by recruiting the exchange factor ARNO, but also can trigger actin assembly directly in coordination with Rac1. Formins constitute another family of evolutionarily conserved cytoskeleton nucleators. Their hallmark is the FH2 domain that promotes the nucleation and elongation of linear actin filaments but can also associate to microtubules. Formins usually make for large families and in *Arabidopsis thaliana* the family has 21 members. While formins are known to form homodimers through their FH2 domain, heterodimerization has been seldom reported. Kollárová et al. [8] investigate two previously uncharacterized plant formins, AtFH13 and AtFH14, and although they show distinct and only partially overlapping patterns of subcellular localization, they are capable of heterodimerizing, a finding not reported previously in plant formins. As important as nucleation in the process of actin remodeling are mechanisms like severing, depolymerization, and regeneration of actin monomers, to which cyclase associated proteins (CAPs) contribute in complex ways. Two isoforms of CAP exist in mammalian cells, but while CAP1 has been extensively studied biochemically, CAP2 has never been. Purde et al. [9] show in their study that the N-terminal domain of both isoforms enhances cofilin-mediated severing and depolymerization of actin filaments. By studying the association status of CAPs, the authors noted that these activities are directly proportional to the degree of oligomerization, with monomers being less effective than tetramers.

About one hundred ABPs contribute to organize the cytoskeleton at the cell cortex, a dense meshwork associated with the plasma membrane. This cortical network is important for the generation of tension needed to maintain cell shape and polarity and to make cell motility possible. Ezrin, radixin, and moesin proteins are among the ABPs that regulate the organization of cortical actin filaments. García-Ortiz and Serrador [10] review the main biochemical mechanisms involved in the regulation of members of this family and their contribution to leukocyte biology, with a focus on the phagocytic cup and the immune synapse. A particular example of the roles of the cortical actin cytoskeleton is the first division of the *Caenorhabditis elegans* embryo, a model of asymmetric cell division that integrates microfilaments, microtubules, and complex signal cues. Samandar Eweis and Plastino [11] review recent research on the roles of the actin cytoskeleton in this crucial stage of the morphogenesis of the worm embryo, with a focus on the processes of symmetry breaking, cortical flows that help establish polarity, and contractile ring formation and positioning.

Having fundamental roles in a plethora of cellular processes, it comes to no surprise that defects in actin and associated proteins have been found to be associated with various diseases. Humans express six actin genes, some of them in a tissue-specific manner, giving rise to highly similar proteins. Disease-causing mutations have been reported for each of the six genes. The most common mutations result in conditions like nemaline myopathy, aortic aneurysms, and cardiomyopathy. Parker et al. [12] review the mutations reported in the human actin genes, their potential consequences for actin function, and the challenges that actins pose for experimental studies. Actin is the major cytoskeletal component of dendritic spines, small protrusions along dendrites, which in the mammalian brain harbor the postsynaptic compartment of glutamatergic excitatory synapses. The actin cytoskeleton contributes decisively to maintaining the dendritic spine architecture and modulating its remodeling. Synaptic dysfunction driven by amyloid β is characteristic of the neurodegenerative disorder Alzheimer's disease. Pelucchi et al. [13] review the role of the actin cytoskeleton in the spine shaping, the participation of actin and actin remodeling proteins in the endocytosis mechanisms implicated in amyloid generation and receptor trafficking, and the evidence supporting the implication of the actin cytoskeleton in synaptic failure.

In the skeletal muscle cell, the transmembrane protein dysferlin facilitates calcium-dependent aggregation and fusion of vesicles during repair of the plasma membrane, at which point it interacts with proteins involved in actin remodeling. Mutations in dysferlin cause a group of muscular dystrophies called dysferlinopathies. Báez-Matus et al. [14] investigate the potential effects of alterations in dysferlin expression on actin dynamics, more specifically G-actin incorporation to filaments. They use immortalized myoblast cell lines derived from dysferlinopathy patients or normal myoblasts in which the dysferlin gene has been silenced and conclude that dysferlin is important for the regulation of actin remodeling.

Many bacterial pathogens have developed the ability to manipulate the actin remodeling machinery to facilitate their own uptake by the host cell and subsequent proliferation and invasion of other cells within the organism. A particular example is the obligate intracellular bacterium *Chlamydia trachomatis*. This organism uses aspects of actin remodeling to induce its own uptake by the host epithelial cell, to create a replicative niche, and, in some cases, to promote its egress from the infected cell, as discussed by Caven and Carabeo [15] in their review.

Overall, the 15 contributions that make up this Special Issue highlight the fundamental roles of the actin cytoskeleton in cellular processes relevant to health and disease. The combination of molecular genetics, biophysics, and advanced imaging techniques in a variety of cell types and model organisms will ensure that exciting discoveries will continue to be made in this field in years to come.

Funding: This research received no external funding.

Conflicts of Interest: The author declares no conflict of interest.

References

1. Liu, Y.; Mollaeian, K.; Shamim, M.; Ren, J. Effect of F-actin and microtubules on cellular mechanical behavior studied using atomic force microscope and an image recognition-based cytoskeleton quantification approach. *Int. J. Mol. Sci.* **2020**, *21*, 392. [CrossRef] [PubMed]
2. Tokuraku, K.; Kuragano, M.; Uyeda, T. Long-range and directional allostery of actin filaments plays important roles in various cellular activities. *Int. J. Mol. Sci.* **2020**, *21*, 3209. [CrossRef] [PubMed]
3. Pacentine, I.; Chatterjee, P.; Barr-Gillespie, P. Stereocilia rootlets: Actin-based structures that are essential for structural stability of the hair bundle. *Int. J. Mol. Sci.* **2020**, *21*, 324. [CrossRef] [PubMed]
4. Holliday, L.; Faria, L.; Rody, W. Actin and actin-associated proteins in extracellular vesicles shed by osteoclasts. *Int. J. Mol. Sci.* **2020**, *21*, 158. [CrossRef] [PubMed]
5. Fabrice, T.; Fiedler, T.; Studer, V.; Vinet, A.; Brogna, F.; Schmidt, A.; Pieters, J. Interactome and F-actin interaction analysis of *Dictyostelium discoideum* coronin A. *Int. J. Mol. Sci.* **2020**, *21*, 1469. [CrossRef] [PubMed]
6. Riley, D.; Khalil, J.; Pieters, J.; Naseem, K.; Rivero, F. Coronin 1 is required for integrin β2 translocation in platelets. *Int. J. Mol. Sci.* **2020**, *21*, 356. [CrossRef] [PubMed]
7. Singh, V.; Davidson, A.; Hume, P.; Koronakis, V. Arf6 can trigger wave regulatory complex-dependent actin assembly independent of Arno. *Int. J. Mol. Sci.* **2020**, *21*, 2457. [CrossRef] [PubMed]
8. Kollárová, E.; Baquero Forero, A.; Stillerová, L.; Přerostová, S.; Cvrčková, F. *Arabidopsis* class II formins AtFH13 and AtFH14 can form heterodimers but exhibit distinct patterns of cellular localization. *Int. J. Mol. Sci.* **2020**, *21*, 348. [CrossRef] [PubMed]
9. Purde, V.; Busch, F.; Kudryashova, E.; Wysocki, V.; Kudryashov, D. Oligomerization affects the ability of human cyclase-associated proteins 1 and 2 to promote actin severing by cofilins. *Int. J. Mol. Sci.* **2019**, *20*, 5647. [CrossRef] [PubMed]
10. García-Ortiz, A.; Serrador, J. ERM proteins at the crossroad of leukocyte polarization, migration and intercellular adhesion. *Int. J. Mol. Sci.* **2020**, *21*, 1502. [CrossRef] [PubMed]

11. Samandar Eweis, D.; Plastino, J. Roles of actin in the morphogenesis of the early *Caenorhabditis elegans* embryo. *Int. J. Mol. Sci.* **2020**, *21*, 3652. [CrossRef] [PubMed]
12. Parker, F.; Baboolal, T.; Peckham, M. Actin mutations and their role in disease. *Int. J. Mol. Sci.* **2020**, *21*, 3371. [CrossRef] [PubMed]
13. Pelucchi, S.; Stringhi, R.; Marcello, E. Dendritic spines in Alzheimer's disease: How the actin cytoskeleton contributes to synaptic failure. *Int. J. Mol. Sci.* **2020**, *21*, 908. [CrossRef] [PubMed]
14. Báez-Matus, X.; Figueroa-Cares, C.; Gónzalez-Jamett, A.; Almarza-Salazar, H.; Arriagada, C.; Maldifassi, M.; Guerra, M.; Mouly, V.; Bigot, A.; Caviedes, P.; et al. Defects in G-actin incorporation into filaments in myoblasts derived from dysferlinopathy patients are restored by dysferlin C2 domains. *Int. J. Mol. Sci.* **2020**, *21*, 37. [CrossRef] [PubMed]
15. Caven, L.; Carabeo, R. Pathogenic puppetry: Manipulation of the host actin cytoskeleton by *Chlamydia trachomatis*. *Int. J. Mol. Sci.* **2020**, *21*, 90. [CrossRef] [PubMed]

© 2020 by the author. Licensee MDPI, Basel, Switzerland. This article is an open access article distributed under the terms and conditions of the Creative Commons Attribution (CC BY) license (http://creativecommons.org/licenses/by/4.0/).

Article

Effect of F-actin and Microtubules on Cellular Mechanical Behavior Studied Using Atomic Force Microscope and an Image Recognition-Based Cytoskeleton Quantification Approach

Yi Liu [1], Keyvan Mollaeian [1], Muhammad Huzaifah Shamim [2] and Juan Ren [1,*]

[1] Department of Mechanical Engineering, Iowa State University, Ames, IA 50011, USA; yil1@iastate.edu (Y.L.); keyvanm@iastate.edu (K.M.)
[2] Department of Electrical and Computer Engineering, Rice University, Houston, TX 77005, USA; mhs.huzaifah@gmail.com
* Correspondence: juanren@iastate.edu; Tel.: +1-515-294-1805

Received: 29 November 2019; Accepted: 3 January 2020; Published: 8 January 2020

Abstract: Cytoskeleton morphology plays a key role in regulating cell mechanics. Particularly, cellular mechanical properties are directly regulated by the highly cross-linked and dynamic cytoskeletal structure of F-actin and microtubules presented in the cytoplasm. Although great efforts have been devoted to investigating the qualitative relation between the cellular cytoskeleton state and cell mechanical properties, comprehensive quantification results of how the states of F-actin and microtubules affect mechanical behavior are still lacking. In this study, the effect of both F-actin and microtubules morphology on cellular mechanical properties was quantified using atomic force microscope indentation experiments together with the proposed image recognition-based cytoskeleton quantification approach. Young's modulus and diffusion coefficient of NIH/3T3 cells with different cytoskeleton states were quantified at different length scales. It was found that the living NIH/3T3 cells sense and adapt to the F-actin and microtubules states: both the cellular elasticity and poroelasticity are closely correlated to the depolymerization degree of F-actin and microtubules at all measured indentation depths. Moreover, the significance of the quantitative effects of F-actin and microtubules in affecting cellular mechanical behavior is depth-dependent.

Keywords: cell mechanics; F-actin; microtubules; image recognition-based cytoskeleton quantification; AFM

1. Introduction

Cellular cytoskeleton, composed of F-actin (actin filaments), microtubules and intermediate filaments, is a highly cross-linked and dynamic network present in all cells cytoplasm [1–3]. Studies have shown that cytoskeletal morphology directly controls the cellular mechanical behavior [1,4]. As one of the major components of the cytoskeleton, F-actin performs its primary function on cell cycling control, amoeba movement, cell shape change, cell contractility and mechanical stability [5,6]. Microtubules provide a platform for cellular cargo transportation including macromolecular assembly, organelles and secretory movement [7,8]. It has been widely demonstrated that both F-actin and microtubules can reorganize their network structures to control the cellular mechanical properties through the assembly and disassembly when the extracellular environment changes [9–12]. Therefore, quantitative results on how the F-actin and microtubules affect the cellular mechanical properties may provide in-depth understandings of the cellular adaptive response to external stimuli, and intracellular transduction mechanisms. Although great efforts have been devoted to investigating the quantitative relation between the cellular cytoskeleton network and the cell mechanical properties, comprehensive quantification results involving cytoskeleton morphology and mechanical parameters are still lacking.

Tseng et al. (2005) added α-actin to living cells and showed that the stiffness of cells with more α-actin was significantly larger than that of the original cells [13]. Brangwynne et al. (2006) used fluorescent images together with macroscopic rods to investigate the effect of microtubules, it was found that the buckling wavelength of microtubules reduced dramatically to increase the sustainable compressive forces of microtubules in cells [14]. By using the microfluidic device, Schaedel et al. (2015) demonstrated that microtubules had self-healing properties and their ductile structure enables the cell adaptation to external mechanical stresses [15]. These aforementioned studies indicate that there indeed exist correlations between the morphology of either F-actin or the microtubules and cellular mechanism. However, they did not compare the effects of F-actin and microtubules in affecting cellular mechanical behaviors [13–15].

By using the atomic force microscope (AFM), Rotsch et al. (2000) investigated the correlation between the cell elasticity and fluorescence images of cells treated with multiple drugs for disrupting or stabilizing the cytoskeleton structure [16]. Haga et al. (2000) used force mapping mode of AFM to measure the cellular elasticity, and then analyzed the correlation between the distribution of cellular cytoskeleton and elastic moduli [17]. S.kasas et al. (2005) investigated the superficial and deep changes of cellular mechanical properties due to the cytoskeleton disassembly using AFM and finite element simulation [18]. CAMSAP3-ACF7, which is able to keep the length and orientation of F-actin and microtubules, was used by Ning et al. (2016) to study the impact of the morphology of cellular cytoskeleton on regulating the cellular adhesion and cell migration [19]. The researches mentioned above were proposed for showing the relation between the cytoskeleton morphology and cell mechanical behavior. However, these studies only either focused on cellular elasticity [16,17], or selected one indentation depth with a fixed treatment concentration in AFM experiments, therefore could not provide quantitative details of cytoskeleton impact on the cellular mechanics at different length scales [18,19]. Therefore, the cellular poroelasticity quantification is missing and the length scale of the effects of F-actin and microtubules has not been reported as well.

Therefore, in this study, we report the quantitative investigation on the effects of F-actin and microtubules in affecting both the elasticity and poroelasticity at different indentation depths. The contribution of this study is two-fold: (1) In order to quantify the cytoskeleton morphology, an image recognition-based cytoskeleton quantification (IRCQ) approach was developed which quantifies both the F-actin and microtubules morphologies using their fluorescent intensity, respectively; (2) the quantitative effects of F-actin and microtubules in affecting the cellular elasticity and poroelasticity were investigated. Specifically, AFM indentation experiments were performed to quantify both the cellular Young's modulus and diffusion coefficient at different depths for the cells treated with F-actin inhibitor (latrunculin B) and microtubule inhibitor (nocodazole), respectively. The cytoskeleton treatments were designed that the F-actin and microtubules were inhibited at similar degrees, and the treatment results were verified using the proposed IRCQ approach. Then the cellular mechanical behavior was measured for each treatment using the AFM indentation data and the effects of F-actin and microtubules were compared and analyzed.

2. Materials and Methods

2.1. Cell Preparation

2.1.1. Cell Culture and Treatment

Primary mouse embryonic fibroblast cells (NIH/3T3) were seeded in six-well plates (ThermoFisher Scientific, Waltham, MA, USA) and 35 mm tissue culture dishes (Azzota Scientific, DE, USA) for fluorescent intensity quantification and AFM indentation experiments, respectively, using Dulbecco's Modified Eagle's Medium (ATCC, Rockville, MD, USA), together with 10% (V/V) Calf Bovine Serum (Sigma, St. Louis, MO, USA) and 1% (V/V) penicillin-streptomycin (Gibco, Grand Island,

NY, USA). The cell culture vessels were maintained in the incubator at the temperature of 37° and humidified atmosphere of 5% CO_2. The cultured cells were ready after 24 h.

To investigate the different cytoskeletal states of F-actin and microtubules, the cells were treated with latrunculin B (George Town, Cayman Islands) and nocodazole (Belgium, USA), respectively. Living 3T3 cells were divided into two groups for the actin and microtubule treatments, respectively. The cellular F-actin were inhibited using latrunculin B at the final concentration of 0 nM (control), 10 nM, 30 nM, 40 nM, 60 nM, 75 nM, and 100 nM in the aforementioned cell culture medium. The cellular microtubules were treated with nocodazole at the final concentration of 0 nM (control), 10 nM, 30 nM, 50 nM, 75 nM, 100 nM, and 200 nM in cell culture medium. The cells were treated for 30 min in the incubator before the AFM measurements.

2.1.2. Immunofluorescence

To observe the cytoskeletal morphology, F-actin and microtubules were stained using immunofluorescence. 4% paraformaldehyde (Alfa Aesar, Ward Hill, MA, USA) diluted in PBS was used to fix the NIH/3T3 cells in the incubator for 10 min. 0.1% Triton-X (Fisher Scientific, Fair Lawn, NJ, USA) was then applied for permeabilization of the cell membrane at room temperature for 10 min.

(i) **F-actin.** To observe the F-actin, the untreated fixed cells were stained using 100 nM working stock of Actin-stain^{TM} 555 phalloidin (Cytoskeleton Inc, Denver, CO, USA), which could bind to and visualize F-actin [20], and incubated at room temperature in dark for 30 min.

(ii) **Microtubules.** The observe the microtubules, the untreated fixed cells were blocked with 5% BSA (Fisher Scientific, Fair Lawn, NJ, USA) and kept in the refrigerator for 12 h. The cells were then incubated using Alpha-Tubulin (Acetylated) Recombinant Mouse Monoclonal Antibody (Fisher Scientific, Fair Lawn, NJ, USA) at 1 µg/mL in 1% BSA at room temperature for 3 h. To label the microtubules, Alexa Fluor 488 Rabbit Anti-Mouse IgG Secondary Antibody (Fisher Scientific, Fair Lawn, NJ, USA) at dilution of 1:400 in PBS was used for 30 min at room temperature.

During the staining process, the cells were rinsed three times with PBS after each step.

2.2. Fluorescence Microscope

An AxioObserve Z1 inverted optical microscope equipped with a sola light engine (Lumencor, Beaverton, OR, USA) was used to obtain the fluorescent images of F-actin and microtubules. The microscope was controlled by a Zeiss 780 confocal microscope system (Zeiss, Oberkochen, Germany). The fluorescent images were taken in 10 s using the same light strength and exposure time for preventing the light bleaching effect and obtaining the images under the same imaging conditions.

2.3. F-actin and Microtubules Quantification

2.3.1. Image Pre-Processing

To process the fluorescent images of the untreated and treated cells, the original RGB images were converted to grayscale with the brightness range from 0~255 for each pixel [21]. To minimize the background color effect, the pixel brightness lower than the image average brightness was mandatorily set as zero. To quantify the morphologies (i.e., quantity) of F-actin and microtubules, an image recognition-based cytoskeleton quantification (IRCQ) approach was proposed and applied in the image processing.

2.3.2. Image Recognition-Based Cytoskeleton Quantification Approach

In the previous study, an image recognition-based F-actin quantification (IRAQ) approach was proposed to quantify both the F-actin orientation and intensity simultaneously [22]. In IRAQ, Canny and Sobel edge detectors, as well as the Matlab filling tools were utilized in filament skeletonization and cell area detection. However, compared to F-actin, determined by the structure, the microtubules show dense labeled fluorescent spots rather than clear fibrous cross-network in the fluorescence images (see Figure 1). Therefore, quantifying the orientation deviation of microtubules is meaningless. Moreover, the image skeletonization processing in IRAQ is not feasible for microtubules intensity quantification. Overall, the brightness intensity quantification algorithm designed in IRAQ is not suitable for microtubules due to the significant structural difference between F-actin and microtubules. Therefore, an image recognition-based cytoskeleton quantification (IRCQ) for quantifying the intensity of both the actin-cytoskeleton and microtubules was proposed. IRCQ uses the breadth-first search (BFS) instead of edge detector and filling tools to quantify the brightness intensity of F-actin and microtubules.

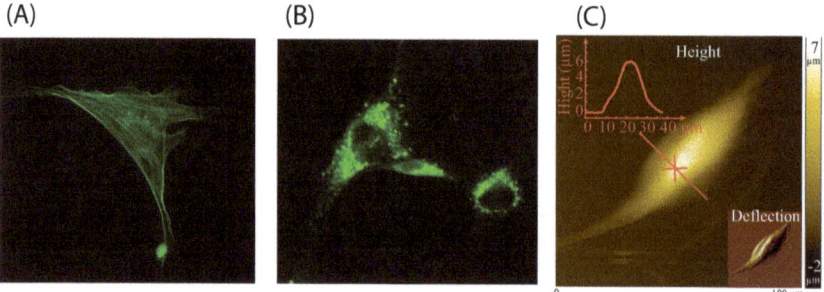

Figure 1. The fluorescent images of (**A**) F-actin and (**B**) microtubules in control NIH/3T3 cells, respectively. (**C**) AFM topography image of a NIH/3T3 cell, where the red cross denotes the poroelasticity measurement.

Breadth-first search (BFS) is a common searching algorithm for large unknown graph data structures [23]. BFS starts from a root node of the searching tree and explores all of the neighbor nodes incident to the source node. It keeps moving toward the next-depth neighbor nodes until all nodes in the graph have been visited exactly once. BFS uses the opposite strategy compare to the depth-first search, which explores as far as possible along one branch before backtracking and expands the next branch [24].

In IRCQ, BFS algorithm is used to quantify the intensity of cellular cytoskeleton by calculating the total pixel brightness over the detected cell area. The designed algorithm starts from the first pixel at the upper-left corner of the grayscale image (see Figure 2). If the pixel brightness is larger than or equal to the prechosen threshold ϵ, the cell area counter is increased by one (remains the same otherwise) and the corresponding brightness is added to the cell brightness counter, and the brightness of the surrounding pixels are then checked recursively. Each checked pixel is marked as "Read" to prevent redetection. This process continues until no surrounding pixels are brighter than the threshold (i.e., the boundary of a cell is detected). Next, the algorithm moves toward the next "Unread" pixel in the image pixel matrix and recreates new cell area and brightness counters, respectively, and the above "checking" process is repeated. Finally, we could obtain at least one detected cell area. To eliminate the unwanted staining spots on the image, the original cell images can be cropped into several ones such that each new image only contains one single cell. Then only the largest cell area detected in each new image. is chosen as the quantification target. The BFS algorithm in IRCQ is shown as Algorithm 1. The average F-actin intensity (AAI) and the average microtubule intensity (AMI) are both quantified as,

$$I = \frac{B}{C \times r}, \qquad (1)$$

where I is the average intensity, C and B are the maximum cell area count and the corresponding brightness, respectively. r is pixel area of the obtained fluorescent images. The relative intensity percentage change, Δ, is quantified as,

$$\Delta = \frac{I_0 - I_i}{I_0 - I_m} \times 100\%, \tag{2}$$

where I_0 and I_m are the average intensity of untreated and fully treated (i.e., the morphology does not change if the treatment strength is further increased.) cells, respectively. I_i is the average intensity of cells treated with certain treatment concentration i.

Algorithm 1: BFS algorithms in IRCQ

    Data: A fluorescent single image with pixel matrix
    Result: Area and brightness of a detected cell
1   ϵ = threshold;
2   $n = 0$;
3   Start from the first upper-left pixel P of the input image;
4   **while** *there exist "Unread" pixels* **do**
5   Create a cell area counter $C_n = 0$;
6   Create a cell brightness counter $B_n = 0$;
7   Mark P as "Read";
8   Initialize queue Q with P;
9   **while** *Q is not empty* **do**
10   Poll front of $Q \to q$;
11   **for** *all neighbors p of q* **do**
12   **if** *p is "Unread" & its brightness $B_p \geq \epsilon$* **then**
13                 $C_n + 1 \to C_n$;
14                 $B_n + B_p \to B_n$;
15   Add p to the end of Q;
16   Mark p as "Read";
17   **end**
18   **end**
19   **end**
20   The next "Unread" pixel $\to P$;
21      $n + 1 \to n$;
22   **end**
23   **return** *Cell area count* $C = max\{C_k\}, (k = 1, 2, ..., n)$ *& the corresponding cell brightness* B.

2.4. AFM Measurement

The AFM indentation experiments were performed in the aforementioned cell treatment medium at room temperature using Bruker BioScope Resolve AFM system (Santa Barbara, CA, USA) integrated with an inverted optical microscope (Olympus, IX73, Tokyo, Japan). Glass bead/sphere AFM probe (Novascan, IA, USA) with the radius of 2.5 µm was used, and its cantilever spring constant of 0.03 N/m was acquired using the thermal tune approach. To minimize the nucleus effect, the cells were indented at the location away from the top during the experiments. To minimize the limited cell thickness and substrate effects, the target indentations were selected as 650, 1000, 1300 nm, which were less than a quarter of the cell height at 7 ± 1 µm [3,25]. The reason of performing AFM measurement at different desired indentations is to study the length scale of the effect of cytoskeleton morphology on cellular mechanical behavior [26,27]. To quantify the cell elasticity and poroelasticity, the AFM indentation procedure reported in [4] was applied. Specifically, cells were indented at the speed of 20 µm/s until the desired indentations were reached (indenting process), and the probe was then kept resting on the cell at that position for one second to obtain the force-relaxation curve (force-relaxation process). For each treatment concentration, the AFM experiment was performed on at least 8 cells for each designed indentation depth.

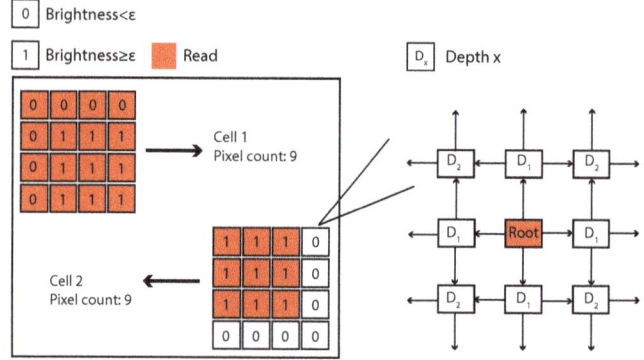

Figure 2. Illustrative demonstration of BFS algorithm.

2.5. Mechanical Property Quantification

The AFM probe-cell interaction force, $F(t)$, was quantified as,

$$F(t) = k \times S \times d(t), \tag{3}$$

where k is the cantilever spring constant, $d(t)$ is the cantilever deflection increase, S is the deflection sensitivity. The indentation depth, $\delta(t)$ is calculated as,

$$\delta(t) = z(t) - d(t), \tag{4}$$

where $z(t)$ is piezo displacement with respect to its initial probe-cell contact position.

2.5.1. Elasticity

During the indenting process, the used AFM probe was spherical and its indenting speed at 20 μm/s was significantly faster than the intracellular fluid efflux rate [28]. Therefore, the measured cell can be treated as incompressible. The Young's modulus, E, of cells then could be quantified using Hertzian model as,

$$F(t) = \frac{4}{3}\frac{E}{1-\nu^2}r^{\frac{1}{2}}\delta^{\frac{3}{2}}(t), \tag{5}$$

where $\nu = 0.5$ is the incompressible Poisson's ratio, $r = 2.5$ μm is the AFM probe radius.

2.5.2. Poroelasticity

During the force-relaxation process, the AFM probe was resting on the sample cells after the targeted indentation depth was reached. The intracellular efflux occurs to equilibrate the unbalanced intracellular pressure caused by the fast deformation of the cell morphology generated in the indenting process. Therefore, the probe-cell contact force, $F(t)$, was decreasing without changing the AFM displacement. Additionally, The AFM imaging performed on NIH/3T3 cells (see Figure 1) shows that the cell size (>30 μm) was significantly larger than the spherical probe radius (=2.5 μm). Therefore, the probe-cell interaction could be treated as a poroelastic half-space indented by a sphere indenter during the force-relaxation process. The diffusion coefficient, D, was quantified using the empirical mathematical model as follows [28,29],

$$\frac{F(t) - F_f}{F_i - F_f} = 0.491 \exp^{-0.908\sqrt{\frac{Dt}{a^2}}} + 0.509 \exp^{-1.679\frac{Dt}{a^2}}, \tag{6}$$

where F_i and δ_i are the cell-probe interaction force and the indentation depth at the end of the probe indenting process(i.e., the beginning of the force-relaxation process). F_f is the fully relaxed force obtained at the end of the relaxation process (i.e., one second after the indenting process in this paper). $a = \sqrt{r\delta_i}$ is the probe-cell contact size.

3. Results and Discussion

Although previous studies have shown the effect of the F-actin and microtubules on the cellular mechanical properties, quantitative analysis on how F-actin and microtubules affect the intracellular elasticity and poroelasticity has not yet been investigated. Therefore, using the proposed IRCQ, we quantified the correlation between the cytoskeleton morphology and cellular mechanical behavior (elasticity and poroelasticity).

3.1. F-actin and Microtubules Average Intensity Quantification

F-actin of NIH/3T3 cells were treated with latrunculin B at the concentrations of 0, 10, 30, 40, 60, 75, and 100 nM. Nocodazole with concentrations of 0, 10, 30, 50, 75, 100, and 200 nM were chosen in the microtubule treatments as well. After the treatments, the IRCQ approach was then applied to quantify the average F-actin intensity (i.e., actin filament intensity, AAI) and average microtubule intensity (AMI) of the cells. Fifty images of each treatment concentration were taken using a fluorescent microscope, respectively. The obtained images were pre-processed as mentioned in Section 2.3.1. AAI and AMI were quantified according to Equation (1), the results are shown in Table 1 and Figure 3. The treatment concentrations were chosen (after trials) such that the corresponding relative AAI and AMI changes (i.e., Δ) were at similar levels, respectively.

Table 1. Quantification results of average F-actin intensity (AAI), average microtubule intensity (AMI), and corresponding percentage change (Δ) at different treatment concentrations.

Latrunculin B (nM)	AAI (mean ± S.E./μm^2)	Δ (%)	Nocodazole (nM)	AMI (mean ± S.E./μm^2)	Δ (%)
0	375.1 ± 172.99	0.00	0	904.58 ± 194.41	0.00
10	363.5 ± 122.78	18.89	10	891.32 ± 213.50	19.49
30	351.3 ± 111.83	38.76	30	875.49 ± 145.98	42.75
40	338.7 ± 123.74	62.00	50	865.07 ± 142.82	58.10
60	322.7 ± 133.10	88.78	75	845.67 ± 193.25	86.57
75	313.7 ± 73.98	100.00	100	836.53 ± 151.28	100.00
100	315.7 ± 89.22	–	200	830.23 ± 186.67	–

The AAI results clearly show that the average intensity of F-actin is negatively correlated with the latrunculin B concentration when less than 100 nM. As AAI barely changed when the treatment concentration was further increased to 100 nM indicating that the F-actin were fully depolymerized at the concentration of 75 nM. Thus the mean value of AAI = 313.7/μm^2 was used as I_m in Equation (percentage) when calculating the relative intensity change. The AMI quantification results demonstrate that the average intensity of microtubules decreases with the nocodazole concentration increase (see Table 1). Note that the reduction of AMI becomes much less significant when the treatment concentration doubled from 100 nM compared to other treatments. This indicates that the treatment became saturated if the treatment concentration was beyond 100 nM. Therefore, the microtubules were fully depolymerized at the nocodazole concentration of 100 nM. Then the percentage change (Δ) of

AMI was calculated using Equation (2), in which $I_m = 836.53 / \mu m^2$. Example cell cytoskeleton images and the intensity quantification results are shown in Figure 3.

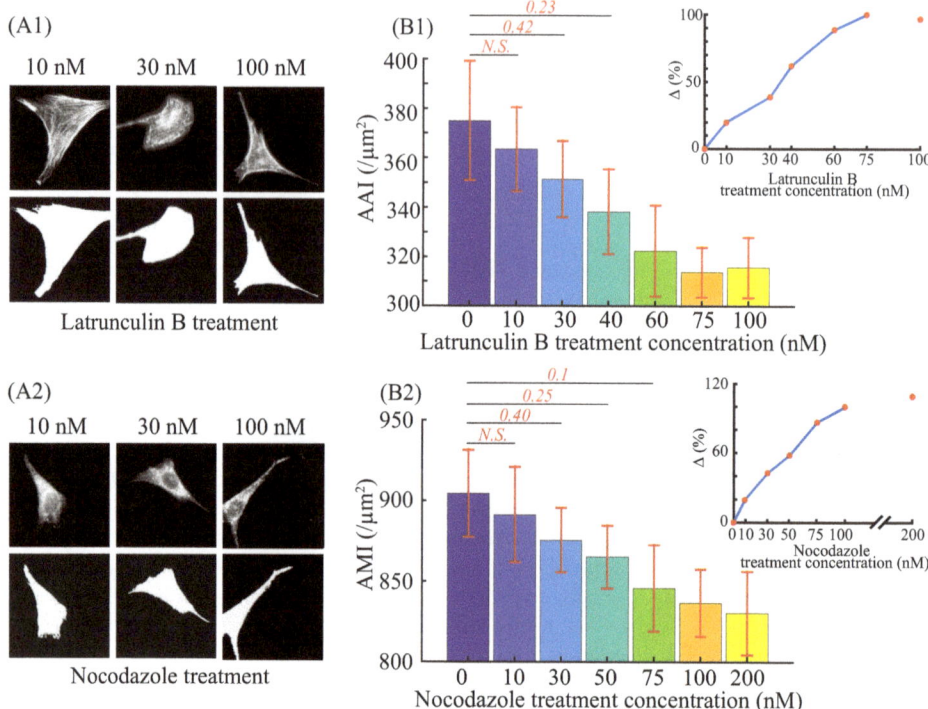

Figure 3. Example F-actin and microtubule images of cells treated with (**A1**) latrunculin B and (**A2**) nocodazole, respectively, together with their corresponding detected cell area (lower rows, respectively). (**B1,B2**) are the AAI and the AMI together with their corresponding relative change Δ, respectively. The error bars represent the standard errors. $n = 50$. Student's *t*-test was performed to analyze the statistical difference: for each treatment concentration, data were compared with respect to the untreated ones. A $p < 0.05$ was yielded for each comparison, unless otherwise denoted in the figure (with *p* values in red bold italic font).

3.2. Elasticity and Poroelasticity Quantification

The AFM experiments designed with three target indentation depths (i.e., 650, 1000, 1300 nm) were performed on at least 8 cells for each aforementioned treatment concentration. Since the substrate stiffness of cell culture dish was at least three orders of magnitude higher than cells', and the desired target indentation was less than a quarter of the cell height, the substrate influence could be ignored in cellular mechanical property quantification [3,25]. Therefore, the results shown in Figure 4 indeed represent the mechanical response of the indented cells. In general, the experimental results show significant changes in both the cellular elasticity (Young's modulus, E) and poroelasticity (diffusion coefficient, D) with the depolymerization of F-actin and microtubules, respectively.

As shown in Figure 4, Young's modulus and diffusion coefficient were decreased and increased by 78.37∼89.53% and 182.34∼263.17%, respectively, with the increase of the treatment concentrations for each indentation depth. These changes are consistent with the previous studies that the cellular F-actin and microtubules provide cells with mechanical support and driving forces for movement [30,31]. Specifically, the depolymerization of F-actin and microtubules reduces the strength of the cytoskeletal network, which leads to the weakening of the supporting ability to resist the external force stimuli. Moreover, the cytoskeleton depolymerization resulted in increased pore size of the cross-linked cytoskeleton network, therefore, the diffusion coefficient was increased as previous studies have shown the cytoplasmic pore size is the dominant factor in affecting the cellular poroelasticity: the larger the cytoskeleton pore size, the higher the cellular diffusion coefficient [32,33].

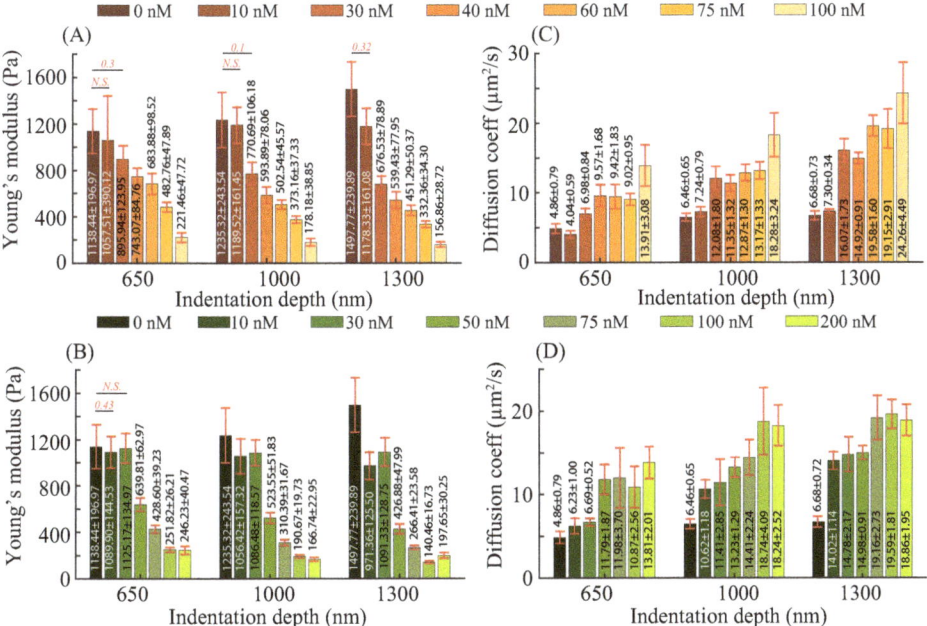

Figure 4. (**A,B**) Young's modulus and (**C,D**) diffusion coefficient of NIH/3T3 cells treated with different treatment concentrations of (**A,C**) latrunculin B and (**B,D**) nocodazole, respectively, quantified at different indentation depths at the indenting velocity of 20 μm/s. $n \geq 8$. Error bar: standard error. Student's t-test was performed to analyze the statistical difference: for each indentation, data were compared with respect to the untreated ones at the same indentation. A $p < 0.05$ was yielded for each comparison, unless otherwise denoted in the figure (with p values in red bold italic font).

As can be seen in Figure 4A,B, Young's modulus E of the control (untreated) cells increased 31.56%, when the indentation depth increased from 650 to 1300 nm, and the increases were 11.43% for cells treated with latrunculin B concentration at 10 nM. However, for the rest of latrunculin treatments ≥ 30 nM, with the increase of the indentation depth, E decreased by 24.49~34.01% (see Figure 4A), and the decreases was 3.01~44.22%, for all the nocodazole concentrations (see Figure 4B). Moreover, the diffusion coefficient D increased by 37.5~112.44%, when the cells were indented with the indentation depth from 650~1300 nm under the latrunculin B treatments (see Figure 4C), and it increased by 36.61~125.14% for the nocodazole treatments (see Figure 4D). Two observations can be made from these results: (1) The NIH/3T3 cells were not homogeneous in terms of elasticity and poroelasticity; (2) This heterogeneity was changed as once the latrunculin B and nocodazole treatment concentrations were increased. The former agrees with previous studies that mammalian cells have a multilayered structure of its cytoplasm, and the cytoplasm heterogeneity (especially the cytoskeleton heterogeneity) is involved in affecting the intracellular elasticity and poroelasticity [4,18,34]. With the increase of the indentation depth, deeper layers of the cytoskeleton were excited and deformed, and higher Young's modulus was observed for the untreated cells. This agrees with previous findings that the deeper layers of the cytoskeleton are stiffer [10,35]. However, once the treatment concentration was high enough, the F-actin and microtubule structures in the deeper layers were sufficiently depolymerized, thus lower Young's modulus was observed at deeper indentations. Meanwhile, a more significant increase in the diffusion coefficient was observed. These findings are consistent with previous studies that F-actin and microtubules containing signal molecules that could regulate the formation of the focal adhesion [9,36]. Specifically, because of the relatively high treatment concentrations, the depolymerization of deeper layer F-actin and microtubules could lead to unstabilized focal adhesion, which reduced the cell-substrate bonding strength and cells' contractibility. Therefore, higher lateral expansion of cells induced by the deeper indentation depth resulted in lower Young's modulus. Meanwhile, the enlarged expansion volume increased the pore size in the cross-linked cytoskeleton network, causing the higher diffusion coefficient. Therefore, the change of the cellular elasticity and poroelasticity became more significant as the cytoskeleton treatment strength increased.

3.3. Effects of F-actin and Microtubule on Elastic and Poroelastic Behavior of Cells

In order to investigate the significance of F-actin and microtubules in affecting the cellular mechanical behavior, the relative cytoskeleton morphology change (Δ of F-actin and microtubules, respectively) vs. cellular elasticity E and poroelasticity D relations were investigated for each measured indentation depth, respectively. Meanwhile, the relative changes of E and D with respect to the values measured form the control under each indentation depth were also quantified. The results are shown in Figure 5.

As can be seen in Figure 5(A1–A3), for the three measured indentation depths, the dominance of F-actin and microtubules in affecting cellular elasticity depends on their morphology quantity changes, respectively. For the smallest indentation (650 nm), the effect of F-actin is more significant when the intensity (i.e., F-actin and microtubule quantity) decrease percentage Δ is relatively small, however, this difference is overturned as the increase of the cytoskeleton treatment concentration (i.e., the F-actin and microtubule were inhibited more significantly). As shown in previous studies, F-actin are approximately three hundred times less resistant than microtubules subject to mechanical forces [37,38], and they have a presence of higher concentrations than microtubules that promotes the assembly of highly organized cytoskeleton [10]. Also, F-actin affect cellular mechanics at superficial layers [10]. Therefore, a small decrease of the F-actin quantity (<60%) could weaken the actin structure stiffness greatly as F-actin were softer and the fibrous actin alignment was disturbed as well [37,39], which further led to the notable change of the cellular elasticity even at the smallest indentation depth. However, at the same depolymerization degree, as the microtubules were more resistant, the change in cell elasticity caused by the low degree of depolymerization of microtubules was less significant compared to the former case. As the morphology quantity decrease became more and more significant (>60%), the microtubules–the stiff support of the cytoskeleton structure–were sufficiently depolymerized and this led to a dramatic drop of the cellular elasticity. Therefore, although the cellular elasticity is more sensitive to the small decrease of the F-actin (signaled as the slope of the curves in Figure 5(A1–A3)) when the morphology variation was small, microtubules are more dominant in affecting E. The dominant effect of microtubules became more significant as the indentation depth was increased: E became more sensitive to the decrease of the microtubule quantity even when Δ was small, and the decrease of E was more dramatic as the increase of Δ for the microtubule treatments. The reason for the more obvious effect of microtubules in E at deeper indentation can be explained using the result reported previously that microtubules are concentrated at the deeper layer of cells compared to superficial layers [3,4].

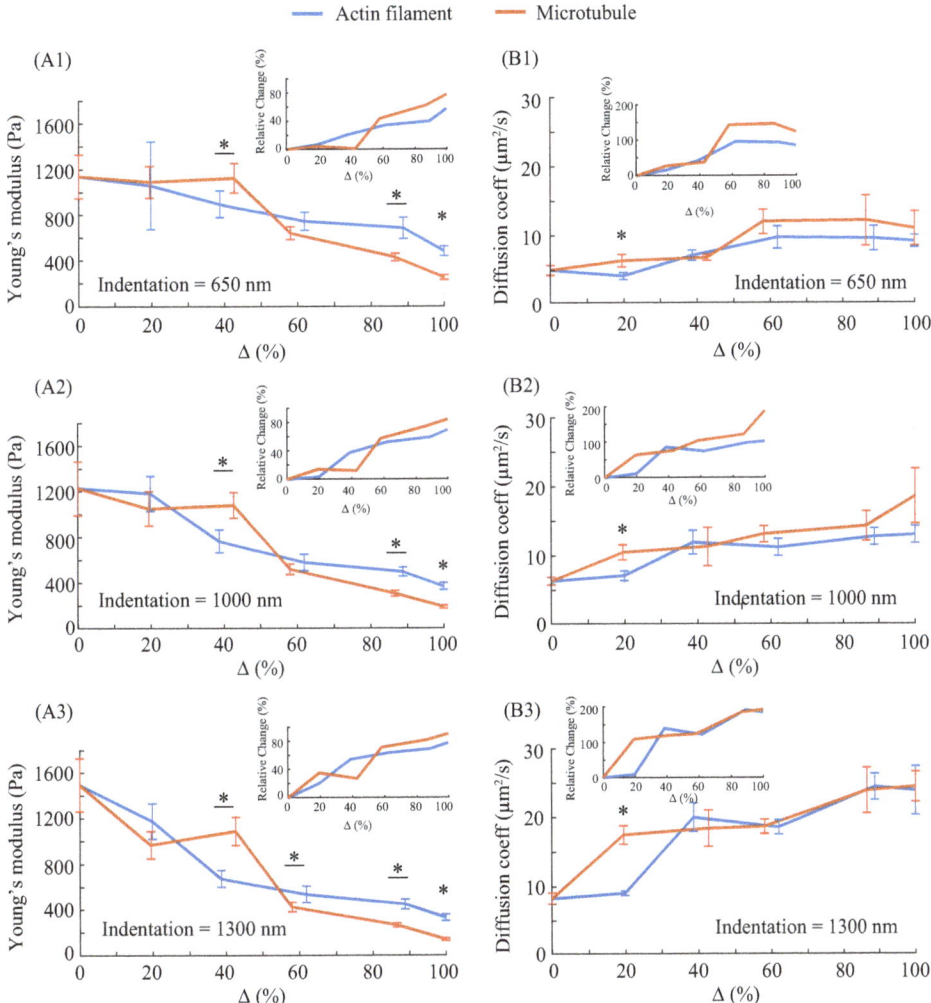

Figure 5. The cytoskeleton morphology intensity decrease percentage (Δ) vs. the cellular (**A1**–**A3**) Young's modulus and (**B1**–**B3**) diffusion coefficient of NIH/3T3 cells quantified at the indentation depths of 650, 1000, and 1300 nm, respectively. $n \geq 8$. Error bar: standard error. Student's t-test was performed to analyze the statistical difference: for each similar treatment concentration, data from the two drug treatments were compared. ∗: $p < 0.05$.

As shown in Figure 5(B1–B3), the diffusion coefficient (D) is more sensitive to the change of microtubule quantity (signaled by the values in Figure 5(B1–B3)), and the increase of D induced by the depolymerization of microtubules is more significant than that induced by F-actin depolymerization at all three measured indentation depths for small depolymerization degrees (<40%). The results agree with the reported studies that the mesh-like morphology of microtubules is highly dynamic, known as dynamic instability, compared to F-actin that have branched dendritic network, or parallel bundles [40]. Specifically, microtubules underwent rapid depolymerization and reorganization during the nocodazole treatments and AFM indentation process, thus it resulted in a more significant and sensitive change of the porous morphology of the cytoskeleton compare to the F-actin. As previous studies have shown that the cytoplasmic pore size is more dominant than elasticity in affecting the cellular poroelasticity (i.e., diffusion coefficient) [3,4], thereby the increase of diffusion coefficient induced by microtubules treatments is more notable than that caused by F-actin treatments. As the morphology quantity decrease became more and more significant, the effect difference on D between microtubules and F-actin became unremarkable (when $\Delta > 40\%$). As aforementioned, F-actin have a higher concentration and much less rigid than microtubules [10,40]. Therefore, when the parallel cross-linked actin structure was sufficiently depolymerized (with the reduction in quantity > 40%), the effect of actin treatment on D was similar to that of the microtubules at all three indentation depth. Compare to the results of cellular elasticity, it is clear that a small degree of depolymerization of microtubules affects the cytoskeleton morphology (e.g., pore size) more significantly than its effect on the cellular elasticity.

Therefore, the AFM indentation experiments have shown that the cellular elasticity is more sensitive to the variation of F-actin at small ranges, but microtubules are still dominant in affecting the elasticity of the cellular structure and the dominant role becomes more significant at deeper layers. Meanwhile, depolymerization of microtubules always has a more significant effect on cellular poroelasticity at all measured depths.

4. Conclusions

In this study, the effect of both F-actin and microtubules on cellular mechanical behavior was investigated. To obtain quantitative comparison, cytoskeleton treatments were performed on NIH/3T3 cells such that F-actin and microtubules were depolymerized at similar levels with the aid of the proposed IRCQ. The cellular elasticity and poroelasticity were quantified using AFM indentation measurements. It was found that the cellular Young's modulus decreased monotonically with the treatment concentration for each measured indentation depth. The trend of the cellular elasticity heterogeneity (i.e., elasticity vs. indentation depth) was changed once the F-actin and/or microtubules were sufficiently depolymerized. Although the cell Young's modulus is more sensitive to the reduction of F-actin at superficial layers, microtubules are more dominant in affecting cellular elasticity. Moreover, the cellular poroelasticity is very sensitive to the change of the microtubule structure, even at the low degree of depolymerization and small indentations.

Author Contributions: Conceptualization, J.R., Y.L.; Methodology, Y.L., K.M., J.R.; Software, Y.L., K.M., M.H.S.; Validation, Y.L., K.M., M.H.S.; Formal analysis, Y.L., K.M., M.H.S., J.R.; Investigation, Y.L., K.M.; Data curation, Y.L., K.M., M.H.S., J.R.; All authors participated in writing the manuscript. All authors have read and agreed to the published version of the manuscript.

Funding: This work was supported by the National Science Foundation (NSF) [CMMI-1634592] and [CMMI-1751503], and Iowa State University.

Acknowledgments: The authors thank Xuefeng Wang for providing NIH/3T3 cells.

Conflicts of Interest: The authors declare no conflict of interest.

References

1. Lichtenstein, N.; Geiger, B.; Kam, Z. Quantitative analysis of cytoskeletal organization by digital fluorescent microscopy. *Cytom. Part A* **2003**, *54*, 8–18. [CrossRef]
2. Pullarkat, P.A.; Fernández, P.A.; Ott, A. Rheological properties of the eukaryotic cell cytoskeleton. *Phys. Rep.* **2007**, *449*, 29–53. [CrossRef]
3. Moeendarbary, E.; Valon, L.; Fritzsche, M.; Harris, A.R.; Moulding, D.A.; Thrasher, A.J.; Stride, E.; Mahadevan, L.; Charras, G.T. The cytoplasm of living cells behaves as a poroelastic material. *Nat. Mater.* **2013**, *12*, 253. [CrossRef] [PubMed]
4. Mollaeian, K.; Liu, Y.; Bi, S.; Ren, J. Atomic force microscopy study revealed velocity-dependence and nonlinearity of nanoscale poroelasticity of eukaryotic cells. *J. Mech. Behav. Biomed. Mater.* **2018**, *78*, 65–73. [CrossRef] [PubMed]

5. Salbreux, G.; Charras, G.; Paluch, E. Actin cortex mechanics and cellular morphogenesis. *Trends Cell Biol.* **2012**, *22*, 536–545. [CrossRef] [PubMed]
6. Blanchoin, L.; Boujemaa-Paterski, R.; Sykes, C.; Plastino, J. Actin dynamics, architecture, and mechanics in cell motility. *Physiol. Rev.* **2014**, *94*, 235–263. [CrossRef]
7. Horio, T.; Hotani, H. Visualization of the dynamic instability of individual microtubules by dark-field microscopy. *Nature* **1986**, *321*, 605. [CrossRef]
8. Wang, N.; Naruse, K.; Stamenović, D.; Fredberg, J.J.; Mijailovich, S.M.; Tolić-Nørrelykke, I.M.; Polte, T.; Mannix, R.; Ingber, D.E. Mechanical behavior in living cells consistent with the tensegrity model. *Proc. Natl. Acad. Sci. USA* **2001**, *98*, 7765–7770. [CrossRef]
9. Enomoto, T. Microtubule disruption induces the formation of actin stress fibers and focal adhesions in cultured cells: Possible involvement of the rho signal cascade. *Cell Struct. Funct.* **1996**, *21*, 317–326. [CrossRef]
10. Fletcher, D.A.; Mullins, R.D. Cell mechanics and the cytoskeleton. *Nature* **2010**, *463*, 485. [CrossRef]
11. Lin, C.H.; Thompson, C.A.; Forscher, P. Cytoskeletal reorganization underlying growth cone motility. *Curr. Opin. Neurobiol.* **1994**, *4*, 640–647. [CrossRef]
12. Pollard, T.D.; Borisy, G.G. Cellular motility driven by assembly and disassembly of actin filaments. *Cell* **2003**, *112*, 453–465. [CrossRef]
13. Tseng, Y.; Kole, T.P.; Lee, J.S.; Fedorov, E.; Almo, S.C.; Schafer, B.W.; Wirtz, D. How actin crosslinking and bundling proteins cooperate to generate an enhanced cell mechanical response. *Biochem. Biophys. Res. Commun.* **2005**, *334*, 183–192. [CrossRef]
14. Brangwynne, C.P.; MacKintosh, F.C.; Kumar, S.; Geisse, N.A.; Talbot, J.; Mahadevan, L.; Parker, K.K.; Ingber, D.E.; Weitz, D.A. Microtubules can bear enhanced compressive loads in living cells because of lateral reinforcement. *J. Cell Biol.* **2006**, *173*, 733–741. [CrossRef] [PubMed]
15. Schaedel, L.; John, K.; Gaillard, J.; Nachury, M.V.; Blanchoin, L.; Théry, M. Microtubules self-repair in response to mechanical stress. *Nat. Mater.* **2015**, *14*, 1156. [CrossRef]
16. Rotsch, C.; Radmacher, M. Drug-induced changes of cytoskeletal structure and mechanics in fibroblasts: An atomic force microscopy study. *Biophys. J.* **2000**, *78*, 520–535. [CrossRef]
17. Haga, H.; Sasaki, S.; Kawabata, K.; Ito, E.; Ushiki, T.; Sambongi, T. Elasticity mapping of living fibroblasts by AFM and immunofluorescence observation of the cytoskeleton. *Ultramicroscopy* **2000**, *82*, 253–258. [CrossRef]
18. Kasas, S.; Wang, X.; Hirling, H.; Marsault, R.; Huni, B.; Yersin, A.; Regazzi, R.; Grenningloh, G.; Riederer, B.; Forro, L.; et al. Superficial and deep changes of cellular mechanical properties following cytoskeleton disassembly. *Cell Motil. Cytoskelet.* **2005**, *62*, 124–132. [CrossRef]
19. Ning, W.; Yu, Y.; Xu, H.; Liu, X.; Wang, D.; Wang, J.; Wang, Y.; Meng, W. The CAMSAP3-ACF7 complex couples noncentrosomal microtubules with actin filaments to coordinate their dynamics. *Dev. Cell* **2016**, *39*, 61–74. [CrossRef]
20. Cooper, J.A. Effects of cytochalasin and phalloidin on actin. *J. Cell Biol.* **1987**, *105*, 1473–1478. [CrossRef]
21. Ojala, T.; Pietikainen, M.; Maenpaa, T. Multiresolution gray-scale and rotation invariant texture classification with local binary patterns. *IEEE Trans. Pattern Anal. Mach. Intell.* **2002**, *24*, 971–987. [CrossRef]
22. Liu, Y.; Mollaeian, K.; Ren, J. An Image Recognition-Based Approach to Actin Cytoskeleton Quantification. *Electronics* **2018**, *7*, 443. [CrossRef]
23. Beamer, S.; Asanović, K.; Patterson, D. Direction-optimizing breadth-first search. *Sci. Program.* **2013**, *21*, 137–148. [CrossRef]
24. Tarjan, R. Depth-first search and linear graph algorithms. *SIAM J. Comput.* **1972**, *1*, 146–160. [CrossRef]
25. Chen, J. Nanobiomechanics of living cells: A review. *Interface Focus* **2014**, *4*, 20130055. [CrossRef]
26. Schillers, H.; Wälte, M.; Urbanova, K.; Oberleithner, H. Real-time monitoring of cell elasticity reveals oscillating myosin activity. *Biophys. J.* **2010**, *99*, 3639–3646. [CrossRef]
27. Mollaeian, K.; Liu, Y.; Ren, J. Investigation of Nanoscale Poroelasticity of Eukaryotic Cells Using Atomic Force Microscopy. In Proceedings of the ASME 2017 Dynamic Systems and Control Conference, Tysons, VA, USA, 11–13 October 2017; p. V001T08A005.
28. Mollaeian, K.; Liu, Y.; Bi, S.; Wang, Y.; Ren, J.; Lu, M. Nonlinear Cellular Mechanical Behavior Adaptation to Substrate Mechanics Identified by Atomic Force Microscope. *Int. J. Mol. Sci.* **2018**, *19*, 3461. [CrossRef]
29. Hu, Y.; Zhao, X.; Vlassak, J.J.; Suo, Z. Using indentation to characterize the poroelasticity of gels. *Appl. Phys. Lett.* **2010**, *96*, 121904. [CrossRef]
30. Janmey, P.A. Mechanical properties of cytoskeletal polymers. *Curr. Opin. Cell Biol.* **1991**, *3*, 4–11. [CrossRef]

31. Pollard, T.D.; Cooper, J.A. Actin, a central player in cell shape and movement. *Science* **2009**, *326*, 1208–1212. [CrossRef]
32. Nikaido, H.; Rosenberg, E.Y. Effect on solute size on diffusion rates through the transmembrane pores of the outer membrane of Escherichia coli. *J. Gener. Physiol.* **1981**, *77*, 121–135. [CrossRef] [PubMed]
33. Potma, E.O.; de Boeij, W.P.; Bosgraaf, L.; Roelofs, J.; van Haastert, P.J.; Wiersma, D.A. Reduced protein diffusion rate by cytoskeleton in vegetative and polarized dictyostelium cells. *Biophys. J.* **2001**, *81*, 2010–2019. [CrossRef]
34. Fuhrmann, A.; Staunton, J.; Nandakumar, V.; Banyai, N.; Davies, P.; Ros, R. AFM stiffness nanotomography of normal, metaplastic and dysplastic human esophageal cells. *Phys. Biol.* **2011**, *8*, 015007. [CrossRef] [PubMed]
35. Yeung, T.; Georges, P.C.; Flanagan, L.A.; Marg, B.; Ortiz, M.; Funaki, M.; Zahir, N.; Ming, W.; Weaver, V.; Janmey, P.A. Effects of substrate stiffness on cell morphology, cytoskeletal structure, and adhesion. *Cell Motil. Cytoskelet.* **2005**, *60*, 24–34. [CrossRef] [PubMed]
36. Wehrle-Haller, B.; Imhof, B.A. Actin, microtubules and focal adhesion dynamics during cell migration. *Int. J. Biochem. Cell Biol.* **2003**, *35*, 39–50. [CrossRef]
37. Gittes, F.; Mickey, B.; Nettleton, J.; Howard, J. Flexural rigidity of microtubules and actin filaments measured from thermal fluctuations in shape. *J. Cell Biol.* **1993**, *120*, 923–934. [CrossRef]
38. Wen, Q.; Janmey, P.A. Polymer physics of the cytoskeleton. *Curr. Opin. Solid State Mater. Sci.* **2011**, *15*, 177–182. [CrossRef]
39. Butt, T.; Mufti, T.; Humayun, A.; Rosenthal, P.B.; Khan, S.; Khan, S.; Molloy, J.E. Myosin motors drive long range alignment of actin filaments. *J. Biol. Chem.* **2010**, *285*, 4964–4974. [CrossRef]
40. Gardel, M.L.; Kasza, K.E.; Brangwynne, C.P.; Liu, J.; Weitz, D.A. Mechanical response of cytoskeletal networks. *Methods Cell Biol.* **2008**, *89*, 487–519.

© 2020 by the authors. Licensee MDPI, Basel, Switzerland. This article is an open access article distributed under the terms and conditions of the Creative Commons Attribution (CC BY) license (http://creativecommons.org/licenses/by/4.0/).

Review

Long-Range and Directional Allostery of Actin Filaments Plays Important Roles in Various Cellular Activities

Kiyotaka Tokuraku [1,*], **Masahiro Kuragano** [1] **and Taro Q. P. Uyeda** [2]

1. Department of Applied Sciences, Muroran Institute of Technology, Muroran, Hokkaido 050-8585, Japan; gano@mmm.muroran-it.ac.jp
2. Department of Physics, Faculty of Science and Engineering, Waseda University, Tokyo 169-8555, Japan; t-uyeda@waseda.jp
* Correspondence: tokuraku@mmm.muroran-it.ac.jp

Received: 1 April 2020; Accepted: 30 April 2020; Published: 1 May 2020

Abstract: A wide variety of uniquely localized actin-binding proteins (ABPs) are involved in various cellular activities, such as cytokinesis, migration, adhesion, morphogenesis, and intracellular transport. In a micrometer-scale space such as the inside of cells, protein molecules diffuse throughout the cell interior within seconds. In this condition, how can ABPs selectively bind to particular actin filaments when there is an abundance of actin filaments in the cytoplasm? In recent years, several ABPs have been reported to induce cooperative conformational changes to actin filaments allowing structural changes to propagate along the filament cables uni- or bidirectionally, thereby regulating the subsequent binding of ABPs. Such propagation of ABP-induced cooperative conformational changes in actin filaments may be advantageous for the elaborate regulation of cellular activities driven by actin-based machineries in the intracellular space, which is dominated by diffusion. In this review, we focus on long-range allosteric regulation driven by cooperative conformational changes of actin filaments that are evoked by binding of ABPs, and discuss roles of allostery of actin filaments in narrow intracellular spaces.

Keywords: long-range allostery; actin; actin-binding protein; cofilin; cooperativity; drebrin; filamin; fimbrin; gelsolin; myosin; tropomyosin

1. Introduction

Actin filaments play many important roles in eukaryotic cells by interacting with various actin-binding proteins (ABPs). To realize these activities, the localization of ABPs is spatially and temporally regulated in cells. For example, myosin II is localized only in the posterior region in a migrating *Dictyostelium* cell, even though the anterior region is rich in actin filaments [1,2]. On the other hand, cofilin is localized to lamellipodia in the anterior region, and recycles actin for continuous polymerization at the leading edge [3]. This different localization of ABPs is one driving force for cell migration. Nonmigratory cells have other types of actin-based structures involved in cell-cell and cell-substrate adhesion such as adherens junctions, tight junctions and focal adhesions, as well as more specialized structures such as microvilli and stereocilia. Again, each of those structures depends on spatially-regulated interactions of actin filaments with specific sets of ABPs. Although the regulation of ABP localization is generally explained by local biochemical signaling such as phosphorylation and changes in the concentration of signaling molecules, not all aspects of ABP regulation can be explained by local biochemical signaling [4], and the overall picture is still unclear.

Recent developments in structural biology have revealed dynamic structural polymorphism in pure actin filaments [5,6]. Furthermore, various ABPs are now known to induce cooperative conformational changes of actin filaments [7], thereby controlling the binding of ABPs to actin protomers not in direct contact with the bound ABP in vitro, which we call long-range allostery. In addition to local biochemical signals, long-range allostery in actin filaments may be involved in controlling the localization of ABPs in micrometer-scale intracellular spaces. In this review, we focus on the spatial and temporal allosteric regulation of ABP localization by cooperative conformational changes of actin filament cables.

2. Cooperative Binding of ABPs

For decades, cooperative binding of various ABPs to actin filaments, accompanied by cooperative conformational changes in filaments, has been reported. Such ABPs include myosin II [8–23], tropomyosin [24–27], cofilin [4,28–41], drebrin [42,43], fimbrin [44–46], α-actinin [47,48], filamin [49], α-catenin [50], gelsolin [51,52] and formin [53]. In recent years, the interaction between different ABPs via conformational changes of actin filaments, including a mutually exclusive interaction, has also been reported. In this section, we will list ABPs that have been reported to bind cooperatively with actin filaments. They are classified by one of three binding styles—side-binding, cross-linking, and end-binding.

2.1. Side-Binding ABPs

2.1.1. Myosin II and V

Myosin [54] is a family of motor proteins that convert chemical energy of ATP into motion energy. There are a large number of reports of cooperative conformational changes within actin filaments induced by the binding of myosin [8,9]. Nearly fifty years ago, Oosawa et al. first reported the cooperative interaction between heavy meromyosin (HMM) of myosin II and actin filaments, observing that the binding of HMM increased the fluorescence intensity of labeled actin, with the effect being saturated in the presence of a 1:20 molar ratio of HMM to actin protomers [10]. Miki et al. also reported that the maximal change in fluorescence in actin filaments occurred when only one HMM molecule was bound per 50 actin protomers or when one subfragment 1 (S1) molecule was bound per 25 actin protomers [11]. There are many similar reports on the cooperative interaction between myosin II and actin filaments [12–15]. Cooperative binding of myosin II and actin filaments was observed by electron microscopy [16–20]. For, example, Orlova and Egelman reported very large cooperativity in the binding of HMM to actin filaments with Ca^{2+} bound at the high-affinity metal binding site [20]. Cooperative binding was also reported by fluorescence microscopic observation using GFP-fused HMM (HMM-GFP) derived from *Dictyostelium* myosin II [21]. When HMM-GFP was mixed with actin filaments with either Ca^{2+} or Mg^{2+} at the high-affinity metal binding site, HMM-GFP formed clusters along both forms of actin filaments. The density of HMM molecules in the cluster along Ca^{2+}-actin filaments was higher than that along Mg^{2+}-actin filaments [21]. The weak cooperative binding to Mg^{2+}-actin filaments required low, submicromolar concentrations of ATP [21]. The growth of HMM-GFP clusters along Mg^{2+}-actin filaments, which is the physiological form [22], was unidirectional, as determined by real-time fluorescence microscopy [23]. Cooperative cluster formation was observed along Mg^{2+}-actin filaments that were loosely immobilized on positively charged lipid bilayers but not when tightly immobilized by biotin-avidin linkage, suggesting that the myosin-induced cooperative conformational changes in actin filaments involves a change in the helical pitch [23]. Kozuka et al. examined the dynamic polymorphism of actin filaments by using single molecule intramolecular Förester resonance energy transfer (FRET) imaging, noting that actin protomer switches between low- and high-FRET efficiency states [6]. The high-FRET efficiency state is favored when actin filaments interact with myosin V in the presence of ATP [6], suggesting that myosin V can also induce a cooperative conformational change in actin protomers.

2.1.2. Tropomyosin

Tropomyosin [55] is a regulatory protein found in muscle and non-muscle cells. In striated muscle, this protein functions in association with actin filaments and troponin to confer calcium sensitivity to the actomyosin ATPase [56,57]. Cooperative effects between tropomyosin and actin filaments have recently been reported. For example, proteolytic modification within the DNase-binding loop of actin increased the rate of subunit exchange along actin filaments and tropomyosin binding almost completely suppressed the increase in subunit exchange [25]. The effect was cooperative, with half-maximal inhibition observed at about a 1:50 molar ratio of tropomyosin:actin [25]. In mammals, >40 tropomyosin isoforms can be generated through alternative splicing from four *tropomyosin* genes. Interestingly, the binding of these isoforms is segregated to different actin filament structures in nonmuscle cells, presumably because of cooperative interactions [27].

2.1.3. Cofilin

Cofilin and the closely related actin-depolymerizing factor (ADF) control actin polymerization and depolymerization in a pH-sensitive manner, and inhibit interaction of actin filaments with myosin and tropomyosin [58,59]. Cofilin also binds to actin filaments cooperatively [28–31] and forms tight clusters [4,32–38]. Severing of actin filaments by cofilin often occurs at or near the boundaries between cofilin clusters and the bare zones, presumably because of structural discontinuities at those sites [32,36,38]. Actin filaments bound with cofilin clusters were supertwisted by 25% [32–35], and high-speed atomic force microscopy (HS-AFM) revealed that the supertwisted conformation propagates unidirectionally to the neighboring bare zone on the pointed-end side of the cluster [32]. On the other hand, fluorescence microscopy revealed that cofilin clusters grow bidirectionally [38]. The cause of this discrepancy is currently unknown, but it may be due to the difference in resolution between AFM and fluorescence microscopy or to differences in the method of immobilizing actin filaments to the substrate. The cooperative binding of cofilin to actin filaments has been analyzed using actin mutants as well. For example, actin of the G146V mutant fails to bind cofilin. Intriguingly, binding of cofilin to copolymers of wild type and mutant actins was strongly inhibited even when the ratio of wild type actin and G146V actin in copolymers was 9:1, suggesting that the cooperative conformational change propagates about 10 molecules of actin protomers [39]. K336I mutant actin also cooperatively inhibited binding of cofilin to wild type actin in a copolymer [41]. Moreover, HS-AFM demonstrated that approximately a half helical pitch, which contains 14 actin protomers, of the bare zone on the pointed end side of a cofilin cluster is supertwisted [32]. Recent cryoelectron microscopic analysis suggested that only one or two actin protomers adjacent to cofilin clusters show maximum binding cooperativity [40], and again, the source of discrepancy between this and the HS-AFM observation is currently unknown. On the other hand, differential scanning calorimetry demonstrated that one molecule of cofilin affects the structure of about 100 actin protomers [60]. These results suggest that cofilin induces a wide range (~100 protomers) of structural changes in actin filaments, which may be a different type of cooperative conformational change from that which occurs in a narrower range (1–10 protomers) involving changes in the helical pitch.

2.1.4. Drebrin

Drebrin [61] is an actin-binding protein in the brain that regulates synaptic plasticity of neuronal dendrites [62]. Drebrin A induces cooperative change in the helical structure of actin filaments and cooperatively binds to filaments [42,43]. AFM analysis showed that drebrin A-induced structural and mechanical remodeling in actin filaments involves significant changes in helical twisting and filament stiffness [43]. Drebrin A N-terminal fragment (1-300) containing an actin-binding domain (ABD) and an ADF homology domain forms clusters (average cluster size was ~107 nm or ~2.6 actin filament helical repeats) along actin filaments [42].

2.2. Cross-Linking ABPs

2.2.1. Fimbrin

Fimbrin [63], referred to as plastin in humans, belongs to a superfamily of actin cross-linking proteins that share calponin homology (CH) domains. In addition to fimbrin, the actin cross-linking proteins with CH domains include α-actinin, filamin, spectrin, dystrophin and ABP-120 [64–67]. The N-terminal CH domains of fimbrin induced a conformational change in actin filaments [44–46]. Fimbrin inhibits tropomyosin binding to actin filaments [46]. It would be interesting to test if this inhibition is due to simple competition of the binding sites on actin protomers, or due to the fimbrin-induced cooperative conformational changes in actin filaments.

2.2.2. α-Actinin

α-Actinin [68] forms a dimer and cross-links actin filament cables to form parallel bundles. α-Actinin also enhances nucleotide exchange of bound actin filaments, and this activity was maximal when α-actinin was added at a 1:49 molar ratio of α-actinin to actin protomers [47], strongly suggesting the cooperative impact of α-actinin binding to many actin protomers.

2.2.3. Filamin

Filamin [69] is also a dimeric actin filament cross-linking protein [70]. The meshwork of actin filaments cross-linked by filamin is involved in many cellular activities such as cell migration, chemotaxis and mechanosensing [71–74]. The ABD of filamin selectively binds to actin filaments in the rear of migrating cells and contributes to the posterior localization of filamin [49]. The speed of translocation was much faster than that of the retrograde flow of cortical actin filaments, suggesting that the ABD of filamin diffusing in the cytoplasm recognizes a certain feature of the actin filament structure in the rear of the cell and selectively binds to it [49].

Further detailed studies are needed to obtain direct evidence for long-range allostery involving ABDs of cross-linking ABPs.

2.3. End-Binding ABPs

2.3.1. Gelsolin

Gelsolin [75] was discovered as a calcium-dependent regulatory protein that controls cytoplasmic actin gel-sol transformation. Low resolution electron microscopy revealed that the binding of one molecule of gelsolin at the barbed end of an actin filament changes the structure of all actin protomers in a filament, representing a case of extremely long-range cooperative conformational change in actin filaments [51]. The amino-terminal half of gelsolin, G1-3, and full-length gelsolin both quenched the pyrene fluorescence of actin, and the extent of quenching and stoichiometry were identical between G1-3 and gelsolin. In contrast, severing of actin filaments by G1-3 was much less efficient than by full-length gelsolin. Quantitative experiments suggested cooperative interactions in which the binding of two G1-3 molecules in close proximity leads to cooperative severing of the polymer, thus increasing severing efficiency [52].

2.3.2. Formin

Formin dimer [76] is a ubiquitous actin filament nucleator that progressively elongates filaments by incorporating actin monomers complexed with profilin [77]. One formin dimer at the barbed end can affect the dynamic properties of the entire filament. Analyses of the results obtained at various formin/actin concentration ratios indicated that at least 160 actin protomers are affected by the binding of a single formin dimer to the barbed end of a filament [53].

3. Allosteric Interaction between ABPs and Actin Filaments Involved in the Formation and Function Mechanism of the Actin-Based Machinery

The actin-based motile machineries, which include stress fibers, lamellipodia, filopodia, invadopodia, muscle, and contractile rings, are formed by the interaction between actin filaments and specific ABPs (Figure 1). As mentioned above, the interaction of ABPs with actin filaments induces structural changes of filaments, which regulates the interaction between ABPs. Thus, allosteric interactions involving multiple ABPs may control the formation and function of actin-based machineries, but research in this area has only just begun. This section describes the formation and function mechanisms and characteristics of these machineries, focusing on the stress fibers, lamellipodia, and filopodia (Figure 1), which have been studied to some extent already.

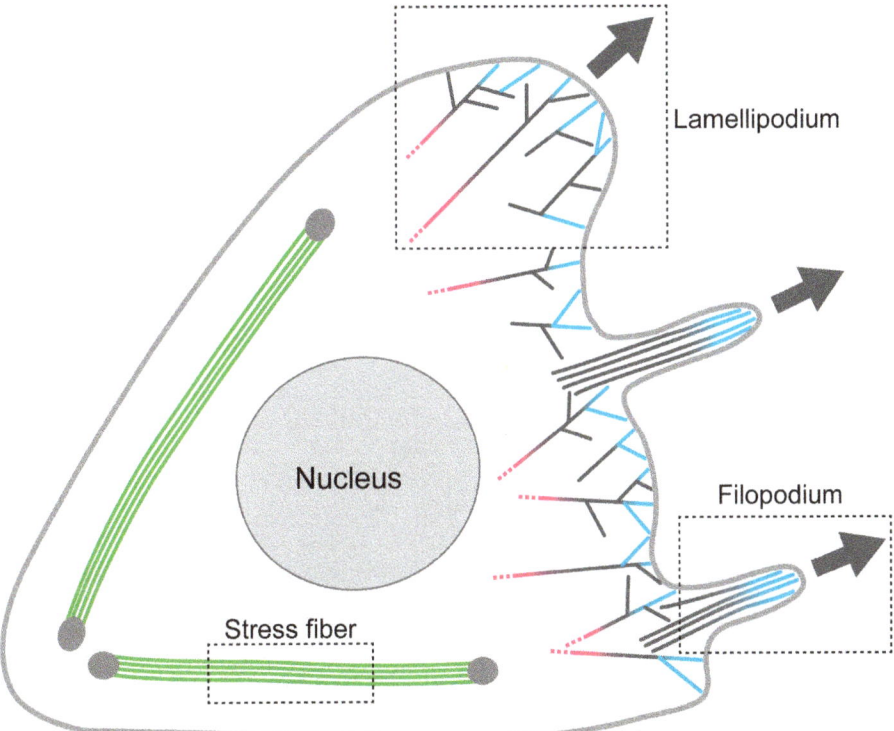

Figure 1. Possible contribution of allosteric interaction between actin-binding proteins (ABPs) and actin filaments to the formation and functions of actin-based machineries, including stress fiber, lamellipodium, and filopodium. Green filaments indicate actin filaments along which cooperative structural changes are evoked by myosin II. Blue filaments in lamellipodium and filopodia are polymerized by formin bound to the barbed end near the cell membrane [78]. Formin would induce a cooperative structural change, but those structural changes do not reach all the way to the pointed ends of filaments (gray sections of the filaments). The binding of cofilin to the gray regions induced another type of structural change (magenta filaments), so that the actin filaments are depolymerized in regions deeper inside the cell. The detailed mechanism driven by cooperative structural changes in actin filaments constituting the stress fibers, lamellipodia and filopodia will be explained in the following sections.

3.1. Stress Fibers

Stress fibers are contractile force-generating bundled structures mainly composed of actin filaments and myosin II filaments [79], and are found in nonmuscle cells such as fibroblasts, endothelial cells, myofibroblasts and epithelial cells. The coordination of stress fibers and cell adhesion controls cell migration and cell morphogenesis. Stress fibers contain proteins with various functions, such as actin crosslinking proteins and kinases, in addition to actin filaments and myosin II [80]. The formation and contraction of stress fibers are controlled by the coordinated function of these proteins. The cooperative interaction between actin filaments and ABPs presumably plays a role in the correct assembly of stress fibers. For example, if the actin-depolymerizing protein, cofilin, enters a stress fiber, actin filaments constituting the stress fiber will be severed and its structure will be disrupted.

Cofilin binding supertwists the helix of actin filaments by about 25%, and this conformational change is propagated to a neighboring bare zone [32–35]. On the other hand, the helix slightly untwists when the myosin motor domain binds in the rigor state [81,82]. This allosteric regulation by cooperative conformational changes of actin filaments may contribute to mutually exclusive binding of cofilin and myosin II [4] and/or tropomyosin [26] to actin filaments. However, the presence of ATP was essential for the mutually exclusive binding of myosin II and cofilin [4], suggesting that the structural change of actin filaments induced by the active myosin heads is different from that induced by rigor crossbridges, and is important for long-range allostery. Additionally, cofilin cannot bind to tensed actin filaments in vivo and in vitro [83]. Since interactions with myosin II in the presence of ATP generates tension, myosin II can generate long-range allostery in actin filaments through two mechanisms, i.e., a direct structural effect and the generation of tension, which would synergistically inhibit the binding of cofilin to actively contracting stress fibers. This would partially explain why stress fibers are disassembled when myosin II is pharmacologically inhibited [84,85] or when the deformable substrate to which cells adhere is relaxed [80,83,86].

Stress fibers also include tropomyosin [87] and α-actinin [88], which interact cooperatively with actin filaments [25,47]. Tropomyosin 4 recruits myosin II to stress fiber precursors, and tropomyosin 1, 2/3 and 5NM1/2 stabilize the actin filaments of stress fibers [89]. Although details of these mechanisms are unknown, it is possible that the cooperative interaction between myosin II and tropomyosin 4 or α-actinin is involved. In addition to the mutually exclusive effects between myosin II and cofilin, these positive cooperative effects may control the proper self-assembly of the actin filament-based molecular machineries. Actin structural changes caused by the binding of myosin II, tropomyosin and α-actinin are all transmitted up to about 50 actin protomers ahead, implying that these ABPs induce structural changes of actin protomers using a similar mechanism. It is likely that the cooperative structural changes of actin filaments evoked by the binding of these ABPs and the tension generated by myosin II are involved in the resistance of stress fibers to cofilin (Figure 2), but also to recruit myosin II. This speculation is supported by the observation that the motor domain alone [90] or S1 [91] of the IIB isoform of nonmuscle myosin II binds to stress fibers in normal rat kidney (NRK) cells or fibroblasts, in the absence of any known biochemical regulation. Moreover, the motor domain or S1 of myosin II is localized to tensed, myosin II-containing actin structures, such as contractile rings in dividing cells and the posterior cortex of migrating cells, in Dictyostelium [92]. This latter result is noteworthy in that it suggested that myosin II-induced and/or tension-induced conformational changes in actin filaments are sufficient to recruit myosin II, since Dictyostelium does not have tropomyosin.

In the NRK study by Tang and Ostap, unlike the motor domain of myosin II, the motor domain of myosin I did not localize to stress fibers. Furthermore, the response to tropomyosin binding to actin filaments was different between myosin I and IIB. Interestingly, S1 of the IIA isoform of nonmuscle myosin II did not bind to stress fibers [91]. It is therefore unlikely that the motor domains of all classes of myosin, or even different isoforms within a specific class, induce similar cooperative conformational changes to actin filaments, or recognize a similar structure of actin filaments for cooperative binding. Each isoform or class of myosin appears to be optimized in terms of inducing and/or recognizing cooperative conformational changes in actin filaments.

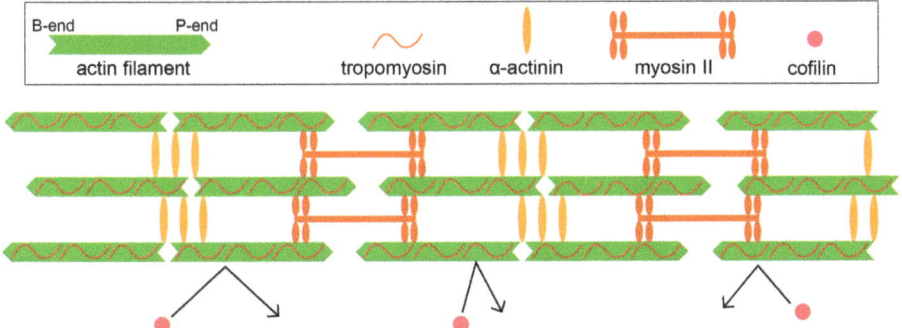

Figure 2. Stress fiber composed of tropomyosin, α-actinin, and myosin II, and the proposed mechanism for how cofilin is excluded. The binding of myosin II (and probably tropomyosin and α-actinin) to actin filaments induces a cooperative structural change of filaments accompanied by untwisting of the helix (green filaments). In addition, stress fibers are also tensed by the movement of myosin II. The untwisting and tension of the actin filaments presumably involve constructing a stress fiber that repels cofilin.

3.2. Lamellipodia

Lamellipodia are broad and flat protrusions along the leading edge of a migrating cell, including neural crest derived cells in several species, keratocytes and macrophages in *Drosophila melanogaster* [93]. They are also composed of actin filaments and various ABPs, such as formin, profilin, the ARP2/3 complex and cofilin [78]. Lamellipodia drive cell migration by the force of actin polymerization at the leading edge. Therefore, it is important that G-actin is recruited to the tip of filaments to maintain continuous polymerization. G-actin is supplied by recycling from old filaments that interact with cofilin [94].

HS-AFM observation revealed that shortening of the helical pitch of actin filaments induced by cofilin binding propagate to the pointed-end side of the cofilin cluster, and that cofilin clusters grow unidirectionally toward the pointed-end of the filament [32]. Severing by cofilin was often observed near the boundaries between cofilin clusters and bare zones or within each cofilin cluster [32]. Cluster growth on the pointed-end side of the initially bound cofilin molecule may accumulate strain within the cluster, resulting in severing of actin filaments at the boundary between the cluster and the bare region and/or inside the cluster. In lamellipodia, the pointed end and barbed end of actin filaments are oriented toward the nucleus and leading edge of the cell, respectively. Formin is bound to the barbed end of actin filaments [78]. Helical rotation of formin decreases the helical twist of an actin filament if there is rotational friction between the formin molecule and the surroundings and between the actin filament and the surroundings. This would confer resistance to cofilin [95]. Cofilin interacts with actin filaments in sections near the pointed end and depolymerizes filaments because structural changes due to formin bound at the barbed end are probably not transmitted to the sections close to the pointed end due to the branch involving the ARP2/3 complex [78,96] (Figure 3). This structural effect, together with the differential distribution of ATP-, ADP•Pi- and ADP-bound protomers along the filament [97,98], which is not discussed in this article, may allow cofilin to efficiently depolymerize actin filaments deeper inside a cell, which are no longer necessary for the generation of force along the leading edge of a lamellipodium.

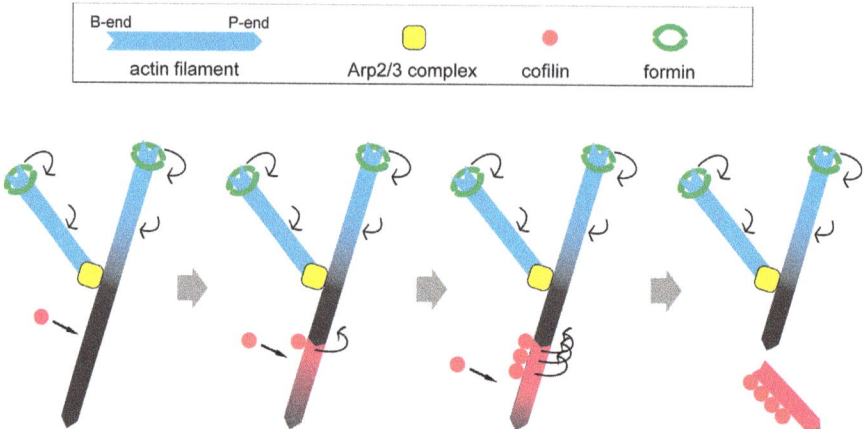

Figure 3. Severing of formin-bound actin filament by cofilin in lamellipodia. Formin bound to the barbed end of an actin filament induces untwisting of the filament (blue), but the effect does not reach the pointed end of filaments (grey) deeper inside the cells because the untwisting is presumably blocked at the branching point where ARP2/3 complex is connected. Cofilin binds to the grey section of the filament unaffected by formin and causes cooperative structural changes with supertwisting of filaments (magenta). Cofilin clusters grow in the magenta area, and the actin filament is severed.

3.3. Filopodia

Filopodia are thin (diameter 0.1-0.3 μm), finger-like projections containing parallel bundles of actin filaments, and function as an antenna for detecting the cell environment [78,99]. The major actin filament-bundling protein in filopodia is fascin [100]. The homologous molecules of fascin have been discovered in a wide range of species, from insects to mammals [101]. The distribution of fascin throughout the actin filament bundles in filopodia [102,103] suggests that fascin can bind to actin filaments independently of formin binding and stabilize filopodia (Figure 4). The crosslinking of actin filaments by fascin presumably suppresses the propagation of structural changes of filaments, including not only untwisting by formin but also supertwisting by cofilin. In fact, Hirakawa et al. reported that cofilin was difficult to cooperatively bind to actin filaments tightly immobilized on a glass surface using biotin-avidin linkages [23]. It is likely that filopodia show resistance to cofilin due to the untwisting of filaments by formin and the crosslinking of filaments by fascin (Figure 4).

It has not been reported whether fascin induces cooperative structural changes of actin filaments, but a small-angle x-ray scattering study revealed that fascin bundles actin into a continuous spectrum of intermediate twist states [104]. Fascin and α-actinin can work in concert to generate enhanced cell stiffness [105], and myosin V selectively moves to the tip of filopodia along the actin filament bundled by fascin [106]. These results imply that there is also an unknown cooperative interaction between fascin and actin filaments.

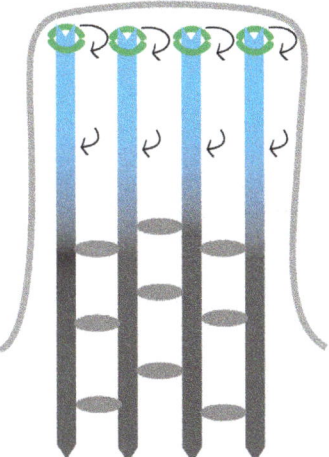

Figure 4. Extension and stabilization of filopodia. Similar to lamellipodia, in filopodia, formin binds to the barbed end of actin filaments and untwists the filaments. Since fascin can bind to actin filaments regardless of this structural change, filaments are bundled in filopodia. The formin-induced untwisting is presumably blocked and not propagated to the pointed-end of filaments by the immobilization by fascin. Cofilin cannot bind not only to the untwisted segments near the tip of filopodia, but also to the bundled sections, since those bundled sections cannot be supertwisted.

3.4. Other Actin-Based Machineries

Muscle is the most representative and well-studied actin-based machinery. Proteins constituting myofibrils are classified into contractile, regulatory, and structural proteins. Actin and myosin II, which are contractile proteins, polymerize to form thin and thick filaments, respectively, and muscles can contract as they slide on each other [54]. Tropomyosin and the troponin complex, which are regulatory proteins, coordinate with each other to switch the myosin activity [107]. Connectin [108] (titin [109]), nebulin [110], tropomodulin [111], α-actinin [68], Fhod3 [112] and others, which are structural proteins, are involved in myofibril formation and maintenance, elasticity and elongation [113].

The contractile ring is a well-known actin-based machinery with an important function in eukaryotic cell division [114]. The contractile ring is also composed of contractile proteins, actin filaments and myosin II, and associated proteins such as formin [115], anillin [116] and septin [117]. α-Actinin and fimbrin cooperate with myosin II when the contractile ring is organized [118].

Muscles and contractile rings are categorized as actomyosin-based machinery, along with stress fibers, because their power source is actomyosin-based motors. We suspect that in these machineries, ABP-induced cooperative structural changes in actin filaments also cause allosteric regulatory effects on other ABP interactions. Further research is needed to uncover the relationships between these allosteric regulatory mechanisms.

Elongation of invadopodia of cancer cells requires microtubules and vimentin intermediate filaments in addition to filopodia machinery [119]. We recently discovered that microtubule elongation along actin filaments induced by microtubule-associated protein-4 contributes to the formation of cellular protrusions [120]. These results suggest the possibility that cancer invasion is regulated by complex and dynamic cell functions expressed by the cooperative interaction between the actin-based machinery and the microtubule-based machinery.

4. Significance of Propagation Distance and Directionality of Long-Range Allostery

Cofilin binds to actin filaments and induces structural changes accompanied by supertwisting in a relatively narrow range of around 10 actin protomers [32,39]. As a result, dense cofilin clusters are formed in a narrow section of the actin filament, and filaments that accumulate strain are severed at the end or inside of the cluster [32]. Such a narrow range of transmission of conformational change presumably results in dense clustering of cofilin, leading to severing of actin filaments by the cluster. In a moderate range of about 50 actin protomers, transmission of conformational changes induced by myosin II [11], tropomyosin [25] and α-actinin [47] may be involved in the formation and stabilization of actin-based machineries such as stress fibers, muscles and contractile rings. Finally, there are cases of structural changes over a much longer distance. For example, more than 160 actin protomers are affected by formin [53], and this would be useful to prevent cofilin binding to actin filaments in a wide, anterior area of lamellipodia.

Directionality of the propagation of structural changes may also play an important role in functional expression of actin-based machineries. For example, the unidirectionality of the cooperative interaction of myosin II with actin filaments [23] presumably plays a role in the unidirectional motor properties of myosin. The mechanism of the cooperative structural change of actin filaments is not well understood, but unidirectional propagation of structural change may be an inherent property of polar cables. In other words, unidirectional propagation of cooperative structural changes may be a consequence rather than being of physiological significance.

As described above, the information transmitted by a structural change in actin filament cables may be important for the formation and function of the actin-based machinery in cells (Figure 3). In addition to the stress fibers and lamellipodia discussed in this review, actin filaments and various ABPs interact with each other in filopodia and the cell cortex, and long-range allostery may play important roles in the expression of their functions. In the future, studies focusing on the types, directions, and propagation distances of cooperative actin filament structural changes induced by the binding of ABPs will elucidate how actin-based machineries are assembled and function.

Similar long-range allostery has been discovered in microtubules as well. Muto et al. reported that kinesin-1 binding to microtubules in the presence of ATP causes a long-range structural change of microtubules, increasing their affinity for kinesin toward the plus end [121]. Shima et al. reported that kinesin-1 binding induces conformation switching of microtubules to increase the affinity for kinesin-1 [122]. They concluded that the biased transport triggered by this positive feedback loop would specify a future axon [122].

5. Future Perspectives

Future studies are needed to reveal detailed allosteric interactions between actin filaments and ABPs whose cooperative interactions with actin filaments are not well analyzed, including drebrin, fimbrin, filamin and gelsolin. Those studies will provide important information to unveil the entire picture of how various ABPs control the formation and function of actin-based machineries through long-range allosteric regulation driven by cooperative conformational changes of actin filaments.

Mutation of the Lys-336 residues of α-actin, which is located near the ATP binding site, to Ile or Asp causes congenital myopathy [123,124]. Umeki et al. reported that K336I actin forms apparently normal cofilaments with wild type actin, but the interactions between the cofilaments and α-actinin, cofilin, and myosin II are impaired in vitro [41]. Noguchi et al. also revealed that the G146V mutation, which causes dominant lethality in yeast, cooperatively inhibits cofilin binding in vitro [39]. These results support the idea that ABP-dependent cooperative structural changes in actin protomers, or the long-range allostery, play important roles in essential cellular activities and that the defect in the long-range allostery can cause disease. Clearly, more work is needed to understand the relationship between diseases and actin's long-range allostery.

6. Conclusions

Actin filaments express some types of long-range allostery caused by an interaction with ABPs, which have different propagation distances of cooperative structural change. These cooperative structural changes influence each other, so that actin-based machineries can be properly formed and function. Research to elucidate the function of the actin-based machinery through the cooperative structural change of actin filaments is still in its infancy. This long-range allostery, found not only in actin filaments but also in microtubules, will be the key to understanding differences between the biological and artificial machineries. That is, artificial machines composed of gears, motors, etc., are designed without considering the flexibility and cooperativity of the parts. In contrast, in biological machineries, the functions realized by the high degree of cooperativity of parts themselves play important roles in the operation mechanism. Protein molecules are by no means simply smaller versions of parts of artificial machines.

Author Contributions: Conceptualization, K.T. and T.Q.P.U.; writing—original draft preparation, K.T.; writing—review and editing, M.K. and T.Q.P.U. All authors have read and agreed to the published version of the manuscript.

Funding: This research was funded by The Japan Society for the Promotion of Science (KAKENHI, JP24117008 grand awarded to T.Q.P.U.; KAKENHI, JP16K14704 grand awarded to K.T.).

Conflicts of Interest: The authors declare no conflicts of interest.

Abbreviations

ABD	actin-binding domain
ABP	actin-binding protein
ADF	actin-depolymerizing factor
CH	calponin-homology
FRET	fluorescence resonance energy transfer
HMM	heavy meromyosin
HMM-GFP	GFP-fused HMM
HS-AFM	high-speed atomic force microscopy

References

1. Moores, S.L.; Sabry, J.H.; Spudich, J.A. Myosin dynamics in live Dictyostelium cells. *Proc. Natl. Acad. Sci. USA* **1996**, *93*, 443–446. [CrossRef] [PubMed]
2. Yumura, S.; Mori, H.; Fukui, Y. Localization of actin and myosin for the study of ameboid movement in Dictyostelium using improved immunofluorescence. *J. Cell Biol.* **1984**, *99*, 894–899. [CrossRef] [PubMed]
3. Aizawa, H.; Fukui, Y.; Yahara, I. Live dynamics of Dictyostelium cofilin suggests a role in remodeling actin latticework into bundles. *J. Cell Sci.* **1997**, *110*, 2333–2344. [PubMed]
4. Ngo, K.X.; Umeki, N.; Kijima, S.T.; Kodera, N.; Ueno, H.; Furutani-Umezu, N.; Nakajima, J.; Noguchi, T.Q.P.; Nagasaki, A.; Tokuraku, K.; et al. Allosteric regulation by cooperative conformational changes of actin filaments drives mutually exclusive binding with cofilin and myosin. *Sci. Rep.* **2016**, *6*, 35449. [CrossRef] [PubMed]
5. Galkin, V.E.; Orlova, A.; Schroder, G.F.; Egelman, E.H. Structural polymorphism in F-actin. *Nat. Struct. Mol. Biol.* **2010**, *17*, 1318–1323. [CrossRef] [PubMed]
6. Kozuka, J.; Yokota, H.; Arai, Y.; Ishii, Y.; Yanagida, T. Dynamic polymorphism of single actin molecules in the actin filament. *Nat. Chem. Biol.* **2006**, *2*, 83–86. [CrossRef] [PubMed]
7. Uyeda, T.Q.P.; Ngo, K.X.; Kodera, N.; Tokuraku, K. Uni-directional Propagation of Structural Changes in Actin Filaments. In *The Role of Water in ATP Hydrolysis Energy Transduction by Protein Machinery*; Suzuki, M., Ed.; Springer: Singapore, 2018; pp. 157–177.
8. Oosawa, F. Macromolecular assembly of actin. In *Muscle and Non-muscle Motility*; Stracher, A., Ed.; Academic Press: New York, NY, USA, 1983; pp. 151–216.

9. Egelman, E.H.; Orlova, A. New insights into actin filament dynamics. *Curr. Opin. Struct. Biol.* **1995**, *5*, 172–180. [CrossRef]
10. Oosawa, F.; Fujime, S.; Ishiwata, S.; Mihashi, K. Dynamic property of F-actin and thin filament. *Cold Spring Harbor Symp. Quant. Biol.* **1972**, *37*, 277–285. [CrossRef]
11. Miki, M.; Wahl, P.; Auchet, J.C. Fluorescence anisotropy of labeled F-actin: Influence of divalent cations on the interaction between F-actin and myosin heads. *Biochemistry* **1982**, *21*, 3661–3665. [CrossRef]
12. Tawada, K. Physicochemical studies of F-actin-heavy meromyosin solutions. *Biochim. Biophys. Acta* **1969**, *172*, 311–318. [CrossRef]
13. Fujime, S.; Ishiwata, S. Dynamic study of F-actin by quasielastic scattering of laser light. *J. Mol. Biol.* **1971**, *62*, 251–265. [CrossRef]
14. Loscalzo, J.; Reed, G.H.; Weber, A. Conformational change and cooperativity in actin filaments free of tropomyosin. *Proc. Natl. Acad. Sci. USA* **1975**, *72*, 3412–3415. [CrossRef] [PubMed]
15. Thomas, D.D.; Seidel, J.C.; Gergely, J. Rotational dynamics of spin-labeled F-actin in the sub-millisecond time range. *J. Mol. Biol.* **1979**, *132*, 257–273. [CrossRef]
16. Craig, R.; Szent-Gyorgyi, A.G.; Beese, L.; Flicker, P.; Vibert, P.; Cohen, C. Electron microscopy of thin filaments decorated with a Ca^{2+}-regulated myosin. *J. Mol. Biol.* **1980**, *140*, 35–55. [CrossRef]
17. Frado, L.L.; Craig, R. Electron microscopy of the actin-myosin head complex in the presence of ATP. *J. Mol. Biol.* **1992**, *223*, 391–397. [CrossRef]
18. Walker, M.; White, H.; Belknap, B.; Trinick, J. Electron cryomicroscopy of acto-myosin-S1 during steady-state ATP hydrolysis. *Biophys. J.* **1994**, *66*, 1563–1572. [CrossRef]
19. Woodrum, D.T.; Rich, S.A.; Pollard, T.D. Evidence for biased bidirectional polymerization of actin filaments using heavy meromyosin prepared by an improved method. *J. Cell Biol.* **1975**, *67*, 231–237. [CrossRef]
20. Orlova, A.; Egelman, E.H. Cooperative rigor binding of myosin to actin is a function of F-actin structure. *J. Mol. Biol.* **1997**, *265*, 469–474. [CrossRef]
21. Tokuraku, K.; Kurogi, R.; Toya, R.; Uyeda, T.Q.P. Novel mode of cooperative binding between myosin and Mg^{2+}-actin filaments in the presence of low concentrations of ATP. *J. Mol. Biol.* **2009**, *386*, 149–162. [CrossRef]
22. Kitazawa, T.; Shuman, H.; Somlyo, A.P. Calcium and magnesium binding to thin and thick filaments in skinned muscle fibres: Electron probe analysis. *J. Muscle Res. Cell Motil.* **1982**, *3*, 437–454. [CrossRef]
23. Hirakawa, R.; Nishikawa, Y.; Uyeda, T.Q.P.; Tokuraku, K. Unidirectional growth of heavy meromyosin clusters along actin filaments revealed by real-time fluorescence microscopy. *Cytoskeleton (Hoboken)* **2017**, *74*, 482–489. [CrossRef] [PubMed]
24. Butters, C.A.; Willadsen, K.A.; Tobacman, L.S. Cooperative interactions between adjacent troponin-tropomyosin complexes may be transmitted through the actin filament. *J. Biol. Chem.* **1993**, *268*, 15565–15570. [PubMed]
25. Khaitlina, S.; Tsaplina, O.; Hinssen, H. Cooperative effects of tropomyosin on the dynamics of the actin filament. *FEBS Lett.* **2017**, *591*, 1884–1891. [CrossRef] [PubMed]
26. Christensen, J.R.; Hocky, G.M.; Homa, K.E.; Morganthaler, A.N.; Hitchcock-DeGregori, S.E.; Voth, G.A.; Kovar, D.R. Competition between Tropomyosin, Fimbrin, and ADF/Cofilin drives their sorting to distinct actin filament networks. *Elife* **2017**, *6*, 6. [CrossRef] [PubMed]
27. Gateva, G.; Kremneva, E.; Reindl, T.; Kotila, T.; Kogan, K.; Gressin, L.; Gunning, P.W.; Manstein, D.J.; Michelot, A.; Lappalainen, P. Tropomyosin Isoforms Specify Functionally Distinct Actin Filament Populations In Vitro. *Curr. Biol.* **2017**, *27*, 705–713. [CrossRef] [PubMed]
28. De La Cruz, E.M. Cofilin binding to muscle and non-muscle actin filaments: Isoform-dependent cooperative interactions. *J. Mol. Biol.* **2005**, *346*, 557–564. [CrossRef] [PubMed]
29. Hawkins, M.; Pope, B.; Maciver, S.K.; Weeds, A.G. Human actin depolymerizing factor mediates a pH-sensitive destruction of actin filaments. *Biochemistry* **1993**, *32*, 9985–9993. [CrossRef]
30. Hayakawa, K.; Sakakibara, S.; Sokabe, M.; Tatsumi, H. Single-molecule imaging and kinetic analysis of cooperative cofilin-actin filament interactions. *Proc. Natl. Acad. Sci. USA* **2014**, *111*, 9810–9815. [CrossRef]
31. Hayden, S.M.; Miller, P.S.; Brauweiler, A.; Bamburg, J.R. Analysis of the interactions of actin depolymerizing factor with G- and F-actin. *Biochemistry* **1993**, *32*, 9994–10004. [CrossRef]
32. Ngo, K.X.; Kodera, N.; Katayama, E.; Ando, T.; Uyeda, T.Q.P. Cofilin-induced unidirectional cooperative conformational changes in actin filaments revealed by high-speed atomic force microscopy. *Elife* **2015**, *4*, 04806. [CrossRef]

33. Galkin, V.E.; Orlova, A.; Kudryashov, D.S.; Solodukhin, A.; Reisler, E.; Schroder, G.F.; Egelman, E.H. Remodeling of actin filaments by ADF/cofilin proteins. *Proc. Natl. Acad. Sci. USA* **2011**, *108*, 20568–20572. [CrossRef] [PubMed]
34. Galkin, V.E.; Orlova, A.; Lukoyanova, N.; Wriggers, W.; Egelman, E.H. Actin depolymerizing factor stabilizes an existing state of F-actin and can change the tilt of F-actin subunits. *J. Cell Biol.* **2001**, *153*, 75–86. [CrossRef] [PubMed]
35. McGough, A.; Pope, B.; Chiu, W.; Weeds, A. Cofilin changes the twist of F-actin: Implications for actin filament dynamics and cellular function. *J. Cell Biol.* **1997**, *138*, 771–781. [CrossRef] [PubMed]
36. Suarez, C.; Roland, J.; Boujemaa-Paterski, R.; Kang, H.; McCullough, B.R.; Reymann, A.C.; Guerin, C.; Martiel, J.L.; De la Cruz, E.M.; Blanchoin, L. Cofilin tunes the nucleotide state of actin filaments and severs at bare and decorated segment boundaries. *Curr. Biol.* **2011**, *21*, 862–868. [CrossRef] [PubMed]
37. Umeki, N.; Hirose, K.; Uyeda, T.Q.P. Cofilin-induced cooperative conformational changes of actin subunits revealed using cofilin-actin fusion protein. *Sci. Rep.* **2016**, *6*, 20406. [CrossRef] [PubMed]
38. Wioland, H.; Guichard, B.; Senju, Y.; Myram, S.; Lappalainen, P.; Jegou, A.; Romet-Lemonne, G. ADF/Cofilin Accelerates Actin Dynamics by Severing Filaments and Promoting Their Depolymerization at Both Ends. *Curr. Biol.* **2017**, *27*, 1956–1967 e1957. [CrossRef]
39. Noguchi, T.Q.P.; Toya, R.; Ueno, H.; Tokuraku, K.; Uyeda, T.Q.P. Screening of novel dominant negative mutant actins using glycine targeted scanning identifies G146V actin that cooperatively inhibits cofilin binding. *Biochem. Biophys. Res. Commun.* **2010**, *396*, 1006–1011. [CrossRef]
40. Huehn, A.R.; Bibeau, J.P.; Schramm, A.C.; Cao, W.; De La Cruz, E.M.; Sindelar, C.V. Structures of cofilin-induced structural changes reveal local and asymmetric perturbations of actin filaments. *Proc. Natl. Acad. Sci. USA* **2020**, *117*, 1478–1484. [CrossRef]
41. Umeki, N.; Shibata, K.; Noguchi, T.Q.P.; Hirose, K.; Sako, Y.; Uyeda, T.Q.P. K336I mutant actin alters the structure of neighbouring protomers in filaments and reduces affinity for actin-binding proteins. *Sci. Rep.* **2019**, *9*, 5353. [CrossRef]
42. Sharma, S.; Grintsevich, E.E.; Hsueh, C.; Reisler, E.; Gimzewski, J.K. Molecular cooperativity of drebrin1-300 binding and structural remodeling of F-actin. *Biophys. J.* **2012**, *103*, 275–283. [CrossRef]
43. Sharma, S.; Grintsevich, E.E.; Phillips, M.L.; Reisler, E.; Gimzewski, J.K. Atomic force microscopy reveals drebrin induced remodeling of f-actin with subnanometer resolution. *Nano Lett* **2011**, *11*, 825–827. [CrossRef] [PubMed]
44. Galkin, V.E.; Orlova, A.; Cherepanova, O.; Lebart, M.C.; Egelman, E.H. High-resolution cryo-EM structure of the F-actin-fimbrin/plastin ABD2 complex. *Proc. Natl. Acad. Sci. USA* **2008**, *105*, 1494–1498. [CrossRef] [PubMed]
45. Hanein, D.; Matsudaira, P.; DeRosier, D.J. Evidence for a conformational change in actin induced by fimbrin (N375) binding. *J. Cell Biol.* **1997**, *139*, 387–396. [CrossRef] [PubMed]
46. Skau, C.T.; Kovar, D.R. Fimbrin and tropomyosin competition regulates endocytosis and cytokinesis kinetics in fission yeast. *Curr. Biol.* **2010**, *20*, 1415–1422. [CrossRef]
47. Craig-Schmidt, M.C.; Robson, R.M.; Goll, D.E.; Stromer, M.H. Effect of alpha-actinin on actin structure. Release of bound nucleotide. *Biochim. Biophys. Acta* **1981**, *670*, 9–16. [CrossRef]
48. Galkin, V.E.; Orlova, A.; Salmazo, A.; Djinovic-Carugo, K.; Egelman, E.H. Opening of tandem calponin homology domains regulates their affinity for F-actin. *Nat. Struct. Mol. Biol.* **2010**, *17*, 614–616. [CrossRef]
49. Shibata, K.; Nagasaki, A.; Adachi, H.; Uyeda, T.Q.P. Actin binding domain of filamin distinguishes posterior from anterior actin filaments in migrating Dictyostelium cells. *Biophys Physicobiol* **2016**, *13*, 321–331. [CrossRef]
50. Hansen, S.D.; Kwiatkowski, A.V.; Ouyang, C.Y.; Liu, H.; Pokutta, S.; Watkins, S.C.; Volkmann, N.; Hanein, D.; Weis, W.I.; Mullins, R.D.; et al. Alpha E-catenin actin-binding domain alters actin filament conformation and regulates binding of nucleation and disassembly factors. *Mol. Biol. Cell* **2013**, *24*, 3710–3720. [CrossRef]
51. Orlova, A.; Prochniewicz, E.; Egelman, E.H. Structural dynamics of F-actin: II. Cooperativity in structural transitions. *J. Mol. Biol.* **1995**, *245*, 598–607. [CrossRef]
52. Selden, L.A.; Kinosian, H.J.; Newman, J.; Lincoln, B.; Hurwitz, C.; Gershman, L.C.; Estes, J.E. Severing of F-actin by the amino-terminal half of gelsolin suggests internal cooperativity in gelsolin. *Biophys. J.* **1998**, *75*, 3092–3100. [CrossRef]
53. Papp, G.; Bugyi, B.; Ujfalusi, Z.; Barko, S.; Hild, G.; Somogyi, B.; Nyitrai, M. Conformational changes in actin filaments induced by formin binding to the barbed end. *Biophys. J.* **2006**, *91*, 2564–2572. [CrossRef] [PubMed]

54. Huxley, H.E. The mechanism of muscular contraction. *Science* **1969**, *164*, 1356–1365. [CrossRef] [PubMed]
55. Bailey, K. Tropomyosin: A new asymmetric protein component of muscle. *Nature* **1946**, *157*, 368. [CrossRef] [PubMed]
56. Ebashi, S.; Endo, M. Calcium ion and muscle contraction. *Prog. Biophys. Mol. Biol.* **1968**, *18*, 123–183. [CrossRef]
57. Weber, A.; Murray, J.M. Molecular control mechanisms in muscle contraction. *Physiol. Rev.* **1973**, *53*, 612–673. [CrossRef]
58. Nishida, E.; Maekawa, S.; Sakai, H. Cofilin, a protein in porcine brain that binds to actin filaments and inhibits their interactions with myosin and tropomyosin. *Biochemistry* **1984**, *23*, 5307–5313. [CrossRef]
59. Yonezawa, N.; Nishida, E.; Sakai, H. pH control of actin polymerization by cofilin. *J Biol Chem* **1985**, *260*, 14410–14412.
60. Bobkov, A.A.; Muhlrad, A.; Pavlov, D.A.; Kokabi, K.; Yilmaz, A.; Reisler, E. Cooperative effects of cofilin (ADF) on actin structure suggest allosteric mechanism of cofilin function. *J. Mol. Biol.* **2006**, *356*, 325–334. [CrossRef]
61. Shirao, T.; Kojima, N.; Kato, Y.; Obata, K. Molecular cloning of a cDNA for the developmentally regulated brain protein, drebrin. *Brain Res.* **1988**, *464*, 71–74. [CrossRef]
62. Lin, Y.C.; Koleske, A.J. Mechanisms of synapse and dendrite maintenance and their disruption in psychiatric and neurodegenerative disorders. *Annu. Rev. Neurosci.* **2010**, *33*, 349–378. [CrossRef]
63. Bretscher, A.; Weber, K. Fimbrin, a new microfilament-associated protein present in microvilli and other cell surface structures. *J. Cell Biol.* **1980**, *86*, 335–340. [CrossRef] [PubMed]
64. Dubreuil, R.R. Structure and evolution of the actin crosslinking proteins. *Bioessays* **1991**, *13*, 219–226. [CrossRef] [PubMed]
65. Hartwig, J.H.; Kwiatkowski, D.J. Actin-binding proteins. *Curr. Opin. Cell Biol.* **1991**, *3*, 87–97. [CrossRef]
66. Matsudaira, P. Modular organization of actin crosslinking proteins. *Trends Biochem. Sci.* **1991**, *16*, 87–92. [CrossRef]
67. Otto, J.J. Actin-bundling proteins. *Curr. Opin. Cell Biol.* **1994**, *6*, 105–109. [CrossRef]
68. Maruyama, K.; Ebashi, S. Alpha-actinin, a new structural protein from striated muscle. II. Action on actin. *J. Biochem.* **1965**, *58*, 13–19. [CrossRef]
69. Wang, K.; Ash, J.F.; Singer, S.J. Filamin, a new high-molecular-weight protein found in smooth muscle and non-muscle cells. *Proc. Natl. Acad. Sci. USA* **1975**, *72*, 4483–4486. [CrossRef]
70. Stossel, T.P.; Condeelis, J.; Cooley, L.; Hartwig, J.H.; Noegel, A.; Schleicher, M.; Shapiro, S.S. Filamins as integrators of cell mechanics and signalling. *Nat. Rev. Mol. Cell Biol.* **2001**, *2*, 138–145. [CrossRef]
71. Byfield, F.J.; Wen, Q.; Leventhal, I.; Nordstrom, K.; Arratia, P.E.; Miller, R.T.; Janmey, P.A. Absence of filamin A prevents cells from responding to stiffness gradients on gels coated with collagen but not fibronectin. *Biophys. J.* **2009**, *96*, 5095–5102. [CrossRef]
72. Lynch, C.D.; Gauthier, N.C.; Biais, N.; Lazar, A.M.; Roca-Cusachs, P.; Yu, C.H.; Sheetz, M.P. Filamin depletion blocks endoplasmic spreading and destabilizes force-bearing adhesions. *Mol. Biol. Cell* **2011**, *22*, 1263–1273. [CrossRef]
73. Luo, T.; Mohan, K.; Iglesias, P.A.; Robinson, D.N. Molecular mechanisms of cellular mechanosensing. *Nat. Mater.* **2013**, *12*, 1064–1071. [CrossRef] [PubMed]
74. Sun, C.; Forster, C.; Nakamura, F.; Glogauer, M. Filamin-A regulates neutrophil uropod retraction through RhoA during chemotaxis. *PLoS ONE* **2013**, *8*, e79009. [CrossRef] [PubMed]
75. Yin, H.L.; Stossel, T.P. Control of cytoplasmic actin gel-sol transformation by gelsolin, a calcium-dependent regulatory protein. *Nature* **1979**, *281*, 583–586. [CrossRef] [PubMed]
76. Mass, R.L.; Zeller, R.; Woychik, R.P.; Vogt, T.F.; Leder, P. Disruption of formin-encoding transcripts in two mutant limb deformity alleles. *Nature* **1990**, *346*, 853–855. [CrossRef] [PubMed]
77. Courtemanche, N. Mechanisms of formin-mediated actin assembly and dynamics. *Biophys. Rev.* **2018**, *10*, 1553–1569. [CrossRef] [PubMed]
78. Ridley, A.J. Life at the leading edge. *Cell* **2011**, *145*, 1012–1022. [CrossRef]
79. Kreis, T.E.; Birchmeier, W. Stress fiber sarcomeres of fibroblasts are contractile. *Cell* **1980**, *22*, 555–561. [CrossRef]
80. Tojkander, S.; Gateva, G.; Lappalainen, P. Actin stress fibers—Assembly, dynamics and biological roles. *J. Cell Sci.* **2012**, *125*, 1855–1864. [CrossRef]

81. Tsaturyan, A.K.; Koubassova, N.; Ferenczi, M.A.; Narayanan, T.; Roessle, M.; Bershitsky, S.Y. Strong binding of myosin heads stretches and twists the actin helix. *Biophys. J.* **2005**, *88*, 1902–1910. [CrossRef]
82. Holmes, K.C.; Angert, I.; Kull, F.J.; Jahn, W.; Schroder, R.R. Electron cryo-microscopy shows how strong binding of myosin to actin releases nucleotide. *Nature* **2003**, *425*, 423–427. [CrossRef]
83. Hayakawa, K.; Tatsumi, H.; Sokabe, M. Actin filaments function as a tension sensor by tension-dependent binding of cofilin to the filament. *J. Cell Biol.* **2011**, *195*, 721–727. [CrossRef] [PubMed]
84. Chrzanowska-Wodnicka, M.; Burridge, K. Rho-stimulated contractility drives the formation of stress fibers and focal adhesions. *J. Cell Biol.* **1996**, *133*, 1403–1415. [CrossRef] [PubMed]
85. Ono, S.; Abe, H.; Obinata, T. Stimulus-dependent disorganization of actin filaments induced by overexpression of cofilin in C2 myoblasts. *Cell Struct. Funct.* **1996**, *21*, 491–499. [CrossRef] [PubMed]
86. Engler, A.J.; Sen, S.; Sweeney, H.L.; Discher, D.E. Matrix elasticity directs stem cell lineage specification. *Cell* **2006**, *126*, 677–689. [CrossRef]
87. Lazarides, E. Tropomyosin antibody: The specific localization of tropomyosin in nonmuscle cells. *J. Cell Biol.* **1975**, *65*, 549–561. [CrossRef]
88. Lazarides, E.; Burridge, K. Alpha-actinin: Immunofluorescent localization of a muscle structural protein in nonmuscle cells. *Cell* **1975**, *6*, 289–298. [CrossRef]
89. Tojkander, S.; Gateva, G.; Schevzov, G.; Hotulainen, P.; Naumanen, P.; Martin, C.; Gunning, P.W.; Lappalainen, P. A molecular pathway for myosin II recruitment to stress fibers. *Curr. Biol.* **2011**, *21*, 539–550. [CrossRef]
90. Tang, N.; Ostap, E.M. Motor domain-dependent localization of myo1b (myr-1). *Curr. Biol.* **2001**, *11*, 1131–1135. [CrossRef]
91. Kuragano, M.; Uyeda, T.Q.P.; Kamijo, K.; Murakami, Y.; Takahashi, M. Different contributions of nonmuscle myosin IIA and IIB to the organization of stress fiber subtypes in fibroblasts. *Mol. Biol. Cell* **2018**, *29*, 911–922. [CrossRef]
92. Uyeda, T.Q.P.; Iwadate, Y.; Umeki, N.; Nagasaki, A.; Yumura, S. Stretching actin filaments within cells enhances their affinity for the myosin II motor domain. *PLoS ONE* **2011**, *6*, e26200. [CrossRef]
93. Krause, M.; Gautreau, A. Steering cell migration: Lamellipodium dynamics and the regulation of directional persistence. *Nat. Rev. Mol. Cell Biol.* **2014**, *15*, 577–590. [CrossRef] [PubMed]
94. Van Rheenen, J.; Condeelis, J.; Glogauer, M. A common cofilin activity cycle in invasive tumor cells and inflammatory cells. *J. Cell Sci.* **2009**, *122*, 305–311. [CrossRef] [PubMed]
95. Mizuno, H.; Tanaka, K.; Yamashiro, S.; Narita, A.; Watanabe, N. Helical rotation of the diaphanous-related formin mDia1 generates actin filaments resistant to cofilin. *Proc. Natl. Acad. Sci.* **2018**, *115*, E5000–E5007. [CrossRef]
96. Goley, E.D.; Welch, M.D. The ARP2/3 complex: An actin nucleator comes of age. *Nat. Rev. Mol. Cell Biol.* **2006**, *7*, 713–726. [CrossRef] [PubMed]
97. Carlier, M.F.; Laurent, V.; Santolini, J.; Melki, R.; Didry, D.; Xia, G.X.; Hong, Y.; Chua, N.H.; Pantaloni, D. Actin depolymerizing factor (ADF/cofilin) enhances the rate of filament turnover: Implication in actin-based motility. *J. Cell Biol.* **1997**, *136*, 1307–1322. [CrossRef] [PubMed]
98. Pollard, T.D.; Borisy, G.G. Cellular motility driven by assembly and disassembly of actin filaments. *Cell* **2003**, *112*, 453–465. [CrossRef]
99. Mattila, P.K.; Lappalainen, P. Filopodia: Molecular architecture and cellular functions. *Nat. Rev. Mol. Cell Biol.* **2008**, *9*, 446–454. [CrossRef]
100. Machesky, L.M.; Li, A. Fascin: Invasive filopodia promoting metastasis. *Commun. Integr. Biol.* **2010**, *3*, 263–270. [CrossRef]
101. Adams, J.C. Roles of fascin in cell adhesion and motility. *Curr. Opin. Cell Biol.* **2004**, *16*, 590–596. [CrossRef]
102. Nakagawa, H.; Terasaki, A.G.; Suzuki, H.; Ohashi, K.; Miyamoto, S. Short-term retention of actin filament binding proteins on lamellipodial actin bundles. *FEBS Lett.* **2006**, *580*, 3223–3228. [CrossRef]
103. Cohan, C.S.; Welnhofer, E.A.; Zhao, L.; Matsumura, F.; Yamashiro, S. Role of the actin bundling protein fascin in growth cone morphogenesis: Localization in filopodia and lamellipodia. *Cell Motil. Cytoskeleton* **2001**, *48*, 109–120. [CrossRef]
104. Shin, H.; Purdy Drew, K.R.; Bartles, J.R.; Wong, G.C.; Grason, G.M. Cooperativity and frustration in protein-mediated parallel actin bundles. *Phys. Rev. Lett.* **2009**, *103*, 238102. [CrossRef] [PubMed]

105. Tseng, Y.; Kole, T.P.; Lee, J.S.; Fedorov, E.; Almo, S.C.; Schafer, B.W.; Wirtz, D. How actin crosslinking and bundling proteins cooperate to generate an enhanced cell mechanical response. *Biochem. Biophys. Res. Commun.* **2005**, *334*, 183–192. [CrossRef] [PubMed]
106. Nagy, S.; Rock, R.S. Structured post-IQ domain governs selectivity of myosin X for fascin-actin bundles. *J. Biol. Chem.* **2010**, *285*, 26608–26617. [CrossRef] [PubMed]
107. Gordon, A.M.; Homsher, E.; Regnier, M. Regulation of contraction in striated muscle. *Physiol. Rev.* **2000**, *80*, 853–924. [CrossRef] [PubMed]
108. Maruyama, K. Connectin, an elastic protein from myofibrils. *J. Biochem.* **1976**, *80*, 405–407. [CrossRef]
109. Wang, K.; McClure, J.; Tu, A. Titin: Major myofibrillar components of striated muscle. *Proc. Natl. Acad. Sci. USA* **1979**, *76*, 3698–3702. [CrossRef]
110. Wang, K.; Ramirez-Mitchell, R. A network of transverse and longitudinal intermediate filaments is associated with sarcomeres of adult vertebrate skeletal muscle. *J. Cell Biol.* **1983**, *96*, 562–570. [CrossRef]
111. Fowler, V.M. Tropomodulin: A cytoskeletal protein that binds to the end of erythrocyte tropomyosin and inhibits tropomyosin binding to actin. *J. Cell Biol.* **1990**, *111*, 471–481. [CrossRef]
112. Taniguchi, K.; Takeya, R.; Suetsugu, S.; Kan, O.M.; Narusawa, M.; Shiose, A.; Tominaga, R.; Sumimoto, H. Mammalian formin fhod3 regulates actin assembly and sarcomere organization in striated muscles. *J. Biol. Chem.* **2009**, *284*, 29873–29881. [CrossRef]
113. Alberts, B. *Molecular Biology of the Cell*, 6th ed.; Garland Science, Taylor and Francis Group: New York, NY, USA, 2015; p. 1.
114. Schwayer, C.; Sikora, M.; Slovakova, J.; Kardos, R.; Heisenberg, C.P. Actin Rings of Power. *Dev. Cell* **2016**, *37*, 493–506. [CrossRef] [PubMed]
115. Pelham, R.J.; Chang, F. Actin dynamics in the contractile ring during cytokinesis in fission yeast. *Nature* **2002**, *419*, 82–86. [CrossRef] [PubMed]
116. Field, C.M.; Alberts, B.M. Anillin, a contractile ring protein that cycles from the nucleus to the cell cortex. *J. Cell Biol.* **1995**, *131*, 165–178. [CrossRef] [PubMed]
117. Kinoshita, M.; Kumar, S.; Mizoguchi, A.; Ide, C.; Kinoshita, A.; Haraguchi, T.; Hiraoka, Y.; Noda, M. Nedd5, a mammalian septin, is a novel cytoskeletal component interacting with actin-based structures. *Genes Dev.* **1997**, *11*, 1535–1547. [CrossRef] [PubMed]
118. Laporte, D.; Ojkic, N.; Vavylonis, D.; Wu, J.Q. alpha-Actinin and fimbrin cooperate with myosin II to organize actomyosin bundles during contractile-ring assembly. *Mol. Biol. Cell* **2012**, *23*, 3094–3110. [CrossRef]
119. Schoumacher, M.; Goldman, R.D.; Louvard, D.; Vignjevic, D.M. Actin, microtubules, and vimentin intermediate filaments cooperate for elongation of invadopodia. *J. Cell Biol.* **2010**, *189*, 541–556. [CrossRef]
120. Doki, C.; Nishida, K.; Saito, S.; Shiga, M.; Ogara, H.; Kuramoto, A.; Kuragano, M.; Nozumi, M.; Igarashi, M.; Nakagawa, H.; et al. Microtubule elongation along actin filaments induced by microtubule-associated protein 4 contributes to the formation of cellular protrusions. *J. Biochem.* **2020**. [CrossRef]
121. Muto, E.; Sakai, H.; Kaseda, K. Long-range cooperative binding of kinesin to a microtubule in the presence of ATP. *J. Cell Biol.* **2005**, *168*, 691–696. [CrossRef]
122. Shima, T.; Morikawa, M.; Kaneshiro, J.; Kambara, T.; Kamimura, S.; Yagi, T.; Iwamoto, H.; Uemura, S.; Shigematsu, H.; Shirouzu, M.; et al. Kinesin-binding-triggered conformation switching of microtubules contributes to polarized transport. *J. Cell Biol.* **2018**, *217*, 4164–4183. [CrossRef]
123. Laing, N.G.; Dye, D.E.; Wallgren-Pettersson, C.; Richard, G.; Monnier, N.; Lillis, S.; Winder, T.L.; Lochmuller, H.; Graziano, C.; Mitrani-Rosenbaum, S.; et al. Mutations and polymorphisms of the skeletal muscle alpha-actin gene (ACTA1). *Hum. Mutat.* **2009**, *30*, 1267–1277. [CrossRef]
124. Sparrow, J.C.; Nowak, K.J.; Durling, H.J.; Beggs, A.H.; Wallgren-Pettersson, C.; Romero, N.; Nonaka, I.; Laing, N.G. Muscle disease caused by mutations in the skeletal muscle alpha-actin gene (ACTA1). *Neuromuscul. Disord.* **2003**, *13*, 519–531. [CrossRef]

© 2020 by the authors. Licensee MDPI, Basel, Switzerland. This article is an open access article distributed under the terms and conditions of the Creative Commons Attribution (CC BY) license (http://creativecommons.org/licenses/by/4.0/).

Review

Stereocilia Rootlets: Actin-Based Structures That Are Essential for Structural Stability of the Hair Bundle

Itallia Pacentine, Paroma Chatterjee and Peter G. Barr-Gillespie *

Oregon Hearing Research Center & Vollum Institute, Oregon Health & Science University, Portland, OR 97239, USA; pacentin@ohsu.edu (I.P.); chatterp@ohsu.edu (P.C.)
* Correspondence: gillespp@ohsu.edu; Tel.: +1-503-494-2936

Received: 12 December 2019; Accepted: 1 January 2020; Published: 3 January 2020

Abstract: Sensory hair cells of the inner ear rely on the hair bundle, a cluster of actin-filled stereocilia, to transduce auditory and vestibular stimuli into electrical impulses. Because they are long and thin projections, stereocilia are most prone to damage at the point where they insert into the hair cell's soma. Moreover, this is the site of stereocilia pivoting, the mechanical movement that induces transduction, which additionally weakens this area mechanically. To bolster this fragile area, hair cells construct a dense core called the rootlet at the base of each stereocilium, which extends down into the actin meshwork of the cuticular plate and firmly anchors the stereocilium. Rootlets are constructed with tightly packed actin filaments that extend from stereocilia actin filaments which are wrapped with TRIOBP; in addition, many other proteins contribute to the rootlet and its associated structures. Rootlets allow stereocilia to sustain innumerable deflections over their lifetimes and exemplify the unique manner in which sensory hair cells exploit actin and its associated proteins to carry out the function of mechanotransduction.

Keywords: rootlet; actin; stereocilia; hair cell

1. Introduction

Eukaryotic cells use actin as a basic building block of the cytoskeleton. Actin monomers bind end-to-end to one another and form filaments, which grow through a polymerization reaction that is controlled by ATP [1]. All actin filaments are polarized, with filament ends that are named for their appearance when decorated with a myosin motor fragment; polymerization occurs about ten times faster at the barbed (or plus) end than it does at the pointed (or minus) end. Depending on how filaments are organized, actin can be deployed for a variety of cellular processes. Cells use actin to power cell motility or control cytokinesis [2]; cells also exploit actin to adopt and maintain specific cellular structures [3,4], which include lamellipodia, filopodia, microvilli, and stereocilia. Adding another layer of versatility, compartments of actin can form within larger actin frameworks simply by differential organization of the filaments; these specialized compartments can then perform unique functions. One such compartment is the rootlet.

Actin-containing rootlets anchor actin-based structures to the cell's soma, and these rootlets can be found associated with microvilli, stereocilia, and other actin protrusions [5–12]. By their appearance, rootlets appear to secure those actin protrusions to the rest of the cell. Confusingly, the term "rootlet" is also used to describe anchors for microtubule-based processes, including cilia and flagella [13–17]. Cilia and flagella emanate apically from the basal body, an organizing structure for microtubules, whereas rootlets project from the basal body deep into the cell [18]. Ciliary rootlets are thought to be formed by rootletin [19,20], which is also a structural component of basal bodies and centrosomes. Microtubule-associated rootlets are not thought to have any molecular similarities to the rootlets of actin-based processes.

Both types of rootlets share several key features, including an internal organization that is distinct from surrounding structures, a role in anchoring projecting structures of cells, and specialized binding proteins. We focus here on the essential role of the rootlet in the stereocilia of sensory hair cells.

2. Hair Cells Mediate Hearing and Balance

Hair cells are specialized neuroepithelial cells that are responsible for detecting sound and head movements [21]. Hair cells of the auditory system are functionally similar to those of the vestibular system, capable of responding to deflections of their apical hair bundles by external forces in both cases. Auditory and vestibular stimuli are uniquely conveyed to hair cells, accounting for the differential sensitivity of the two systems to differing external stimuli. Hair bundles each contain 20–300 stereocilia per hair cell [22], arranged in rows of increasing height. Adjacent stereocilia are connected to each other by a variety of linkages, the most famous of which are the tip links [23], which gate the mechanoelectrical transduction (MET) channels [24]. Deflection of the hair bundle leads to sliding of individual stereocilia relative to their neighbors, which in turn increases tension on tip links, activating the channels. MET channels are nonselective cation channels, and by admitting K^+ from the special ionic environment that is exposed to the apical surface (the endolymph), channel opening leads to hair cell depolarization. In turn, depolarization activates neurotransmitter release at the base of the hair cell, leading to excitation of auditory or vestibular nerves.

3. Compartments of Actin in Sensory Hair Cells

Hair cells compartmentalize actin into three main structures: the stereocilia, a cuticular plate, and the rootlets (Figure 1), each of which has a specialized role in the overall function of the hair cell. These compartments are defined by type of actin filament and how these filaments are arranged.

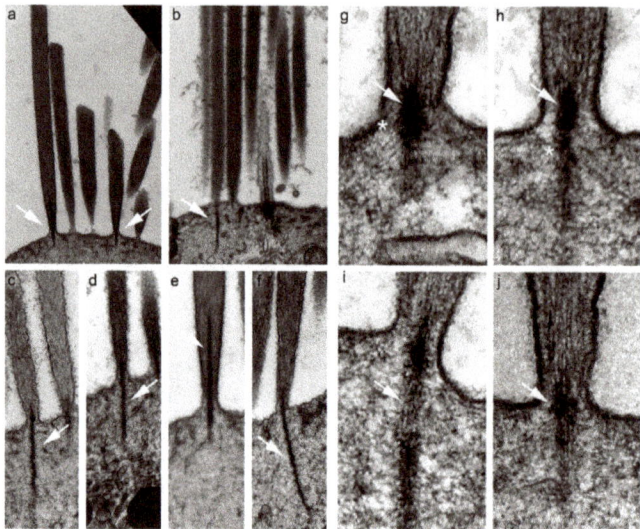

Figure 1. Rootlet structure as visualized with transmission electron microscopy (TEM). Each panel shows an image of a rootlet in a mouse utricle hair cell. Arrows indicate rootlets in each panel; arrows point to upper rootlets in (**a,e**), and to lower rootlets in (**b–d,f,i**). Arrows point to the rootlet at the insertion of the stereocilium into the soma in (**g,h,j**). The apparent length and location of a rootlet depends on the orientation of the thin section used for imaging, as well as the orientation of the rootlet within the stereocilia and cuticular plate. Asterisks in **e** and **f** indicate filaments that appear to connect the rootlet to the membrane and cuticular plate. Ages: (**a,b**) P12; (**c**) P5; (**d–f**) P12; (**g–j**) P5. Panel full widths: (**a,b**) 3000 nm; (**c–f**) 700 nm; (**g–j**) 350 nm.

The most visually striking compartment of a hair cell is the stereocilium, a finger-like extension from the apex of the soma. Stereocilia, which are comprised of parallel actin filaments that are heavily crosslinked, are very rigid and do not bend except at the base [25]. All stereocilia actin filaments are oriented in the same direction, with the barbed end pointed away from the cell [26]. The number of actin filaments decreases from hundreds or even thousands along the shaft to only a few dozen at the base of the stereocilium where it inserts into the apical soma [5]. This reduction in filament number shapes the base into a narrow taper. This taper allows the stereocilium to pivot [27], which is the mechanical movement that is required to activate a hair cell and give rise to hearing and balance [21].

The second compartment is the cuticular plate, which is located below the stereocilium in the apical soma [28]. The cuticular plate is a bowl-shaped mesh of actin, constructed of filaments with no directionality [7]; pointed and barbed ends are randomly oriented, and are crosslinked with proteins like spectrin [29], LMO7 [30], and XIRP2 [31,32]. This meshwork of filaments provides a solid structure similar to the way entwined twigs allow a bird nest to hold its shape, and makes an ideal anchoring point for a third actin structure, the rootlet.

4. Morphology of the Rootlet

Hair cell rootlets were first noted in 1965 in transmission electron microscopy (TEM) images, where they appear as dark, splinter-shaped densities associated with stereocilia [33]. Rootlets are osmiophilic and therefore appear as heavily stained structures in conventional TEM micrographs (Figure 1). A single rootlet appears as a core within a stereocilium, starting halfway down or lower, and extends down through the tapered region and into the cell soma, anchoring into the cuticular plate (Figure 2a). The rootlet thus forms a bridge that connects the stereocilium and cuticular plate. Rootlets contain uniquely organized actin [34–36], with a higher density of filaments than in either the stereocilia or the cuticular plate.

Rootlets begin to develop after the initial stereocilia lengthening phase [37]. In the chick auditory organ, rootlets begin to appear at embryonic day 13 (E13), forming simultaneously with the taper [6]. In some cases, stereocilia actin filaments appear to merge into the rootlet, suggesting that formation of the rootlet may contribute to shaping of the taper [38]. By E14, an osmiophilic plug is visible at the stereocilia insertion; rootlet filaments continue to grow within the cuticular plate for several more days, forming the lower rootlet. In the mouse auditory organ (the cochlea), rootlets begin to develop around E15, and the cuticular plate then forms around the rootlet by E18 [39]. Others have not observed rootlet development in mice until after P1, with the upper rootlet (the dense core at the base of stereocilia) appearing first, and then extending down through the taper into the cuticular plate [40]. These developmental differences may reflect the lack of a strict morphological definition for the rootlet.

Detailed TEM images have revealed the shape and structure of the rootlet. In longitudinal sections, the rootlet is composed of parallel filaments, similar to but more densely packed than in the stereocilium [34,35,41]. When examined by TEM after staining with tannic acid, which can act as a negative stain, the filaments stain relatively lightly but show a periodic banding pattern, occurring at approximately 36 nm intervals [35]. As 36 nm is the actin helix half-repeat, the periodicity at which actin subunits are optimally oriented for interaction with binding proteins [42], these results suggest the presence of a regular structure binding the rootlet together. In alligator lizards, rootlets are demonstrably composed of actin [5,7], but the identity of filaments in mammalian rootlets was initially unclear. While the dimensions and organization of the filaments resembled actin paracrystals [34,35], in early experiments with mammals, rootlets did not label with S1 myosin fragments [25,26,41]. Tight packing of the filaments likely impeded entry by S1 myosin fragments; however, thin sectioning exposed the rootlet filaments, which could then be labeled with anti-actin antibodies [36]. Mammals have six genes for actin [43], two of which—*ACTB* and *ACTG1*—are expressed in non-muscle cells. Immunogold labeling revealed that rootlets are primarily comprised of ACTB oriented with barbed ends up, exactly like the stereocilia, though they label sparsely for ACTG1 too [38,44].

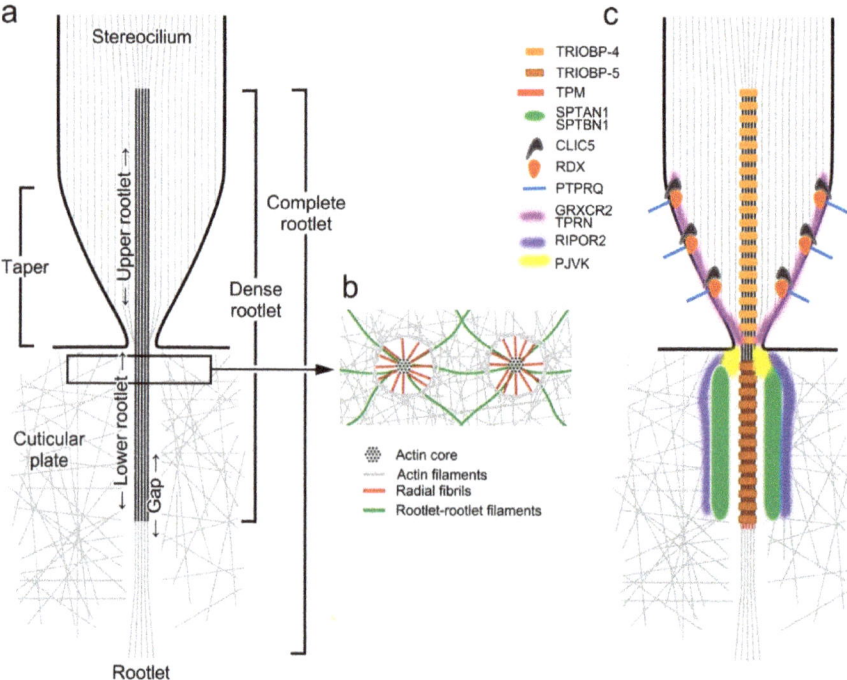

Figure 2. Diagrams illustrating rootlet structure. (**a**) Actin filament structure only. In a stereocilium, spacing of actin filaments (gray) is maintained by actin–actin crosslinkers (not shown). Crosslinkers disappear in the taper region, where most filaments terminate on or near the plasma membrane in a systematic way, forming the taper. By contrast, the central dozen or two filaments are gathered together to form the rootlet, which eventually penetrates into the cell and extends into the cuticular plate. The rootlet stains darkly with osmium tetraoxide, forming the dense rootlet, which accounts for the majority of the rootlet visible in TEM. Rootlets have an upper portion in the stereocilium ('upper rootlet') and a lower portion in the cuticular plate ('lower rootlet'). In the cuticular plate, there is a gap between the filaments of the rootlet and the meshwork of the cuticular plate. (**b**) Rootlet connecting filaments. In cross-sections of the rootlet near the apical surface (see box in **a**), several types of filaments can be identified. Radial fibrils (red) extend from the core of the rootlet to the surrounding cuticular plate, while rootlet–rootlet filaments, visible only after detergent and EDTA extraction prior to fixation, appear to interconnect rootlets. (**c**) Proteins of the rootlet. TRIOBP-4 is exclusively associated with the upper rootlet, while TRIOBP-5 is associated with the lower rootlet. Both TRIOBP splice forms have multiple actin binding domains, which apparently allow TRIOBP to wrap around and bundle the rootlet filaments. TRIOBP-5 has additional domains (separately encoded by the TRIOBP-1 splice form) that may connect TRIOBP-5 to surrounding structures. The spectrin isoforms SPTAN1 and SPTBN1 apparently form a sheath around the lower rootlet, appearing as rings in confocal horizontal sections. CLIC5, RDX, and PTPRQ form a membrane complex in the taper region; they may also bind TPRN, which is also associated with the taper and rootlet. GRXCR2 maintains TPRN in the taper region. The precise localization of PJVK is not clear. RIPOR2 (FAM65B) also forms rings around the rootlet, but it is unclear whether the rings persist along the lower portion of the rootlet.

Rootlets vary in their morphology as they traverse the stereocilia and cuticular plate. In alligator lizards, lower rootlets are composed of filament ribbons held in a circular pattern by fibrils that are 3 nm in diameter, forming a hollow tubular structure [5,7]. These fibrils do not label with S1 myosin fragments, suggesting that they are not actin [5,7]. However, in horizontal sections of mammals, the rootlet is composed of approximately 8 nm diameter filaments that are arranged in a hexagonal pattern in both the upper and lower rootlet [34,35]. Some lower rootlets have a hollow center [34,35], particularly those from the first and second row stereocilia [38]. The filaments stain lightly with tannic acid and are outlined by electron-dense material [41]. With a center-to-center spacing of about 8 nm [34,40], close to the diameter of the filaments themselves [35], it is unlikely that actin binding proteins are present between the actin filaments of mammalian rootlets.

In the mammalian inner ear, horizontal TEM sections show that rootlets within the stereocilium (upper rootlets) have a dense core surrounded by a dense ring, with a gap of non-dense material in between [38]. In outer hair cells, the rootlet is widest just above the entry point to the cuticular plate [38]. The rootlet diameter corresponds to stereocilia height, with longer stereocilia having thicker rootlets [38]. As the rootlet passes down through the taper region, the dense ring converges on the rootlet, so that lower rootlets appear as a solitary dense core with a non-dense gap circling it [38]. In the non-dense gap surrounding lower rootlets, thin fibrils (3–5 nm in diameter) extend outward radially from the rootlet, connecting it to the meshwork of the cuticular plate (Figure 2b) [34,35,38,41,45]. These radial fibrils are most numerous at the apex of the cuticular plate, with the number and length of fibrils decreasing as the rootlet plunges deeper [45]. Other filaments form rootlet–rootlet connectors within the cuticular plate [34]. Neither the radial fibrils nor the rootlet–rootlet connectors decorate with S1 myosin fragments, which are specific for actin filaments, suggesting that these filaments are comprised of proteins distinct from actin [41,45].

In the mouse auditory organ, rootlet position and length vary by hair cell subtype, location of the cell along the sensory epithelium, and by individual rows of stereocilia. In all hair cell types, the total length of a rootlet correlates with the height of its associated stereocilium; the longest rootlets in an individual cell are associated with first row stereocilia, the tallest row [38,40]. Loss of rootlets does not affect stereocilia height, nor does loss of stereocilia affect lower rootlet length, indicating that independent mechanisms control the lengths of these two structures [40]. The longest rootlets, those associated with first row stereocilia, sometimes punch through the cuticular plate entirely [38] and have been observed to take a sharp 110° turn to merge with filaments of the striated organelle [46]. The striated organelle is a cytoskeletal lattice located near the plasma membrane below the cuticular plate; it is only found in mammalian inner hair cells and vestibular hair cells [47–49]. In outer hair cells, the apical soma folds over to form a lip that contains the outer edges of the cuticular plate, and lateral rootlets within this lip have been observed to angle sideways and directly contact the lateral wall [38]. At the base of rootlets within the cuticular plate, some rootlets become thin, flat, or crescent-shaped, and the terminal points can sometimes splay [38].

Another important measure is how far the upper rootlet extends into the stereocilium; greater penetration into the stereocilium may indicate additional strength. In all apical mouse auditory hair cells, both inner and outer, upper rootlets rise about a third of the way up the stereocilia, whereas in basal outer hair cells, they rise about halfway up the stereocilia [38]. This measurement has not been reported for basal inner hair cells. In addition, how deep the rootlet extends into the cuticular plate may reflect the stresses imposed on the stereocilium. In one study [38], reconstructed serial vertical TEM sections were used to measure the lengths of lower rootlets, and this length was compared to the height of their stereocilia as a ratio (stereocilia height: lower rootlet length). In inner hair cells, regardless of location, there was a consistent drop in this ratio across stereocilia rows. This progressive drop in ratio across the stereocilia rows also occurred in apical outer hair cells, where the first, second and third row ratios were 2.7:1 2:1, and 1.7:1. This shifting ratio may suggest that the length of lower rootlets has an upper limit, possibly set by the size of the cuticular plate. Basal outer hair cells showed a uniquely consistent ratio across stereocilia rows; in all rows, the height of the stereocilium was about equal to the length of the lower rootlet (ratio about 1:1).

It is intriguing that there are differences in the morphology of rootlets when comparing basal and apical hair cells. Mammalian auditory hair cells are frequency-tuned, with apical cells responding best to low frequencies and basal cells responding to high frequencies. In basal outer cells, the longer upper and lower rootlets (relative to stereocilia height) may reflect a greater need to stiffen and strengthen stereocilia in high-frequency cells, which are subjected to many more cycles of stimulation. These differences in rootlet organization among hair cell populations could contribute to frequency selection.

Rootlets can bend in the taper region of the stereocilia [38] where the stereocilia pivot; bending makes functional sense, as activation of the hair cell relies on the stereocilia pivoting. Strikingly, however, rootlets are bent even at rest—the lower rootlet can be angled as much 7° off from its upper section and corresponding stereocilium [50]. Furthermore, the direction of this angle differs across stereocilia rows. The angles of rootlets in tall stereocilia are negative (bent towards shorter stereocilia), and those in short stereocilia are positive (bent towards taller stereocilia). This bending suggests that rootlets contribute to forces that cause the stereocilia to maintain contact with their neighbors by leaning into each other, taller toward shorter and vice-versa. Previously, it was thought that the curve of the cuticular plate into a shallow bowl was the only contributor to this resting tendency for stereocilia tips to contact one another [51]. Further investigation revealed that while rootlets may contribute to the resting lean of stereocilia, they are likely not required for maintaining this contact during deflections. During deflections, the bundle integrity is instead attributed to side and ankle linkages between neighboring stereocilia [50]. Internal tension within the hair bundle caused by motor control over the tip links could lead to bending of rootlets [52], as could tension that pulls rootlets apart from each other, especially if they are connected by elastic linkages. Regardless, the observation that rootlets are bent at rest and presumably impart internal tension to the bundle suggests that forces that keep stereocilia spaced at discrete intervals are also very important.

5. Rootlet Protein Composition

While prior studies utilized TEM imaging of the stereocilia rootlets to determine their structure, reports of the protein composition of the rootlet—beyond actin—have only recently emerged (Figure 2c). Two groups of proteins associated with rootlets can be distinguished: those that are internal components, and those that associate externally with the rootlet. Although the filamentous component of the rootlet is actin, several other proteins have been localized to the rootlets directly. An early study examined the localization of contractile proteins in the mammalian auditory hair cells [36]; antibodies against tropomyosin, an actin binding protein, labeled the rootlet region of both outer and inner hair cells. Although tropomyosin's precise localization relative to the rootlet actin density was unclear, later immunogold labeling experiments confirmed tropomyosin's presence within stereocilia rootlets [38]. Tropomyosin binds within the groove of the actin helix and prevents binding of some proteins to actin [53]; in the rootlet, tropomyosin may control access of actin binding proteins to the rootlet structural filaments.

A key structural component of rootlets is TRIOBP (TRIO and F-actin binding protein), an actin-associated protein [54]; highlighting its significance for the auditory system, mutations in *TRIOBP* cause hereditary hearing loss in humans [55,56]. The mouse *Triobp* gene encodes three isoforms: TRIOBP-5 is full-length, while TRIOBP-4 only encodes the N-terminus and TRIOBP-1 the C-terminus [57,58]. TRIOBP-1 is ubiquitously expressed, whereas the expression of TRIOBP-4 and TRIOBP-5 is mainly restricted to the eye and inner ear [55,56]. In vitro, TRIOBP-4 tightly bundles actin into rootlet-like structures [40,58]. While loss of TRIOBP-1 causes embryonic lethality, mice expressing TRIOBP-1 but lacking both TRIOBP-4 and TRIOBP-5 have fragile stereocilia that never develop rootlets, and that are prone to damage [40,58]. Loss of just TRIOBP-5 causes dysmorphic rootlets, with lower rootlets that are thin and wispy or absent, and upper rootlets that are elongated and widened [58].

Immunolabeling demonstrated that TRIOBP-4 is present primarily in the upper rootlet, while TRIOBP-5 expression is restricted to the lower rootlet [40,58]. A GFP fusion with TRIOBP-1 localizes to rootlets [59], suggesting that the C-terminus of TRIOBP-5 (which contains the complete TRIOBP-1 sequence) interacts with binding partners while the N-terminus (which contains actin-bundling domains) interacts with the rootlet actin filaments. TRIOBP expression is reduced in the absence of LIM-only protein 7 (LMO7), a primary component of the cuticular plate [30]. Consistent with loss of TRIOBP-5, LMO7-deficient mice also have abnormal rootlet morphology, and suffer progressive hearing loss [30].

Pejvakin (PJVK) also localizes to rootlets. While mice lacking PJVK suffer from profound hearing loss, their rootlets appear normal, and instead hair bundle morphology is disrupted [59]. TRIOBP co-localizes with PJVK, and TRIOBP-1 interacts with the C-terminus of PJVK [59]. As the exons encoding TRIOBP-1 are also found in TRIOBP-5, it is likely that PJVK is a binding partner of TRIOBP-5. The functional role of PJVK in rootlets is unclear, however.

Taperin (TPRN) may also play a role in rootlet function. TPRN is primarily concentrated at the taper region of the stereocilia and associated with non-syndromic hearing loss in humans [60]. Loss of TPRN in mice leads to the disruption of the stereociliary rootlet and eventual loss of stereocilia, which results in hearing loss [61]. Visualization of TPRN using stochastic optical reconstruction microscopy (STORM) suggested that TPRN is present in the core of the taper, where the rootlet resides [62]. Although TPRN immunoreactivity does not appear to be associated with the lower rootlet within the cuticular plate, $Tprn^{-/-}$ mice have rootlets that are unusually curved and that have hollow central regions surrounded by dense rings on the periphery [61]. TPRN can have profound effects on stereocilia actin structure; when the *Grxcr2* gene is disrupted, TPRN mislocalizes to upper stereocilia shafts, and stereocilia are both longer and profoundly disrupted [63]. The tight restriction of TPRN to stereocilia bases is thus functionally important.

Experiments using immunogold labeling suggest that rootlets contain the actin-binding protein spectrin [38]. Spectrins usually oligomerize as αβ tetramers, and the predominant isoforms expressed in mouse hair cells are SPTAN1 and SPTBN1 [64]. A recent study imaged both SPTAN1 and SPTBN1 using super-resolution fluorescence microscopy [65], and showed that spectrin is not a structural component of rootlets, but rather surrounds the lower rootlet. While spectrin is initially present throughout the cuticular plate in both inner and outer hair cells, by P14, spectrin condenses into ring-like structures that surround lower rootlets, extending several hundred nanometers down into the cuticular plate [65]. The spectrin structures are hollow cylinders or sheaths that surround the rootlets. The late formation of the spectrin sheaths suggests that this protein has a role in maintenance of rootlets. Spectrin sheaths are not found in the third row of inner hair cells [65]; since these stereocilia pivot when the hair bundle is deflected, spectrin cannot be responsible for the ability of the dense rootlet to bend in the taper region. Instead, it suggests a role in strengthening the connection between the lower rootlet and the cuticular plate. The stereocilia of the third row are narrow compared with those of the first and second rows, so the extra support provided by spectrin may only be necessary for thicker stereocilia. This conclusion is supported by spectrin distribution in vestibular hair cells. In these hair cells with thinner stereocilia, spectrin forms a meshwork in the cuticular plate instead of ensheathing the lower rootlets [65].

Mice lacking SPTBN1 display severe deafness, which highlights the importance of spectrin sheaths for mammalian hearing [65]. SPTBN1 is also required for proper localization of SPTAN1, and so SPTAN1 is also missing from cuticular plates in $Sptbn1^{-/-}$ mice [65]. TPRN expression is disrupted by the loss of SPTBN1 [65], although the altered distribution reflects the disorganization of stereocilia in $Sptbn1^{-/-}$ mice rather than a change in TPRN distribution within a stereocilium. Spectrin also plays an important role in the localization of RIPOR2, also known as FAM65B, around the rootlet [65]. RIPOR2 is expressed in a distinct ring around the edges of the taper region, with no labeling in the center [62]; in $Sptbn1^{-/-}$ mice, RIPOR2 is no longer associated with the stereocilia insertions. RIPOR2, like spectrin, apparently surrounds rootlets, and its altered distribution in $Sptbn1^{-/-}$ mice suggests that spectrin anchors it there. While it is unclear whether RIPOR2 has a distinct function in rootlets, its presence influences the localization of TPRN; in *Ripor2* knockouts, TPRN no longer localizes to rootlets [62]. Similarly, TPRN localization to the base of stereocilia is dependent on the protein CLIC5. TPRN and CLIC5 form a complex with RDX and MYO6 [66], and the three proteins other than TPRN are concentrated in the taper region but are not thought to associate directly with rootlets.

In the frog saccule, immunoelectron microscopy revealed that MYO1C, an unconventional myosin, was located in a ring where stereocilia insert into the soma [67]. Additionally, these authors also showed that isolated stereocilia display a concentration of the unconventional myosin MYO6 at their tapered ends [67]. Combined with its known association with TPRN, MYO6 may be involved in rootlet formation or function. Whether myosin molecules can crawl along exposed rootlet actin filaments is unclear; both MYO1C and MYO6 may be bound to rootlet actin filaments but are unable to move further along the filaments, trapping them at the base of stereocilia.

In the auditory organ of guinea pigs, TEM with immunogold labeling demonstrated that calmodulin (CALM) is enriched in the rootlet region of both outer and inner hair cells [68]. These results suggest that a component of the rootlet may bind CALM and be regulated by Ca^{2+}, although no known CALM binding proteins have yet been localized to rootlets.

6. Function of the Rootlet

Projecting structures often have rootlets or similar structures, and it is likely that the fundamental function of a rootlet is to anchor the projection. A dense core that literally roots into the principal cytoskeleton may provide stability and strength to otherwise flimsy structures. Hair cell stereocilia undoubtedly rely on rootlets to resist damage, as stereocilia that lack rootlets are more susceptible to long-term damage after deflections [40]. Rootlets could resist damage either by strengthening the stereocilia insertion or by increasing stereocilia stiffness, preventing large damaging deflections. Stereocilia that lack rootlets are 2–4× more flexible during deflections by fluid-jet stimulation; however, once BAPTA treatment removed the contribution of extracellular linkages between stereocilia, stereocilia without rootlets were then 3–10× more flexible than controls [40]. Since neither TEM nor scanning electron microscopy revealed extra bending in stereocilia lacking rootlets, the change in flexibility must arise from alterations to the mechanical properties of the taper region rather than to changes in the stereocilia core rigidity [40]. Even with its specialized reinforcing structures, or perhaps because it is the first line of defense to mechanical stress, the rootlet is the hair cell structure that is most often damaged by sound stimuli that cause permanent threshold shifts [69]. Damaging noise exposure can cause upper rootlets to shorten, which may reduce bundle stiffness, or even to disconnect entirely from the lower rootlet [69]. If rootlets are prevented from developing, stereocilia form and reach normal heights but eventually fuse to their neighbors and degenerate [40]. These observations argue for an essential structural role for rootlets.

Do rootlets contribute to the features of hair cell MET? Hair cells are able to sense sound and vibrations because stereocilia pivot at their bases and slide with respect to each other within the hair bundle; accordingly, detection of motion by stereocilia is the basis of hearing and balance. Experiments and models show that the stiffness of the rootlets (also known as the pivot springs) contributes to overall stiffness of the stereocilia [52,70,71]. The rootlets provide an opposing force that allows the gating spring, the elastic element that controls opening of MET channels, to remain extended at rest. The magnitude of this opposing force can be measured directly in experiments where gating springs are severed with BAPTA iontophoresis while a hair bundle is under displacement clamp [72,73]. Significant rootlet stiffness is also required for the negative bundle stiffness observed in the frog saccule [74,75]. Rootlets thus play a critical role in the mechanics of stereocilia deflection, but are not required to perform MET itself. When applying a displacement stimulus to hair cells using a glass probe, the maximum current elicited from the MET channel was unchanged in *Triobp* mutants, which lack rootlets. Moreover, MET in these mutants did not differ from that of their wild-type siblings with regard to the current–displacement relationship, fast adaptation rate and extent, and slow adaptation rate and extent [40]. Application of a force stimulus to a bundle, for example using a fluid-jet stimulus, reveals the essential function of rootlets; in *Triobp* mutant mice, bundle stiffness was greatly diminished [40]. More in-depth characterization of mechanotransduction using a flexible-fiber force stimulus should reveal the consequences of alteration of the rootlets for hair bundle function.

7. Conclusions and Perspectives

In sensory hair cells, rootlets play an essential role in ensuring that the pivot point of stereocilia, the locus of greatest mechanical stress, is durable. Rootlet actin filaments, tightly bundled by TRIOBP and surrounded by hollow spectrin cylinders, must resist significant life-long forces.

Many aspects of rootlets remain mysterious. TEM imaging reveals a rich abundance of filaments that both comprise the rootlet and connect it to other structures, but the identities of the proteins making up these filaments remain largely unknown. For example, which protein comprises the rootlet–rootlet connectors in the lower rootlet, or the radial fibrils? The localization of spectrin makes it a plausible candidate, and in some published images rootlet–rootlet connectors or radial fibrils appear to be labeled by spectrin antibodies [65]. High-resolution immunogold labeling of SPTAN1 and SPTBN1 could determine if the filaments observed in TEM images are indeed spectrin. We still do not know the protein component of the dark ring surrounding the upper rootlet, and there are no good candidates.

The unique mechanical needs for the stereocilia pivot point of the stereocilium stimulate questions regarding the molecular structure of the rootlet. Stereocilia can pivot to extreme degrees, and the rootlet pivots with it. How is it possible for the rootlet to form sharp angles when it is comprised of rigid actin filaments? Rootlets damaged by loud sound often fracture at the pivot point, confirming that this is a weak area [69,76]. Do the actin filaments of the upper and lower rootlets terminate at the pivot point? If so, are they held in place by unknown proteins that provide flexibility? Or is the actin specially organized at this single point so as to allow filaments to bend without breaking? If the bundled rootlet actin filaments can slide with respect to each other (shearing), then differential movements of the filaments can accommodate pivoting stereocilia (Figure 3). Note however that this model, where only rootlet filaments slide but stereocilia filaments do not, contradicts Tilney's hypothesis that all stereocilia actin filaments shear during a stimulus [77].

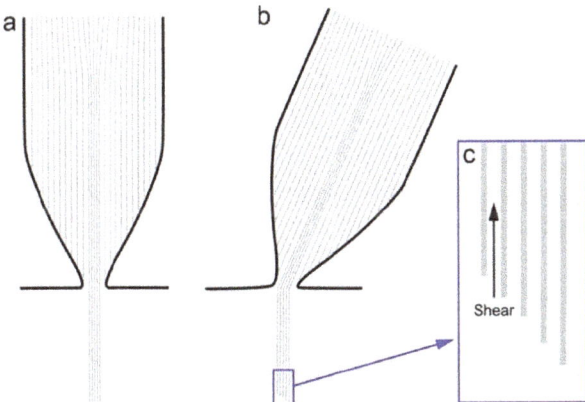

Figure 3. Hypothesis for rootlet filament movement during stereocilia pivoting. (**a**) Rootlet actin filaments are mechanically similar to 1/4" steel cables—flexible but inextensible. Multiple rootlet filaments are bundled together, probably with TRIOBP. (**b**) If rootlet filaments are capable of shearing (sliding with respect to each other), then during stereocilia pivoting, filaments furthest from the direction of the pivoting (left in this diagram) will shear more than those closest to the direction of pivoting. (**c**) Magnified view of the soma ends of rootlets illustrating differential shear of rootlet filaments.

The unique cytoskeletal structure of the stereocilia rootlet requires further investigation, both to determine its structural features and to identify the proteins that are required for its assembly, structure, and function. Because of the difficulty in visualizing the three-dimensional structure of the rootlet using conventional TEM methods, the focused ion beam-scanning electron microscopy (FIB-SEM) approach taken by Katsuno and colleagues [58] should be a superior method for rootlet characterization [78], as is electron cryo-tomography [79]. Both methods allow for three-dimensional reconstruction of structures in a volume once segmentation has been carried out. Localization of proteins using antibodies is possible with each technique, but is difficult.

Like other actin domains in hair cells, the rootlet is a rich source of intriguing cytoskeletal mysteries. Further investigation into the rootlet's structure is essential and now possible given the superior three-dimensional imaging tools available. Moreover, continued identification of new rootlet components, as well as more accurate localization of all components, is necessary to determine the rootlet's molecular structure and mechanism of assembly. Finally, a deeper understanding of the rootlet's function is needed; does it serve to strengthen the stereocilia insertion or simply to stiffen the pivot point? Answering these and related questions will contribute to our deepening understanding of the stereocilia rootlet.

Funding: Research in the authors' laboratory was funded by National Institutes of Health grants R01 DC002368, R01 DC011034, and R01014427.

Acknowledgments: We thank Jocelyn Krey for comments on this manuscript.

Conflicts of Interest: The authors declare no conflict of interest.

Abbreviations

TEM Transmission electron microscopy
MET Mechanoelectrical transduction

References

1. Pollard, T.D.; Borisy, G.G. Cellular motility driven by assembly and disassembly of actin filaments. *Cell* **2003**, *112*, 453–465. [CrossRef]
2. Pollard, T.D.; Cooper, J.A. Actin, a central player in cell shape and movement. *Science* **2009**, *326*, 1208–1212. [CrossRef] [PubMed]
3. Chhabra, E.S.; Higgs, H.N. The many faces of actin: Matching assembly factors with cellular structures. *Nat. Cell Biol.* **2007**, *9*, 1110–1121. [CrossRef] [PubMed]
4. Michelot, A.; Drubin, D.G. Building distinct actin filament networks in a common cytoplasm. *Curr. Biol.* **2011**, *21*, R560–R569. [CrossRef]
5. Tilney, L.G.; DeRosier, D.J.; Mulroy, M.J. The organization of actin filaments in the stereocilia of cochlear hair cells. *J. Cell Biol.* **1980**, *86*, 244–259. [CrossRef]
6. Tilney, L.G.; DeRosier, D.J. Actin filaments, stereocilia, and hair cells of the bird cochlea. IV. How the actin filaments become organized in developing stereocilia and in the cuticular plate. *Dev. Biol.* **1986**, *116*, 119–129. [CrossRef]
7. DeRosier, D.J.; Tilney, L.G. The structure of the cuticular plate, an in vivo actin gel. *J. Cell Biol.* **1989**, *109*, 2853–2867. [CrossRef]
8. Cotanche, D.A. Development of hair cell stereocilia in the avian cochlea. *Hear. Res.* **1987**, *28*, 35–44. [CrossRef]
9. Hirokawa, N.; Tilney, L.G. Interactions between actin filaments and between actin filaments and membranes in quick-frozen and deeply etched hair cells of the chick ear. *J. Cell Biol.* **1982**, *95*, 249–261. [CrossRef]
10. Duckert, L.G.; Rubel, E.W. Ultrastructural observations on regenerating hair cells in the chick basilar papilla. *Hear. Res.* **1990**, *48*, 161–182. [CrossRef]
11. Blest, A.D.; De Couet, H.G.; Sigmund, C. The cytoskeleton of microvilli of leech photoreceptors. A stable bundle of actin microfilaments. *Cell Tissue Res.* **1983**, *234*, 9–16. [CrossRef]
12. Hirokawa, N.; Tilney, L.G.; Fujiwara, K.; Heuser, J.E. Organization of actin, myosin, and intermediate filaments in the brush border of intestinal epithelial cells. *J. Cell Biol.* **1982**, *94*, 425–443. [CrossRef]
13. Horridge, G.A. Statocysts of medusae and evolution of stereocilia. *Tissue Cell* **1969**, *1*, 341–353. [CrossRef]
14. Salisbury, J.L.; Floyd, G.L. Calcium-induced contraction of the rhizoplast of a quadriflagellate green alga. *Science* **1978**, *202*, 975–977. [CrossRef]
15. Salisbury, J.L.; Baron, A.; Surek, B.; Melkonian, M. Striated flagellar roots: Isolation and partial characterization of a calcium-modulated contractile organelle. *J. Cell Biol.* **1984**, *99*, 962–970. [CrossRef]

16. Wolfrum, U. Cytoskeletal elements in arthropod sensilla and mammalian photoreceptors. *Biol. Cell* **1992**, *76*, 373–381. [CrossRef]
17. Chen, J.V.; Kao, L.R.; Jana, S.C.; Sivan-Loukianova, E.; Mendonça, S.; Cabrera, O.A.; Singh, P.; Cabernard, C.; Eberl, D.F.; Bettencourt-Dias, M.; et al. Rootletin organizes the ciliary rootlet to achieve neuron sensory function in Drosophila. *J. Cell Biol.* **2015**, *211*, 435–453. [CrossRef]
18. Worley, L.G.; Fischbeinn, E.; Shapiro, J.E. The structure of ciliated epithelial cells as revealed by the electron microscope and in phase contrast. *J. Morphol.* **1953**, *92*, 545–577. [CrossRef]
19. Yang, J.; Liu, X.; Yue, G.; Adamian, M.; Bulgakov, O.; Li, T. Rootletin, a novel coiled-coil protein, is a structural component of the ciliary rootlet. *J. Cell Biol.* **2002**, *159*, 431–440. [CrossRef]
20. Yang, J.; Gao, J.; Adamian, M.; Wen, X.H.; Pawlyk, B.; Zhang, L.; Sanderson, M.J.; Zuo, J.; Makino, C.L.; Li, T. The ciliary rootlet maintains long-term stability of sensory cilia. *Mol. Cell Biol.* **2005**, *25*, 4129–4137. [CrossRef]
21. Hudspeth, A.J. How the ear's works work. *Nature* **1989**, *341*, 397–404. [CrossRef]
22. Roberts, W.M.; Howard, J.; Hudspeth, A.J. Hair cells: Transduction, tuning, and transmission in the inner ear. *Annu. Rev. Cell. Biol.* **1988**, *4*, 63–92. [CrossRef]
23. Pickles, J.O.; Comis, S.D.; Osborne, M.P. Cross-links between stereocilia in the guinea pig organ of Corti, and their possible relation to sensory transduction. *Hear. Res.* **1984**, *15*, 103–112. [CrossRef]
24. Fettiplace, R.; Kim, K.X. The physiology of mechanoelectrical transduction channels in hearing. *Physiol. Rev.* **2014**, *94*, 951–986. [CrossRef] [PubMed]
25. Flock, A.; Cheung, H.C. Actin filaments in sensory hairs of inner ear receptor cells. *J. Cell Biol.* **1977**, *75*, 339–343. [CrossRef] [PubMed]
26. Flock, A.; Cheung, H.C.; Flock, B.; Utter, G. Three sets of actin filaments in sensory cells of the inner ear. Identification and functional orientation determined by gel electrophoresis, immunofluorescence, and electron microscopy. *J. Neurocytol.* **1981**, *10*, 133–147. [CrossRef]
27. Flock, A.; Flock, B.; Murray, E. Studies on the sensory hairs of receptor cells in the inner ear. *Acta Otolaryngol.* **1977**, *83*, 85–91. [CrossRef] [PubMed]
28. Pollock, L.M.; McDermott, B.M. The cuticular plate: A riddle, wrapped in a mystery, inside a hair cell. *Birth Defects Res. C Embryo Today* **2015**, *105*, 126–139. [CrossRef] [PubMed]
29. Slepecky, N.B.; Ulfendahl, M. Actin-binding and microtubule-associated proteins in the organ of Corti. *Hear. Res.* **1992**, *57*, 201–215. [CrossRef]
30. Du, T.T.; Dewey, J.B.; Wagner, E.L.; Cui, R.; Heo, J.; Park, J.J.; Francis, S.P.; Perez-Reyes, E.; Guillot, S.J.; Sherman, N.E.; et al. LMO7 deficiency reveals the significance of the cuticular plate for hearing function. *Nat. Commun.* **2019**, *10*, 1117. [CrossRef]
31. Francis, S.P.; Krey, J.F.; Krystofiak, E.S.; Cui, R.; Nanda, S.; Xu, W.; Kachar, B.; Barr-Gillespie, P.G.; Shin, J.B. A short splice form of Xin-actin binding repeat containing 2 (XIRP2) lacking the Xin repeats is required for maintenance of stereocilia morphology and hearing function. *J. Neurosci.* **2015**, *35*, 1999–2014. [CrossRef] [PubMed]
32. Scheffer, D.I.; Zhang, D.S.; Shen, J.; Indzhykulian, A.; Karavitaki, K.D.; Xu, Y.J.; Wang, Q.; Lin, J.J.; Chen, Z.Y.; Corey, D.P. XIRP2, an actin-binding protein essential for inner ear hair-cell stereocilia. *Cell Rep.* **2015**, *10*, 1811–1818. [CrossRef] [PubMed]
33. Flock, A.; Duvall, A.J. The ultrastructure of the kinocilium of the senosry cells in the inner ear and lateral line organs. *J. Cell Biol.* **1965**, *25*, 1–8. [CrossRef] [PubMed]
34. Itoh, M.; Nakashima, T. Structure of the hair rootlets on cochlear sensory cells by tannic acid fixation. *Acta Otolaryngol.* **1980**, *90*, 385–390. [CrossRef] [PubMed]
35. Itoh, M. Preservation and visualization of actin-containing filaments in the apical zone of cochlear sensory cells. *Hear. Res.* **1982**, *6*, 277–289. [CrossRef]
36. Slepecky, N.; Chamberlain, S.C. Immunoelectron microscopic and immunofluorescent localization of cytoskeletal and muscle-like contractile proteins in inner ear sensory hair cells. *Hear. Res.* **1985**, *20*, 245–260. [CrossRef]
37. Tilney, L.G.; Tilney, M.S.; DeRosier, D.J. Actin filaments, stereocilia, and hair cells: How cells count and measure. *Ann. Rev. Cell Biol.* **1992**, *8*, 257–274. [CrossRef]
38. Furness, D.N.; Mahendrasingam, S.; Ohashi, M.; Fettiplace, R.; Hackney, C.M. The dimensions and composition of stereociliary rootlets in mammalian cochlear hair cells: Comparison between high- and

low-frequency cells and evidence for a connection to the lateral membrane. *J. Neurosci.* **2008**, *28*, 6342–6353. [CrossRef]
39. Anniko, M. Cytodifferentiation of cochlear hair cells. *Am. J. Otolaryngol.* **1983**, *4*, 375–388. [CrossRef]
40. Kitajiri, S.; Sakamoto, T.; Belyantseva, I.A.; Goodyear, R.J.; Stepanyan, R.; Fujiwara, I.; Bird, J.E.; Riazuddin, S.; Riazuddin, S.; Ahmed, Z.M.; et al. Actin-bundling protein TRIOBP forms resilient rootlets of hair cell stereocilia essential for hearing. *Cell* **2010**, *141*, 786–798. [CrossRef]
41. Slepecky, N.; Chamberlain, S.C. Distribution and polarity of actin in the sensory hair cells of the chinchilla cochlea. *Cell Tissue Res.* **1982**, *224*, 15–24. [CrossRef]
42. Dominguez, R.; Holmes, K.C. Actin structure and function. *Annu. Rev. Biophys.* **2011**, *40*, 169–186. [CrossRef] [PubMed]
43. Vandekerckhove, J.; Weber, K. At least six different actins are expressed in a higher mammal: An analysis based on the amino acid sequence of the amino-terminal tryptic peptide. *J. Mol. Biol.* **1978**, *126*, 783–802. [CrossRef]
44. Furness, D.N.; Katori, Y.; Mahendrasingam, S.; Hackney, C.M. Differential distribution of beta- and gamma-actin in guinea-pig cochlear sensory and supporting cells. *Hear. Res.* **2005**, *207*, 22–34. [CrossRef] [PubMed]
45. Arima, T.; Uemura, T.; Yamamoto, T. Three-dimensional visualizations of the inner ear hair cell of the guinea pig. A rapid-freeze, deep-etch study of filamentous and membranous organelles. *Hear. Res.* **1987**, *25*, 61–68. [CrossRef]
46. Vranceanu, F.; Perkins, G.A.; Terada, M.; Chidavaenzi, R.L.; Ellisman, M.H.; Lysakowski, A. Striated organelle, a cytoskeletal structure positioned to modulate hair-cell transduction. *Proc. Natl. Acad. Sci. USA* **2012**, *109*, 4473–4478. [CrossRef] [PubMed]
47. Spoendlin, H. The organization of the cochlear receptor. *Fortschr. Hals Nasen Ohrenheilkd.* **1966**, *13*, 1–227. [PubMed]
48. Slepecky, N.; Hamernik, R.; Henderson, D. The consistent occurrence of a striated organelle (Friedmann body) in the inner hair cells of the normal chinchilla. *Acta Otolaryngol.* **1981**, *91*, 189–198. [CrossRef]
49. Ross, M.D.; Bourne, C. Interrelated striated elements in vestibular hair cells of the rat. *Science* **1983**, *220*, 622–624. [CrossRef]
50. Karavitaki, K.D.; Corey, D.P. Sliding adhesion confers coherent motion to hair cell stereocilia and parallel gating to transduction channels. *J. Neurosci.* **2010**, *30*, 9051–9063. [CrossRef]
51. Hudspeth, A.J. The hair cells of the inner ear. *Sci. Am.* **1983**, *248*, 54–64. [CrossRef] [PubMed]
52. Hudspeth, A.J. Hair-bundle mechanics and a model for mechanoelectrical transduction by hair cells. *Soc. Gen. Physiol. Ser.* **1992**, *47*, 357–370. [PubMed]
53. Gunning, P.W.; Hardeman, E.C.; Lappalainen, P.; Mulvihill, D.P. Tropomyosin—Master regulator of actin filament function in the cytoskeleton. *J. Cell Sci.* **2015**, *128*, 2965–2974. [CrossRef] [PubMed]
54. Seipel, K.; O'Brien, S.P.; Iannotti, E.; Medley, Q.G.; Streuli, M. Tara, a novel F-actin binding protein, associates with the Trio guanine nucleotide exchange factor and regulates actin cytoskeletal organization. *J. Cell Sci.* **2001**, *114*, 389–399.
55. Shahin, H.; Walsh, T.; Sobe, T.; Abu Sa'ed, J.; Abu Rayan, A.; Lynch, E.D.; Lee, M.K.; Avraham, K.B.; King, M.C.; Kanaan, M. Mutations in a novel isoform of TRIOBP that encodes a filamentous-actin binding protein are responsible for DFNB28 recessive nonsyndromic hearing loss. *Am. J. Hum. Genet.* **2006**, *78*, 144–152. [CrossRef]
56. Riazuddin, S.; Khan, S.N.; Ahmed, Z.M.; Ghosh, M.; Caution, K.; Nazli, S.; Kabra, M.; Zafar, A.U.; Chen, K.; Naz, S.; et al. Mutations in TRIOBP, which encodes a putative cytoskeletal-organizing protein, are associated with nonsyndromic recessive deafness. *Am. J. Hum. Genet.* **2006**, *78*, 137–143. [CrossRef]
57. Bao, J.; Bielski, E.; Bachhawat, A.; Taha, D.; Gunther, L.K.; Thirumurugan, K.; Kitajiri, S.; Sakamoto, T. R1 motif is the major actin-binding domain of TRIOBP-4. *Biochemistry* **2013**, *52*, 5256–5264. [CrossRef]
58. Katsuno, T.; Belyantseva, I.A.; Cartagena-Rivera, A.X.; Ohta, K.; Crump, S.M.; Petralia, R.S.; Ono, K.; Tona, R.; Imtiaz, A.; Rehman, A.; et al. TRIOBP-5 sculpts stereocilia rootlets and stiffens supporting cells enabling hearing. *JCI Insight* **2019**, *4*, 128561. [CrossRef]
59. Kazmierczak, M.; Kazmierczak, P.; Peng, A.W.; Harris, S.L.; Shah, P.; Puel, J.L.; Lenoir, M.; Franco, S.J.; Schwander, M. Pejvakin, a candidate stereociliary rootlet protein, regulates hair cell function in a cell-autonomous manner. *J. Neurosci.* **2017**, *37*, 3447–3464. [CrossRef]

60. Rehman, A.U.; Morell, R.J.; Belyantseva, I.A.; Khan, S.Y.; Boger, E.T.; Shahzad, M.; Ahmed, Z.M.; Riazuddin, S.; Khan, S.N.; Riazuddin, S.; et al. Targeted capture and next-generation sequencing identifies C9orf75, encoding taperin, as the mutated gene in nonsyndromic deafness DFNB79. *Am. J. Hum. Genet.* **2010**, *86*, 378–388. [CrossRef]
61. Men, Y.; Li, X.; Tu, H.; Zhang, A.; Fu, X.; Wang, Z.; Jin, Y.; Hou, C.; Zhang, T.; Zhang, S.; et al. Tprn is essential for the integrity of stereociliary rootlet in cochlear hair cells in mice. *Front. Med.* **2018**, *13*, 690–704. [CrossRef] [PubMed]
62. Zhao, B.; Wu, Z.; Müller, U. Murine Fam65b forms ring-like structures at the base of stereocilia critical for mechanosensory hair cell function. *Elife* **2016**, *5*, e14222. [CrossRef] [PubMed]
63. Liu, C.; Luo, N.; Tung, C.Y.; Perrin, B.J.; Zhao, B. GRXCR2 regulates taperin localization critical for stereocilia morphology and hearing. *Cell Rep.* **2018**, *25*, 1268–1280.e4. [CrossRef] [PubMed]
64. Scheffer, D.I.; Shen, J.; Corey, D.P.; Chen, Z.Y. Gene expression by mouse inner ear hair cells during development. *J. Neurosci.* **2015**, *35*, 6366–6380. [CrossRef] [PubMed]
65. Liu, Y.; Qi, J.; Chen, X.; Tang, M.; Chu, C.; Zhu, W.; Li, H.; Tian, C.; Yang, G.; Zhong, C.; et al. Critical role of spectrin in hearing development and deafness. *Sci. Adv.* **2019**, *5*, eaav7803. [CrossRef]
66. Salles, F.T.; Andrade, L.R.; Tanda, S.; Grati, M.; Plona, K.L.; Gagnon, L.H.; Johnson, K.R.; Kachar, B.; Berryman, M.A. CLIC5 stabilizes membrane-actin filament linkages at the base of hair cell stereocilia in a molecular complex with radixin, taperin, and myosin VI. *Cytoskeleton (Hoboken)* **2014**, *71*, 61–78. [CrossRef]
67. Hasson, T.; Gillespie, P.G.; Garcia, J.A.; MacDonald, R.B.; Zhao, Y.; Yee, A.G.; Mooseker, M.S.; Corey, D.P. Unconventional myosins in inner-ear sensory epithelia. *J. Cell Biol.* **1997**, *137*, 1287–1307. [CrossRef]
68. Furness, D.N.; Karkanevatos, A.; West, B.; Hackney, C.M. An immunogold investigation of the distribution of calmodulin in the apex of cochlear hair cells. *Hear. Res.* **2002**, *173*, 10–20. [CrossRef]
69. Liberman, M.C. Chronic ultrastructural changes in acoustic trauma: Serial-section reconstruction of stereocilia and cuticular plates. *Hear. Res.* **1987**, *26*, 65–88. [CrossRef]
70. Howard, J.; Hudspeth, A.J. Compliance of the hair bundle associated with gating of mechanoelectrical transduction channels in the bullfrog's saccular hair cell. *Neuron* **1988**, *1*, 189–199. [CrossRef]
71. Pickles, J.O. A model for the mechanics of the stereociliar bundle on acousticolateral hair cells. *Hear. Res.* **1993**, *68*, 159–172. [CrossRef]
72. Jaramillo, F.; Hudspeth, A.J. Displacement-clamp measurement of the forces exerted by gating springs in the hair bundle. *Proc. Natl. Acad. Sci. USA* **1993**, *90*, 1330–1334. [CrossRef] [PubMed]
73. Tobin, M.; Chaiyasitdhi, A.; Michel, V.; Michalski, N.; Martin, P. Stiffness and tension gradients of the hair cell's tip-link complex in the mammalian cochlea. *Elife* **2019**, *8*, e43473. [CrossRef] [PubMed]
74. Martin, P.; Mehta, A.D.; Hudspeth, A.J. Negative hair-bundle stiffness betrays a mechanism for mechanical amplification by the hair cell. *Proc. Natl. Acad. Sci. USA* **2000**, *97*, 12026–12031. [CrossRef] [PubMed]
75. Kim, J. Unconventional mechanics of lipid membranes: A potential role for mechanotransduction of hair cell stereocilia. *Biophys. J.* **2015**, *108*, 610–621. [CrossRef] [PubMed]
76. Tilney, L.G.; Saunders, J.C.; Egelman, E.; DeRosier, D.J. Changes in the organization of actin filaments in the stereocilia of noise-damaged lizard cochleae. *Hear. Res.* **1982**, *7*, 181–197. [CrossRef]
77. Tilney, L.G.; Egelman, E.H.; DeRosier, D.J.; Saunders, J.C. Actin filaments, stereocilia, and hair cells of the bird cochlea. II. Packing of actin filaments in the stereocilia and in the cuticular plate and what happens to the organization when the stereocilia are bent. *J. Cell. Biol.* **1983**, *96*, 822–834. [CrossRef]
78. Narayan, K.; Subramaniam, S. Focused ion beams in biology. *Nat. Methods* **2015**, *12*, 1021. [CrossRef]
79. Pfeffer, S.; Mahamid, J. Unravelling molecular complexity in structural cell biology. *Curr Opin. Struct. Biol.* **2018**, *52*, 111. [CrossRef]

© 2020 by the authors. Licensee MDPI, Basel, Switzerland. This article is an open access article distributed under the terms and conditions of the Creative Commons Attribution (CC BY) license (http://creativecommons.org/licenses/by/4.0/).

Review

Actin and Actin-Associated Proteins in Extracellular Vesicles Shed by Osteoclasts

L. Shannon Holliday [1,*], Lorraine Perciliano de Faria [2] and Wellington J. Rody Jr. [3]

1. Department of Orthodontics, College of Dentistry, University of Florida, Gainesville, FL 32610, USA
2. Department of Biomaterials and Oral Biology, School of Dentistry, University of São Paulo, São Paulo 01000, Brazil; LPercilianodeFaria@dental.ufl.edu
3. Department of Orthodontics and Pediatric Dentistry, Stony Brook University School of Dental Medicine, Stony Brook, NY 11794, USA; Wellington.Rody@stonybrookmedicine.edu
* Correspondence: sholliday@dental.ufl.edu

Received: 21 November 2019; Accepted: 16 December 2019; Published: 25 December 2019

Abstract: Extracellular vesicles (EVs) are shed by all eukaryotic cells and have emerged as important intercellular regulators. EVs released by osteoclasts were recently identified as important coupling factors in bone remodeling. They are shed as osteoclasts resorb bone and stimulate osteoblasts to form bone to replace the bone resorbed. We reported the proteomic content of osteoclast EVs with data from two-dimensional, high resolution liquid chromatography/mass spectrometry. In this article, we examine in detail the actin and actin-associated proteins found in osteoclast EVs. Like EVs from other cell types, actin and various actin-associated proteins were abundant. These include components of the polymerization machinery, myosin mechanoenzymes, proteins that stabilize or depolymerize microfilaments, and actin-associated proteins that are involved in regulating integrins. The selective incorporation of actin-associated proteins into osteoclast EVs suggests that they have roles in the formation of EVs and/or the regulatory signaling functions of the EVs. Regulating integrins so that they bind extracellular matrix tightly, in order to attach EVs to the extracellular matrix at specific locations in organs and tissues, is one potential active role for actin-associated proteins in EVs.

Keywords: exosome; microvesicle; microfilament; integrins; bone remodeling; myosins; actin-related protein; proteomics; extracellular vesicles

1. Introduction

Extracellular vesicles (EVs) are 30–150 nm in diameter vesicles that are released by eukaryotic cells and function in intercellular signaling [1,2]. The term EVs encompasses exosomes and microvesicles (Figure 1) [3]. Exosomes develop as inward buds into endocytic compartments, which pinch off into the lumen of the compartment, resulting in the formation of multivesicular bodies. Multivesicular bodies can then fuse with the plasma membrane to shed the exosomes from the cell. Microvesicles bud off directly from the plasma membrane. The two types of EVs have similar size, composition, and regulatory functions and are difficult to distinguish in extracellular vesicle populations, although some articles suggest that microvesicles may be on average larger and may have some differing components [4]. In addition, some non-vesicular particles are probably often isolated in EV preps, including exomeres and lipoproteins [5]. Unless the type of vesicle being studied is known, which is usually not the case at the present time, the term EVs is preferred [3].

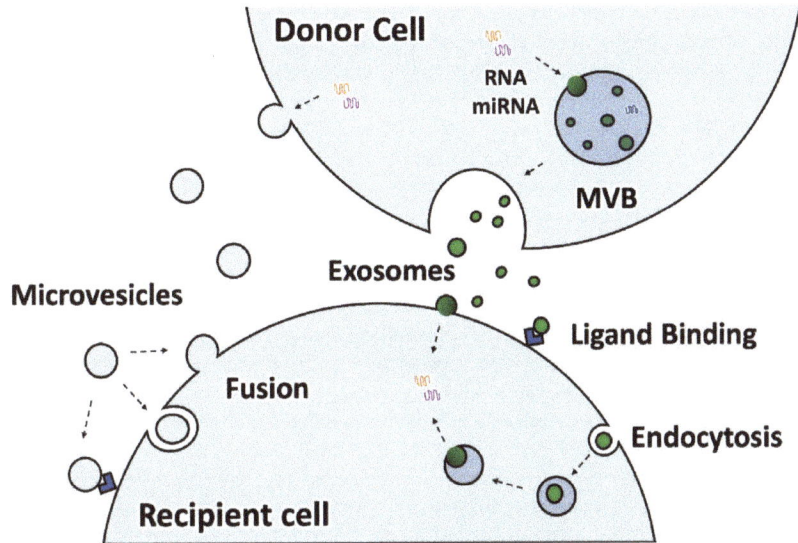

Figure 1. Extracellular vesicles include exosomes which are derived from multivesicular bodies (MVB) and microvesicles (ectosomes) which bud directly from the plasma membrane. Both may bind surface receptors of target cells to stimulate signaling pathways, or to fuse with the plasma membrane or membranes of endocytic compartments. Fusion releases their luminal contents into the cytosol of the target cell, and membrane proteins into either the plasma membrane or endocytic membrane.

Exosomes were first identified and characterized due to their role in the removal of the transferrin receptor from reticulocytes as they differentiated [6,7]. For many years, exosomes were mostly thought of as "garbage bags", although evidence that EVs could present antigen appeared during the 1990s [8]. In 2007, landmark articles showed that exosomes carried mRNAs and microRNAs, and could fuse with target cells to introduce the functional RNAs into the cytosol [9,10]. The concept of EVs being able to regulate target cells acting at different regulatory levels stimulated the EV field. Subsequently, much evidence has accumulated that by transferring microRNAs, EVs modulate target cell protein expression. For example, two groups reported that microRNA 214-3p is found in EVs from osteoclasts, and is transferred to osteoblasts, where it inhibits osteoblast formation by reducing the expression of regulatory proteins [11,12]. Despite the plethora of articles supporting the hypothesis that microRNAs in EVs are crucial to their regulatory function, some studies have cast doubt on whether sufficient numbers of microRNAs are present in EVs to suppress mRNA translation [13].

For EVs to bind and stimulate a target cell, either from the outside through traditional signal transduction pathways, or after fusing, the EVs must interact with the cell. Osteoclast EVs serve as a model for the sorts of interactions and regulation that have been found in EVs in general.

In osteoclasts, three potential modes of interaction have been identified, namely semaphorin 4D in EVs binding plexin-B1 on osteoblasts [11], ephrin-B2 in EVs binding ephB4 [12], and receptor activator of nuclear factor kappa B (RANK) in EVs binding RANK-ligand (RANKL) [14]. Semaphorin 4D and ephrinB2 were reported to be osteoclast-derived signaling factors, before their detection in EVs [15,16]. Both were found to tether EVs to the surface of osteoblasts prior to the fusion that delivers microRNA-214-3p from the EVs lumen to the cytosol of the osteoblast.

RANK is part of the central signaling pathway in bone remodeling [17,18]. RANKL, which is found on osteoblasts and osteocytes, binds RANK on the surface of osteoclasts to stimulate a signaling pathway, resulting in the activation of nuclear factor kappa B and nuclear factor of activated T cells 1 signaling. These pathways are required for osteoclasts to differentiate from multipotent hematopoietic precursors to osteoclasts, and for the bone resorbing activity by mature osteoclasts. A third protein, osteoprotegerin, is a soluble protein secreted by osteoblasts that binds RANKL and serves as a competitive inhibitor of binding between RANKL and RANK. The detection of RANK-containing EVs (RANK-EVs) suggested immediately the possibility that they might function like osteoprotegerin as competitive inhibitors of binding between RANKL and cellular RANK. Consistent with this idea, when added to calcitriol-stimulated mouse marrow, in which calcitriol stimulates the coordinated differentiation of osteoblasts, which produce RANKL, and osteoclasts, RANK-EVs reduced osteoclast formation [14]. Quantitative assessment of the data, however, suggested that there were likely too few RANK-EVs to trigger the observed reduction in osteoclast numbers by simple competitive inhibition, and another regulatory mechanism was likely in play [19].

A solution was presented when it was shown that RANK-EVs stimulate a reverse RANKL signaling pathway through mammalian target of rapamycin and runt-related transcription factor 2, which drive cells of the osteoblastic lineage toward bone formation [20]. This idea is also consistent with findings that osteoblasts and osteoclasts are rarely in direct contact in vivo. In addition, increasingly strong evidence shows that osteocytes, terminally differentiated cells of the osteoblastic lineage which are buried in the bone, are the source of most of the RANKL that stimulates bone resorption in vivo [21–24]. Hence, RANK-EVs produced by osteoclasts in vivo would be free to bind RANKL on osteoblasts to stimulate bone formation, thus providing an elegant mechanism for coupling bone resorption to bone formation [20].

Although this provides a satisfying explanation for regulation by RANK-EVs, it leaves open the question of whether these EVs fuse with osteoblasts and whether other components, like microRNAs, may be present in RANK-EVs. Another crucial question regarding RANK-EVs is how their shedding is regulated. More RANK-EVs were shed from osteoclasts resorbing bone compared with dentine or when they were quiescent on plastic [25]. Understanding the function of regulatory EVs from osteoclasts requires more complete knowledge of their composition.

2. Proteomics Show that EVs Shed by Osteoclasts Are Rich in Actin and Actin-Associated Proteins

We performed two-dimensional, high resolution liquid chromatography/mass spectrometry of EVs isolated from osteoclasts resorbing bone, dentine or quiescent on plastic [25]. The morphology of osteoclasts on bone and dentine was indistinguishable, while those on plastic lacked actin rings and ruffled membranes. The quantitative and qualitative data we obtained identified actin as the most abundant protein of EVs resorbing bone, and various actin-associated proteins as being very abundant. The levels of actin associated proteins referred to in the manuscript, except integrins, which will be presented separately, are shown in Figure 2. These numbers are normalized to actin in EVs from bone resorbing osteoclasts.

Osteoclasts are highly specialized to resorb bone [26]. To do this, they form resorption compartments, each composed of an actin ring and a ruffled plasma membrane, or ruffled border (Figure 3). Actin rings are composed of arrays of podosomes (also called invadopodia) [27]. These dynamic structures are very similar to the podosomes found in other cell types [28], but to meet the needs of resorbing osteoclasts, they are joined by interconnecting actin filaments to form the cohesive actin ring [29]. The actin ring is associated with a very tight contact between the osteoclast membrane and bone, the sealing zone, which segregates an extracellular resorption compartment. The ruffle plasma membrane is surrounded by the actin ring and is packed with vacuolar H^+-ATPase (V-ATPase), which pumps protons out of the cell into the resorption compartment [30,31]. This lowers the pH to about 5.0, which solubilizes bone mineral, and provides an environment where the acid cysteine proteinase, cathepsin K, which is secreted by the osteoclasts, degrades the organic matrix of the bone [32,33]. Because of the prominent role of the actin cytoskeleton in osteoclasts, it is perhaps not surprising that EVs would be loaded with abundant actin and associated proteins. Nevertheless, to understand extracellular vesicles, the potential roles of the most abundant components, which include actin and associated proteins, must be addressed. Even if they have no functional role, that too must be understood, and that would be a clue regarding the formation of EVs and their functions. As a first step in this process, in this article we will discuss what is found, and what is not found, in EVs and the questions that are raised.

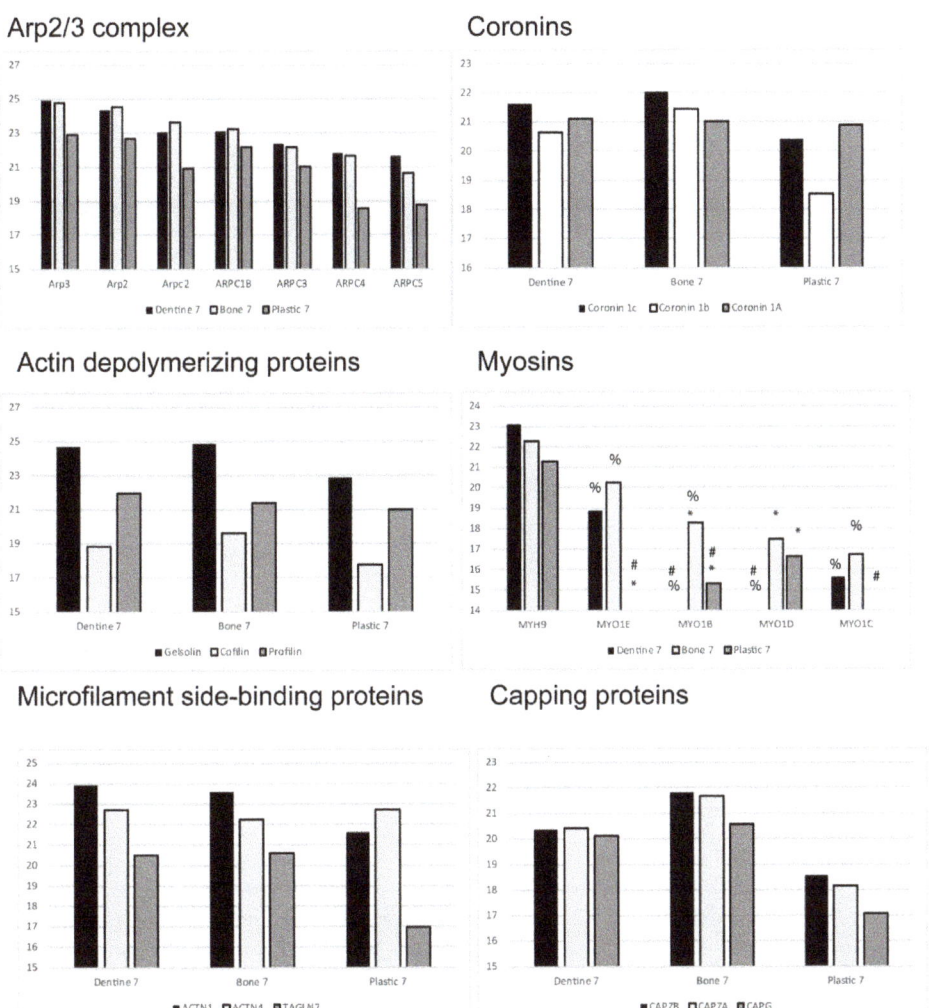

Figure 2. The relative abundance of the actin associate proteins discussed in the current manuscript. These numbers are normalized to actin in extracellular vesicles from conditioned media form osteoclasts resorbing bone at Days 4–7. The raw Day 4–7 data were published previously [25]. Abbreviations: MYH (myosin heavy chain), MYO (myosin), ACTN (alpha-actinin), TAGLN (transgelin), CAP (capping protein). Statistical analysis was performed from Z-scores as described in reference 25. Proteins with Z-scores greater than 1.65 or smaller than 1.65 were considered significantly different. The following symbols are used; # indicates different from bone; % indicates different form plastic; * indicates different from dentine.

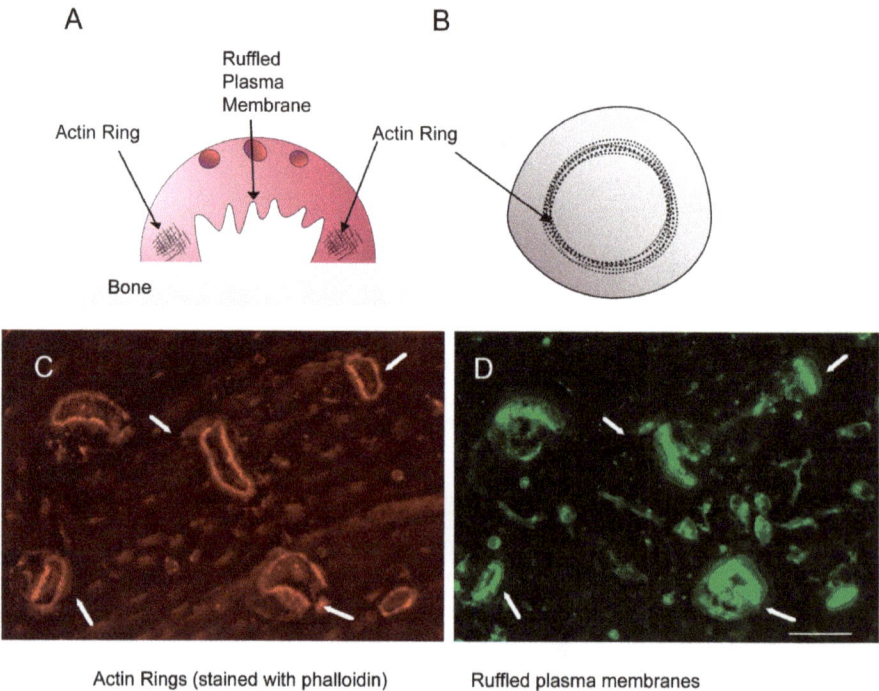

Figure 3. Osteoclasts are highly specialized cells that migrate through mineralized tissue. (**A**) A schematic of a resorbing osteoclast from the side. (**B**) Schematic of cell from side. (**C**) Immunofluorescence micrograph of actin rings stained with Texas Red-phalloidin (red). (**D**) Immunofluorescence micrograph of same cells stained with a rabbit polyclonal anti-E-subunit of V-ATPase antibody followed by an Alexa 488-tagged anti-rabbit IgG secondary antibody to detect ruffled plasma membranes (green). Arrows show examples of actin rings and ruffled membranes, with arrows pointing to same cells in the two panels. The scale bar for the micrographs is 25 µm.

In the discussion that follows, we will assume that actin and associated proteins are evenly distributed among EVs. This is unlikely to be the case. However, this simplification makes it easier to organize the quantitative data and given that actin and associated proteins are very abundant in EVs from various sources, it may be a reasonable approximation. It is possible that these proteins are concentrated in a subset of EVs, for example, into microvesicles and not exosomes. As we go through the different types of actin-associated proteins found in osteoclast EVs, we will provide a brief review of their known functions in cells and discuss how the specific proteins may be involved in the formation and/or function of EVs.

3. Possible Roles for the Actin Cytoskeleton in the Formation of EVs

As EVs form from the budding of a membrane either into an endocytic compartment or from the plasma membrane, the actin cytoskeleton may play a role in the process, and indeed may be included in EVs due to its role in EV formation. There is some data supporting this idea for the formation of microvesicles (also called ectosomes) [34]. The actin cytoskeletal protein filamin A was shown to be involved in regulating the incorporation of tissue factor into EVs [35,36]. It remains uncertain, however, whether elements of the actin cytoskeleton are required for EV formation. The endosomal sorting complexes required for transport (ESCRT) complex [37], Rab27 [38], and neutral sphingomyelinases [39,40] have all been implicated in EV formation.

At least three ways can be envisioned by which the actin cytoskeleton could be involved in EV formation. Actin filaments may be involved by pushing against the membrane, harnessing the force generated by actin polymerization to form a bud [41–43]. Myosin mechanoenzymes could move components to sites where EVs are forming [44]. Actin–myosin contraction could be involved in sealing buds that form [45] (Figure 4).

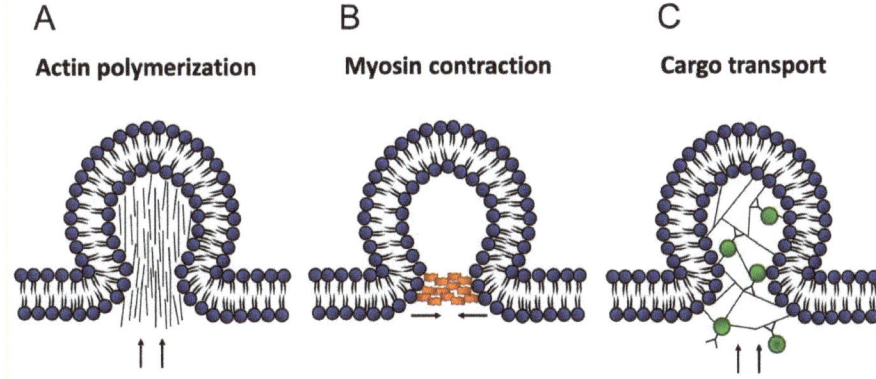

Figure 4. Three ways that the actin cytoskeleton might be involved in extracellular vesicle (EV) formation. (**A**) Microfilaments may push against membranes using force generated by polymerization of actin. (**B**) Myosin motors may materials to the site of EV formation for packaging. (**C**) Myosin contraction may play a role in the scission of the EV. In each of these cases, elements of the actin cytoskeleton may be trapped in the EVs as the result of the cytoskeleton's involvement in the EV formation process. Arrows denote the direction filaments push into membrane in actin piloymerization, or the direction of cargo transport.

Crawling motility is powered by force generated by actin polymerization to push against membranes [46], so the notion of actin polymerization pushing against membranes to facilitate bud formation is plausible. Likewise, myosins for delivering materials [47] and for the scission of membranes, as in cytokinesis where myosin II works in conjunction with elements of ESCRT [48], is well established. In addition, myosin contraction, downstream of ADP-ribosylation factor 6 (ARF6) signaling, has been implicated in the formation of microvesicles [49]. A conventional myosin II and ARF6 are detected in osteoclast EVs [25].

Testing these ideas is more difficult, since actin polymerization and myosins are involved in a wide array of cell processes. To formally test this idea likely will require use of a cell-free model of multivesicular body formation [50]. Ideally a model could be developed that incorporates a minimal number of purified components, in a similar way to that in which the force generation mechanism of actin polymerization has been characterized [51,52].

4. Are Microfilament Dynamics Possible in EVs?

Actin exists as either a 42 kD monomer or as polymers of the monomer called microfilaments [53]. Typically in cells, about half of the total actin is polymerized into microfilaments. Polymerization and depolymerization in the cytosol occur constantly and are tightly controlled [54]. Is actin polymerized or unpolymerized in EVs and do transitions between monomers and polymers occur? Filaments within subpopulations of EVs leading to non-spherical EVs, and perhaps dynamics of the filaments have been reported [55,56]. This provides some empirical support for the idea that dynamic cytoskeletal systems may exist in at least some EVs. As we will discuss, some, but not all, of the components normally associated with actin dynamics are detected in EVs. However, cytoskeletal dynamics in the cytosol of cells is tied to a high rate of adenosine triphosphate (ATP) hydrolysis [57]. There is evidence that ATP can be generated by glycolytic enzymes in certain types of EVs, prostasomes, and seminal plasma exosomes [58,59]. However, it has not been shown that the ability to generate ATP through glycolysis occurs in the lumen of EVs in general.

Actin can bind either ATP, ADP-Pi, or ADP. ATP-actin polymerizes much better and is the species that usually (or always) enters filaments in cells. Soon afterward, the actin-bound ATP is broken down, first to ADP-Pi-actin, then ADP-actin. This occurs as a result of its presence in the microfilament [57]. When filaments depolymerize, ADP-actin typically is released and it is then recharged into ATP-actin from the cytosolic free ATP pool by interaction with profilin, which binds actin monomers and causes a change in conformation that accelerates nucleotide exchange [60]. This in practice means conversion of ADP-actin to ATP-actin, since ATP is much more abundant in the cytosol than ADP.

EVs isolated by standard methods, including EVs from osteoclasts, have abundant glycolytic enzymes [25]. However, recent evidence suggests that at least a portion of the glycolytic enzymes normally detected in EV preparations are components of non-vesicular structures called exomeres [61]. Even if EVs are rich in glycolytic enzymes, no membrane transporters for glucose or other sugars were detected in osteoclast EVs [25]. It is possible that such transporters are present at low levels, which may be enough to supply glucose or other raw materials for glycolysis. Nevertheless, it has not yet been demonstrated that osteoclast EVs, or most other EVs, have the energy producing resources to support traditional actin dynamics. This also does not rule out some sort of non-traditional polymerization scheme. For example, high concentrations of ADP-actin will polymerize [62]. However, unless a mechanism for ATP generation or the import of ATP into EVs is identified, serious concerns about the plausibility of actin dynamics in EVs are warranted.

Finally, isolated EVs are typically stored in media lacking ATP or free sugars. This would mean the EVs that are usually studied by electron microscopy or nanoparticle tracking lack an energy source. It may be worth studying EVs under conditions where ATP might be available or could be, in principle, generated.

One mechanism by which actin polymerization is triggered in cells is through the action of a class of proteins called formins [63]. None of this family were detected in EVs from osteoclasts.

The actin related protein 2/3 (Arp2/3) complex is the basis of the other general mechanism for triggering actin polymerization in cells. It contains one copy each of seven different proteins including Arp2 and Arp3, which are close relatives of actin [64–66]. Arp2/3 is activated to stimulate actin polymerization by interactions with members of the Wiskott–Aldrich syndrome protein (WASP) family of proteins [67]. WASP proteins are activated by Rho-class GTPases [54]. In osteoclast EVs, all of the elements of the Arp2/3 complex are very abundant and, in the stoichiometry, expected for intact complexes (Figure 2). We estimate about 1 copy per 7 actin monomers. Surprisingly, we only detected traces of two members of the WASP family of proteins, WASP and WAVE2. It is not clear that the abundant Arp2/3 complex could be activated to stimulate actin polymerization in EVs.

In principle, Arp2/3 complex could enter EVs independent of actin, associated with the "slow-growing" end of actin filaments, or as part of a branched actin network. The incorporation of the Arp2/3 complex into branched actin networks can involve interactions with cortactin and n-WASP (one of the WASP family) [68]. Cortactin is upregulated as osteoclasts differentiate, is expressed at high levels in mature osteoclasts, and it is required for the formation of actin rings [69–71]. However, it was not detected in osteoclast EVs, even in trace amounts.

Three isoforms of coronin were abundant in osteoclast EVs [72]. Coronin binds and locks Arp2/3 into an inactive complex in the absence of preexisting filaments, but links Arp2/3 to existing filaments, and can either protect them from or enhance the activity of cofilin [73], a protein that binds to the side of actin filaments, disassembles filaments and binds actin monomers. It is abundant in osteoclast EVs (Figure 2). Coronin binds ATP-actin with 47-times higher affinity that ADP-actin. It protects microfilaments from cofilin when it is bound to ATP-actin but does not protect ADP-actin. As discussed above, it seems likely that most of the actin in EVs is ADP-actin. Therefore, although EVs may have initially have had polymerized actin in them, for example, if actin polymerization has a role in EV formation, very soon microfilaments would be expected to be disassemble, aided by the activities of cofilin, gelsolin, and profilin [74,75], which are abundant, unless they are protected by stabilizing proteins. Coronin may associate with Arp2/3 to maintain it in the inactive state. In this scenario, ADP-actin would be unlikely to spontaneously polymerize as it requires a higher concentration to polymerize than ATP-actin, and the abundant profilin and cofilin, which sequester actin from polymerization, would bind ADP-actin monomers and keep the free actin concentration low. It is worth considering that if such an EV was to fuse with a target cell, many of the necessary raw materials for actin polymerization would be locally enriched, requiring only the abundant free ATP in the cytosol, and a WASP family member to trigger rapid local polymerization. Such an event could mechanistically help explain the observation that EVs can contribute to persistent cell movement through extracellular matrix [76,77]. The fusion of EVs associated with the matrix with the migrating cell could provide concentrated packages of cytoskeletal components at the leading edge, in addition to signaling molecules.

Molecule interacting with CasL (Mical) catalyzes redox reactions using microfilaments as a substrate [78]. This makes the filaments more susceptible to depolymerization by cofilin [79]. Mical1 is present in osteoclast EVs.

Based on this analysis, it seems unlikely that actin assembly/disassembly cycles occur in EVs in the manner of the cytosol, at least in most EVs, most of the time. There is probably neither the ATP to support polymerization nor the regulatory proteins to promote polymerization. In addition, any new filaments assembled would be rapidly disassembled by depolymerizing proteins, unless they were protected and stabilized. As we will see, proteins that could protect and stabilize filaments are present. It is therefore possible that microfilaments enter EVs as the EVs are formed and are stabilized to remain polymerized in the EVs.

5. Stable Microfilaments in EVs?

Before examining whether stable microfilaments might exist in osteoclast EVs, it is worth considering the possible length of a microfilament that could be contained in an osteoclast EV, which are on average about 50 nm in outside diameter. Structurally, each monomer in a microfilament is associated with an increase in length of about 3 nm, and microfilaments do not bend very much [53]. This constrains the possible length of microfilaments in most osteoclast EVs to be no more than about 30 nm in length (10 monomers). Even in a large, 150 nm-diameter EV, the size limit is about 50 monomers. In cells, microfilaments often achieve lengths of a micron or more, and contain hundreds of monomers Unless microfilaments deform the EVs, only relatively short microfilaments are physically possible in EVs.

In cells, subsets of microfilaments are stable, depolymerizing slowly or not at all [41]. These microfilaments are bound by proteins that interact with ADP-actin in filaments and protect the filaments from the actions of depolymerizing proteins like cofilin. Proteins also stabilize microfilaments by reducing depolymerization through the binding of free filament ends and blocking actin dynamics at that end. Osteoclast EVs contain a number of abundant proteins from this category.

The following proteins that bind to, and block the fast-growing ends of filaments are abundant in osteoclast EVs, gelsolin, the alpha and beta subunits of capZ and capG [80]. The slow growing end of actin filaments has a higher critical concentration [53]. It is possible that capped stabilized filaments could be incorporated into EVs and stabilized as short filaments. Such short filaments would fit in the lumen of EVs and might be appropriate for stabilizing protein complexes that interact with and regulate the binding state of integrins, as will be discussed in more detail below. Vasodilator-stimulated phosphoprotein (VASP) is an end tracking protein that has also been reported to protect microfilaments from depolymerization by gelsolin [81]. It is detected in EVs with modest abundance.

Tropomyosins bind the sides of microfilaments and protect them from degradation by cofilin [82]. Different isoforms of tropomyosin provide differing levels of protection [83]. Three tropomyosins are detected in osteoclast EVs though none are abundant. Tropomyosin 3 is the most abundant of the three and provides the greatest protection against cofilin severing. Plastins, including L-plastin, which is abundant in the EVs, compete with tropomyosins and cofilin for binding the microfilaments. They synergize with cofilin, under some conditions, to displace tropomyosin [82].

Transgelin-2 is a microfilament-binding protein that protects microfilaments from cofilin-mediated depolymerization, and it is a tumor-suppressor [84–88]. This small protein is abundant in EVs, and in principle could be involved in stabilizing a significant amount of the actin as microfilaments. Little is known about transgelin-2 in osteoclasts or EVs.

6. Myosins in EVs

By far the most abundant myosin in osteoclast EVs is the conventional non-muscle myosin IIa (MYH9). This is a common myosin to be found in EVs from other sources and was identified as a potential exosomal biomarker for inflamed trigeminal satellite glial cells [89]. In osteoclasts, both myosin IIa and myosin IIb are expressed, and myosin IIa is associated with the actin rings. Myosin IIb is associated with the non-actin ring actin cytoskeleton and was detected in trace amounts in EVs from osteoclasts resorbing bone [90].

The unconventional myosin 1E was the second most abundant myosin detected. It has not been studied in osteoclasts but was recently identified as a regulator of phosphatidylinositol signaling and actin polymerization in posodomes in mouse embryonic fibroblasts [91]. Several other myosin I isoforms (myosin 1B, myosin 1D, myosin 1C, myosin 1F, and myosin 1G) were found in trace amounts. Likewise, traces of myosin Va, myosin IIb, myosin XIV, myosin XVIIIA, myosin XV, and myosin VIIB were detected. Myosins are large proteins. We detected 50–80 peptides from myosin IIa in our proteomic analysis. We only detected only 2–4 peptides from the trace myosins, usually in only one sample [25].

As with the discussion of actin dynamics, a major limiting factor of potential myosin activity in EVs is whether ATP for the motor would be present. If not, it would likely be locked in ATP-free rigor. The presence of abundant myosin IIa could implicate it in EV formation. It will be important to test whether myosin contraction is required for exosome or microvesicle formation.

7. Integrins and Regulators of Integrin Binding in EVs

Integrins in EVs have garnered considerable attention for their ability to connect EVs to specific sites in the extracellular matrix [92–95]. The activity of integrins in EVs has been linked to promoting metastasis of cancer cells and to pulmonary tissue destruction [76,96,97]. More generally, it is thought that integrins are at least partially responsible for observed organotropism [98,99]. For example, EVs collected from osteoclasts in cell culture, then injected into the tail vein of mice, were shown to preferentially hone to bone, and EVs from other types of cells to hone preferentially to their organ of origin [11]. An important mechanistic concern remains to be addressed. Integrins, which are composed of an alpha and beta subunit, in the plasma membrane can exist in three conformations, only one of which binds the extracellular matrix with high affinity [100]. These conformational states are regulated by proteins that bind directly or indirectly to the cytosolic (or in the case of EVs luminal) domain of the integrins. These include talin, filamin, vinculin, and L-plastin, which are all abundant in osteoclast EVs.

The most abundant integrins found in osteoclast EVs are alphaV beta3, alphaM beta2, and alpha2 beta1 [25]. The proteomic experiment performed was designed to compare osteoclasts resorbing bone or dentine and osteoclasts quiescent on plastic. AlphaV beta3 was more abundant on EVs from osteoclasts resorbing bone than on those resorbing dentine, while the reverse was true for alphaM beta2 (Figure 5). Our data suggest that there is more beta2 than alphaM, it is likely that very small amounts of alphaL and/or alphaX, other partners of beta2, are also present, but too little to detect. AlphaM/X/L beta 2 have all been reported to be present in osteoclasts or in osteoclast precursors [101–105]. It is possible that EVs containing these integrins are from precursor cells in the culture and are being shed during osteoclast differentiation.

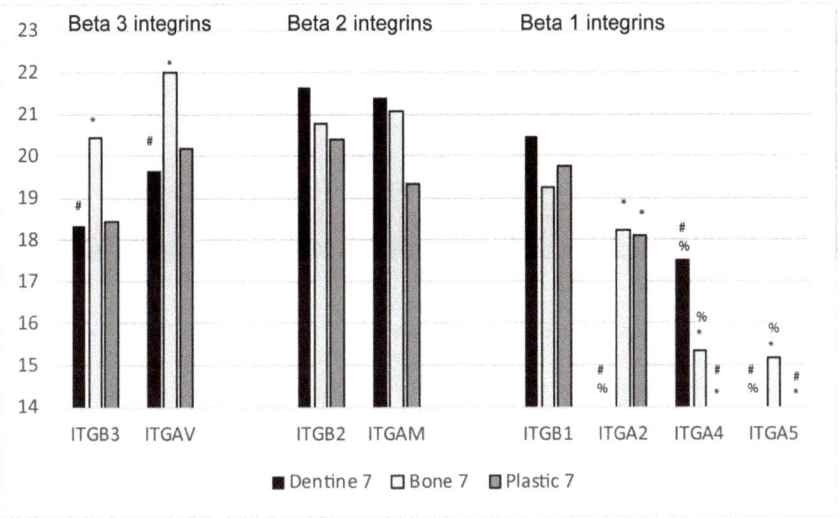

Figure 5. The relative abundance of the integrins (ITGB is integrin beta; ITGA is integrin alpha) found in osteoclast extracellular vesicles. AlphaV beta3 were found at higher levels in EVs from osteoclasts resorbing bone, compared with dentine or quiescent osteoclasts. Beta2 was found at higher levels in EVs from dentine. Our data suggest that beta1 integrin is found at higher levels in EVs from dentine. Alpha2 beta1 is probably found in all samples, although alpha2 was not detected in the EVs from dentine it was not detected in Day 7 dentine. Alpha4 beta1 may be found at higher levels in osteoclasts resorbing dentine and alpha5 beta1 may be more prominent in osteoclasts resorbing bone. Statistical analysis was performed from Z-scores as described in reference 25. Proteins with Z-scores greater than 1.65 or smaller than −1.65 were considered significantly different. The following symbols are used; # indicates different from bone; % indicates different form plastic; * indicates different from dentine.

Three binding partners for beta1 integrin were found, alpha2, which was most abundant, alpha4 and alpha5. Interestingly alpha4 was more prominent in EVs from osteoclasts resorbing dentine while alpha5 was more prominent in EVs from osteoclasts resorbing bone. The binding specificity for the three partially overlaps but has key differences. Alpha2 beta1 binds collagen, laminin, and thrombospondin. Alpha4 beta1 binds thrombospondin, osteopontin, fibronectin, and mucosal vascular addressin cell adhesion molecule (MAdCAM). MAdCAM is a ligand found on mucosal endothelial cells to direct cells into inflamed tissues [106]. Alpha5 beta1 binds fibronectin and osteopontin.

The roots of deciduous teeth are resorbed physiologically during normal tooth. Pathophysiologic root resorption usually involves resorption of the roots of permanent teeth [107]. One example is root resorption associated with orthodontic force application [108]. During orthodontic procedures a small amount of root resorption often occurs, but not enough to cause short- or long-term difficulties with tooth retention. In some patients, for reasons that are not clear, the application of mechanical force triggers massive root resorption that endangers the health and retention of the tooth. Although dentine and bone are both mineralized matrices with similar compositions, our study showed that there are subtle differences in the response of the osteoclasts to bone versus dentine. A crucial difference was the shedding of RANK-containing EVs, which occurred at lower levels from osteoclasts resorbing dentine compared with bone [25].

The difference in the predominant integrins being shed may also be illuminating. AlphaM/L/X beta2 binds fibrinogen, Factor X, iC3b, and intercellular adhesion molecules, which are soluble ligands associated with inflammation or cell surface receptors of cell-cell interactions [109]. It is possible that the beta2 integrin containing EVs could play anti-inflammatory roles by competitively inhibiting interactions between pro-inflammatory ligands with the integrins on the cell. The relative abundance of alpha4 in EVs from osteoclasts resorbing dentine, could support the idea that these EVs may be acting in an anti-inflammatory manner, competitively inhibiting integrin interactions that are involved in inflammatory responses. AlphaV beta3 is important for osteoclast activation and binds denatured collagen, like that present during bone resorption [110–114]. It will be of great interest to determine if alphaV beta3 integrins are found in EVs that also carry RANK. If so, this could be a mechanism for targeting the RANK EVs to the sites where new bone formation is required. We believe that studies to determine the composition of EVs with different integrins are of vital importance. We hypothesize that specific EVs will have one type of integrin and that regulatory factors will be associated with specific EVs.

As described above, there is considerable evidence for integrins targeting EVs to specific locations in the organism. Integrins serve as an information conduit. When they bind an external ligand, not only does it serve, in some cases, as a means to secure adhesion with the extracellular matrix, or in some cases, another cell, but it also triggers signaling within the cell. A complex of proteins that bind the cytosolic region of the integrin accumulate, recruit the actin cytoskeleton and trigger signal transduction pathways [115]. Integrins also serve as a conduit for inside-out signaling [116–118]. To bind external ligands with high affinity, the integrin must be in a specific conformation. This is accomplished by the binding of specific actin-associated proteins (Figure 6). For integrins in EVs to bind extracellular matrix, as has been described, it must be in a conformation that enables that binding, and that is achieved by the interaction with proteins including filamin. Likewise, other binding proteins of the cytosolic (luminal) domain of integrins can lock them into a conformation where they cannot bind their ligands in the outside world.

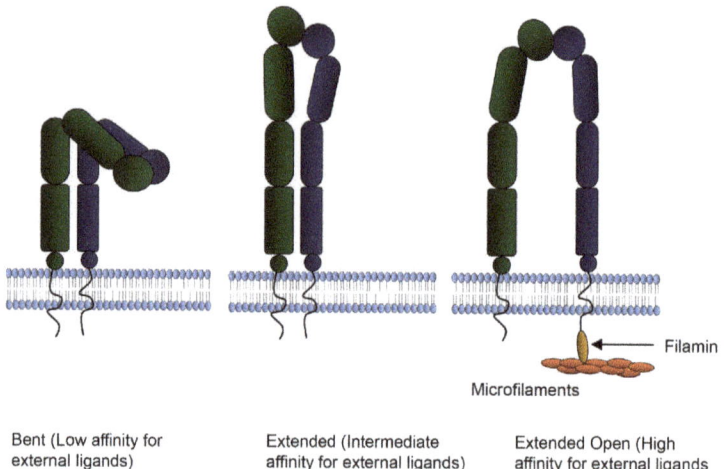

Figure 6. Integrins can take on multiple conformations. In the bent conformation the integrin does not bind external ligands. In extended conformation, integrins bind ligands with intermediate affinity. In the extended open conformation integrins bind ligands with high affinity. Attachment of talin to the cytosolic (luminal) domain of the beta3 integrin can promote the high affinity bonding conformation. In the schematic, alpha integrins are on the left, beta integrins are on right. As described in text, filamin and other integrin–associated proteins can regulate the conformation of integrins by associating with the cytosolic domain.

For EVs to utilize integrins for targeting, they must also have elements necessary to maintain them in their high affinity binding conformation. Our proteomic analysis of osteoclast EVs suggests that the proteins necessary to maintain active integrins are present, and so the idea of a membrane cytoskeleton of actin-associated proteins in EVs is plausible. Experiments are crucial to identify the binding state of integrins in EVs and the overall compositions of those EVs. Affinity isolation of the EVs using the integrin-ligand interaction has been accomplished for EVs from B-cells [92].

When the fusion of EVs with target cells is considered, it is usually thought of as delivering luminal components, like microRNAs, to the cytosol of the target cell [119,120]. Another consequence of the fusion of EVs is the delivery of membrane components, including integrins [121,122]. In a series of studies, data has been presented that suggests that EVs deliver integrins from a cell of origin to a target cell, and that this is an element of the machinery that allows cancer cells to precondition the local environment of a site of metastasis [93,121]. With osteoclast EVs, both alphaV beta3 and alpha2 beta1 integrins have been shown to have critical regulatory functions in osteoblasts [123–125]. It is plausible that fusion of osteoclast EVs, containing these integrins with osteoblasts could enhance signaling responses that are intrinsic to osteoblasts. This is particularly intriguing in the scenario where integrins first attach EVs, with signal like RANK on their surface, to areas where bone has been resorbed, then contact and fuse with osteoblasts. This could both stimulate osteoblastogenesis through reverse RANKL signaling and enhance the response by donating integrins that would increase adherence and adherence based signaling in the osteoblasts.

8. Summary

Actin and various actin-associated proteins are among the most abundant components that make up EVs, including EVs shed by osteoclasts. It is not known whether these proteins are present because they were part of the mechanism by which EVs are made, or whether they are involved in the regulatory function of EVs. Although EVs likely lack sufficient ATP to support dynamic filaments, and key elements of the machinery that stimulate actin polymerization are missing, it is possible that as EVs fuse with target cells, and the incorporation of concentrated cytoskeletal components could enhance cellular responses like directed cell movement. Integrins have been reported to be vital for targeting EVs to the extracellular matrix. Actin-associated proteins, and perhaps microfilaments in EVs may function to maintain integrins in the conformation that binds the extracellular matrix tightly. EVs have only recently emerged as important agents of intercellular signaling. Understanding the roles of the abundant actin cytoskeletal proteins is crucial for the long-term goal of fully understanding signaling by EVs and making use of that understanding to develop new therapeutic approaches.

Author Contributions: Conceptualization, L.S.H., L.P.d.F. and W.J.R.J.; methodology, L.S.H., L.P.d.F. and W.J.R.J.; formal analysis, L.S.H., L.P.d.F. and W.J.R.J.; resources, L.S.H. and W.J.R.J.; data curation, L.S.H. and W.J.R.J.; writing—original draft preparation, L.S.H.; writing—review and editing, W.J.R.J.; visualization, L.P.d.F.; funding acquisition, L.S.H. and W.J.R.J. All authors have read and agreed to the published version of the manuscript.

Funding: This research was funded by NIH/NIDCR R21 DE023900 (LSH) and NIH/NIDCR R03 DE027504 (W.J.R.J.).

Conflicts of Interest: The authors declare no conflict of interest.

Abbreviations

EV	Extracellular vesicle
RANK	Receptor activator of nuclear factor kappa B
RANKL	Receptor activator of nuclear factor kappa B-ligand
ESCRT	Endosomal sorting complexes required for transport
ATP	Adenosine triphosphate
ADP-Pi	Adenosine diphosphate with attached phosphate group
ADP	Adenosine diphosphate

WASP	Wiskott-Aldrich syndrome protein
Mical	Molecule interacting with CasL
MadCAM	Mucosal addressin cell adhesion molecule
ITGB	Integrin beta
ITGA	Integrin alpha
MYH	Myosin heavy chain
MYO	Myosin
ACTN	Alpha actinin
TAGLN	Transgelin
CAP	Capping protein

References

1. Mathieu, M.; Martin-Jaular, L.; Lavieu, G.; Thery, C. Specificities of secretion and uptake of exosomes and other extracellular vesicles for cell-to-cell communication. *Nat. Cell Biol.* **2019**, *21*, 9–17. [CrossRef] [PubMed]
2. Tkach, M.; Thery, C. Communication by Extracellular Vesicles: Where We Are and Where We Need to Go. *Cell* **2016**, *164*, 1226–1232. [CrossRef]
3. Witwer, K.W.; Thery, C. Extracellular vesicles or exosomes? On primacy, precision, and popularity influencing a choice of nomenclature. *J. Extracell. Vesicles* **2019**, *8*, 1648167. [CrossRef]
4. Raposo, G.; Stoorvogel, W. Extracellular vesicles: Exosomes, microvesicles, and friends. *J. Cell Biol.* **2013**, *200*, 373–383. [CrossRef]
5. Zhang, H.; Freitas, D.; Kim, H.S.; Fabijanic, K.; Li, Z.; Chen, H.; Mark, M.T.; Molina, H.; Martin, A.B.; Bojmar, L.; et al. Identification of distinct nanoparticles and subsets of extracellular vesicles by asymmetric flow field-flow fractionation. *Nat. Cell Biol.* **2018**, *20*, 332–343. [CrossRef] [PubMed]
6. Harding, C.; Heuser, J.; Stahl, P. Receptor-mediated endocytosis of transferrin and recycling of the transferrin receptor in rat reticulocytes. *J. Cell Biol.* **1983**, *97*, 329–339. [CrossRef] [PubMed]
7. Pan, B.T.; Johnstone, R.M. Fate of the transferrin receptor during maturation of sheep reticulocytes in vitro: Selective externalization of the receptor. *Cell* **1983**, *33*, 967–978. [CrossRef]
8. Raposo, G.; Nijman, H.W.; Stoorvogel, W.; Liejendekker, R.; Harding, C.V.; Melief, C.J.; Geuze, H.J. B lymphocytes secrete antigen-presenting vesicles. *J. Exp. Med.* **1996**, *183*, 1161–1172. [CrossRef] [PubMed]
9. Valadi, H.; Ekstrom, K.; Bossios, A.; Sjostrand, M.; Lee, J.J.; Lotvall, J.O. Exosome-mediated transfer of mRNAs and microRNAs is a novel mechanism of genetic exchange between cells. *Nat. Cell Biol.* **2007**, *9*, 654–659. [CrossRef] [PubMed]
10. Deregibus, M.C.; Cantaluppi, V.; Calogero, R.; Lo Iacono, M.; Tetta, C.; Biancone, L.; Bruno, S.; Bussolati, B.; Camussi, G. Endothelial progenitor cell derived microvesicles activate an angiogenic program in endothelial cells by a horizontal transfer of mRNA. *Blood* **2007**, *110*, 2440–2448. [CrossRef]
11. Li, D.; Liu, J.; Guo, B.; Liang, C.; Dang, L.; Lu, C.; He, X.; Cheung, H.Y.; Xu, L.; Lu, C.; et al. Osteoclast-derived exosomal miR-214-3p inhibits osteoblastic bone formation. *Nat. Commun.* **2016**, *7*, 10872. [CrossRef] [PubMed]
12. Sun, W.; Zhao, C.; Li, Y.; Wang, L.; Nie, G.; Peng, J.; Wang, A.; Zhang, P.; Tian, W.; Li, Q.; et al. Osteoclast-derived microRNA-containing exosomes selectively inhibit osteoblast activity. *Cell Discov.* **2016**, *2*, 16015. [CrossRef] [PubMed]
13. Chevillet, J.; Kang, Q.; Ruf, I.K.; Briggs, H.A.; Vojtech, L.N.; Hughes, S.M.; Cheng, H.H.; Arroyo, J.D.; Meredith, E.K.; Gallichotte, E.N.; et al. Quantitative and stoichiometric analysis of the microRNA content of exosomes. *Proc. Natl. Acad. Sci. USA* **2014**, *111*, 14888–14893. [CrossRef] [PubMed]
14. Huynh, N.; VonMoss, L.; Smith, D.; Rahman, I.; Felemban, M.F.; Zuo, J.; Rody, W.J., Jr.; McHugh, K.P.; Holliday, L.S. Characterization of regulatory extracellular vesicles from osteoclasts. *J. Dent. Res.* **2016**, *95*, 673–679. [CrossRef] [PubMed]
15. Negishi-Koga, T.; Shinohara, M.; Komatsu, N.; Bito, H.; Kodama, T.; Friedel, R.H.; Takayanagi, H. Suppression of bone formation by osteoclastic expression of semaphorin 4D. *Nat. Med.* **2011**, *17*, 1473–1480. [CrossRef]
16. Takyar, F.M.; Tonna, S.; Ho, P.W.; Crimeen-Irwin, B.; Baker, E.K.; Martin, T.J.; Sims, N.A. EphrinB2/EphB4 inhibition in the osteoblast lineage modifies the anabolic response to parathyroid hormone. *J. Bone Miner. Res.* **2013**, *28*, 912–925. [CrossRef]

17. Boyce, B.F.; Xing, L. Functions of RANKL/RANK/OPG in bone modeling and remodeling. *Arch. Biochem. Biophys.* **2008**, *473*, 139–146. [CrossRef]
18. Hofbauer, L.C.; Kuhne, C.A.; Viereck, V. The OPG/RANKL/RANK system in metabolic bone diseases. *J. Musculoskelet. Neuronal Interact.* **2004**, *4*, 268–275.
19. Holliday, L.S.; McHugh, K.P.; Zuo, J.; Aguirre, J.I.; Neubert, J.K.; Rody, W.J., Jr. Exosomes: Novel regulators of bone remodeling and potential therapeutic agents for orthodontics. *Orthod. Craniofac. Res.* **2017**, *20*, 95–99. [CrossRef]
20. Ikebuchi, Y.; Aoki, S.; Honma, M.; Hayashi, M.; Sugamori, Y.; Khan, M.; Kariya, Y.; Kato, G.; Tabata, Y.; Penninger, J.M.; et al. Coupling of bone resorption and formation by RANKL reverse signalling. *Nature* **2018**, *561*, 195–200. [CrossRef]
21. Sclerostin regulates RANKL expression in osteocytes. *Bonekey Rep.* **2012**, *1*, 19.
22. Killock, D. Bone: Osteocyte RANKL in bone homeostasis: A paradigm shift? *Nat. Rev. Rheumatol.* **2011**, *7*, 619. [CrossRef] [PubMed]
23. Nakashima, T.; Hayashi, M.; Fukunaga, T.; Kurata, K.; Oh-Hora, M.; Feng, J.Q.; Bonewald, L.F.; Kodama, T.; Wutz, A.; Wagner, E.F.; et al. Evidence for osteocyte regulation of bone homeostasis through RANKL expression. *Nat. Med.* **2011**, *7*, 1231–1234. [CrossRef] [PubMed]
24. Xiong, J.; Onal, M.; Jilka, R.L.; Weinstein, R.S.; Manolagas, S.C.; O'Brien, C.A. Matrix-embedded cells control osteoclast formation. *Nat. Med.* **2011**, *17*, 1235–1241. [CrossRef] [PubMed]
25. Rody, W.J., Jr.; Chamberlain, C.A.; Emory-Carter, A.K.; McHugh, K.P.; Wallet, S.M.; Spicer, V.; Krokhin, O.; Holliday, L.S. The proteome of extracellular vesicles released by clastic cells differs based on their substrate. *PLoS ONE* **2019**, *14*, e0219602. [CrossRef] [PubMed]
26. Teitelbaum, S.L. Osteoclasts: What do they do and how do they do it? *Am. J. Pathol.* **2007**, *170*, 427–435. [CrossRef] [PubMed]
27. Chambers, T.J.; Fuller, K. How are osteoclasts induced to resorb bone? *Ann. N. Y. Acad. Sci.* **2011**, *1240*, 1–6. [CrossRef]
28. Destaing, O.; Petropoulos, C.; Albiges-Rizo, C. Coupling between acto-adhesive machinery and ECM degradation in invadosomes. *Cell Adh. Migr.* **2014**, *8*, 256–262. [CrossRef]
29. King, G.J.; Holtrop, M.E. Actin-like filaments in bone cells of cultured mouse calvaria as demonstrated by binding to heavy meromyosin. *J. Cell Biol.* **1975**, *66*, 445–451. [CrossRef]
30. Blair, H.C.; Teitelbaum, S.L.; Ghiselli, R.; Gluck, S. Osteoclastic bone resorption by a polarized vacuolar proton pump. *Science* **1989**, *245*, 855–857. [CrossRef]
31. Toro, E.J.; Ostrov, D.A.; Wronski, T.J.; Holliday, L.S. Rational Identification of Enoxacin as a Novel V-ATPase-Directed Osteoclast Inhibitor. *Curr. Protein Pept. Sci.* **2012**, *13*, 180–191. [CrossRef] [PubMed]
32. Bromme, D.; Okamoto, K.; Wang, B.B.; Biroc, S. Human cathepsin O_2, a matrix protein-degrading cysteine protease expressed in osteoclasts. Functional expression of human cathepsin O_2 in Spodoptera frugiperda and characterization of the enzyme. *J. Biol. Chem.* **1996**, *271*, 2126–2132. [CrossRef] [PubMed]
33. Gelb, B.D.; Moissoglu, K.; Zhang, J.; Martignetti, J.A.; Bromme, D.; Desnick, R.J. Cathepsin K: Isolation and characterization of the murine cDNA and genomic sequence, the homologue of the human pycnodysostosis gene. *Biol. Chem. Mol. Med.* **1996**, *59*, 200–206. [CrossRef] [PubMed]
34. Tomoshige, S.; Kobayashi, Y.; Hosoba, K.; Hamamoto, A.; Miyamoto, T.; Saito, Y. Cytoskeleton-related regulation of primary cilia shortening mediated by melanin-concentrating hormone receptor 1. *Gen. Comp. Endocrinol.* **2017**, *253*, 44–52. [CrossRef]
35. Collier, M.E.; Maraveyas, A.; Ettelaie, C. Filamin-A is required for the incorporation of tissue factor into cell-derived microvesicles. *Thromb. Haemost.* **2014**, *111*, 647–655.
36. Collier, M.E.W.; Ettelaie, C.; Goult, B.T.; Maraveyas, A.; Goodall, A.H. Investigation of the Filamin A-Dependent Mechanisms of Tissue Factor Incorporation into Microvesicles. *Thromb. Haemost.* **2017**, *117*, 2034–2044. [CrossRef]
37. Colombo, M.; Moita, C.; Van Niel, G.; Kowal, J.; Vigneron, J.; Benaroch, P.; Manel, N.; Moita, L.F.; Thery, C.; Raposo, G. Analysis of ESCRT functions in exosome biogenesis, composition and secretion highlights the heterogeneity of extracellular vesicles. *J. Cell Sci.* **2013**, *126*, 5553–5565. [CrossRef]
38. Ostrowski, M.; Carmo, N.B.; Krumeich, S.; Fanget, I.; Raposo, G.; Savina, A.; Moita, C.F.; Schauer, K.; Hume, A.N.; Freitas, R.P.; et al. Rab27a and Rab27b control different steps of the exosome secretion pathway. *Nat. Cell Biol.* **2010**, *12*, 19–30. [CrossRef]

39. Trajkovic, K.; Hsu, C.; Chiantia, S.; Rajendran, L.; Wenzel, D.; Wieland, F.; Schwille, P.; Brugger, B.; Simons, M. Ceramide triggers budding of exosome vesicles into multivesicular endosomes. *Science* **2008**, *319*, 1244–1247. [CrossRef]
40. Yuyama, K.; Sun, H.; Mitsutake, S.; Igarashi, Y. Sphingolipid-modulated exosome secretion promotes clearance of amyloid-beta by microglia. *J. Biol. Chem.* **2012**, *287*, 10977–10989. [CrossRef]
41. Pollard, T.D. What We Know and Do Not Know About Actin. *Handb. Exp. Pharmacol.* **2017**, *235*, 331–347. [PubMed]
42. Kajimoto, T.; Mohamed, N.N.I.; Badawy, S.M.M.; Matovelo, S.A.; Hirase, M.; Nakamura, S.; Yoshida, D.; Okada, T.; Ijuin, T.; Nakamura, S.I. Involvement of Gβγ subunits of G(i) protein coupled with S1P receptor on multivesicular endosomes in F-actin formation and cargo sorting into exosomes. *J. Biol. Chem.* **2018**, *293*, 245–253. [CrossRef] [PubMed]
43. Footer, M.J.; Kerssemakers, J.W.; Theriot, J.A.; Dogterom, M. Direct measurement of force generation by actin filament polymerization using an optical trap. *Proc. Natl. Acad. Sci. USA* **2007**, *104*, 2181–2186. [CrossRef] [PubMed]
44. DePina, A.S.; Langford, G.M. Vesicle transport: The role of actin filaments and myosin motors. *Microsc. Res. Technol.* **1999**, *47*, 93–106. [CrossRef]
45. Hurley, J.H. ESCRTs are everywhere. *EMBO J.* **2015**, *34*, 2398–2407. [CrossRef] [PubMed]
46. Devreotes, P.N.; Bhattacharya, S.; Edwards, M.; Iglesias, P.A.; Lampert, T.; Miao, Y. Excitable Signal Transduction Networks in Directed Cell Migration. *Ann. Rev. Cell Dev. Biol.* **2017**, *33*, 103–125. [CrossRef]
47. Hammer, J.A., III. The structure and function of unconventional myosins: A review. *J. Muscle Res. Cell Motil.* **1994**, *15*, 1–10. [CrossRef]
48. Wang, K.; Wloka, C.; Bi, E. Non-muscle Myosin-II Is Required for the Generation of a constriction Site for Subsequent Abscission. *iScience* **2019**, *13*, 69–81. [CrossRef]
49. Muralidharan-Chari, V.; Clancy, J.; Plou, C.; Romao, M.; Chavrier, P.; Raposo, G.; D'Souza-Schorey, C. ARF6-regulated shedding of tumor cell-derived plasma membrane microvesicles. *Curr. Biol* **2009**, *19*, 1875–1885. [CrossRef]
50. Sun, W.; Vida, T.A.; Sirisaengtaksin, N.; Merrill, S.A.; Hanson, P.I.; Bean, A.J. Cell-free reconstitution of multivesicular body formation and receptor sorting. *Traffic* **2010**, *11*, 867–876. [CrossRef]
51. Carlier, M.F.; Wiesner, S.; Le, C.C.; Pantaloni, D. Actin-based motility as a self-organized system: Mechanism and reconstitution In Vitro. *Comptes Rendus Biol.* **2003**, *326*, 161–170. [CrossRef]
52. Theriot, J.A.; Rosenblatt, J.; Portnoy, D.A.; Goldschmidt-Clermont, P.J.; Mitchison, T.J. Involvement of profilin in the actin-based motility of L. monocytogenes in cells and in cell-free extracts. *Cell* **1994**, *76*, 505–517. [CrossRef]
53. Korn, E.D. Actin polymerization and its regulation by proteins from nonmuscle cells. *Physiol. Rev.* **1982**, *62*, 672–737. [CrossRef] [PubMed]
54. Campellone, K.G.; Welch, M.D. A nucleator arms race: Cellular control of actin assembly. *Nat. Rev. Mol. Cell Biol.* **2010**, *11*, 237–251. [CrossRef]
55. Lasser, C.; Jang, S.C.; Lotvall, J. Subpopulations of extracellular vesicles and their therapeutic potential. *Mol. Aspects Med.* **2018**, *60*, 1–14. [CrossRef]
56. Zabeo, D.; Cvjetkovic, A.; Lasser, C.; Schorb, M.; Lotvall, J.; Hoog, J.L. Exosomes purified from a single cell type have diverse morphology. *J. Extracell. Vesicles* **2017**, *6*, 1329476. [CrossRef]
57. Korn, E.D.; Carlier, M.F.; Pantaloni, D. Actin polymerization and ATP hydrolysis. *Science* **1987**, *238*, 638–644. [CrossRef]
58. Ronquist, K.G.; Ek, B.; Morrell, J.; Stavreus-Evers, A.; Strom, H.B.; Humblot, P.; Ronquist, G.; Larsson, A. Prostasomes from four different species are able to produce extracellular adenosine triphosphate (ATP). *Biochim. Biophys. Acta* **2013**, *1830*, 4604–4610. [CrossRef]
59. Guo, H.; Chang, Z.; Zhang, Z.; Zhao, Y.; Jiang, X.; Yu, H.; Zhang, Y.; Zhao, R.; He, B. Extracellular ATPs produced in seminal plasma exosomes regulate boar sperm motility and mitochondrial metabolism. *Theriogenology* **2019**, *139*, 113–120. [CrossRef]
60. Pantaloni, D.; Carlier, M.F. How profilin promotes actin filament assembly in the presence of thymosin beta 4. *Cell* **1993**, *75*, 1007–1014. [CrossRef]

61. Zhang, Q.; Higginbotham, J.N.; Jeppesen, D.K.; Yang, Y.P.; Li, W.; McKinley, E.T.; Graves-Deal, R.; Ping, J.; Britain, C.M.; Dorsett, K.A.; et al. Transfer of Functional Cargo in Exomeres. *Cell Rep.* **2019**, *27*, 940–954. [CrossRef] [PubMed]
62. Lal, A.A.; Brenner, S.L.; Korn, E.D. Preparation and polymerization of skeletal muscle ADP-actin. *J. Biol. Chem.* **1984**, *259*, 13061–13065. [PubMed]
63. Zigmond, S.H. Formin-induced nucleation of actin filaments. *Curr. Opin. Cell Biol.* **2004**, *16*, 99–105. [CrossRef] [PubMed]
64. Rotty, J.D.; Wu, C.; Bear, J.E. New insights into the regulation and cellular functions of the ARP2/3 complex. *Nat. Rev. Mol. Cell Biol.* **2013**, *14*, 7–12. [CrossRef]
65. Machesky, L.M.; Gould, K.L. The Arp2/3 complex: A multifunctional actin organizer. *Curr. Opin. Cell Biol.* **1999**, *11*, 117–121. [CrossRef]
66. Machesky, L.M.; Atkinson, S.J.; Ampe, C.; Vandekerckhove, J.; Pollard, T.D. Purification of a cortical complex containing two unconventional actins from Acanthamoeba by affinity chromatography on profilin-agarose. *J. Cell Biol.* **1994**, *127*, 107–115. [CrossRef]
67. Millard, T.H.; Machesky, L.M. The Wiskott-Aldrich syndrome protein (WASP) family. *Trends Biochem. Sci.* **2001**, *26*, 198–199. [CrossRef]
68. Weaver, A.M.; Heuser, J.E.; Karginov, A.V.; Lee, W.L.; Parsons, J.T.; Cooper, J.A. Interaction of cortactin and N-WASp with Arp2/3 complex. *Curr. Biol.* **2002**, *12*, 1270–1278. [CrossRef]
69. Luxenburg, C.; Parsons, J.T.; Addadi, L.; Geiger, B. Involvement of the Src-cortactin pathway in podosome formation and turnover during polarization of cultured osteoclasts. *J. Cell Sci.* **2006**, *119*, 4878–4888. [CrossRef]
70. Tehrani, S.; Faccio, R.; Chandrasekar, I.; Ross, F.P.; Cooper, J.A. Cortactin has an essential and specific role in osteoclast actin assembly. *Mol. Biol. Cell* **2006**, *17*, 2882–2895. [CrossRef]
71. Zalli, D.; Neff, L.; Nagano, K.; Shin, N.Y.; Witke, W.; Gori, F.; Baron, R. The Actin-Binding Protein Cofilin and Its Interaction with Cortactin Are Required for Podosome Patterning in Osteoclasts and Bone Resorption In Vivo and In Vitro. *J. Bone Miner. Res.* **2016**, *31*, 1701–1712. [CrossRef]
72. Chan, K.T.; Creed, S.J.; Bear, J.E. Unraveling the enigma: Progress towards understanding the coronin family of actin regulators. *Trends Cell Biol.* **2011**, *21*, 481–488. [CrossRef] [PubMed]
73. Blanchoin, L.; Amann, K.J.; Higgs, H.N.; Kaiser, D.A.; Marchand, J.B.; Mullins, R.D.; Pollard, T.D. Role of ADF/cofilin, Arp2/3 complex, capping proteins and profilin in the dynamic of branched actin filaments networks. *Mol. Biol. Cell* **2000**, *11*, 178A.
74. Blanchoin, L.; Pollard, T.D.; Mullins, R.D. Interactions of ADF/cofilin, Arp2/3 complex, capping protein and profilin in remodeling of branched actin filament networks. *Curr. Biol.* **2000**, *10*, 1273–1282. [CrossRef]
75. Blanchoin, L.; Mullins, R.D.; Robinson, R.C.; Choe, S.; Pollard, T.D. Acanthamoeba actophorin (ADF/cofilin) depolymerizes actin filaments capped with Arp2/3 and gelsolin as well as only barbed ends capped filaments. Effect of phosphorylation on actophorin interaction with actin. *Mol. Biol. Cell* **1999**, *10*, 24A.
76. Sung, B.H.; Ketova, T.; Hoshino, D.; Zijlstra, A.; Weaver, A.M. Directional cell movement through tissues is controlled by exosome secretion. *Nat. Commun.* **2015**, *6*, 7164. [CrossRef]
77. Sung, B.H.; Weaver, A.M. Exosome secretion promotes chemotaxis of cancer cells. *Cell Adhes. Migr.* **2017**, *11*, 187–195. [CrossRef]
78. Zhou, Y.; Gunput, R.A.; Adolfs, Y.; Pasterkamp, R.J. MICALs in control of the cytoskeleton, exocytosis, and cell death. *Cell Mol. Life Sci.* **2011**, *68*, 4033–4044. [CrossRef]
79. Grintsevich, E.E.; Ge, P.; Sawaya, M.R.; Yesilyurt, H.G.; Terman, J.R.; Zhou, Z.H.; Reisler, E. Catastrophic disassembly of actin filaments via Mical-mediated oxidation. *Nat. Commun.* **2017**, *8*, 2183. [CrossRef]
80. Cooper, J.A.; Schafer, D.A. Control of actin assembly and disassembly at filament ends. *Curr. Opin. Cell Biol.* **2000**, *12*, 97–103. [CrossRef]
81. Bearer, E.L.; Prakash, J.M.; Manchester, R.D.; Allen, P.G. VASP protects actin filaments from gelsolin: An in vitro study with implications for platelet actin reorganizations. *Cell Motil. Cytoskeleton.* **2000**, *47*, 351–364. [CrossRef]
82. Christensen, J.R.; Hocky, G.M.; Homa, K.E.; Morganthaler, A.N.; Hitchcock-DeGregori, S.E.; Voth, G.A.; Kovar, D.R. Competition between Tropomyosin, Fimbrin, and ADF/Cofilin drives their sorting to distinct actin filament networks. *eLife* **2017**, *10*, 6.

83. Gateva, G.; Kremneva, E.; Reindl, T.; Kotila, T.; Kogan, K.; Gressin, L.; Gunning, P.W.; Manstein, D.J.; Michelot, A.; Lappalainen, P. Tropomyosin Isoforms Specify Functionally Distinct Actin Filament Populations In Vitro. *Curr. Biol.* **2017**, *27*, 705–713. [CrossRef] [PubMed]
84. Na, B.R.; Jun, C.D. TAGLN2-mediated actin stabilization at the immunological synapse: Implication for cytotoxic T cell control of target cells. *BMB Rep.* **2015**, *48*, 369. [CrossRef] [PubMed]
85. Na, B.R.; Kim, H.R.; Piragyte, I.; Oh, H.M.; Kwon, M.S.; Akber, U.; Lee, H.S.; Park, D.S.; Song, W.K.; Park, Z.Y.; et al. TAGLN2 regulates T cell activation by stabilizing the actin cytoskeleton at the immunological synapse. *J. Cell Biol.* **2015**, *209*, 143–162. [CrossRef] [PubMed]
86. Hao, R.; Liu, Y.; Du, Q.; Liu, L.; Chen, S.; You, H.; Dong, Y. Transgelin-2 expression in breast cancer and its relationships with clinicopathological features and patient outcome. *Breast Cancer* **2019**, *26*, 776–783. [CrossRef] [PubMed]
87. Sun, Y.; Peng, W.; He, W.; Luo, M.; Chang, G.; Shen, J.; Zhao, X.; Hu, Y. Transgelin-2 is a novel target of KRAS-ERK signaling involved in the development of pancreatic cancer. *J. Exp. Clin. Cancer Res.* **2018**, *37*, 166. [CrossRef]
88. Zhou, Q.; Jiang, X.; Yan, W.; Dou, X. Transgelin 2 overexpression inhibits cervical cancer cell invasion and migration. *Mol. Med. Rep.* **2019**, *19*, 4919–4926. [CrossRef]
89. Vinterhoj, H.S.H.; Stensballe, A.; Duroux, M.; Gazerani, P. Characterization of rat primary trigeminal satellite glial cells and associated extracellular vesicles under normal and inflammatory conditions. *J. Proteom.* **2019**, *190*, 27–34. [CrossRef]
90. Krits, I.; Wysolmerski, R.B.; Holliday, L.S.; Lee, B.S. Differential Localization of Myosin II Isoforms in Resting and Activated Osteoclasts. *Calcif. Tissue Int.* **2002**, *71*, 530–538. [CrossRef]
91. Zhang, Y.; Cao, F.; Zhou, Y.; Feng, Z.; Sit, B.; Krendel, M.; Yu, C.H. Tail domains of myosin-1e regulate phosphatidylinositol signaling and F-actin polymerization at the ventral layer of podosomes. *Mol. Biol. Cell* **2019**, *30*, 622–635. [CrossRef] [PubMed]
92. Clayton, A.; Turkes, A.; Dewitt, S.; Steadman, R.; Mason, M.D.; Hallett, M.B. Adhesion and signaling by B cell-derived exosomes: The role of integrins. *FASEB J.* **2004**, *18*, 977–979. [CrossRef] [PubMed]
93. DeRita, R.M.; Sayeed, A.; Garcia, V.; Krishn, S.R.; Shields, C.D.; Sarker, S.; Friedman, A.; McCue, P.; Molugu, S.K.; Rodeck, U.; et al. Tumor-Derived Extracellular Vesicles Require beta1 Integrins to Promote Anchorage-Independent Growth. *iScience* **2019**, *14*, 199–209. [CrossRef] [PubMed]
94. Guo, Q.; Furuta, K.; Lucien, F.; Gutierrez Sanchez, L.H.; Hirsova, P.; Krishnan, A.; Kabashima, A.; Pavelko, K.D.; Madden, B.; Alhuwaish, H.; et al. Integrin beta1-enriched extracellular vesicles mediate monocyte adhesion and promote liver inflammation in murine NASH. *J. Hepatol.* **2019**, *71*, 1193–1205. [CrossRef] [PubMed]
95. Manou, D.; Caon, I.; Bouris, P.; Triantaphyllidou, I.E.; Giaroni, C.; Passi, A.; Karamanos, N.; Vigetti, D.; Theocharis, A.D. The Complex Interplay between Extracellular Matrix and Cells in Tissues. *Methods Mol. Biol.* **2019**, *1952*, 1–20. [PubMed]
96. Hoshino, D.; Kirkbride, K.C.; Costello, K.; Clark, E.S.; Sinha, S.; Grega-Larson, N.; Tyska, M.J.; Weaver, A.M. Exosome secretion is enhanced by invadopodia and drives invasive behavior. *Cell Rep.* **2013**, *5*, 1159–1168. [CrossRef]
97. Genschmer, K.R.; Russell, D.W.; Lal, C.; Szul, T.; Bratcher, P.E.; Noerager, B.D.; Abdul, R.M.; Xu, X.; Rezonzew, G.; Viera, L.; et al. Activated PMN Exosomes: Pathogenic Entities Causing Matrix Destruction and Disease in the Lung. *Cell* **2019**, *176*, 113–126. [CrossRef]
98. Chen, W.; Hoffmann, A.D.; Liu, H.; Liu, X. Organotropism: New insights into molecular mechanisms of breast cancer metastasis. *NPJ Precis. Oncol.* **2018**, *2*, 4. [CrossRef]
99. Hoshino, A.; Costa-Silva, B.; Shen, T.L.; Rodrigues, G.; Hashimoto, A.; Tesic, M.M.; Molina, H.; Kohsaka, S.; Di, G.A.; Ceder, S.; et al. Tumour exosome integrins determine organotropic metastasis. *Nature* **2015**, *527*, 329–335. [CrossRef]
100. Harburger, D.S.; Calderwood, D.A. Integrin signalling at a glance. *J. Cell Sci.* **2009**, *122*, 159–163. [CrossRef]
101. Hayashi, H.; Nakahama, K.; Sato, T.; Tuchiya, T.; Asakawa, Y.; Maemura, T.; Tanaka, M.; Morita, I. The role of Mac-1 (CD11b/CD18) in osteoclast differentiation induced by receptor activator of nuclear factor-kappaB ligand. *FEBS Lett.* **2008**, *582*, 3243–3248. [CrossRef] [PubMed]

102. Ohtsuji, M.; Lin, Q.; Okazaki, H.; Takahashi, K.; Amano, H.; Yagita, H.; Nishimura, H.; Hirose, S. Anti-CD11b antibody treatment suppresses the osteoclast generation, inflammatory cell infiltration, and autoantibody production in arthritis-prone FcgammaRIIB-deficient mice. *Arthritis Res. Ther.* **2018**, *20*, 25. [CrossRef] [PubMed]
103. Yang, G.; Chen, X.; Yan, Z.; Zhu, Q.; Yang, C. CD11b promotes the differentiation of osteoclasts induced by RANKL through the spleen tyrosine kinase signalling pathway. *J. Cell Mol. Med.* **2017**, *21*, 3445–3452. [CrossRef] [PubMed]
104. Hamzei, M.; Ventriglia, G.; Hagnia, M.; Antonopolous, A.; Bernal-Sprekelsen, M.; Dazert, S.; Hildmann, H.; Sudhoff, H. Osteoclast stimulating and differentiating factors in human cholesteatoma. *Laryngoscope* **2003**, *113*, 436–442. [CrossRef]
105. Ruef, N.; Dolder, S.; Aeberli, D.; Seitz, M.; Balani, D.; Hofstetter, W. Granulocyte-macrophage colony-stimulating factor-dependent CD11c-positive cells differentiate into active osteoclasts. *Bone* **2017**, *97*, 267–277. [CrossRef]
106. Strauch, U.G.; Lifka, A.; Gosslar, U.; Kilshaw, P.J.; Clements, J.; Holzmann, B. Distinct binding specificities of integrins alpha 4 beta 7 (LPAM-1), alpha 4 beta 1 (VLA-4), and alpha IEL beta 7. *Int. Immunol.* **1994**, *6*, 263–275. [CrossRef]
107. Ne, R.F.; Witherspoon, D.E.; Gutmann, J.L. Tooth resorption. *Quintessence Int.* **1999**, *30*, 9–25.
108. Hartsfield, J.K., Jr. Pathways in external apical root resorption associated with orthodontia. *Orthod. Craniofac. Res.* **2009**, *12*, 236–242. [CrossRef]
109. Humphries, J.D.; Byron, A.; Humphries, M.J. Integrin ligands at a glance. *J. Cell Sci.* **2006**, *119*, 3901–3903. [CrossRef]
110. Chellaiah, M.A.; Hruska, K.A. The integrin alpha(v)beta(3) and CD44 regulate the actions of osteopontin on osteoclast motility. *Calcif. Tissue Int.* **2003**, *72*, 197–205. [CrossRef]
111. Duong, L.T.; Lakkakorpi, P.; Nakamura, I.; Rodan, G.A. Integrins and signaling in osteoclast function. *Matrix Biol.* **2000**, *19*, 97–105. [CrossRef]
112. McHugh, K.P.; Hodivala-Dilke, K.; Zheng, M.H.; Namba, N.; Lam, J.; Novack, D.; Feng, X.; Ross, F.P.; Hynes, R.O.; Teitelbaum, S.L. Mice lacking beta 3 integrins are osteosclerotic because of dysfunctional osteoclasts. *J. Clin. Investig.* **2000**, *105*, 433–440. [CrossRef]
113. McHugh, K.P.; Shen, Z.; Crotti, T.N.; Flannery, M.R.; Fajardo, R.; Bierbaum, B.E.; Goldring, S.R. Role of cell-matrix interactions in osteoclast differentiation. *Adv. Exp. Med. Biol.* **2007**, *602*, 107–111. [PubMed]
114. Holliday, L.S.; Welgus, H.G.; Fliszar, C.J.; Veith, G.M.; Jeffrey, J.J.; Gluck, S.L. Initiation of osteoclast bone resorption by interstitial collagenase. *J. Biol. Chem.* **1997**, *272*, 22053–22058. [CrossRef] [PubMed]
115. Horton, E.R.; Humphries, J.D.; James, J.; Jones, M.C.; Askari, J.A.; Humphries, M.J. The integrin adhesome network at a glance. *J. Cell Sci.* **2016**, *129*, 4159–4163. [CrossRef]
116. Faull, R.J.; Ginsberg, M.H. Inside-out signaling through integrins. *J. Am. Soc. Nephrol.* **1996**, *7*, 1091–1097.
117. Shen, B.; Delaney, M.K.; Du, X. Inside-out, outside-in, and inside-outside-in: G protein signaling in integrin-mediated cell adhesion, spreading, and retraction. *Curr. Opin. Cell Biol.* **2012**, *24*, 600–606. [CrossRef]
118. Springer, T.A.; Dustin, M.L. Integrin inside-out signaling and the immunological synapse. *Curr. Opin. Cell Biol.* **2012**, *24*, 107–115. [CrossRef]
119. Bayraktar, R.; Van, R.K.; Calin, G.A. Cell-to-cell communication: microRNAs as hormones. *Mol. Oncol.* **2017**, *11*, 1673–1686. [CrossRef]
120. Kim, K.M.; Abdelmohsen, K.; Mustapic, M.; Kapogiannis, D.; Gorospe, M. RNA in extracellular vesicles. *Wiley Interdiscip. Rev. RNA* **2017**, *8*, e1413. [CrossRef]
121. Fedele, C.; Singh, A.; Zerlanko, B.J.; Iozzo, R.V.; Languino, L.R. The alphavbeta6 integrin is transferred intercellularly via exosomes. *J. Biol. Chem.* **2015**, *290*, 4545–4551. [CrossRef] [PubMed]
122. Singh, A.; Fedele, C.; Lu, H.; Nevalainen, M.T.; Keen, J.H.; Languino, L.R. Exosome-mediated Transfer of alphavbeta3 Integrin from Tumorigenic to Nontumorigenic Cells Promotes a Migratory Phenotype. *Mol. Cancer Res.* **2016**, *14*, 1136–1146. [CrossRef] [PubMed]
123. Cheng, S.L.; Lai, C.F.; Fausto, A.; Chellaiah, M.; Feng, X.; McHugh, K.P.; Teitelbaum, S.L.; Civitelli, R.; Hruska, K.A.; Ross, F.P.; et al. Regulation of alphaVbeta3 and alphaVbeta5 integrins by dexamethasone in normal human osteoblastic cells. *J. Cell Biol. Chem.* **2000**, *77*, 265–276.

124. Hu, H.M.; Yang, L.; Wang, Z.; Liu, Y.W.; Fan, J.Z.; Fan, J.; Liu, J.; Luo, Z.J. Overexpression of integrin a2 promotes osteogenic differentiation of hBMSCs from senile osteoporosis through the ERK pathway. *Int. J. Clin. Exp. Pathol.* **2013**, *6*, 841–852. [PubMed]
125. Phillips, J.A.; Almeida, E.A.; Hill, E.L.; Aguirre, J.I.; Rivera, M.F.; Nachbandi, I.; Wronski, T.J.; van der Meulen, M.C.; Globus, R.K. Role for beta1 integrins in cortical osteocytes during acute musculoskeletal disuse. *Matrix Biol.* **2008**, *27*, 609–618. [CrossRef] [PubMed]

© 2019 by the authors. Licensee MDPI, Basel, Switzerland. This article is an open access article distributed under the terms and conditions of the Creative Commons Attribution (CC BY) license (http://creativecommons.org/licenses/by/4.0/).

Article

Interactome and F-Actin Interaction Analysis of *Dictyostelium discoideum* Coronin A

Tohnyui Ndinyanka Fabrice †, Thomas Fiedler †, Vera Studer †, Adrien Vinet, Francesco Brogna, Alexander Schmidt and Jean Pieters *

Biozentrum, University of Basel, Klingelbergstrasse 50, 4056 Basel, Switzerland; nf.tohnyui@unibas.ch (T.N.F.); tommyfiedler@gmail.com (T.F.); vera_studer@gmx.ch (V.S.); a.vinet@lady-green.com (A.V.); brogna.francesco@hotmail.com (F.B.); alex.schmidt@unibas.ch (A.S.)
* Correspondence: jean.pieters@unibas.ch; Tel.: +41-61-267-14-94
† These authors contributed equally to this work.

Received: 22 December 2019; Accepted: 17 February 2020; Published: 21 February 2020

Abstract: Coronin proteins are evolutionary conserved WD repeat containing proteins that have been proposed to carry out different functions. In *Dictyostelium*, the short coronin isoform, coronin A, has been implicated in cytoskeletal reorganization, chemotaxis, phagocytosis and the initiation of multicellular development. Generally thought of as modulators of F-actin, coronin A and its mammalian homologs have also been shown to mediate cellular processes in an F-actin-independent manner. Therefore, it remains unclear whether or not coronin A carries out its functions through its capacity to interact with F-actin. Moreover, the interacting partners of coronin A are not known. Here, we analyzed the interactome of coronin A as well as its interaction with F-actin within cells and in vitro. Interactome analysis showed the association with a diverse set of interaction partners, including fimbrin, talin and myosin subunits, with only a transient interaction with the minor actin10 isoform, but not the major form of actin, actin8, which was consistent with the absence of a coronin A-actin interaction as analyzed by co-sedimentation from cells and lysates. In vitro, however, purified coronin A co-precipitated with rabbit muscle F-actin in a coiled-coil-dependent manner. Our results suggest that an in vitro interaction of coronin A and rabbit muscle actin may not reflect the cellular interaction state of coronin A with actin, and that coronin A interacts with diverse proteins in a time-dependent manner.

Keywords: *Dictyostelium*; coronin A; interactome analysis; Actin

1. Introduction

The coronin protein family is comprised of a group of evolutionary conserved proteins that are characterized by the presence of a central Tryptophan-Aspartate (WD or WD-40) repeat-containing domain fused via a linker of variable length to a coiled-coil domain that is involved in homo-oligomerization [1,2]. Coronin molecules are widespread in eukaryotes, with a bioinformatic analysis defining over 723 coronin molecules from 358 different eukaryotic species [3]. Notably, while lower eukaryotes such as yeast, amoeba and parasites including *Leishmania*, *Toxoplasma* and *Plasmodium* appear to express one or maximally two coronin molecules [4–6], in higher eukaryotes, multiple coronin molecules are expressed, with up to seven coronins expressed in mammals [1,7–9].

The biological function for many of the coronins within cells or organisms remains unclear. While a number of studies have demonstrated an interaction of coronin molecules with actin in vitro, most of the work linking coronin molecules to F-actin interaction has been performed using recombinantly expressed *Saccharomyces cerevisiae* coronin (Crn1) [4,10]. In vitro, Crn1 was found to co-precipitate with F-actin [11], which is in accordance with the presence of a CA-like (Central region fused to an acidic region) domain in yeast Crn1 [12] that is known to be responsible for interactions with actin and

Arp2/3 [13,14]. However, the CA-like domain is missing in most other coronin molecules, and in fact, it is unclear to what degree yeast Crn1 is a functional homologue of *Dictyostelium* and mammalian coronins [3]. Furthermore, yeast cells lacking Crn1 do not show an obvious phenotype and have no detectable defects in actin-based processes under a variety of different growth conditions [4,15]. In the unicellular parasites *Toxoplasma gondii*, *Plasmodium* and *Leishmania*, coronins appear to play divergent roles; while in *Leishmania*, coronin regulates microtubule remodeling during cytokinesis [16], in *Toxoplasma gondii*, deletion of coronin does not affect a number of actin-dependent processes, although a weak interaction with actin was observed in vitro [6]. Similarly, *Plasmodium* coronin only weakly interacts with actin in vitro [5] and was shown to localize within the cell in a calcium-dependent and actin-independent manner [17].

In mammals, coronin molecules are emerging as multifunctional regulators of diverse physiological processes, and a common molecular function for the different coronins has not been clearly established. Thus far, F-actin modulation has been the common denominator to explain the role of the different coronin proteins; whereas several coronin proteins were shown to bind F-actin in vitro and within cells [18–21], other coronin proteins were specifically shown to neither bind to nor modulate F-actin within cells [22,23]. Notably, for one of the best characterized coronins, mammalian coronin 1 (also known as P57 or TACO, for Tryptophan Aspartate containing Coat protein) [24,25], as well as a number of other coronins, emerging evidence is suggesting that they perform actin-independent functions that include neuronal signaling, T cell homeostasis and the initiation of multicellular differentiation [6,26–30].

Given the above-mentioned conflicting reports on the capacity of coronin proteins to interact with F-actin in vitro and within cells as well as the issue of potential redundancy, for example, in mammals, where multiple coronin molecules can be co-expressed [1], we turned to *Dictyostelium discoideum*, that only expresses a single short coronin, coronin A. Coronin A was initially described as a myosin-actin co-precipitating protein that accumulates at crown-shaped, actin-rich cell protrusions (hence the name 'coronin') although subsequent work showed that 'crowns' are also formed in the absence of coronin A [31,32]. *Dictyostelium* cells lacking coronin A show pleiotropic defects in cytokinesis, uptake of yeast particles as well as motility and migration [33–35]. In addition to coronin A, *Dictyostelium* also expresses a 'tandem' coronin molecule, termed coronin B, and the two coronins appear to have non-redundant functions [36]. *Dictyostelium* cells, which are a unicellular species when sufficient food is available, have the remarkable capacity, upon starvation, to transform into multicellular structures, resulting in spore-bearing fruiting body formation to ensure long-term survival. The developmental program responsible for the transformation from single cells to spores is initiated upon starvation and depends on cell density and food-deprivation factors that induce pulsating release of cyclic Adenosine Monophosphate (cAMP). This cAMP-release induces the upregulation of genes necessary for cAMP production and chemotaxis, driving the initiation of multicellular development [37–39]. Recent work showed that coronin A is responsible for the initiation of the cAMP relay that is required for development upon starvation, but dispensable for cAMP sensing, chemotaxis, and development per se [30]. Together with the finding that F-actin depolymerization does not compromise cAMP-mediated signal transduction, these results suggest that coronin A does not directly modulate F-actin during multicellular development [30]. Instead, F-actin-dependent processes may occur downstream of the coronin A-dependent starvation response, and, in accordance with the role for coronin 1 in mammals and coronin in *Plasmodium*, a prime role for coronin A in *Dictyostelium* may lie within the regulation of cAMP-dependent signal transduction [30].

In this paper, we characterized the interactome of coronin A by affinity purification followed by mass spectrometry as well as analyzing the interaction of *Dictyostelium* coronin A with F-actin within cells and with rabbit muscle F-actin in vitro. We found that while the interactome analysis revealed the co-precipitation of coronin A with a number of actin-interacting proteins as well as a transient interaction with the minor actin10 isoform, within cells, coronin A failed to interact with actin under conditions in which actin robustly interacted with myosin. In accordance with *Dictyostelium* coronin A being dispensable for F-actin modulation, phagocytosis of bacteria and inert beads were unaffected by deletion of coronin A. Together, these data suggest that an interaction of coronin A with the actin cytoskeleton occurs indirectly, and that an in vitro association with rabbit muscle actin may not be indicative for the cellular state of coronin A.

2. Results

2.1. Coronin A-F-Actin Interaction in Vitro and within Cells

Coronin A was originally identified as an actin-myosin interacting protein and has been suggested to play diverse roles in the regulation of a number of actin-dependent processes [40–42]. However, more recent work has suggested that the function for coronin A in initiating multicellular development occurs independently of a role in F-actin reorganization [30]. To investigate an interaction of coronin A with actin, a number of experimental approaches were undertaken to determine the interaction partners of coronin A within *Dictyostelium* as well as to assess the capacity of coronin A to interact with F-actin within cells (with endogenous actin) and in vitro (using rabbit muscle F-actin).

First, to identify the coronin A interactome in an unbiased manner, *corA*$^-$ cells that were transfected with FLAG-tagged coronin A or with non-tagged coronin A (control) and grown in HL5 medium were immunoprecipitated from cell lysates using FLAG affinity chromatography, eluted and interacting proteins analyzed by quantitative mass spectrometry. A total of 47 significantly enriched (log2ratio > 1.5; q-value < 0.05) coronin A interacting proteins were identified (Figure 1A,B and Tables S1 and S2). The most prominently associated proteins included several uncharacterized proteins, metabolic enzymes, tubulin chaperones and a transcription factor (Figure 1A,B). While a > 2-fold enrichment for the actin interacting proteins myosin-K heavy chain, actobindin-B/C, talin-B and fimbrin was observed (Figure 1A,B and Tables S1 and S2), actin was not present in the interactome.

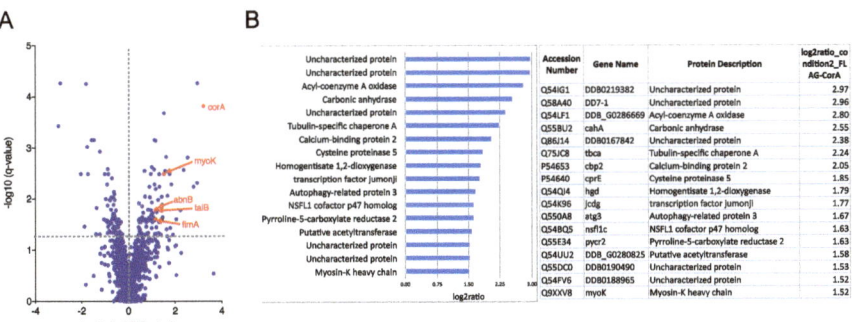

Figure 1. Coronin A interactome analysis. Growing *Dictyostelium* cells were collected, lysed and subjected to AP-MS as described in Materials and Methods. Shown are all interacting proteins with log2ratio > 1.5 and q-value < 0.05. (**A**). Volcano plot; (**B**). Log2ratiorank. See also Tables S1 and S2.

Since cell lysis prior to mass spectrometry analysis may have disrupted any interactions between coronin A and actin, as a second approach, we analyzed an interaction of coronin A with F-actin within cells. To do so, cells were lysed using F-actin stabilization buffer [43–45], followed by analysis of the pellets and supernatants by SDS-PAGE and immunoblotting for actin and coronin A. Since this assay probes the state of F-versus G-actin in situ, care was taken to avoid any dilution factor (see also Materials and Methods). As shown in Figure 2 ('*untreated*'), all of the coronin A immunoreactivity was recovered in the supernatant, suggesting that at steady state, coronin A does not interact with F-actin. To analyze interaction of coronin A with F-actin under conditions in which F-actin is polymerized, cells were either left untreated or incubated with the F-actin polymerizing drug Jasplakinolide. As a control, F-actin was fully depolymerized by the inclusion of Latrunculin A. Cells were then harvested, lysed, and sedimented followed by analysis of soluble proteins as mentioned above. As can be seen in Figure 2 ('Jasp' and 'LatA'), under all conditions of actin polymerization/depolymerization, coronin A was resolved in the supernatant, independent of the polymerization state of actin.

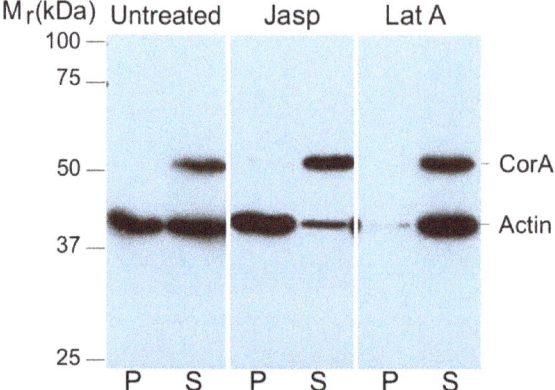

Figure 2. Coronin A-F-actin interaction within cells. Cells were either left untreated (**left**) or treated with Jasplakinolide (**middle**) or Latrunculin A (**right**). The cells were lysed in 200 µL F-actin stabilization buffer and F-actin and G-actin were separated by ultracentrifugation of the lysate as described in Materials and Methods. The pellet was resuspended in the exact same volume as the original lysis volume. Proteins in the supernatant (S) or pellet (P) fractions were separated by SDS-PAGE and immunoblotted for coronin A and actin. Shown are representative results from at least four independent experiments.

In a third approach, to asses coronin A-F-actin interaction, we directly analyzed whether actin co-eluted with FLAG-coronin A following affinity purification. To that end, the purification procedure as described above for FLAG-coronin A was adapted to ensure that F-actin remained intact by replacing the filtration step (that may have resulted in the clearance of F-actin) by homogenization followed by low-speed centrifugation to remove large debris. Cell lysates were subsequently loaded onto the anti-FLAG column, and following elution with the FLAG peptide, fractions were analyzed by SDS-PAGE and immunoblotted using either coronin A antibodies or anti-actin antibodies. As can be seen in Figure 3A,B, all actin eluted in the flow through, without any actin co-eluting in FLAG-coronin A containing fractions. To test whether the absence of actin in FLAG-coronin A-eluted fractions was due to the FLAG tag, we repeated the purification using His-tagged coronin A (Figure 3C,D). In addition, both the lysis and the elution buffer did not contain any NaCl, given the reported sensitivity of the interaction of *Dictyostelium* coronin A with F-actin to NaCl in vitro [31]. As can be seen in Figure 3C,D, no actin co-eluted with His-coronin A-containing fractions and all the detectable actin signal was found in the flow through. As a positive control, a His-tagged myosin-coronin A fusion protein was expressed in *Dictyostelium*, and this fusion protein was purified by metal affinity chromatography as for His-coronin A. In this case (Figure 3E,F), as expected, actin co-eluted in His-tagged myosin-coronin A containing fractions.

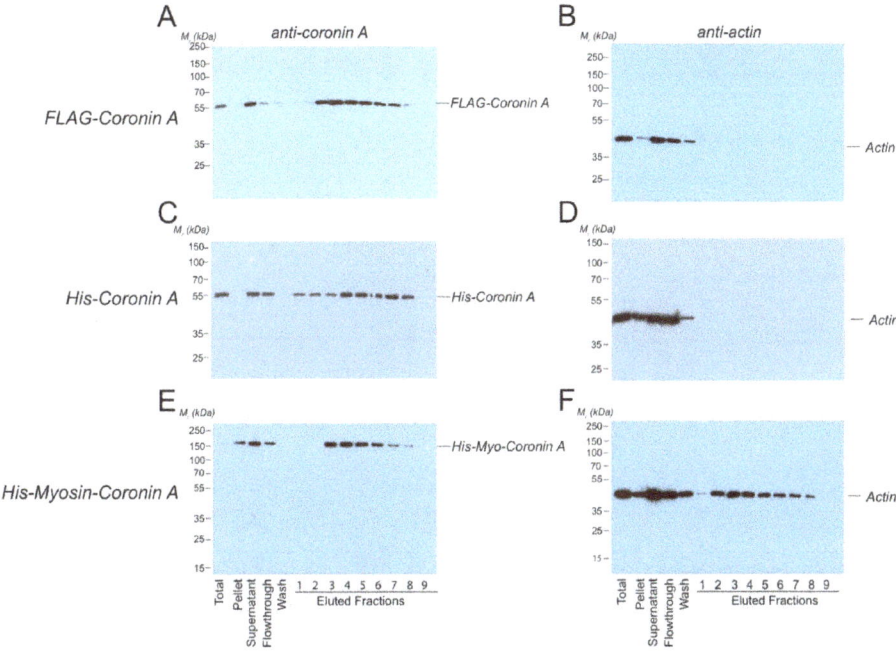

Figure 3. Co-purification of *Dictyostelium* coronin A and actin. *Dictyostelium* cells expressing the constructs indicated were lysed in lysis buffer and homogenized using a glass Tenbroek homogenizer followed by low speed centrifugation. (**A**,**B**). FLAG-CorA was purified using an anti-FLAG column. Fractions were collected, separated by SDS-PAGE, and tested for the presence of coronin A (**A**) and actin (**B**) by Western blotting. (**C**,**D**). Coronin A fused to a Histidine-tag was purified using Nickel beads. Cells were lysed in the absence of NaCl, fractions were collected, separated by SDS-PAGE, and tested for the presence of coronin A (**C**) and actin (**D**) by Western blotting. (**E**,**F**). Coronin A fused to a Histidine-tagged myosin heavy chain fragment was purified using Nickel beads. Fractions were collected, separated by SDS-PAGE, and tested for the presence of coronin A (**E**) and actin (**F**) by Western blotting. Shown are representative results from at least three independent experiments.

Together, the above data suggest that within cells, coronin A failed to interact with actin; instead, actin may interact with coronin A in an indirect manner, possibly via one or more of the interactors defined by mass spectrometry, such as myosin, fimbrin or talin (Figure 1). However, given the published datasets showing the interaction of coronin A, as well as a number of other coronin molecules with (rabbit muscle) F-actin in vitro [4–6,12,31,46,47], we also analyzed the capacity of FLAG-coronin A to interact with purified rabbit muscle F-actin. In addition, given the reported interaction of several coronins with F-actin via their coiled coils [4,18], we included a coronin A mutant lacking the coiled coil domain. We found, in accordance with earlier reports [31], that the interaction of coronin A with rabbit muscle F-actin depended on the ionic strength (Figure 4A), and that coronin A co-pelleted with rabbit muscle F-actin at a concentration of 50 mM NaCl but not at 100 or 150 mM NaCl (Figure 4A). As a control, *S. cerevisiae* Crn1 was employed as it possesses actin interaction domains (in contrast to most other coronins, including *Dictyostelium* coronin A, see [3,12]); as expected, Crn1 readily co-sedimented with rabbit muscle F-actin (Figure 4B). These data suggest that in vitro, coronin A can be co-precipitated with rabbit muscle F-actin under low—but not at elevated ionic—strength conditions.

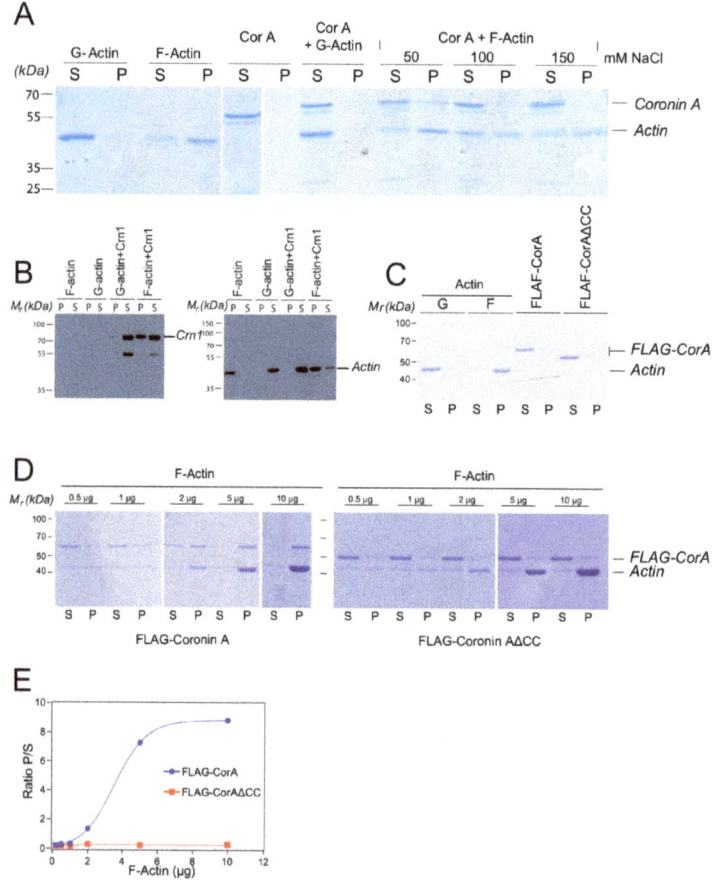

Figure 4. Coronin A-F-actin interaction with muscle F-actin in vitro. (**A**). Purified FLAG-CorA and/or equimolar amounts of rabbit muscle G-actin or F-actin was subjected to ultracentrifugation at 100,000× g, for 1h at 4 °C in the presence of rabbit muscle G-actin and F-actin and different NaCl concentrations. The supernatant was removed and the pellet resuspended in 2× SDS-PAGE sample buffer. Samples were separated by SDS-PAGE and the gel stained using Coomassie blue. Lanes 1–4: controls; lanes 5–8: coronin A and G-actin; lanes 9–14: Sedimentation analysis was performed in the presence of the NaCl concentrations indicated. (**B**). Rabbit muscle F-actin or G-actin were incubated in the absence or presence of *S. cerevisiae* Crn1, incubated for 20 min at room temperature and the samples were then processed as described in the Materials and Methods. Pellets (P) and supernatants (S) were separated by SDS-PAGE and immunoblotted for Crn1 (**left panel**) or actin (**right panel**) as described above. The lower band most likely represents a degradation product of Crn1. (**C**). Purified actin, FLAG-CorA or FLAG-CorAΔCC were analyzed as in A, separated by SDS-PAGE and the gel stained using Coomassie blue. (**D**). Interaction of the indicated amounts of purified FLAG-CorA (top) or FLAG-CorAΔCC (bottom) with rabbit muscle G- and F-actin was carried out as described in the Materials and Methods. Samples were separated by SDS-PAGE and the gel stained using Coomassie blue. (**E**). Plot of the ratio of rabbit muscle F-actin-bound (co-pelleting, P) to non-bound (S) for FLAG-CorA or FLAG-CorAΔCC determined from the mean grey values of the bands from the Coomassie blue stained gels. Curve fitting shows an apparent Kd of 8.9 µg (CI 5–20 µg). Shown are representative results from at least three (two in the case of panels D, E) independent experiments.

To further analyze a potential interaction of coronin A with rabbit muscle F-actin, as well as a possible involvement of the C-terminal coiled coil in this process, FLAG-tagged coronin A or coronin A lacking the coiled coil (FLAG-CorAΔCC) was purified and analyzed for co-sedimentation with rabbit muscle F-actin (Figure 4C). As shown in Figure 4D,E, coronin A co-sedimented with rabbit muscle F-actin in vitro in a saturable manner (Figure 4D,E and left panels). Co-sedimentation was dependent on the presence of the coiled coil, since purified FLAG-tagged coronin A lacking the coiled coil did not co-sediment with rabbit muscle F-actin (Figure 4D,E and right panels).

The here shown in vitro interaction of coronin A with rabbit muscle actin through its coiled coil is in sharp contrast to the absence of an interaction of coronin A with *Dictyostelium* actin (Figures 1–3). Therefore, to further investigate a potential interaction of coronin A and *Dictyostelium* actin that may have a temporal aspect and possibly depends on the presence of the coiled coil, we used affinity precipitation followed by mass spectrometry to analyze the interactome of cells expressing either FLAG-tagged coronin A or coronin A lacking the coiled coil (FLAG-CorAΔCC) at the time points shown in Figure 5. Interestingly, there were only a limited number of common interacting proteins among the top 25 hits across the different time points (Figure 5 and Table S1), suggesting a highly dynamic coronin A interactome, at least at this time resolution. Furthermore, we found that for the different time points, a number of actin-interacting molecules were detected in the coronin A interactome, including myosin (24 h), talin, fimbrin (48 h) and actobindin (110 h). In contrast to the in vitro results showing robust interaction with rabbit muscle actin, of the 31 different actin genes expressed in *Dictyostelium* that encode for 15 different isoforms [48,49] we found only the minor actin-10 form to associate with coronin A in a coiled coil-dependent manner at 48 h, but not at 24 or 110 h. We conclude from these data that while *Dictyostelium* coronin A can be co-pelleted in vitro with rabbit muscle F-actin, an interaction of coronin A with actin within cells or cell lysates is not detected; rather, coronin A was found to interact with a range of proteins, including actin-interacting proteins, in a transient manner.

2.2. Coronin A is Required for the Phagocytosis of Yeast Particles, but not Bacteria and Inert Beads, Independent of the Coiled-Coil Domain

Together the above data suggest that while in vitro, coronin A can interact with rabbit muscle F-actin in a manner dependent on its coiled coil, it fails to directly bind (F-)actin within cells. To further analyze a potential role for coronin A in F-actin-mediated processes, we assessed the rate of phagocytosis, a process highly dependent on F-actin rearrangement [50,51]. Indeed, blocking actin dynamics using cytochalasin strongly reduced bead uptake, similar to internalization at 4 °C (Figure S1). To determine a role for coronin A in phagocytosis, we assessed the capacity of wild type, *corA*-, or as well as *corA*-cells expressing coronin A or the delta coiled coil mutant to ingest a range of fluorescently labelled particles of different surface compositions and sizes including yeast, bacteria and inert beads using fluorescent activated cell sorting (FACS). We first assessed the ingestion of the natural food of *Dictyostelium* (bacteria) such as live *Escherichia coli* expressing the neon green fluorescent protein (Figure 6A,C) and heat-killed *Klebsiella aerogenes* (Figure 6E), but also of carboxylated inert beads of 1 μm, 4.5 μm, and 6 μm diameter (Figure 6B). No major differences were observed in the phagocytosis of bacteria and inert beads between wild type, coronin A-deficient, or cells expressing full length or delta coiled coil coronin A (Figure 6), suggesting that coronin A as well as the coiled coil is dispensable for phagocytosis of bacteria and inert beads. Interestingly, when internalization of heat-killed *Saccharomyces cerevisiae* was analyzed, cells lacking coronin A displayed significantly reduced levels of phagocytosis compared to wild type cells (Figure 6E), which is consistent with earlier work describing a defect in the uptake of yeast particles in the absence of coronin A [34]. Since yeast uptake in *Dictyostelium* is known to depend on receptor-mediated uptake and is characterized by associated activation of signal transduction [52–54], these data are consistent with a role for coronin A in the modulation of signal transduction rather than F-actin rearrangement.

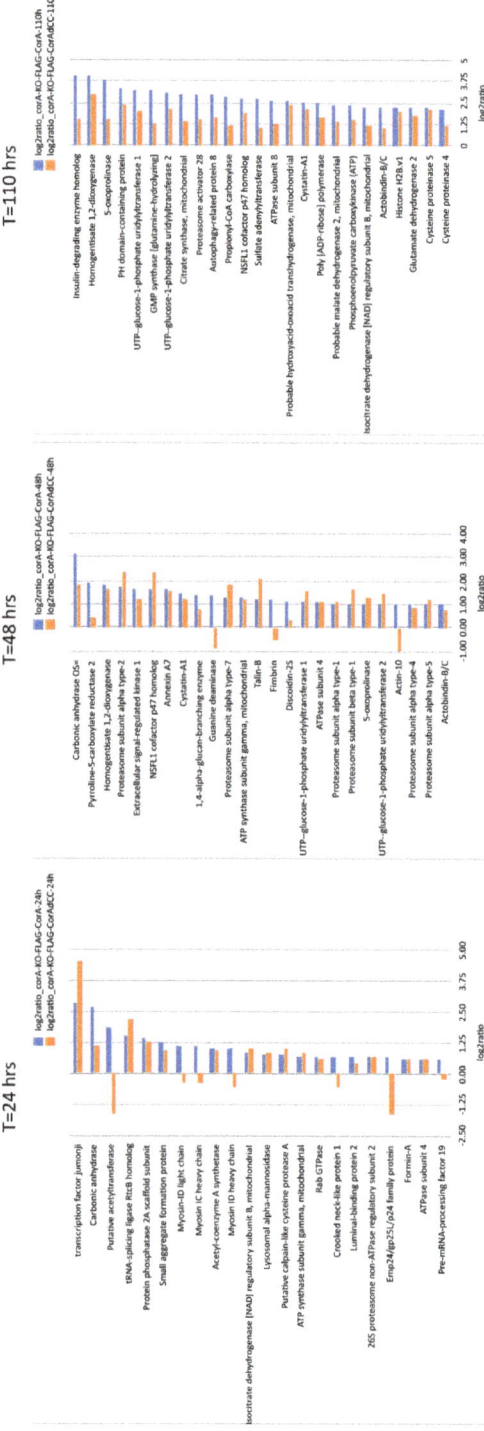

Figure 5. Time-dependent analysis of the coronin A interactome. Cells expressing either FLAG-CorA or FLAG-CorAΔCC were sampled at 24, 48 or 110 h, lysed, and subjected to AP-MS as described in Materials and Methods. The 25 most significant interactors (log2ratio > 1; q-value < 0.05) are shown. For clarity, proteins with unknown function are not represented. Note that the expression of FLAG-CorAΔCC is ~40-fold enriched relative to FLAG-CorA. See also Table S1.

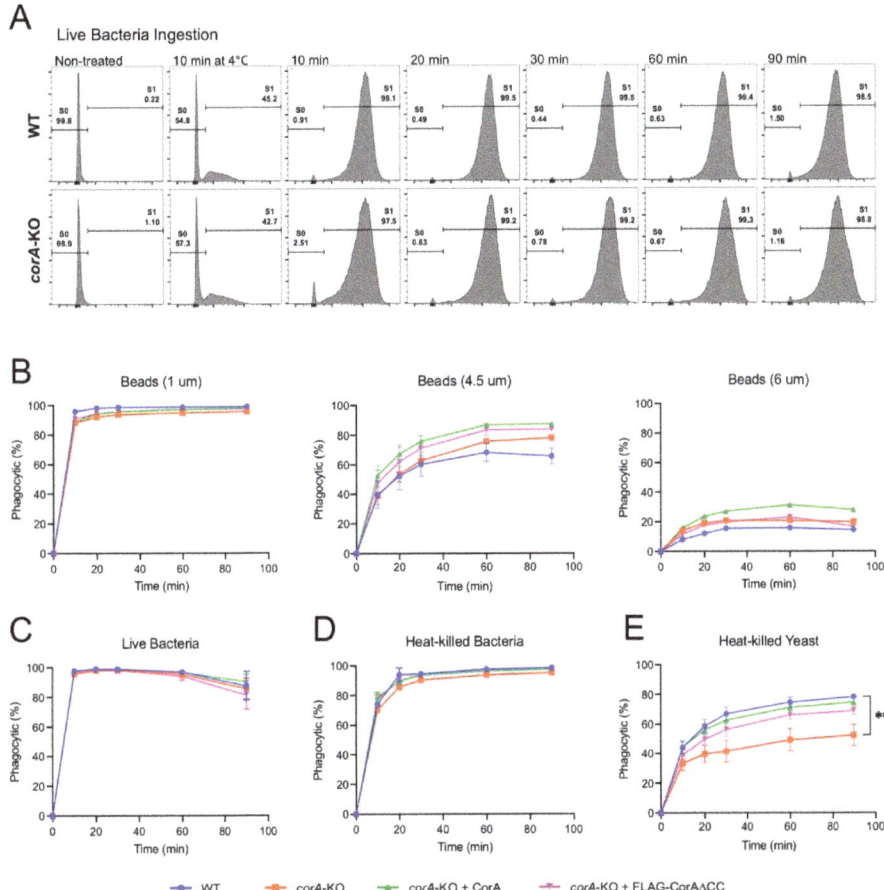

Figure 6. Phagocytosis in the presence and absence of coronin A. (A): Histogram profiles of the distribution and percentage of wild type and $corA^-$ cells that have phagocytosed (S1) live *E. coli* after different time points of incubation. Incubation at 4 °C (control) showed significant reduction of phagocytosis. Shown are representative results from at least 3 independent experiments. (**B–E**): Plot of the percentage of wild type, $corA^-$, $corA^-$ + CorA, $corA^-$ + FLAG-CorA and $corA^-$ + FLAG-CorAΔCC DH1-10 cells that have taken up beads of the indicated sizes (**B**) live *E. coli* (**C**), heat-killed bacteria (**D**), or heat-killed yeast (**E**), respectively (error bar = standard error; ** $p < 0.002$).

Together, these results suggest that coronin A is dispensable for phagocytosis of bacteria and inert beads, but plays a role in yeast particle uptake in a manner that is independent of F-actin interaction.

3. Discussion

Coronins constitute a family of WD repeat containing proteins that are often referred to as F-actin binding and modulating molecules. One reason for this assignment is the fact that coronin A from *Dictyostelium discoideum* was originally identified as a molecule that co-sedimented with an actin/myosin precipitate [31]. However, the evidence for coronin A modulating F-actin within cells is indirect and largely based on the reported phenotypes of *Dictyostelium* lacking coronin A, namely a defect in phagocytosis, chemotaxis and migration [31,32,34,55]. Interestingly, more recent work showed that the defect in chemotaxis and migration in $corA^-$ cells is readily complemented by pulsing the cells with cAMP [30], suggesting that, per se, these processes do not depend on coronin A. Instead, coronin A was found to be responsible for the initiation of multicellular development [30].

Here, we analyzed the interactome of *Dictyostelium* coronin A as well as its interaction with actin. We found that coronin A interacts with diverse proteins in a transient manner and could not find evidence for a direct interaction of coronin A and F-actin within cells or cell lysates. Furthermore, we found a transient association with the minor (<5%) actin10 isoform, but not with actin8, that represents >95% of the cellular actin [48,56]. Since we did detect a number of actin-interacting molecules to co-purify specifically with coronin A, including talin, fimbrin, actobindin and myosin subunits, it is possible that any interaction of coronin A with actin may occur indirectly via these proteins. We also show that the phagocytosis of a range of different cargos was not compromised by the absence of coronin A, with the exception of yeast phagocytosis, which is known to depend on signaling processes [52]. Together these data suggest that coronin A does not directly interacts with/modulates F-actin.

Most previous studies analyzing coronin–actin interaction have employed rabbit muscle actin to demonstrate in vitro co-pelleting with F-actin. Similarly to earlier studies, we confirmed the in vitro interaction of coronin A with muscle F-actin and further found that this interaction was dependent on the coiled coil, which is consistent with the demonstration that the coiled coil in other coronins also possesses low-affinity actin binding sites [11,57]. Since these latter analyses were also performed using in vitro interaction analyses, it is unclear to what degree this reflects an in vivo association with F-actin; while for a number of mammalian coronins, association with F-actin within cells could not be demonstrated [22,23,45], we cannot, however, also given the high degree of homology between *Dictyostelium* and rabbit muscle actin, exclude the possibility that the presence (within cells) or absence (in vitro) of other factors or differential posttranslational modifications determines coronin A-F-actin interaction.

The absence of a direct interaction between *Dictyostelium* coronin A and F-actin within cells as shown here is consistent with the finding that coronin A in *Dictyostelium* is dispensable for several F-actin dependent processes: first, as shown here, the phagocytosis of bacteria and inert beads, an exquisite actin-dependent process [50,51], was unaltered in the absence of coronin A. Second, in *Dictyostelium* cells lacking coronin A, F-actin-dependent processes including folate-mediated chemotaxis as well as chemotaxis upon external cAMP pulsing occurred normally [30]. Rather, the reported phenotypes of *corA⁻ Dictyostelium*, including altered chemotaxis, reduced yeast particle phagocytosis, reduced macropinocytosis as well as defective cytokinesis, all of which largely depend on proper signal transduction [54,58–61], suggests that *Dictyostelium* coronin A may perform a signaling function, consistent with the sequence homology of coronin A with the Gβ subunit of trimeric G proteins as well as the function of mammalian coronin 1 in the modulation of the cAMP protein kinase A pathway [1,26,28,31]. In this light, it is interesting to note that recent work also suggests that the capacity of *Plasmodium* coronin to modulate actin filament turnover occurs in a manner dependent on protein kinase A/cAMP signaling [17].

It is possible that the identified coronin A interacting proteins fimbrin, myosin, talin and actobindin, all of which are known actin interactors [62–65], function as intermediates to link coronin A to the actin cytoskeleton. For example, the interaction with the F-actin cytoskeleton might serve to regulate a dose- and time-dependent availability and/or potentiation of coronin A for signaling processes such as in multicellular development initiated upon starvation [30,66,67]. In such a scenario, an indirect and labile interaction of coronin A with the cytoskeleton would be advantageous, making it quickly available for incorporation into other complexes.

4. Materials and Methods

4.1. Cells, Antibodies, and Growth Conditions

DH1-10 wild-type *Dictyostelium discoideum* cells were acquired from dictybase.org. The *corA*-deficient (*corA⁻*) cells are described elsewhere [30]. Cells were grown in HL-5 media [68] in 100 mL or 500 mL Erlenmeyer flasks at 22 °C with 160 rpm. Anti-coronin A antiserum was described earlier [30]. Anti-yeast Crn1 antiserum was produced in rabbits using recombinant Crn1 (Thermo Fisher). Mouse anti-actin clone 4 was purchased from Millipore. For production of N-terminal FLAG-tagged (DYKDDDDK) coronin A, full length coronin A-encoding cDNA was amplified from DH1-10 genomic DNA by PCR, using primers to insert a thrombin cleavage site to the 5′ end, and BamHI restriction sites to either end of the PCR product. The thrombin cleavage site was optimized for *Dictyostelium* codon usage. Forward primer: 5′ ATTGGATCCTTAGTTCCAAGAGGTTCAATGTCTAAAGTAGTCCGTAGTAG ′3; Reverse primer: 5′ ATTGGATCCTTAGTTGGTGAGTTCTTTGATTTTGGGATCCTTTTTAACG ′3. The PCR product was first subcloned into pCR-BluntII-TOPO vector (Invitrogen), sequenced, digested with BamHI and inserted into the vector pTX-FLAG (dictybase.org). The resulting vector carried a FLAG-tag 10 amino acids upstream of the inserted thrombin cleavage site and coronin A. The expression of the fusion protein is driven by a *Dictyostelium*-actin 15 promoter. For the construction of the coronin A expression vector, the actin15 promoter was synthesized with XbaI restriction sites at both ends and cloned into pUC57 vector (Eurofin genomics); then subcloned into the XbaI site of pBIG (dictybase.org) and checked for correct orientation by sequencing. Then full length coronin A CDS was amplified from DH1-10 genomic DNA by PCR, using primers to introduce BamHI restriction sites on both ends of the PCR product, which was then cloned into the pBIG vector to give pBIG-CorA driven by the actin15 promoter. For ΔCC-CorA, the expression plasmid was generated by synthesizing (Eurofins genomics) BamHI flanked coronin A CDS lacking the last 34 codons, which encode the coiled coil motif, and cloning into the BamHI site of pTX-FLAG vector to generate pTX-FLAG-CorAΔCC.

For the generation of a histidine-tagged (6×) version of coronin A, the coronin A coding region was amplified from DH1-10 genomic DNA with forward (FwCorA HpaI AGAGCGT TAACATGTCTAAAGTAGTCCG) and reverse (RevCorA HpaI AGAGCGTTAACTTAGTTGGT GAGTTCTTTG) primers adding HpaI restriction sites on both ends of the gene. The resulting fragment was ligated into the cloning vector T-easy (Promega) according to the manufacturer's protocol. The vector for production of N-terminal His-tagged (6× His) coronin A was then generated by removing the GFP sequence from the vector pTX-GFP (dictybase.org) with EcoRV and inserting the full length HpaI restricted coronin A sequence in its place via blunt end ligation. For use as a positive control in the actin co-purification experiments, we generated a vector that expresses coronin A as a fusion protein fused to the C terminus of a 6x His-tagged portion of myosin heavy chain capable of binding actin. pDIC2, the vector containing the myosin heavy chain fragment, was a kind gift from Thomas Reubold of the Institute for Biophysical Chemistry at Hannover Medical School [69,70]. The coronin A-encoding gene was excised from the T-easy cloning vector described earlier using the restriction enzyme HpaI to create blunt ends, pDIC2 was linearized with the blunt-cutting restriction enzyme EcoICRI and the coronin A-coding sequence was ligated via blunt end ligation. For the generation of a yeast Crn1 expression vector, Glutathione S-transferase (GST)-fused Crn1 was amplified from an existing vector (pGAT_Crn1); forward primer: 5′GTGTCTGCAGATGTCCCCTATACTAGGTTATTG′3 and reverse primer: 5′CACACTGCAGTCATTTTGACAGTTCGCC′3. GST-Crn1 was then cloned into a p425-TEF yeast expression vector via PstI sites [71].

4.2. Anti-Flag Immunoprecipitation and Mass Spectrometry

Cells (5×10^6, the $corA^-$ transformed with and stably expressing FLAG-CorA or FLAG-CorAΔCC and $corA^-$ expressing CorA as control for unspecific binding) were seeded in triplicates at 10^5 cells/mL, grown in HL-5 media as mentioned above and harvested at 24, 48 and 110 h, washed 2 times with ice cold PBS and lysed with 500 µL Lysis Buffer (low salt TBS (20 mM Tris-HCl pH 8.0, 25 mM NaCl, 5 mM KCl)/1% Triton ×100/HALT from Thermo #1861281) for 30 min on ice with gentle agitation every 5 min and clarified for 15 min at 18,200× g at 4 °C. Monoclonal anti-FLAG-M2 slurry (Sigma, F1804-50UG, 25 µL) was washed twice with low salt TBS and co-incubated with 450 µL of cleared lysate for 90 min at 4 °C in 2 mL microfuge tubes with 360° rotation. Unbound/non-interacting proteins were removed by 3 washes with low salt TBS. Peptide elution occurred in series, first with 100 µL elution buffer 1

(1.6 M urea, 100 mM Ammoniumbicarbonate, 5 µg/mL trypsin) at 37 °C with 1200 rpm for 30 min and then twice, each with 40 µL elution buffer 2 (1.6 M urea, 100 mM Ammoniumbicarbonate, 1 mM TCEP), vortexing and centrifuging, collecting and pooling the supernatant. Eluted proteins were reduced by adding 9 µL TCEP from a 200 mM stock solution to the pooled supernatant (total volume 180 µL) and alkylated with 3.8 µL chloroacetamide (750 mM stock solution) for 1 h at 37 °C, then digested overnight with 0.5 µg of trypsin (Promega USA). Samples were acidified with 150 µL of 5% TFA, pH < 2. Peptides were bound to acetonitrile conditioned C18-columns, washed with 0.1% TFA and eluted with C18-buffer (50% acetonitrile/50% water (v/v) and 0.1% TFA). Eluted peptides were concentrated under vacuum to dryness, then dissolved in and adjusted to 0.2 µg/µL with 0.1% formic acid.

For the LC-MS/MS analysis, the µRPLC-MS system was setup as described previously [72]. Chromatographic separation of peptides was carried out using an EASY nano-LC 1000 system (Thermo Fisher Scientific), equipped with a heated RP-HPLC column (75 µM × 37 cm) packed in-house with 1.9 µM C18 resin (Reprosil-AQ Pur, Dr. Maisch). Aliquots of 1 µg total peptides were analyzed per LC-MS/MS run using a linear gradient ranging from 95% solvent A (0.15% formic acid, 2% acetonitrile) and 5% solvent B (98% acetonitrile, 2% water, 0.15% formic acid) to 30% solvent B over 120 min at a flow rate of 200 nl/min. A mass spectrometry analysis was performed on Q-Exactive HF mass spectrometer equipped with a nano electrospray ion source (both Thermo Fisher Scientific). Each MS1 scan was followed by high-collision-dissociation (HCD) of the 20 most abundant precursor ions with dynamic exclusion for 30 s. The total cycle time was approximately 1–2 s For MS1, 3e6 ions were accumulated in the Orbitrap cell over a maximum time of 100 ms and scanned at a resolution of 120,000 FWHM (at 200 m/z). MS2 scans were acquired at a target setting of 1e5 ions, accumulation time of 50 ms and a resolution of 15,000 FWHM (at 200 m/z). Singly charged ions and ions with unassigned charge state were excluded from triggering MS2 events. The normalized collision energy was set to 28%, the mass isolation window was set to 1.4 m/z and one microscan was acquired for each spectrum.

To determine bait-binding affinities, an MS1 based label-free quantification was carried out. Therefore, the generated raw files were imported into the Progenesis QI for proteomics software (Nonlinear Dynamics, Version 2.0) and analyzed using the default parameter settings. MS/MS-data were exported directly from Progenesis QI for proteomics in mgf format and searched against a decoy database of the forward and reverse sequences of the SwissProt entries of *Dictyostelium discoideum* (www.ebi.ac.uk, release date 2017/10/09) and commonly observed contaminants (in total 26,272 sequences) using MASCOT (Matrix Science, Version 2.4.1). The search criteria were set as follows: full tryptic specificity was required (cleavage after lysine or arginine residues); 3 missed cleavages were allowed; carbamidomethylation (C) was set as fixed modification; oxidation (M) as variable modification. The mass tolerance was set to 10 ppm for precursor ions and 0.02 Da for fragment ions. Results from the database search were imported into Progenesis QI for proteomics and the final peptide measurement list containing the peak areas of all identified peptides, respectively, was exported. This list was further processed and statically analyzed using our in-house developed SafeQuant R script (SafeQuant, https://github.com/eahrne/SafeQuant, [72]). The peptide and protein false discovery rate (FDR) was set to 1% using the number of reverse hits in the dataset. All quantitative analyses were performed in biological triplicates. The resulting details of the proteomics experiments carried out including identification scores, number of peptides quantified, normalized (by sum of all peak intensities) peak intensities, log2 ratios, coefficients of variations and p-values for each quantified protein and sample are displayed in Supplementary Table S1. All raw data and results associated with the manuscript will be deposited into the Proteome X change Consortium via the PRIDE [73] partner repository with the dataset identifier PXD009483 and 10.6019/PXD009483.

For the data shown in Figure 1b (Table S2), the log2ratio for FLAG-CorA samples (pooled of all samples from the different timepoints) was determined against the control sample expressing non-FLAG tagged CorA and background proteins were filtered out using the cutoff of log2ratio > 1.5 and q-value < 0.05; this dataset reflects the growth phase-independent interactome of coronin A. The significantly enriched proteins ranked per their fold enrichment were plotted using Numbers. For the time-dependent analysis in Figure 5, the log2ratio for FLAG-CorA and FLAG-CorAΔCC against the control sample expressing non-FLAG tagged CorA were calculated and the same cutoff set as above, and the samples for each time point (representing the early log, log and early stationary growth phases, i.e., common to exponential growth and stationary phases) were compared to the control of the same time point. The top 25 significantly enriched proteins ranked per their fold enrichment in FLAG-CorA were plotted using Numbers.

4.3. Analysis of F-Actin and G-Actin from Cell Lysates

Coronin A-actin interaction from cell lysates was essentially performed as described [45]. In brief, DH1-10 wild type cells were harvested (4×10^6 cells/sample), washed and resuspended in 0.5 mL starvation buffer B (5 mM Na_2HPO_4, 5 mM KH_2PO_4, 2.5 mM $MgSO_4$, 200 µM $CaCl_2$ [74]). The cells were then exposed to either 9 µM Jasplakinolide (Sigma-Aldrich) for 1 h to induce actin polymerization, 10 µM Latrunculin A (Sigma-Aldrich) for 30 min to induce actin depolymerization, or buffer alone as a control and placed on a shaking platform at 22 °C. The cells were pelleted and washed twice with phosphate-buffered saline (PBS), and then lysed with 200 µL F-actin stabilization buffer (50 mM PIPES pH 7.0, 50 mM NaCl, 5 mM $MgCl_2$, 5 mM ethylene glycol-bis(β-aminoethyl ether)-N,N,N′,N′-tetraacetic acid (EGTA), 5% glycerol, 0.1% Triton X-100, 0.1% Tween 20, 0.1% NonidetP-40, 0.1% β-Mercaptoethanol, 1 mM ATP, protease inhibitor mix [44,75]) on ice for 15 min. The lysates were pre-cleared by centrifugation at 600× g for 5 min and the supernatant was subjected to ultracentrifugation at 150,000× g for 30 min at 4 °C to sediment F-actin. The supernatant was removed and the remaining pellet was resuspended in 100 µL ice cold distilled water containing 10 µM Cytochalasin D (Sigma, St. Louis, MO, USA) for 30 min on ice and occasionally agitated gently by pipetting up and down. The resuspended pellet fraction was then mixed with 100 µL 2× F-actin stabilization buffer to bring the solution to the same volume as the previously removed supernatant fraction. Ten microliter of each supernatant (G-actin) and pellet (F-actin) were analyzed by SDS-PAGE and immunoblotting using anti-actin and anti-coronin A antibodies as described below.

4.4. Protein Purifications

For co-precipitation analysis, FLAG-coronin A and FLAG-CorAΔCC were purified using the M2 Flag affinity gel. In brief, 5×10^8 FLAG-CorA- or FLAG-CorAΔCC-expressing cells were harvested in log growth phase and washed twice with ice cooled TBS (20 mM Tris-HCl pH 8.0, 150 mM NaCl, 5 mM KCl). The cells were then lysed in 4 mL lysis buffer (TBS, 2 mM EDTA, 1% Triton ×100, Protease/Phosphatase Inhibitor from Thermo-Fischer #1861281) on ice for 30 min with gentle agitation every 5 min. The lysate was cleared at 18,000× g for 15 min at 4 °C, filtered through a 0.45 µM filter (Sartorius), then loaded onto 250 µL of M2-anti-FLAG slurry and incubated with rotation in 15 mL Falcon for 90 min at 4 °C. Non-bound protein were washed off 4× (each with 1 mL lysis buffer), and then 8× with 1 mL each of low salt TBS (20 mM Tris-HCl pH 8.0, 25 mM NaCl, 5 mM KCl) at 1000× g, 5 min at 4 °C, while collecting the supernatant and determining the presence of protein at OD280 with an Eppendorf BioSpectrometer® basic (Eppendorf, Hamburg, Germany). By the 9th wash, no protein was detected in the supernatant. Bound proteins were eluted with 250 µL of 3× FLAG peptide (Sigma or GenScript) at 0.2 µg/µL in low salt TBS for 1 h at 4 °C. Purified fractions were run on 10% SDS-PAGE gel, stained with Coomassie® G-250 SimplyBlue™ SafeStain (ThermoFisher) and imaged by scanning with a CanoScan 9000F Mark II scanner (Canon).

G-actin was isolated from rabbit muscle acetone powder (Sigma M6890) as described previously [76]. In brief, 1.5g acetone powder was dissolved in 30 mL buffer G (2 mM imidazole, 0.2 mM ATP, 0.5 mM dithiothreitol (DTT), 0.2 mM $MgCl_2$, pH 7.2–7.4) and stirred on ice for 30 min. The extract was filtered and the remaining acetone powder was extracted with another 30 mL of buffer G for 30 min. Both supernatants were pooled and centrifuged for 20 min at 25,000× g at 4 °C. The supernatant was pooled and supplemented with KCl (1M stock) to a final concentration of 50 mM and $MgCl_2$ to a final concentration of 2 mM and stirred at 4 °C for 1 h. After stirring we slowly added ground KCl powder to reach 0.8 M final concentration and stirred for another 30 min at 4 °C. The solution containing polymerized actin was centrifuged for 1 h at 150,000× g at 4 °C and the resulting pellet was resuspended in 4 mL of buffer A (2.5 mM imidazole, 0.2 mM ATP, 0.2 mM $CaCl_2$, 0.005% NaN_3, 0.2 mM DTT, pH 7.2–7.4) using 18G and 23G syringe needles. The dissolved pellets were dialyzed for 3 days against daily exchanged buffer A, followed by ultracentrifugation at 250,000× g at 4 °C for 1.5 h. The depolymerized actin in the supernatant was further purified using gel filtration (Superdex 200 10/300 GL, GE Healthcare). The purified G-actin was kept dialyzing against buffer A at 4 °C for two weeks.

For the purification of yeast Crn1, the cDNA coding for Crn1 was fused with that of GST as described above. An overnight culture of Y36032_GST-Crn1 yeast (Euroscarf) was diluted in synthetic dropout complete-LEU medium to an OD600 of 0.2. The cells were grown at 30 °C to an OD600 of 0.8–1 in 3l Erlenmeyer flasks on a shaking platform. The cells were harvested at 5000× g for 5 min at 4 °C and washed with ddH_2O. Pellets from a volume of 100 mL original culture were resuspended in 500 µL yeast lysis buffer (20 mM Tris-HCl pH 8.0, 150 mM NaCl, 2 mM EDTA, 0.1% TX-100, 1 mM PMSF, 3 mM DTT, complete protease inhibitor from Roche) and mixed with 500 µL glass beads (Carl Roth: A553.1 Glasperlen 0.25–0.5 mm). The yeast cells were disrupted in a bead vortex for 2 × 30 s. In between vortexing steps, the lysate was placed on ice for 5 min. The supernatant was removed and the remaining material was extracted a second time with 500 µL yeast lysis buffer. Pooled lysates were centrifuged at 17,000× g for 1 min at 4 °C and GST-Crn1 purified on a GSTrap (GE Life Sciences) column followed by incubation with thrombin (4 hrs, 4 °C; 1 unit per 100 µg of protein) to remove the GST-tag. Thrombin was removed using benzamidine beads (50% slurry; Sigma-Aldrich #A7155) and the Crn1 solution was dialyzed (Spectrapor, 10 kDa cutoff) against KMEI buffer (see below) followed by removal of the GST tag using GSTrap from GE and concentrated in an Amicon MWCO 50 kDa device (Millipore, Burlington, MA, USA).

For the co-purification of His-coronin A and His-Myosin-coronin A with actin, the purification was performed as described above with the following changes: cells were lysed in 5 mL imidazole lysis buffer (50 mM Tris-HCl pH 7.5, 20 mM Imidazole, 2 mM Benzamidine, 1 mM EDTA, 0.5% Triton × 100, protease inhibitor mix) on ice for 20 min. The lysate was loaded onto a 300 µL bed of Ni-NTA resin (Qiagen, Hilden, Germany). The loaded column was washed with 40 column volumes of imidazole washing buffer (50 mM Tris-HCl pH 7.5, 40 mM Imidazole, 1 mM EDTA, 2 mM Benzamidine), and bound proteins were eluted with imidazole elution buffer (washing buffer + 260 mM imidazole). Collected fractions were then analyzed by Western blotting as described below.

4.5. Coronin A-F-Actin co-Precipitation Analysis

Coronin A-F-actin co-precipitation analysis was carried out using the FLAG-coronin A or FLAG-CorAΔCC and the procedure according to the Hypermol Actin toolkit (Hypermol, Bielefeld, Germany). In brief, FLAG-CorA and FLAG-CorAΔCC were purified as described above using M2-anti-FLAG resin and dissolved in low salt TBS (20 mM Tris-HCl pH 8.0, 25 mM NaCl, 5 mM KCl). For preparation of rabbit muscle actin, lyophilized G-Actin was reconstituted with 900 µL H_2O to obtain a 1.1 mg/mL stock solution, and left to rehydrate at room temperature for 5 min, and subsequently dialyzed overnight in MonoMix buffer (0.1 mM $CaCl_2$, 0.5 mM DTT, 0.4 mM ATP, 2 mM Tris-HCl, pH 8.2). For F-Actin preparation, the G-actin stock was pre-spun at 100,000× g, 1 h, 4 °C in an Optima TLX ultracentrifuge (TLA55) and the supernatant was used for the co-sedimentation assay. G-actin was mixed in a 1:10 ratio with 10× PolyMix buffer (Hypermol, 1 M KCl, 0.02 M $MgCl_2$, 0.01 M ATP, 0.1 M imidazole, pH 7.4) and left 30 min at room temperature for polymerization. For co-sedimentation, 2 µg prespun (100,000× g, 1 h at 4 °C in an Optima TLX ultracentrifuge (TLA55)) FLAG-CorA or FLAG-CorAΔCC was incubated with different amounts of F-actin for 45 min at room temperature. For higher salt concentrations, FLAG-CorA and G/F-Actin were mixed at equal molar ratios of 1 µM and additional NaCl (from a 5M stock solution in ddH_2O) was added to the samples to final concentrations of 50 mM, 100 mM and 150 mM in a total volume of 40 µL. After incubation, samples were centrifuged at 100,000× g, 1 h, 4 °C in an Optima TLX ultracentrifuge (TLA55, Beckman Coulter). As controls G-actin, F-actin, FLAG-CorA, and FLAG-CorAΔCC were centrifuged separately to assess their solubility. After centrifugation, the supernatants were transferred to 1.5 mL microfuge tubes containing SDS sample buffer (Tris-HCl pH 6.8, 2% SDS, 5% Glycerol, 0.015% DTT, 0.0002% bromophenol blue). The pellet was washed twice with low salt TBS (25 mM NaCl, 20 mM Tris-HCl, pH 7.5) and resuspended in low salt TBS containing SDS sample buffer and transferred into 1.5 mL microfuge tubes. Samples were boiled for 5 min at 95 °C and separated on a 10% SDS-PAGE. The gel was stained using Coomassie G250 SimplyBlue™ SafeStain (ThermoFisher, Waltham, MA, USA). For the sedimentation analysis of yeast Crn1, a stock of freshly purified G-actin was diluted to 4 µM into buffer A. KMEI (10×) actin polymerization buffer was added to yield a 1× concentration (KMEI; 20 mM Imidazole, 50 mM KCl, 1 mM EGTA, 1 mM $MgCl_2$, pH 7.5) and the actin was left to polymerize at RT for 1 h. Thirty µL of F-actin or G-actin were mixed with 10 µL yeast-Crn1 purified as described [4] to a final volume of 40 µL and a final concentration of 500 nM. The mixture was incubated for 30 min at room temperature while shaking. The samples were then subjected to ultracentrifugation at 150,000× g for 30 min at 4 °C. After removal of the supernatant the pellets were resuspended in 40 µL distilled water with 10 µM cytochalasin D and left to stand for 20 min at RT. Both supernatant and pellet were mixed or resuspended in equal amounts of SDS sample buffer. The samples were separated by SDS-PAGE and Western blot was performed as described below. The apparent Kd was obtained by non-linear fitting of the data using Prism (8.3.0) based on duplicate data allowing different values of the maximum P/S ratio for each group.

4.6. Western Blotting

Proteins were separated on 10% Sodium dodecyl sulfate polyacrylamide gel electrophoresis (SDS-PAGE) gels and transferred onto nitrocellulose membranes with semi-dry or wet transfer systems (BioRad, Hercules, CA, USA), depending on the size of proteins to be analyzed. The membranes were stained with Ponceau red protein stain for 15 min, rinsed with ddH_2O and scanned with a CanoScan 9000F Mark II scanner (Canon). The Ponceau red was washed off and the membrane blocked with 5% milk in PBS-Tween20 for 1 h at RT or overnight at 4 °C. The antibodies were diluted in 5% milk PBS-Tween20 at 1:15,000 for anti-coronin A, and 1:5000 for anti-actin. Primary antibody incubation was done either at room temperature for 2 h or overnight at 4 °C, followed by washes and by incubation with horseradish peroxidase (HRP)-coupled secondary antibodies (Southern Biotech). Membranes were developed using SuperSignal PicoWest chemiluminescence substrate (Thermo-Fisher) or WesternBright Quantum HRP substrate (Advansta) and imaged using a Fuji FPM 800A (Fuji, Tokyo, Japan) or Fusion FX7 (VILBER, Paris, France)

4.7. Phagocytosis

For the preparation of particles, harvested log-growth phase *Klebsiela aerogenes* and *S. cerevisiae* (strain NYYO-1, [77]) were washed with and resuspended in KK2 buffer (16 mM KH_2PO_4, 4 mM K_2HPO_4) before heat killing at 80 °C and 65 °C, respectively, for 20 min. Heat-killed bacteria and yeast cells were stained in the dark, respectively, with 10× and 2.5× manufacturer recommended working concentration of CellBrite Fix 640 dye (#30089, Biotium, Freemont, CA, USA) for 30 min at RT. Excess dye solution was removed by centrifugation and labelled cell were resuspended in KK2 buffer. Live bacteria were laboratory strain *E. coli* (DH5α) expressing neon green fluorescent protein grown overnight in liquid broth to stationary growth phase. The live bacteria were a kind gift from Dirk Bumann at the Biozentrum, University of Basel, Basel, Switzerland. For beads, Ø = 1 µm, 3 µm, 4.5 µm or 6 µm fluorescent carboxylate-modified microspheres particles were obtained from Life technologies or PolySciences. For the uptake experiment, 12×10^7 *Dictyostelium* cells were harvested in early log growth and resuspended in 6 mL HL5 medium at a density of 2×10^6 cells/mL in a 2 conical flask and then incubate at 22 °C for 1 h at 160 rpm. For control, cells were pretreated with the actin depolymerizing drug cytochalasin A (C6637 Sigma-Aldrich) to a final concentration of 5 µg/mL for 30 min at 22 °C before the fluorescent particles were added. Particles were added to the *Dictyostelium* cells at MOI of 5, 10, 100, or 200 and incubated at 22 °C with 160 rpm, or at 4 °C (control) and 500 µL samples were collected at 0, 10, 20, 30, 60, and 90 min into 3 mL ice-cold KK2 supplemented with 5 µM NaN$_3$, washed and resuspended in 250 µL ice cold FACS buffer (PBS, 2% FCS, 10 mM EDTA, 0.05% Na-azide) and maintained on ice until measurement [78]. Non-ingested fluorescently labelled bacteria and yeast particles were quenched by adding 0.4% trypan blue at a ratio of 2:1 (trypan blue:sample) and incubated for 10 min prior to analysis. Samples were analyzed by Fluorescent Activated Cell Sorting (FACS) on a BD LSRFortessa (Becton-Dickinson, Franklin Lakes, NJ, USA) and FlowJo Software (flowjo.com).

Supplementary Materials: Supplementary materials can be found at http://www.mdpi.com/1422-0067/21/4/1469/s1. Figure S1: Control for phagocytosis, Table S1: Table S1_Summary Sheet for all time points plus the combined, Table S2: Log2ratios_FLAG-CorA.

Author Contributions: Conceptualization, T.N.F., T.F. and J.P.; methodology, T.F., V.S., T.N.F., A.V., F.B., A.S., J.P.; validation, T.F., V.S., T.N.F., A.V., F.B., A.S.; writing—original draft preparation, T.F., J.P.; writing—review and editing, T.F., V.S., T.N.F., A.V., F.B., A.S.; supervision, J.P.; project administration, J.P.; funding acquisition, J.P., A.S. All authors have read and agreed to the published version of the manuscript.

Funding: This research was funded by the Swiss National Science Foundation, the Novartis Foundation for Bio Medical Research and EMBO long term fellowship to AV.

Acknowledgments: We thank Timothy Sharpe (Biozentrum, University of Basel) for discussions and help in data analysis, Thomas Reubolt, Hannover, for the pDIC2 vector, Dirk Bumann (Biozentrum, University of Basel) for the fluorescent *E. coli*, Cora-Ann Schönenberger and Stefan Drexler for input and advice, and Pierre Cosson (University of Geneva, Switzerland) for continued discussions. This work was funded by the Swiss National Science Foundation, the Novartis Foundation for Bio Medical Research and an EMBO long term fellowships to AV.

Conflicts of Interest: The authors declare no conflict of interest.

References

1. Pieters, J.; Muller, P.; Jayachandran, R. On guard: Coronin proteins in innate and adaptive immunity. *Nat. Rev. Immunol.* **2013**, *13*, 510–518. [CrossRef] [PubMed]
2. de Hostos, E.L. A brief history of the coronin family. *Subcell. Biochem.* **2008**, *48*, 31–40. [PubMed]
3. Eckert, C.; Hammesfahr, B.; Kollmar, M. A holistic phylogeny of the coronin gene family reveals an ancient origin of the tandem-coronin, defines a new subfamily, and predicts protein function. *BMC Evolut. Biol.* **2011**, *11*, 268. [CrossRef]
4. Goode, B.L.; Wong, J.J.; Butty, A.C.; Peter, M.; McCormack, A.L.; Yates, J.R.; Drubin, D.G.; Barnes, G. Coronin promotes the rapid assembly and cross-linking of actin filaments and may link the actin and microtubule cytoskeletons in yeast. *J. Cell Biol.* **1999**, *144*, 83–98. [CrossRef]
5. Tardieux, I.; Liu, X.; Poupel, O.; Parzy, D.; Dehoux, P.; Langsley, G. A Plasmodium falciparum novel gene encoding a coronin-like protein which associates with actin filaments. *FEBS Lett.* **1998**, *441*, 251–256. [CrossRef]
6. Salamun, J.; Kallio, J.P.; Daher, W.; Soldati-Favre, D.; Kursula, I. Structure of Toxoplasma gondii coronin, an actin-binding protein that relocalizes to the posterior pole of invasive parasites and contributes to invasion and egress. *FASEB J.* **2014**, *28*, 4729–4747. [CrossRef]
7. Okumura, M.; Kung, C.; Wong, S.; Rodgers, M.; Thomas, M.L. Definition of family of coronin-related proteins conserved between humans and mice: Close genetic linkage between coronin-2 and CD45-associated protein. *DNA Cell Biol.* **1998**, *17*, 779–787. [CrossRef]
8. Chan, K.T.; Creed, S.J.; Bear, J.E. Unraveling the enigma: Progress towards understanding the coronin family of actin regulators. *Trends Cell Biol.* **2011**, *21*, 481–488. [CrossRef]
9. Pieters, J. Coronin 1 in innate immunity. *Subcell. Biochem.* **2008**, *48*, 116–123.
10. Humphries, C.L.; Balcer, H.I.; D'Agostino, J.L.; Winsor, B.; Drubin, D.G.; Barnes, G.; Andrews, B.J.; Goode, B.L. Direct regulation of Arp2/3 complex activity and function by the actin binding protein coronin. *J. Cell Biol.* **2002**, *159*, 993–1004. [CrossRef]
11. Gandhi, M.; Jangi, M.; Goode, B.L. Functional surfaces on the actin-binding protein coronin revealed by systematic mutagenesis. *J. Biol. Chem.* **2010**, *285*, 34899–34908. [CrossRef] [PubMed]

12. Liu, S.L.; Needham, K.M.; May, J.R.; Nolen, B.J. Mechanism of a concentration-dependent switch between activation and inhibition of Arp2/3 complex by coronin. *J. Biol. Chem.* **2011**, *286*, 17039–17046. [CrossRef] [PubMed]
13. Marchand, J.B.; Kaiser, D.A.; Pollard, T.D.; Higgs, H.N. Interaction of WASP/Scar proteins with actin and vertebrate Arp2/3 complex. *Nat. Cell Biol.* **2001**, *3*, 76–82. [CrossRef]
14. Veltman, D.M.; Insall, R.H. WASP family proteins: Their evolution and its physiological implications. *Mol. Biol. Cell* **2010**, *21*, 2880–2893. [CrossRef] [PubMed]
15. Heil-Chapdelaine, R.A.; Tran, N.K.; Cooper, J.A. The role of Saccharomyces cerevisiae coronin in the actin and microtubule cytoskeletons. *Curr. Biol.* **1998**, *8*, 1281–1284. [CrossRef]
16. Sahasrabuddhe, A.A.; Nayak, R.C.; Gupta, C.M. Ancient Leishmania coronin (CRN12) is involved in microtubule remodeling during cytokinesis. *J. Cell Sci.* **2009**, *122*, 1691–1699. [CrossRef]
17. Bane, K.S.; Lepper, S.; Kehrer, J.; Sattler, J.M.; Singer, M.; Reinig, M.; Klug, D.; Heiss, K.; Baum, J.; Mueller, A.-K.; et al. The Actin Filament-Binding Protein Coronin Regulates Motility in Plasmodium Sporozoites. *PLoS Pathog.* **2016**, *12*, e1005710. [CrossRef]
18. Spoerl, Z.; Stumpf, M.; Noegel, A.A.; Hasse, A. Oligomerization, F-actin interaction, and membrane association of the ubiquitous mammalian coronin 3 are mediated by its carboxyl terminus. *J. Biol. Chem.* **2002**, *277*, 48858–48867. [CrossRef]
19. Huang, W.; Ghisletti, S.; Saijo, K.; Gandhi, M.; Aouadi, M.; Tesz, G.J.; Zhang, D.X.; Yao, J.; Czech, M.P.; Goode, B.L.; et al. Coronin 2A mediates actin-dependent de-repression of inflammatory response genes. *Nature* **2011**, *470*, 414–418. [CrossRef]
20. Nakamura, T.; Takeuchi, K.; Muraoka, S.; Takezoe, H.; Takahashi, N.; Mori, N. A neurally enriched coronin-like protein, ClipinC, is a novel candidate for an actin cytoskeleton-cortical membrane-linking protein. *J. Biol. Chem.* **1999**, *274*, 13322–13327. [CrossRef]
21. Chen, Y.; Ip, F.C.; Shi, L.; Zhang, Z.; Tang, H.; Ng, Y.P.; Ye, W.-C.; Fu, A.K.; Ip, N.Y. Coronin 6 regulates acetylcholine receptor clustering through modulating receptor anchorage to actin cytoskeleton. *J. Neurosci.* **2014**, *34*, 2413–2421. [CrossRef]
22. Cai, L.; Marshall, T.W.; Uetrecht, A.C.; Schafer, D.A.; Bear, J.E. Coronin 1B coordinates Arp2/3 complex and cofilin activities at the leading edge. *Cell* **2007**, *128*, 915–929. [CrossRef] [PubMed]
23. Rybakin, V.; Stumpf, M.; Schulze, A.; Majoul, I.V.; Noegel, A.A.; Hasse, A. Coronin 7, the mammalian POD-1 homologue, localizes to the Golgi apparatus. *FEBS Lett.* **2004**, *573*, 161–167. [CrossRef] [PubMed]
24. Ferrari, G.; Langen, H.; Naito, M.; Pieters, J. A coat protein on phagosomes involved in the intracellular survival of mycobacteria. *Cell* **1999**, *97*, 435–447. [CrossRef]
25. Suzuki, K.; Nishihata, J.; Arai, Y.; Honma, N.; Yamamoto, K.; Irimura, T.; Toyoshima, S. Molecular cloning of a novel actin-binding protein, p57, with a WD repeat and a leucine zipper motif. *FEBS Lett.* **1995**, *364*, 283–288. [CrossRef]
26. Jayachandran, R.; Liu, X.; BoseDasgupta, S.; Müller, P.; Zhang, C.L.; Moshous, D.; Studer, V.; Schneider, J.; Genoud, C.; Fossoud, C.; et al. Coronin 1 regulates cognition and behavior through modulation of cAMP/protein kinase A signaling. *PLoS Biol.* **2014**, *12*, e1001820. [CrossRef] [PubMed]
27. Suo, D.; Park, J.; Harrington, A.W.; Zweifel, L.S.; Mihalas, S.; Deppmann, C.D. Coronin-1 is a neurotrophin endosomal effector that is required for developmental competition for survival. *Nat. Neurosci.* **2014**, *17*, 36–45. [CrossRef]
28. Jayachandran, R.; Gumienny, A.; Bolinger, B.; Ruehl, S.; Lang, M.J.; Fucile, G.; Mazumder, S.; Tchang, V.; Woischnig, A.-K.; Stiess, M.; et al. Disruption of Coronin 1 Signaling in T Cells Promotes Allograft Tolerance while Maintaining Anti-Pathogen Immunity. *Immunity* **2019**, *50*, 152–165.e8. [CrossRef]
29. Haraldsson, M.K.; Louis-Dit-Sully, C.A.; Lawson, B.R.; Sternik, G.; Santiago-Raber, M.L.; Gascoigne, N.R.; Theofilopoulos, A.N.; Kono, D.H. The lupus-related Lmb3 locus contains a disease-suppressing Coronin-1A gene mutation. *Immunity* **2008**, *28*, 40–51. [CrossRef]
30. Vinet, A.F.; Fiedler, T.; Studer, V.; Froquet, R.; Dardel, A.; Cosson, P.; Pieters, J. Initiation of multicellular differentiation in Dictyostelium discoideum is regulated by coronin A. *Mol. Biol. Cell* **2014**, *25*, 688–701. [CrossRef]
31. de Hostos, E.L.; Bradtke, B.; Lottspeich, F.; Guggenheim, R.; Gerisch, G. Coronin, an actin binding protein of Dictyostelium discoideum localized to cell surface projections, has sequence similarities to G protein beta subunits. *EMBO J.* **1991**, *10*, 4097–4104. [CrossRef] [PubMed]

32. de Hostos, E.L.; Rehfuess, C.; Bradtke, B.; Waddell, D.R.; Albrecht, R.; Murphy, J.; Gerisch, G. Dictyostelium mutants lacking the cytoskeletal protein coronin are defective in cytokinesis and cell motility. *J. Cell Biol.* **1993**, *120*, 163–173. [CrossRef] [PubMed]
33. Gerisch, G.; Albrecht, R.; De Hostos, E.; Wallraff, E.; Heizer, C.; Kreitmeier, M.; Muller Taubenberger, A. Actin-associated proteins in motility and chemotaxis of Dictyostelium cells. *Symp. Soc. Exp. Biol.* **1993**, *47*, 297–315. [PubMed]
34. Maniak, M.; Rauchenberger, R.; Albrecht, R.; Murphy, J.; Gerisch, G. Coronin involved in phagocytosis: Dynamics of particle-induced relocalization visualized by a green fluorescent protein Tag. *Cell* **1995**, *83*, 915–924. [CrossRef]
35. Swaminathan, K.; Muller-Taubenberger, A.; Faix, J.; Rivero, F.; Noegel, A.A. A Cdc42- and Rac-interactive binding (CRIB) domain mediates functions of coronin. *Proc. Natl. Acad. Sci. USA* **2014**, *111*, E25–E33. [CrossRef] [PubMed]
36. Shina, M.C.; Müller-Taubenberger, A.; Ünal, C.; Schleicher, M.; Steinert, M.; Eichinger, L.; Müller, R.; Blau-Wasser, R.; Glöckner, G.; Noegel, A.A. Redundant and unique roles of coronin proteins in Dictyostelium. *Cell. Mol. Life Sci.* **2011**, *68*, 303–313. [CrossRef] [PubMed]
37. Devreotes, P. Dictyostelium discoideum: A model system for cell-cell interactions in development. *Science* **1989**, *245*, 1054–1058. [CrossRef]
38. Kimmel, A.R.; Firtel, R.A. cAMP signal transduction pathways regulating development of Dictyostelium discoideum. *Curr. Opin. Genet. Dev.* **1991**, *1*, 383–390. [CrossRef]
39. Weijer, C.J. Morphogenetic cell movement in Dictyostelium. *Semin. Cell Dev. Biol.* **1999**, *10*, 609–619. [CrossRef]
40. de Hostos, E.L. The coronin family of actin-associated proteins. *Trends Cell Biol.* **1999**, *9*, 345–350. [CrossRef]
41. Uetrecht, A.C.; Bear, J.E. Coronins: The return of the crown. *Trends Cell Biol.* **2006**, *16*, 421–426. [CrossRef] [PubMed]
42. Shina, M.C.; Noegel, A.A. Invertebrate coronins. *Subcell. Biochem.* **2008**, *48*, 88–97. [PubMed]
43. Mueller, P.; Quintana, A.; Griesemer, D.; Hoth, M.; Pieters, J. Disruption of the cortical actin cytoskeleton does not affect store operated Ca(2+) channels in human T-cells. *FEBS Lett.* **2007**, *581*, 3557–3562. [CrossRef]
44. Zhang, W.; Wu, Y.; Du, L.; Tang, D.D.; Gunst, S.J. Activation of the Arp2/3 complex by N-WASp is required for actin polymerization and contraction in smooth muscle. *Am. J. Physiol. Cell Physiol.* **2005**, *288*, C1145–C1160. [CrossRef]
45. Mueller, P.; Liu, X.; Pieters, J. Migration and homeostasis of naive T cells depends on coronin 1-mediated prosurvival signals and not on coronin 1-dependent filamentous actin modulation. *J. Immunol.* **2011**, *186*, 4039–4050. [CrossRef] [PubMed]
46. Fukui, Y.; Engler, S.; Inoue, S.; de Hostos, E.L. Architectural dynamics and gene replacement of coronin suggest its role in cytokinesis. *Cell Motil. Cytoskeleton.* **1999**, *42*, 204–217. [CrossRef]
47. Shiow, L.R.; Roadcap, D.W.; Paris, K.; Watson, S.R.; Grigorova, I.L.; Lebet, T.; An, J.; Xu, Y.; Jenne, C.N.; Föger, N.; et al. The actin regulator coronin 1A is mutant in a thymic egress-deficient mouse strain and in a patient with severe combined immunodeficiency. *Nat. Immunol.* **2008**, *9*, 1307–1315. [CrossRef] [PubMed]
48. Fiedler, T.; Ndinyanka Fabrice, T.; Studer, V.; Vinet, A.; Faltova, L.; Sharp, T.; Pieters, J. Homodimerization of coronin A through the C-terminal coiled coil domain is essential for multicellular differentiation of Dictyostelium discoideum in preparation. *Dictybase.org*. (under review).
49. Joseph, J.M.; Fey, P.; Ramalingam, N.; Liu, X.I.; Rohlfs, M.; Noegel, A.A.; Müller-Taubenberger, A.; Schleicher, M. The actinome of Dictyostelium discoideum in comparison to actins and actin-related proteins from other organisms. *PLoS ONE* **2008**, *3*, e2654. [CrossRef]
50. Castellano, F.; Chavrier, P.; Caron, E. Actin dynamics during phagocytosis. *Semin. Immunol.* **2001**, *13*, 347–355. [CrossRef]
51. May, R.C.; Machesky, L.M. Phagocytosis and the actin cytoskeleton. *J. Cell Sci.* **2001**, *114*, 1061–1077.
52. Pan, M.; Xu, X.; Chen, Y.; Jin, T. Identification of a Chemoattractant G-Protein-Coupled Receptor for Folic Acid that Controls Both Chemotaxis and Phagocytosis. *Dev. Cell* **2016**, *36*, 428–439. [CrossRef]
53. Vogel, G.; Thilo, L.; Schwarz, H.; Steinhart, R. Mechanism of phagocytosis in Dictyostelium discoideum: Phagocytosis is mediated by different recognition sites as disclosed by mutants with altered phagocytic properties. *J. Cell Biol.* **1980**, *86*, 456–465. [CrossRef]

54. Riyahi, T.Y.; Frese, F.; Steinert, M.; Omosigho, N.N.; Glöckner, G.; Eichinger, L.; Orabi, B.; Williams, R.S.B.; Noegel, A.A. RpkA, a highly conserved GPCR with a lipid kinase domain, has a role in phagocytosis and anti-bacterial defense. *PLoS ONE* **2011**, *6*, e27311. [CrossRef] [PubMed]
55. Fukui, Y.; de Hostos, E.; Yumura, S.; Kitanishi-Yumura, T.; Inoué, S. Architectural dynamics of F-actin in eupodia suggests their role in invasive locomotion in Dictyostelium. *Exp. Cell Res.* **1999**, *249*, 33–45. [CrossRef]
56. Tunnacliffe, E.; Corrigan, A.M.; Chubb, J.R. Promoter-mediated diversification of transcriptional bursting dynamics following gene duplication. *Proc. Natl. Acad. Sci. USA* **2018**, *115*, 8364–8369. [CrossRef] [PubMed]
57. Chan, K.T.; Roadcap, D.W.; Holoweckyj, N.; Bear, J.E. Coronin 1C harbours a second actin-binding site that confers co-operative binding to F-actin. *Biochem. J.* **2012**, *444*, 89–96. [CrossRef] [PubMed]
58. Berzat, A.; Hall, A. Cellular responses to extracellular guidance cues. *EMBO J.* **2010**, *29*, 2734–2745. [CrossRef]
59. Hacker, U.; Albrecht, R.; Maniak, M. Fluid-phase uptake by macropinocytosis in Dictyostelium. *J. Cell Sci.* **1997**, *110*, 105–112.
60. Nagasaki, A.; Hibi, M.; Asano, Y.; Uyeda, T.Q. Genetic approaches to dissect the mechanisms of two distinct pathways of cell cycle-coupled cytokinesis in Dictyostelium. *Cell Struct. Funct.* **2001**, *26*, 585–591. [CrossRef]
61. Williams, T.D.; Peak-Chew, S.Y.; Paschke, P.; Kay, R.R. Akt and SGK protein kinases are required for efficient feeding by macropinocytosis. *J. Cell Sci.* **2019**, *132*, jcs224998. [CrossRef]
62. Prassler, J.; Stocker, S.; Marriott, G.; Heidecker, M.; Kellermann, J.; Gerisch, G. Interaction of a Dictyostelium member of the plastin/fimbrin family with actin filaments and actin-myosin complexes. *Mol. Biol. Cell* **1997**, *8*, 83–95. [CrossRef] [PubMed]
63. Kollmar, M. Thirteen is enough: The myosins of Dictyostelium discoideum and their light chains. *BMC Genomics* **2006**, *7*, 183. [CrossRef] [PubMed]
64. Friedberg, F.; Rivero, F. Single and multiple CH (calponin homology) domain containing multidomain proteins in Dictyostelium discoideum: An inventory. *Mol. Biol. Rep.* **2010**, *37*, 2853–2862. [CrossRef]
65. Lambooy, P.K.; Korn, E.D. Purification and characterization of actobindin, a new actin monomer-binding protein from Acanthamoeba castellanii. *J. Biol. Chem.* **1986**, *261*, 17150–17155. [PubMed]
66. Howe, A.K. Regulation of actin-based cell migration by cAMP/PKA. *Biochim. Biophys. Acta* **2004**, *1692*, 159–174. [CrossRef]
67. Mattila, P.K.; Batista, F.D.; Treanor, B. Dynamics of the actin cytoskeleton mediates receptor cross talk: An emerging concept in tuning receptor signaling. *J. Cell Biol.* **2016**, *212*, 267–280. [CrossRef]
68. Ashworth, J.M.; Watts, D.J. Metabolism of the cellular slime mould Dictyostelium discoideum grown in axenic culture. *Biochem. J.* **1970**, *119*, 175–182. [CrossRef]
69. Reubold, T.F.; Eschenburg, S.; Becker, A.; Kull, F.J.; Manstein, D.J. A structural model for actin-induced nucleotide release in myosin. *Nat. Struct. Biol.* **2003**, *10*, 826–830. [CrossRef]
70. Fujita-Becker, S.; Reubold, T.F.; Holmes, K.C. The actin-binding cleft: Functional characterisation of myosin II with a strut mutation. *J. Muscle Res. Cell Motil.* **2006**, *27*, 115–123. [CrossRef]
71. Mumberg, D.; Muller, R.; Funk, M. Yeast vectors for the controlled expression of heterologous proteins in different genetic backgrounds. *Gene* **1995**, *156*, 119–122. [CrossRef]
72. Ahrne, E.; Glatter, T.; Vigano, C.; Schubert, C.; Nigg, E.A.; Schmidt, A. Evaluation and Improvement of Quantification Accuracy in Isobaric Mass Tag-Based Protein Quantification Experiments. *J. Proteome Res.* **2016**, *15*, 2537–2547. [CrossRef] [PubMed]
73. Vizcaino, J.A.; Csordas, A.; Del-Toro, N.; Dianes, J.A.; Griss, J.; Lavidas, I.; Mayer, G.; Perez-Riverol, Y.; Reisinger, F.; Ternent, T.; et al. 2016 update of the PRIDE database and its related tools. *Nucleic Acids Res.* **2016**, *44*, D447–D456. [CrossRef] [PubMed]
74. Mueller, P.; Massner, J.; Jayachandran, R.; Combaluzier, B.; Albrecht, I.; Gatfield, J.; Blum, C.; Ceredig, R.; Rodewald, H.-R.; Rolink, A.G.; et al. Regulation of T cell survival through coronin-1-mediated generation of inositol-1,4,5-trisphosphate and calcium mobilization after T cell receptor triggering. *Nat. Immunol.* **2008**, *9*, 424–431. [CrossRef] [PubMed]
75. Eichinger, L.; Rivero, F. Dictyostelium discoideum protocols. In *Methods in Molecular Biology*; Clifton, N.J., Ed.; Humana Press: New York, NY, USA, 2006.
76. Pardee, J.D.; Spudich, J.A. Purification of muscle actin. *Methods Enzymol.* **1982**, *85*, 164–181.

77. Yahara, N.; Ueda, T.; Sato, K.; Nakano, A. Multiple roles of Arf1 GTPase in the yeast exocytic and endocytic pathways. *Mol. Biol. Cell* **2001**, *12*, 221–238. [CrossRef]
78. Sattler, N.; Monroy, R.; Soldati, T. Quantitative analysis of phagocytosis and phagosome maturation. *Methods Mol. Biol.* **2013**, *983*, 383–402.

© 2020 by the authors. Licensee MDPI, Basel, Switzerland. This article is an open access article distributed under the terms and conditions of the Creative Commons Attribution (CC BY) license (http://creativecommons.org/licenses/by/4.0/).

Article

Coronin 1 Is Required for Integrin β2 Translocation in Platelets

David R. J. Riley [1], Jawad S. Khalil [1,2], Jean Pieters [3], Khalid M. Naseem [4] and Francisco Rivero [1,*]

[1] Centre for Atherothrombosis and Metabolic Disease, Hull York Medical School, Faculty of Health Sciences, University of Hull, Hull HU6 7RX, UK; David.riley@hyms.ac.uk (D.R.J.R.); jawad.khalil@bristol.ac.uk (J.S.K.)
[2] School of Physiology, Pharmacology and Neuroscience, Faculty of Life Sciences, University of Bristol, Bristol BS8 1TD, UK
[3] Biozentrum, University of Basel, CH-4056 Basel, Switzerland; jean.pieters@unibas.ch
[4] Leeds Institute for Cardiovascular and Metabolic Medicine, University of Leeds, Leeds LS2 9NL, UK; k.naseem@leeds.ac.uk
* Correspondence: Francisco.rivero@hyms.ac.uk; Tel.: +44-1482-644-633

Received: 22 November 2019; Accepted: 1 January 2020; Published: 5 January 2020

Abstract: Remodeling of the actin cytoskeleton is one of the critical events that allows platelets to undergo morphological and functional changes in response to receptor-mediated signaling cascades. Coronins are a family of evolutionarily conserved proteins implicated in the regulation of the actin cytoskeleton, represented by the abundant coronins 1, 2, and 3 and the less abundant coronin 7 in platelets, but their functions in these cells are poorly understood. A recent report revealed impaired agonist-induced actin polymerization and cofilin phosphoregulation and altered thrombus formation in vivo as salient phenotypes in the absence of an overt hemostasis defect in vivo in a knockout mouse model of coronin 1. Here we show that the absence of coronin 1 is associated with impaired translocation of integrin β2 to the platelet surface upon stimulation with thrombin while morphological and functional alterations, including defects in Arp2/3 complex localization and cAMP-dependent signaling, are absent. Our results suggest a large extent of functional overlap among coronins 1, 2, and 3 in platelets, while aspects like integrin β2 translocation are specifically or predominantly dependent on coronin 1.

Keywords: actin; Arp2/3 complex; cAMP; coronin 1; integrin β2; platelets; thrombin; collagen; prostacyclin

---

1. Introduction

Vascular injury leads to exposure of prothrombotic extracellular matrix proteins, which facilitates the entrapment and activation of platelets through specialized receptors. These interactions contribute to stable adhesion of platelets by generating intracellular signals that lead to shape change, secretion of granules, and activation of integrins. Activation of integrins facilitates the binding of the plasma protein fibrinogen, which subsequently supports platelet aggregation and clot formation, rapidly consolidated by secreted soluble agonists [1]. While this process is critical to hemostatic protection of the vasculature after injury, the rupture of atherosclerotic plaques drives uncontrolled platelet activation that leads to arterial thrombosis and clinical events such as myocardial infarction and stroke.

Platelet activation is the result of multiple integrated signaling cascades that ultimately drive remodeling of the platelet cytoskeleton and sustain the morphological changes required for adhesion, spreading, aggregation, and secretion at the sites of vascular damage [2]. The cytoskeleton is also the target of inhibitory signaling pathways regulated by cyclic nucleotides that balance the activating pathways and prevent thrombus formation [3]. Coronins are a family of evolutionarily conserved regulators of the actin cytoskeleton turnover represented by seven members in mammals They have been grouped into three classes based on phylogenetic and functional criteria [4,5]. Class I includes Coronins 1, 2, 3, and 6 (also called 1A, 1B, 1C, and 1D) that associate with the actin cytoskeleton, localize at the leading edge of migrating cells, and participate in various signaling processes. Class II includes Coro4 and 5 (also called 2A and 2B), involved in focal adhesion turnover, reorganization of the cytoskeleton, and cell migration. The class III coronin (Coro7) has an unusual structure and plays a role in Golgi morphology maintenance. We have reported that class I coronins coronin 1, 2, and 3 are abundant in both human and mouse platelets, whereas coronin 7 is also present in human and mouse platelets in very low amounts and class II coronins are apparently absent [6].

Coronin 1 (coronin-1A or Coro1, also known as P57 or Tryptophan Aspartate containing COat protein (TACO)) [7,8] participates in the modulation of a number of processes through protein–protein interactions. For example, it modulates cyclic adenosine monophosphate (cAMP) signaling in neurons through interaction with the Gαs subunit of heterotrimeric G proteins [9], neutrophil adhesion through interaction with the cytoplasmic tail of integrin β2 [10], and the activity of the small GTPase Rac1 [11]. Coro1 also participates in a number of other cellular processes including NADPH oxidase complex regulation, calcium signaling, vesicle trafficking, and apoptosis [12–16].

Coro1 is abundantly expressed in cells of the hematopoietic lineage, where it is essential for the survival of naïve T cells [16–19], but little is known about its role in platelets. We have shown that Coro1 is mainly a cytosolic protein, but a significant amount associates to membranes in an actin-independent manner. It rapidly translocates to the detergent-insoluble cytoskeleton upon platelet stimulation with thrombin or collagen. Along with Coro2 and 3, it accumulates at the cell cortex and actin nodules [6]. Stocker et al. reported the absence of an overt hemostasis defect in vivo in a knockout mouse model of Coro1. Detailed examination revealed impaired agonist-induced actin polymerization and cofilin phosphoregulation and altered thrombus formation in vivo as salient phenotypes [20]. Here we extend Stocker et al. report by an in-depth characterization of platelet function exploring additional aspects. Our data show that the absence of Coro1 is associated with impaired translocation of integrin β2 to the platelet surface upon stimulation with thrombin but otherwise does not result in noticeable morphological and functional alterations, including Arp2/3 complex localization and cAMP-dependent signaling. This mild phenotype suggests a complex picture in which class I coronins might share roles extensively in platelets.

2. Results

2.1. Absence of Coro1 Is Not Compensated by Increased Coro3

To gain insight into the roles of Coro1 in platelet function, we undertook the characterization of a previously described *Coro1a* knockout (KO) model [15]. We confirmed the absence of the protein in platelet lysates of homozygous KO mice by Western blot analysis and observed that heterozygous mouse platelets expressed approximately half of the amount of the protein present in wild type (WT) mouse platelets (Figure 1A). Coro1 KO mice have been reported to exhibit unaffected hematological parameters, including platelet counts, indicating that hematopoiesis is not affected [17,20]. The size of Coro1 KO platelets was comparable to that of WT platelets as estimated from the forward light scatter in flow cytometry experiments ($p = 0.8164$, Student's *t*-test) (Figure 1B).

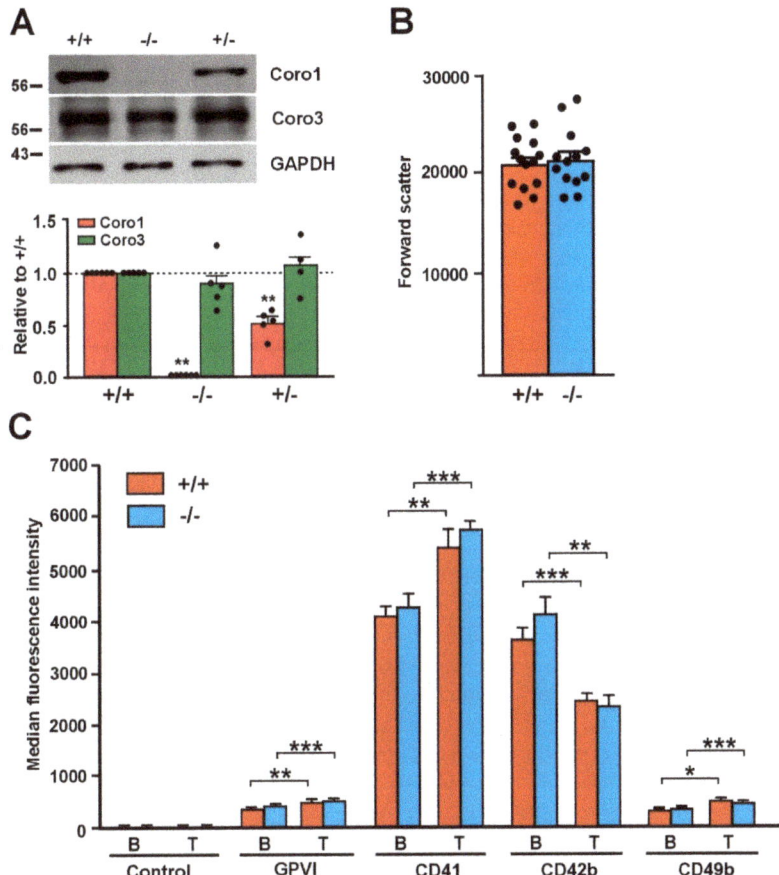

Figure 1. Relative size and receptor expression in *Coro1a* deficient platelets. (**A**) Absence of Coro1 in *Coro1a* deficient platelets and no obvious compensation by Coro3. Platelet lysates were resolved by SDS-PAGE, blotted and probed with specific antibodies for the indicated proteins. GAPDH was used for normalization. Data represent mean ± standard error of the mean (SEM) of 4–6 independent experiments. ** $p < 0.01$; Mann–Whitney U-test. Full blots are shown in Supplemental Figure S1; (**B**) Relative size of *Coro1a* deficient platelets. Mean platelet volume was estimated in platelet-rich plasma (PRP) by mean forward light scatter area using flow cytometry. Data represent mean ± SEM of 13–14 independent experiments. No statistically significant differences were found, Student's *t*-test; (**C**) Surface receptor expression in *Coro1a* deficient platelets. Platelet surface receptors were determined in PRP by flow cytometry both in basal conditions (B) and upon stimulation with 0.1 U/mL thrombin for 20 min at 37 °C (T). Data represent mean ± SEM of 7–16 independent experiments. * $p < 0.05$; ** $p < 0.01$; *** $p < 0.001$; paired Student's *t*-test between basal and stimulated conditions. No statistically significant differences were found between wild type and knockout, nonpaired Student's *t*-test.

2.2. Receptor Expression Is Not Affected in Coro1 Deficient Platelets

We assessed the expression of characteristic surface platelet receptors (GPVI, CD41, CD42b, and CD49b) by flow cytometry both in unstimulated and in thrombin-stimulated platelets. Thrombin stimulation caused a significant increase in the expression of GPVI, CD41 (integrin αIIb), and CD49b (integrin α2) (20–40%) and a significant decrease in the expression of CD42b (GP1b) (32–43%), the latter due to cleavage and internalization of the GP1b/IX/V complex [21]. Both basal and thrombin-stimulated receptor expression levels were comparable in Coro1 WT and KO platelets (Figure 1C).

2.3. Translocation of Integrin β2 Is Impaired in the Absence of Coro1

Coro1 interacts with the cytoplasmic tail of integrin β2 and regulates its function in neutrophils [10]. Although less abundant than integrins β1 and β3, integrin β2 (CD18) is expressed in murine platelets [22–25] and has also been described in human platelets, where expression increases upon thrombin stimulation [26]. This prompted us to investigate whether Coro1 deficiency would have an effect on this integrin. We used flow cytometry to assess the levels of expression of CD18 both in resting and in thrombin stimulated platelets and observed that in resting platelets the levels of CD18 were higher, although statistically not significant, in WT platelets (940 ± 70 median fluorescence intensity) than in KO platelets (783 ± 51; $p = 0.1016$). However, upon thrombin stimulation expression increased significantly in WT platelets to 1562 ± 158 ($p = 0.0032$ relative to basal) but only modestly in KO platelets (to 986 ± 110; $p = 0.0915$ relative to basal, $p = 0.0123$ relative to WT) (Figure 2A,B). The impaired translocation of CD18 in KO platelets can be visualized in immunostained platelets (Figure 2C).

Integrin β2 main ligand is intercellular adhesion molecule-1 (ICAM-1), a glycoprotein expressed in endothelial cells and leukocytes. We used fluorescence microscopy to investigate the effect of Coro1 absence on platelet adhesion and spreading on surfaces coated with 5 mg/mL native BSA, a surrogate method of assessing binding to ICAM-1, both basally and upon stimulation with 0.1 U/mL thrombin [27,28]. On average, similar numbers of WT and KO resting platelets adhered to coverslips (116.7 ± 14.0 and 120.8 ± 7.1, respectively). Resting platelets of both strains attached to the BSA-coated surface but most did not appear to spread, presenting a round morphology and covering a small area (approximately 9 μm^2) (Figure 2D–F). Thrombin stimulation prior to seeding resulted in more than twice the numbers of adhering platelets (280.3 ± 17.2 in WT vs. 276.9 ± 24.0 in KO). Most stimulated platelets presented a well spread round morphology with stress fibers, although some had a spiky morphology, and covered an area of approximately 21 μm^2. No obvious differences were apparent in cell area between WT and KO platelets (Figure 2D–F). To investigate whether stimulation with lower thrombin doses would reveal any subtle difference in spreading between WT and KO platelets, we performed a set of experiments basally and upon stimulation with 0.05 and 0.025 U/mL thrombin. We observed that both doses resulted in numbers of adhering platelets similar to those obtained with 0.1 U/mL: 309.3 ± 16.8 in the WT vs. 282.0 ± 13.7 in the KO with 0.05 U/mL and 296.3 ± 22.4 in the WT vs. 320.8 ± 38.9 in the KO with 0.025 U/mL. The areas of the spread platelets were also in a range similar (20–21 μm^2) to those observed with 0.1 U/mL thrombin. This indicates that low thrombin doses (0.025 U/mL) are sufficient to elicit full spreading on native BSA and Coro1 is dispensable for this response.

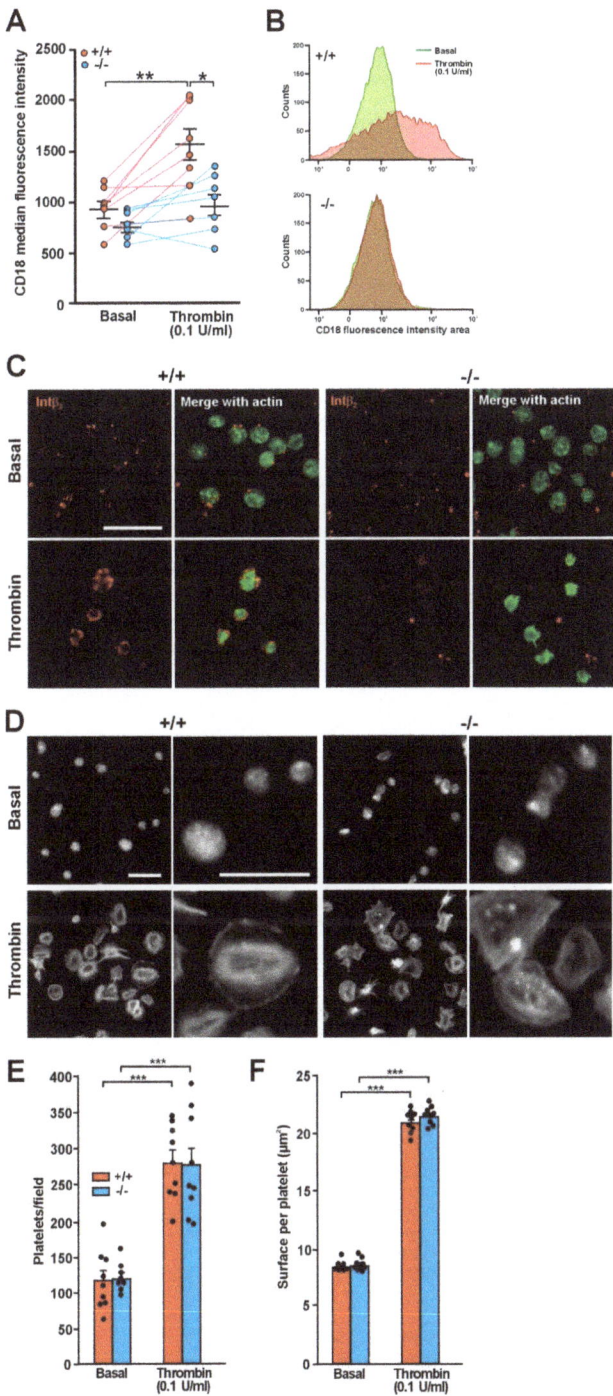

Figure 2. Impaired translocation of integrin β2 in *Coro1a* deficient platelets. (**A**) Platelet surface integrin β2 (CD18) was determined in PRP by flow cytometry both in basal conditions and upon stimulation

with 0.1 U/mL thrombin for 20 min at 37 °C. Individual data and the mean ± SEM of 7–8 independent experiments are shown. * $p < 0.05$; ** $p < 0.01$; paired Student's t-test between basal and stimulated conditions. Nonpaired Student's t-test between wild type (WT) and knockout (KO); (**B**) Representative flow cytometry data of platelet surface CD18 distribution in basal conditions and upon thrombin stimulation; (**C**) Washed platelets were stimulated in suspension with 0.1 U/mL thrombin, fixed with 4% paraformaldehyde (PFA) and spun on poly-L-lysine coated coverslips. The permeabilization step was omitted and the cells were stained with an anti-integrin β2 antibody followed by an Alexa568-coupled secondary antibody (red) and counterstained with fluorescein isothiocyanate (FITC)-phalloidin for filamentous actin (green). Images were acquired with a fluorescence microscope equipped with a structured illumination attachment and deconvolved. Scale bar represents 10 µm; (**D**) Adhesion of Coro1 KO and WT platelets to native bovine serum albumin (BSA). Washed platelets were stimulated with 0.1 U/mL thrombin and immediately allowed to attach to glass coverslips coated with 5 mg/mL of native BSA. Adherent platelets were fixed with 4% PFA, permeabilized with 0.3% Triton X-100, and stained with tetramethylrhodamine isothiocyanate (TRITC)-phalloidin. Images of random areas were acquired with a fluorescence microscope. Examples of platelets at two magnifications are shown. Scale bars represent 10 µm; (**E**) Number of platelets adhering to BSA. 5 fields each 31,560 µm^2 from 9 independent experiments were scored per condition. Data represent mean ± SEM. Number of platelets was significantly higher upon thrombin stimulation (** $p < 0.01$, paired Student's t-test). No significant differences were found between WT and KO platelets both resting and stimulated (unpaired Student's t-test); (**F**) Surface coverage per platelet calculated by thresholding using ImageJ. Data represent mean ± SEM from 9 independent experiments and 600–1200 platelets per condition for each experiment. Platelet surface was significantly higher upon thrombin stimulation (*** $p < 0.001$, paired Student's t-test). No significant differences were found between WT and KO platelets, both resting and stimulated (unpaired Student's t-test).

2.4. Effect of Coro1 Deficiency on Integrin αIIbβ3 Activation and Granule Secretion

We assessed the potential effects of Coro1 deficiency on integrin αIIbβ3 activation with the activation state-specific antibody JON/A by flow cytometry. Stimulation with a wide range of agonists (thrombin, collagen-related peptide (CRP), as well as adenosine diphosphate (ADP) and the thromboxane analog U46619 alone or in combination) caused activation of αIIbβ3, in the case of thrombin and CRP in a dose-dependent manner (Figure 3A). However, we were not able to detect any significant differences in JON/A levels between Coro1 KO and WT platelets, indicating that Coro1 is dispensable for αIIbβ3 activation.

We next explored whether Coro1 KO platelets have a defect in granule secretion. To monitor alpha and dense granule secretion, we induced P-selectin and CD63 expression, respectively, by the same agonists as in the αIIbβ3 activation experiment. In both cases, thrombin produced a clear dose–response effect, CRP had little effect and ADP and U46619 had a synergistic effect in both WT and KO platelets (Figure 3B,C). None of the conditions tested revealed any statistically significant difference between both populations, suggesting that Coro1 is dispensable for granule secretion.

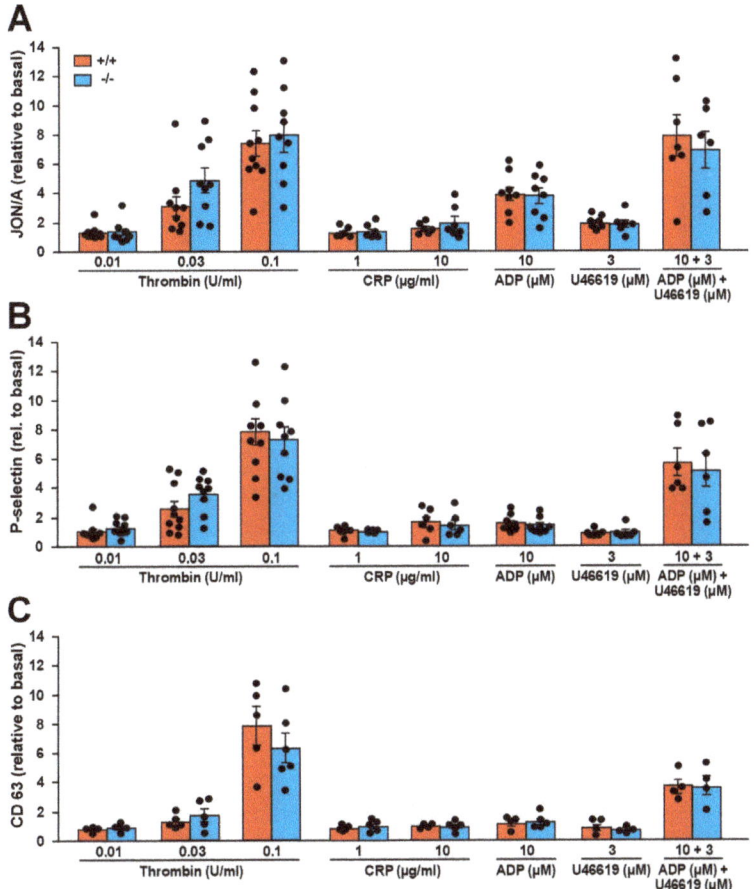

Figure 3. Integrin activation and secretion in *Coro1a* deficient platelets. Integrin activation (**A**), P-selectin exposure (**B**), and CD63 exposure (**C**) were determined in PRP upon stimulation with the indicated doses of agonists for 20 min at 37 °C and subsequent flow cytometry analysis. The data (median fluorescence intensity) represent the mean ± SEM of 5–9 independent experiments expressed relative to basal (unstimulated) platelets. No statistically significant differences were found between WT and KO, Student's *t*-test.

2.5. Effect of Coro1 Deficiency on Platelet Aggregation and Spreading

A functioning actin cytoskeleton remodeling is critical for platelet aggregation and for adhesion and spreading on extracellular matrix proteins. We next investigated the implications of Coro1 deficiency for those processes. Stocker et al. reported subtle defects in aggregation induced by low doses of collagen using impedance-based aggregometry on whole blood [20]. We applied light transmission aggregometry on washed platelets using a range of doses of thrombin (0.0125–0.1 U/mL), collagen (1–10 µg/mL), and CRP (3–10 µg/mL). All three agonists elicited, as expected, a dose-dependent aggregation response, which was comparable in both WT and KO platelets at all doses (Figure 4). The aggregation velocity, calculated as the slope of the aggregation curve, was also dose-dependent for all three agonists. We only observed a statistically significant alteration in the response to high-dose thrombin, with KO platelets showing a marginally higher percentage of aggregation (91.7 vs. 82.4, $p = 0.0420$) and a moderately higher velocity (3.29 vs. 4.30, $p = 0.0137$, Student's *t*-test) compared to WT platelets. We did not observe any statistically significant difference between WT and KO platelets at any dose of collagen or CRP.

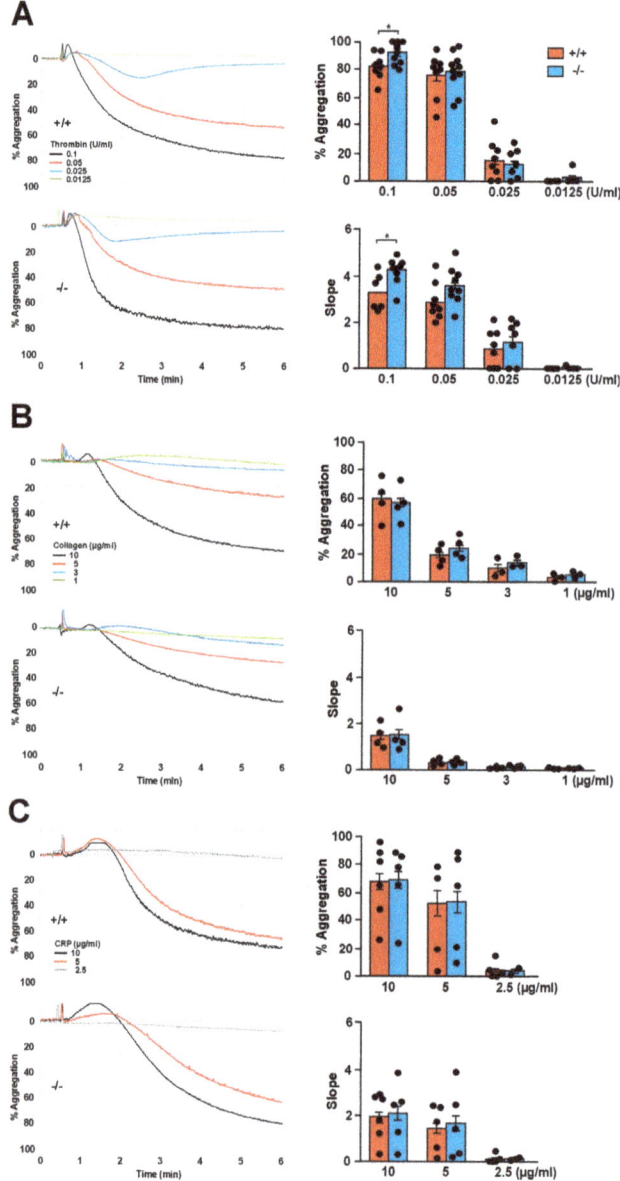

Figure 4. Aggregation in *Coro1a* deficient platelets. Washed platelets (2.0×10^8 platelets/mL) were stimulated with the indicated doses of thrombin (**A**), collagen (**B**), or collagen-related peptide (CRP) (**C**) and aggregation was recorded for 6 min in a Chrono-Log aggregometer. Representative traces are shown on the left. Bar diagrams show percentage of maximum aggregation within 5 min of stimulation and slope as calculated from the linear part of the aggregation trace. Data are mean ± SEM of 4–10 independent experiments. * $p < 0.05$, Student's *t*-test for thrombin; no significant differences were found with collagen and CRP, Mann–Whitney U-test.

The effect of Coro1 absence on platelet adhesion and spreading was further investigated on surfaces coated with collagen (100 µg/mL) or fibrinogen (100 µg/mL) by fluorescence microscopy. On average, slightly more platelets per observation field adhered on fibrinogen; however, there were no statistically significant differences in the numbers of platelets adhering to either surface between the WT and the KO platelets (60.4 ± 4.5 vs. 61.4 ± 5.1 on fibrinogen and 48.3 ± 4.6 vs. 52.3 ± 7.4 on collagen) (Figure 5A, B). Irrespective of genotype, platelets covered a slightly larger surface on collagen (13.43 ± 1.07 μm^2 in the WT vs. 15.83 ± 0.85 μm^2 in the KO) than on fibrinogen (10.75 ± 0.74 μm^2 in the WT vs. 11.89 ± 0.71 μm^2 in the KO) (Figure 5C). Characteristically, on fibrinogen, most platelets showed abundant filopods and actin nodules whereas on collagen most displayed stress fibers, however, no differences in the morphology were apparent between WT and KO platelets in any of the matrices.

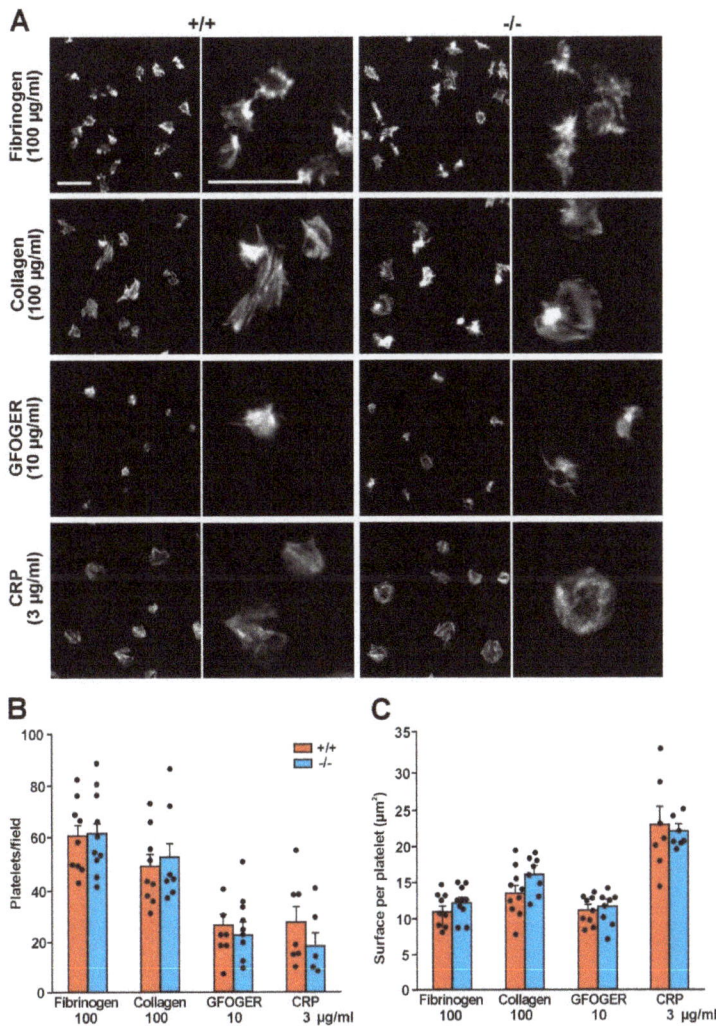

Figure 5. Absence of Coro1 does not impair platelet spreading. (**A**) Adhesion of washed platelets to glass coverslips coated with the indicated concentration of collagen, fibrinogen, Gly-Phe-Hyp-Gly-Glu-Arg (GFOGER), or CRP. Adherent platelets were fixed with 4% PFA, permeabilized with 0.3% Triton X-100,

and stained with TRITC-phalloidin. Images were acquired with a fluorescence microscope equipped with a structured illumination attachment and deconvolved. Examples of platelets at two magnifications are shown. Scale bars represent 10 µm; (**B**) Number of platelets adhering to the indicated concentrations of collagen, fibrinogen, GFOGER, or CRP. 5 fields each 12,500 µm^2 from 5–10 independent experiments were scored per condition. Data represent mean ± SEM. No significant differences were found between WT and KO platelets for any condition (Mann–Whitney U-test); (**C**) Surface coverage per platelet calculated by thresholding using ImageJ. Data represent mean ± SEM from 5–10 independent experiments and 250–1000 platelets per condition for each experiment. No significant differences were found between WT and KO platelets for any condition (Mann–Whitney U-test).

We investigated any subtle effect of Coro1 ablation on adhesion to specific collagen receptors using coverslips coated with peptides that discriminate between receptors, Gly-Phe-Hyp-Gly-Glu-Arg (GFOGER) (for α2β1 integrin), or CRP (for GPVI). Approximately 50% less platelets adhered on GFOGER and CRP compared to collagen and the trend was similar in both WT and KO platelets (Figure 5B). Surface coverage was lower on GFOGER (10.98 ± 0.62 µm^2 in the WT vs. 11.43 ± 0.81 µm^2 in the KO) and higher on CRP (22.92 ± 2.24 µm^2 in the WT vs. KO 21.96 ± 0.77 µm^2 in the KO) compared to collagen (Figure 5C). On CRP, platelets morphologically resembled the ones on collagen, whereas on GFOGER they were discoid and spiky. Again, there were neither statistically significant differences in surface coverage between WT and KO platelets nor noticeable morphological differences in collagen receptor-specific matrices.

2.6. Coro1 Is Dispensable for Arp2/3 Complex Localization

We have shown that in platelets Coro1 co-immunoprecipitates and colocalizes with components of the Arp2/3 complex and that the activity of the complex is necessary for extension of lamellipodia and accumulation of Coro1 at the cell cortex of spreading platelets [6]. We used Coro1 deficient platelets to address whether Coro1 is necessary for Arp2/3 complex localization by allowing them to spread on fibrinogen upon activation with 0.1 U/mL thrombin (Figure 6A). Virtually all unstimulated platelets spread on fibrinogen and showed abundant filopods and actin nodules. In those platelets, the Arp2/3 component ARPC2 (p34-Arc, component of the Arp2/3 complex) displayed a diffuse distribution and some accumulation at actin nodules. Upon thrombin stimulation, approximately 92% of platelets adopted a well spread circular shape with stress fibers and neat accumulation of actin and ARPC2 at the cell cortex. These patterns of platelet morphology and ARPC2 distribution were indistinguishable in both WT and KO platelets (Figure 6B), indicating that Coro1 is dispensable for Arp2/3 localization and lamellipodia formation upon platelet stimulation. Using the same approach, we explored whether activation of the Arp2/3 complex is required for spreading and for cortical localization of ARPC2 upon thrombin stimulation (Figure 6A). We treated wild type and Coro1 deficient platelets with the Arp2/3 complex inhibitor CK666 (50 µM) prior to stimulation with thrombin and seeding. While still capable of adhering to fibrinogen, virtually all unstimulated platelets remained discoid with inconspicuous ARPC2 distribution. Upon thrombin stimulation, approximately 80% of platelets remained discoid, whereas the rest spread, however, their morphology was irregular, their F-actin staining was weaker compared to untreated platelets, and ARPC2 was almost never found at the cell cortex. This shows that activation of the Arp2/3 complex is required for its cortical localization and for efficient spreading, irrespective of the presence of coronin 1.

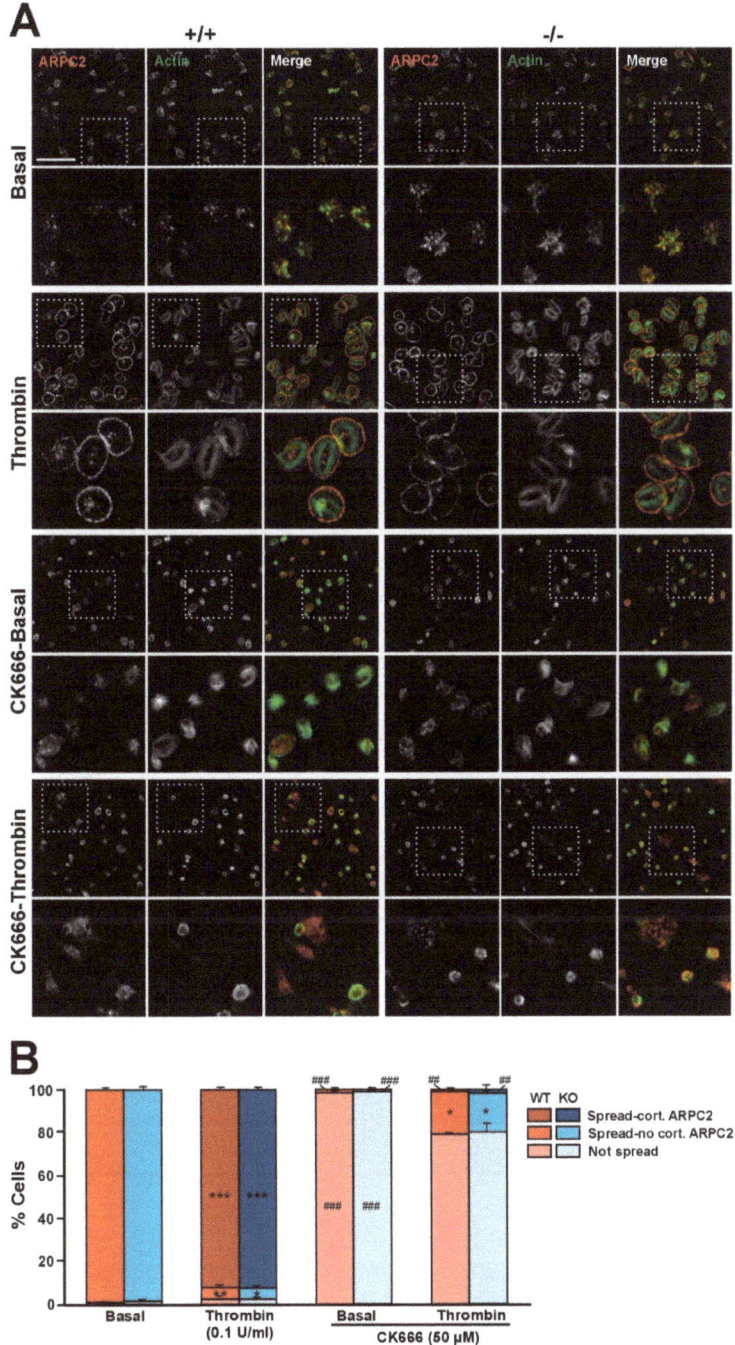

Figure 6. Coro1 is dispensable for Arp2/3 complex localization. (**A**) Localization of ARPC2 in resting and thrombin-stimulated platelets. Washed platelets were stimulated with 0.1 U/mL thrombin and immediately allowed to attach to glass coverslips coated with 100 µg/mL of fibrinogen. A population

of platelets was treated with the Arp2/3 complex inhibitor CK666 (50 µM) for 30 min at 37 °C prior to thrombin stimulation. Adherent platelets were fixed with 4% PFA, permeabilized with 0.3% Triton X-100, stained with an anti-ARPC2 antibody followed by an Alexa568-coupled secondary antibody (red) and counterstained with FITC-phalloidin for filamentous actin (green). Images were acquired with a fluorescence microscope equipped with a structured illumination attachment and deconvolved. Examples of platelets at two magnifications are shown. Boxes mark the enlarged regions. Scale bar represents 25 µm; boxes are 27 × 27 µm; (**B**) Platelet morphology. Platelets were assigned to one of three classes based on spreading and ARPC2 distribution (cortical or not). 5 fields each 36,670 µm^2 from 4 independent experiments were scored per condition. Data are shown as percentage of platelets of each class and represent mean ± SEM. * $p < 0.05$, ** $p < 0.01$, *** $p < 0.001$ relative to the corresponding basal condition. ## $p < 0.01$, ### $p < 0.001$ relative to platelets not treated with CK666 of the same condition. Symbols are placed inside their corresponding bars. No significant differences were found between WT and KO platelets for any condition (Kruskal–Wallis test).

2.7. Absence of Coro1 Does Not Affect cAMP Signaling

Jayachandran et al. have shown that Coro1 interacts with Gαs and modulates the cAMP signaling pathway in neurons and T cells [9,29]. In platelets the cAMP pathway can be triggered by exposure to prostacyclin (PGI2), whose receptor is coupled to heterotrimeric G proteins containing the Gαs subunit, resulting in dampening of the ability to respond to thrombin stimulation. We have shown that in platelets Coro1 is able to immunoprecipitate and colocalize with Gαs [6], prompting us to investigate the functionality of the cAMP pathway in Coro1 deficient platelets. We monitored the activity of the cAMP pathway by detection of vasodilator-stimulated phosphoprotein (VASP) phosphorylation at Ser157. Treatment with a low (5 nM) and a high (100 nM) dose of PGI2 resulted in a dose-dependent increase in the amount of pVASP-S157 in both WT and KO platelets. No statistically significant differences were observed between both genotypes (Figure 7A). We used flow cytometry to quantify the effect of PGI2 on thrombin-stimulated integrin αIIbβ3 activation and granule secretion. Platelets were pretreated with 100 nM PGI2 prior to stimulation with 0.1 U/mL thrombin. As already shown in Figure 3, stimulation with thrombin caused activation of integrin β3 as well as P-selectin and CD63 expression, whereas PGI2 itself did not elicit any response. Treatment with 100 nM PGI2 prior to thrombin stimulation completely abolished those responses both in WT and KO platelets (Figure 7B). Collectively, our results indicate that Coro1 is dispensable for Gαs-dependent modulation of the cAMP pathway in platelets.

2.8. Absence of Coro1 Does Not Impair Hemostasis

To evaluate the influence of *Coro1a* deletion on hemostasis, we examined tail bleeding (Figure 8). Both Coro1 KO and WT animals showed a comparable average bleeding time (1.99 ± 0.19 min in the KO vs. 1.71 ± 0.20 min in the WT). In these experiments, two WT and two KO mice out of 15 per genotype re-bled within one minute of cessation of bleeding.

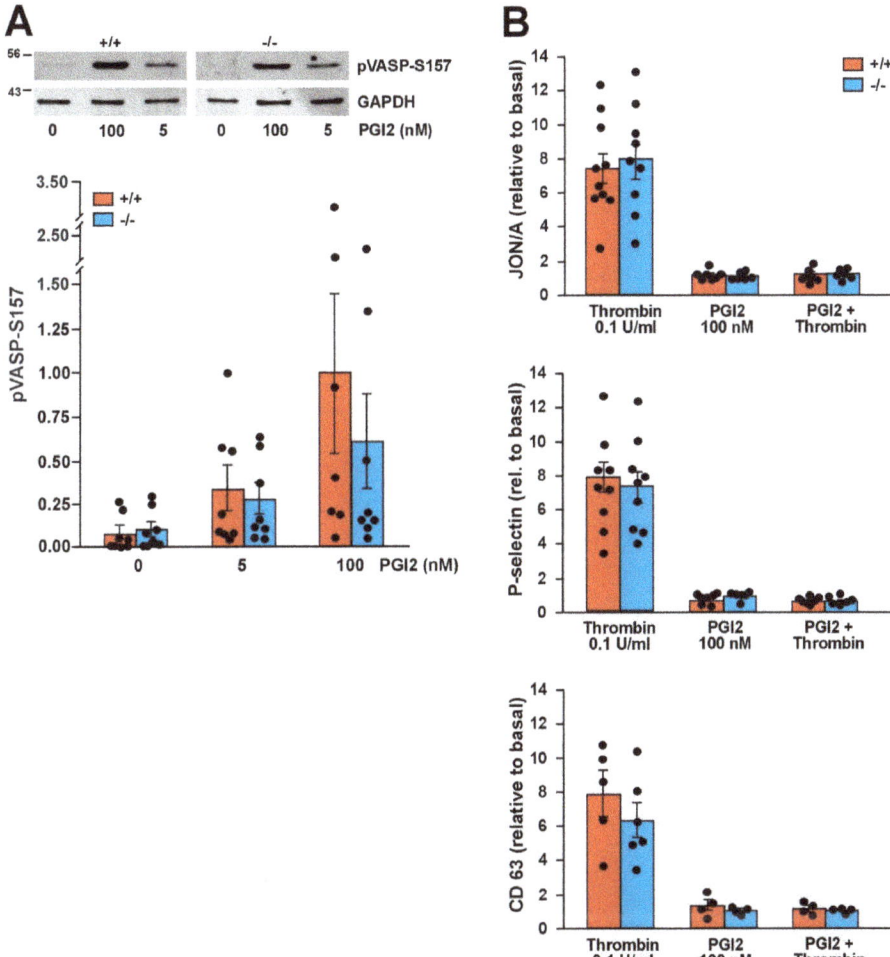

Figure 7. The cAMP pathway is not affected in *Coro1a* deficient platelets. (**A**) Phosphorylation of vasodilator-stimulated phosphoprotein (VASP) upon prostacyclin (PGI2) stimulation for 5 min at the indicated doses. Platelet lysates were resolved by SDS-PAGE, blotted, and probed with specific antibodies for pVASP-Ser157. GAPDH was used for normalization. Representative blots and bar diagrams showing mean ± SEM of 7 independent experiments. Full blots are shown in Supplemental Figure S2; (**B**) Integrin activation, P-selectin exposure, and CD63 exposure upon stimulation with thrombin prior to PGI2. Platelets in PRP were treated with 100 nM PGI2 for 5 min prior to stimulation with 0.1 U/mL thrombin and subsequently analyzed by flow cytometry. This set of experiments was carried out simultaneously with the ones presented in Figure 3. The data (median fluorescence intensity) are expressed relative to basal (unstimulated) platelets. The data represent the mean ± SEM of 4–9 independent experiments. No significant differences were found between WT and KO in any of the assays (Student's *t*-test or Mann–Whitney U-test).

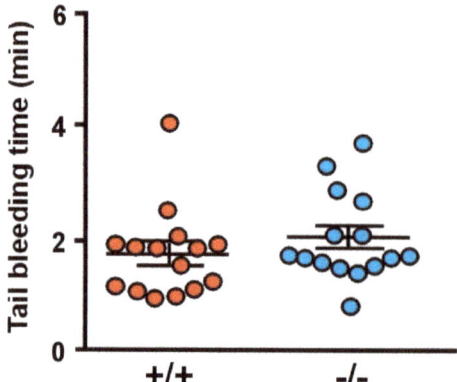

Figure 8. Tail bleeding time. Tests were performed by cutting off 2 mm of the tail tip and immediately placing the tail in PBS at 37 °C. The time until hemostasis was recorded for up to 10 min and re-bleeding monitored for 60 s beyond hemostasis. Data represent mean ± SEM of 15 animals. No significant differences were found between WT and KO (Student's *t*-test).

3. Discussion

The availability of animal models has significantly contributed to elucidate the roles in platelet function of cytoskeleton proteins, which usually cannot be targeted pharmacologically [30]. Here we present a functional characterization of Coro1, an abundant class I coronin, in a KO mouse model. The salient phenotype of Coro1 deficient platelets is the impaired translocation of integrin β2 to the cell surface upon thrombin stimulation, in the absence of any alteration in a range of morphological and functional tests. Our study broadly confirms a recent report by Stocker et al. and explores aspects not covered there [20]. However, we failed to observe some of the mild defects reported by Stocker et al., namely increased relative platelet size and adhesion receptor expression, decreased platelet spreading area upon stimulation with thrombin or collagen, and decreased velocity of aggregation in response to low-dose collagen [20]. These divergent outcomes could be tracked back to methodological differences, size of experimental populations, and statistical analysis. For example, Stocker et al. used an impedance-based method on whole blood to study aggregometry, whereas we used light transmission aggregometry on washed platelets. Impedance-based aggregometry on whole blood captures responses that depend on interactions with leukocytes and red blood cells and is, therefore, closer to the physiological situation, however, it is considered insensitive to low levels of platelet activation. Light transmission aggregometry, by contrast, requires more manipulations but takes platelets at face value, without the influence of variations in hematocrit and cellular content [31,32]. The different methods might have influenced platelet reactivity, causing opposite outcomes, although the differences between WT and KO were always small.

While translocation and activation of integrin αIIbβ3 were not affected in the Coro1 KO platelets, we observed impaired translocation of integrin β2 by deletion of Coro1, suggesting that Coro1 is specifically implicated in the regulation of this integrin in platelets. Integrin β2, a component of lymphocyte function-associated antigen 1 (LFA-1) when associated with integrin αL, is one of the 6 integrins expressed in mouse platelets and the fourth most abundant [25]. β2 integrins are important for polymorphonuclear neutrophil adhesion to the endothelium and subsequent events, like extravasation [33]. Coro1 is critical for these processes because it interacts with the cytoplasmic tail of integrin β2 and regulates the accumulation of activated integrin in focal zones of adherent cells [10]. In platelets, LFA-1 has not been extensively investigated. Platelets from mice deficient in integrin β2 are characterized by a shorter lifespan, reduced adhesion to the endothelium in response to tumor necrosis factor (TNF), and caspase activation [24]. Stocker et al. reported a normal lifespan of Coro1 deficient platelets, suggesting that this coronin is not the only protein responsible for the regulation of integrin β2 [20]. Similarly, we did not observe any defective adhesion and spreading on an ICAM-1 surrogate matrix, suggesting that Coro1 deficient platelets retain sufficient binding capacity through LFA-1 and/or other mechanisms. In line with this observation, Stocker et al. reported unaffected accumulation of neutrophils within arterial thrombi in Coro1 deficient platelets [20]. However, the role of integrin β2 in platelet–leukocyte interaction is difficult to dissect due to concurrent and more prevalent mechanisms mediating those interactions [34] and to the fact that the interactions mediated by LFA-1 and ICAM family molecules are reciprocal: both are present simultaneously in platelets and leukocytes. A rigorous attempt at exploring this aspect would require the generation of a platelet-specific Coro1 knockout model combined with platelet-specific deletion of ICAM-2, the adhesion molecule isoform present in the platelet membrane [25].

Jayachandran et al. have uncovered the role of Coro1 in modulating the cAMP signaling pathway in excitatory neurons, where deficiency of the protein resulted in the loss of excitatory synapses and a range of neurobehavioral disabilities [9]. Coro1 interacts with Gαs in a stimulus-dependent manner, leading to increased cAMP production [9]. Moreover, the association of Coro1 with Gαs is regulated by cyclin-dependent kinase 5 (CDK5)-mediated phosphorylation of Coro1 on two particular threonine residues [35]. Furthermore, Coro1 regulates cAMP signaling in T cells [29] whereas the homolog in *Dictyostelium discoideum* regulates cAMP-dependent initiation of multicellular aggregation [36] and the homolog in the fungus *Magnaporthe oryzae* interacts with a Gαs subunit to regulate cAMP production and pathogenicity [37]. In platelets, Gαs activation and subsequent cAMP production are coupled to binding of PGI2 to its G protein-coupled receptor. Although Coro1 is able to co-immunoprecipitate Gαs in platelets [6], absence of Coro1 does not appear to be detrimental to the production of cAMP, as demonstrated by the ability of Coro1 deficient platelets to phosphorylate VASP and block the effects of thrombin stimulation when exposed to PGI2. The role of Coro1 in cAMP regulation is, however, complex. In T cells depletion of Coro1 results in reduced production of cAMP, however, cAMP levels are increased due to a compensatory decrease in phosphodiesterase 4 (PDE4) levels [29]. Further research would be needed to clarify whether Coro1 regulates the cAMP pathway in platelets and, if so, through which molecular mechanisms. PDE4 is absent [38] but CDK5 is present both in human and mouse platelets [25,39], although the role of the latter in platelets has not been addressed so far. In addition, we have reported the presence of Coro2 and Coro3 in immunocomplexes with Gαs [6], suggesting that these two class I coronins may compensate for the absence of Coro1 for regulation of the cAMP pathway. Functional compensation by other class I coronins might also explain the retained ability of the Arp2/3 complex to accumulate at the cell cortex and enable the formation of lamellipodia and consequently spreading. We have shown that Coro2 and 3 can be found in immunocomplexes with ARPC2 and accumulate at the cell cortex of spread platelets [6].

In summary, we propose that class I coronins display a large extent of functional overlap in platelets. This would explain the absence of a strong phenotype in most platelet functional assays while aspects like integrin β2 translocation reported by us and the formation of F-actin and cofilin dephosphorylation in response to agonists reported by others [20] are specifically or more strongly dependent on Coro1 function. This is not uncommon among components of the actin cytoskeleton, where examples abound [30]. Thus, disruption of the Arp2/3 complex regulators cortactin and its homolog HS1 does not cause any noticeable alteration in platelet function, indicating that their roles might be fulfilled by other proteins [40]. Similarly, disruption of the formin mDia results in no major platelet phenotype, pointing at functional compensation by other formins present in platelets [41]. Future studies toward the elucidation of coronin function in platelets will, therefore, require the generation of mouse models lacking two or three class I coronins in order to arrive at a complete picture of the shared and unique roles of these proteins.

4. Materials and Methods

4.1. Reagents

Primary antibodies against following proteins were used: Coro1 (ab56820 and ab72212), β-actin (ab20272) from Abcam (Cambridge, UK); Coro3 (K6-444 hybridoma supernatant) [42]; β3-integrin (HC93 sc-14009) and Gαs (sc-823) from Santa Cruz Biotechnology (Heidelberg, Germany); phosphor-VASP(Ser157) (#3111) from Cell Signaling Technology (Leiden, The Netherlands); GAPDH (6C5-CB1001) from Calbiochem/Merck (Watford, UK); p34-Arc/ARPC2 (07-227) from Millipore/Merck. Secondary antibodies Alexa Fluor 568-conjugated anti-rat or anti-rabbit immunoglobulins (Molecular Probes, Thermo Fisher Scientific, Altrincham, UK) were used for immunofluorescence. IRDye 680 or IRDye 800 anti-mouse and anti-rabbit immunoglobulins (LI-COR Biosciences, Lincoln, NE, USA) were used for Western blot. Human fibrinogen was from Enzyme Research (Swansea, UK), collagen (Kollagenreagens Horm) was from Takeda (Osaka, Japan), and PGI2 was from Cayman Chemical (Ann Arbor, MI, USA). Phosflow Lyse/Fix Buffer and P-selectin were from BD Biosciences (Oxford, UK). Gly-Phe-Hyp-Gly-Glu-Arg (GFOGER) and collagen-related peptide (CRP) were from Cambridge University (Cambridge, UK). U46619 was from Enzo (Exeter, UK). CK666 was from Tocris Bioscience (Abingdon, UK). Thrombin, ADP and FITC, or TRITC-conjugated phalloidin were from Merck (Dorset, UK). Other reagents were from Merck unless otherwise indicated.

4.2. Experimental Animals

C57Bl/6 mice with a homozygous targeting of the *Coro1a* gene have been previously described [15] and are available from The Jackson Laboratory (JAX stock no. 030203). The animals were kept in the animal facility of the University of Hull using standard conditions. All animal work was performed in accordance with UK Home Office regulations, UK Animals (Scientific Procedures) Act of 1986, under the Home Office project license no. PPL 70/8253 (2 January 2015). Age-matched WT littermates were used as controls in all experiments. Twelve to twenty-week-old animals were used for experiments.

4.3. Mouse Platelet Preparation

Blood was taken by cardiac puncture into acid citrate dextrose (ACD) (29.9 mM trisodium citrate, 113.8 mM glucose, 72.6 mM NaCl, and 2.9 mM citric acid, pH 6.4) or sodium citrate (109 mM tri-sodium citrate pH 7.4) and centrifuged at 100× g for 5 min. The platelet-rich plasma (PRP) was collected in a separate tube, modified Tyrode's buffer was added to the pellet, and the procedure repeated to increase the platelet yield. For washed platelet preparation, the PRP was pelleted at 800× g for 6 min and platelets resuspended in modified Tyrode's buffer and allowed to rest for 30 min at 37 °C prior to experiments.

4.4. Western Blot

Lysates were prepared from washed platelet suspensions by mixing with one volume of 2× Laemmli buffer. Proteins were resolved by SDS-polyacrylamide gel electrophoresis (PAGE) and blotted onto polyvinylidene difluoride (PVDF) membrane. The membrane was incubated with the relevant primary antibody and the corresponding fluorochrome-labeled secondary antibody and visualized and quantified with an LI-COR Odyssey CLx Imaging System (LI-COR Biosciences).

4.5. Flow Cytometry

PRP was prepared in sodium citrate and stimulated with thrombin (in the presence of 10 μM Gly-Pro-Arg-Pro-NH$_2$), CRP, ADP, or U46619 for 20 min at 37 °C in the presence of FITC-conjugated anti-P-selectin (BD Biosciences), PE-conjugated JON/A (Emfret, Würzburg, Germany) and APC/Cy7-conjugated anti-CD63 (Biolegend) antibodies. Platelets were subsequently fixed and analyzed by fluorescence-activated cell sorting (FACS) using an LSRFortessa cell analyzer (BD Biosciences) and FlowJo software.

For receptor expression studies, PRP was incubated with FITC-conjugated antibodies directed against surface membrane glycoproteins GP1b (CD42b), GPVI, integrin α2 (CD49b) (Emfret, Eibelstadt, Germany), integrin αIIb (CD41) (BD Biosciences, Oxford, UK), or PE-conjugated antibodies against integrin β2 (CD18) (Biolegend). Receptor expression was also studied upon stimulation with 0.1 U/mL thrombin for 20 min at 37 °C in the presence of 10 µM Gly-Pro-Arg-Pro-NH$_2$. Platelets were subsequently analyzed by FACS.

4.6. Aggregation, Spreading, and Immunostaining

Platelet aggregation in response to agonists was recorded in washed platelets under constant stirring conditions (1000 rpm) for 7 min at 37 °C using light transmission aggregometry with a CHRONO-LOG 490 aggregometer (CHRONO-LOG, Havertown, PA, USA). Washed platelets in suspension were fixed with an equal volume of ice-cold 4% paraformaldehyde (PFA) and spun at 350× g for 10 min on poly-L-lysine (0.01% in PBS) coated coverslips. Platelets were stained for 1 h at room temperature with the indicated primary antibodies followed by the corresponding secondary antibodies and fluorescently labeled phalloidin diluted in PBG (0.5% bovine serum albumin (BSA), 0.05% fish gelatin in PBS). For adhesion studies, coverslips were coated overnight at 4 °C with fibrinogen, collagen, CRP, or GFOGER in PBS at the concentrations indicated and blocked with 5 mg/mL heat-denatured fatty acid free BSA for 1 h before the experiment. For adhesion on native BSA, coverslips were coated overnight at 4 °C with 5 mg/mL fatty acid free BSA in 0.05 M sodium bicarbonate buffer pH 9 [27]. Washed platelets were allowed to spread for 45 min at 37 °C, fixed with 4% PFA for 10 min, permeabilized with 0.3% Triton X-100 for 5 min, and stained as described above for platelets in suspension. Platelets were imaged by fluorescence microscopy using a Zeiss ApoTome.2 equipped with an AxioCam 506 and Zeiss Plan-Apochromat 63× and 100× NA 1.4 objectives. Platelets were manually counted, and the surface coverage area was analyzed by thresholding using ImageJ.

4.7. Tail Bleeding Assay

Mice were anesthetized with 50 mg/kg ketamine and 1 mg/kg medetomidine. The tail was cut off at 2 mm from the tip and immediately immersed in 37 °C PBS. Bleeding time between the cut and cessation of bleeding for at least one minute was monitored by visual inspection until hemostasis for up to 10 min.

4.8. Statistical Analysis

Experimental data were analyzed by GraphPad Prism v6.0 (La Jolla, CA, USA). Data are presented as mean ± standard error of the mean (SEM) of at least 4 independent experiments. Normality was assessed by the Shapiro–Wilk test. Differences between groups were assessed using the appropriate parametric or nonparametric test and statistical significance was taken at $p \leq 0.05$.

Supplementary Materials: The following are available online at http://www.mdpi.com/1422-0067/21/1/356/s1, Figure S1: Full length blots corresponding to Figure 1A; Figure S2: Full length blots corresponding to Figure 7A. The fourth lane of each genotype corresponds to an experimental condition not included in the article.

Author Contributions: Conceptualization, J.P., K.M.N., and F.R.; formal analysis, D.R.J.R.; funding acquisition, J.P., K.M.N., and F.R.; investigation, D.R.J.R. and J.S.K.; project administration, F.R.; supervision, F.R.; validation, D.R.J.R.; writing—original draft, F.R.; writing—review and editing, J.P. and F.R. All authors have read and agreed to the published version of the manuscript.

Funding: This research was funded by the British Heart Foundation, grant number FS/15/46/31606, the Hull York Medical School (F.R.), and grants from the Swiss National Science Foundation and the Canton of Basel (J.P.). D.R.J.R. was a recipient of a British Heart Foundation scholarship. J.S.K. was a recipient of a University of Hull PhD scholarship.

Acknowledgments: The authors would like to thank Angelika Noegel (University of Cologne, Germany) and Christoph Clemen (University of Bochum, Germany) for kindly providing antibodies, Michael Stiess (Biozentrum, Basel, Switzerland) for support and discussions, and the J. Andrew Grant Fund for Cardiovascular Research for support.

Conflicts of Interest: The authors declare no conflict of interest. The funders had no role in the design of the study; in the collection, analyses, or interpretation of data; in the writing of the manuscript; or in the decision to publish the results.

Abbreviations

ACD	acid citrate dextrose
ADP	adenosine diphosphate
ARPC2	p34-Arc, component of the Arp2/3 complex
BSA	bovine serum albumin
cAMP	cyclic adenosine monophosphate
CDK5	cyclin-dependent kinase 5
CDx	cluster of differentiation x
CRP	collagen related peptide
FACS	fluorescence-activated cell sorting
FITC	fluorescein isothiocyanate
GEFOGER	Gly-Phe-Hyp-Gly-Glu-Arg
ICAM-1	intercellular adhesion molecule 1
KO	knockout
LFA-1	lymphocyte function-associated antigen 1
PAGE	polyacrylamide gel electrophoresis
PBS	phosphate buffered saline
PDE4	phosphodiesterase 4
PFA	paraformaldehyde
PGI2	prostacyclin
PRP	platelet-rich plasma
SEM	standard error of the mean
TNF	tumor necrosis factor
TRITC	tetramethylrhodamine isothiocyanate
VASP	vasodilator-stimulated phosphoprotein
WT	wild type

References

1. Tomaiuolo, M.; Brass, L.F.; Stalker, T.J. Regulation of platelet activation and coagulation and its role in vascular injury and arterial thrombosis. *Interv. Cardiol. Clin.* **2017**, *6*, 1–12. [CrossRef] [PubMed]
2. Hartwig, J.H. The platelet cytoskeleton. In *Platelets*, 3rd ed.; Michelson, A.D., Ed.; Academic Press: London, UK, 2013; pp. 145–168.
3. Smolenski, A. Novel roles of cAMP/cGMP-dependent signaling in platelets. *J. Thromb. Haemost.* **2012**, *10*, 167–176. [CrossRef] [PubMed]
4. Chan, K.T.; Creed, S.J.; Bear, J.E. Unraveling the enigma: Progress towards understanding the coronin family of actin regulators. *Trends Cell Biol.* **2011**, *21*, 481–488. [CrossRef] [PubMed]
5. Pieters, J.; Muller, P.; Jayachandran, R. On guard: Coronin proteins in innate and adaptive immunity. *Nat. Rev. Immunol.* **2013**, *13*, 23765056. [CrossRef] [PubMed]
6. Riley, D.R.J.; Khalil, J.S.; Naseem, K.M.; Rivero, F. Biochemical and immunocytochemical characterization of coronins in platelets. *Platelets* **2019**, *4*, 1–12. [CrossRef]
7. Suzuki, K.K.; Nishihata, J.; Arai, Y.; Honma, N.; Yamamoto, K.; Irimura, T.; Toyoshima, S. Molecular cloning of a novel actin-binding protein, p57, with a WD repeat and a leucine zipper motif. *FEBS Lett.* **1995**, *364*, 283–288. [CrossRef]
8. Ferrari, G.; Langen, H.; Naito, M.; Pieters, J. A coat protein on phagosomes involved in the intracellular survival of mycobacteria. *Cell* **1999**, *97*, 435–447. [CrossRef]

9. Jayachandran, R.; Liu, X.; Bosedasgupta, S.; Müller, P.; Zhang, C.-L.; Moshous, D.; Studer, V.; Schneider, J.; Genoud, C.; Fossoud, C.; et al. Coronin 1 regulates cognition and behavior through modulation of cAMP/protein kinase A signaling. *PLoS Biol.* **2014**, *12*, e1001820. [CrossRef]
10. Pick, R.; Begandt, D.; Stocker, T.J.; Salvermoser, M.; Thome, S.; Böttcher, R.T.; Montanez, E.; Harrison, U.; Forné, I.; Khandoga, A.G.; et al. Coronin 1A, a novel player in integrin biology, controls neutrophil trafficking in innate immunity. *Blood* **2017**, *130*, 847–858. [CrossRef]
11. Ojeda, V.; Castro-Castro, A.; Bustelo, X.R. Coronin1 proteins dictate Rac1 intracellular dynamics and cytoskeletal output. *Mol. Cell. Biol.* **2014**, *34*, 3388–3406. [CrossRef]
12. Combaluzier, B.; Mueller, P.; Massner, J.; Finke, D.; Pieters, J. Coronin 1 is essential for IgM-mediated Ca^{2+} mobilization in B cells but dispensable for the generation of immune responses in vivo. *J. Immunol.* **2009**, *182*, 1954–1961. [CrossRef] [PubMed]
13. Grogan, A.; Reeves, E.; Keep, N.; Wientjes, F.; Totty, N.F.; Burlingame, A.L.; Hsuan, J.J.; Segal, A.W. Cytosolic phox proteins interact with and regulate the assembly of coronin in neutrophils. *J. Cell Sci.* **1997**, *110*, 3071–3081. [PubMed]
14. Moriceau, S.; Kantari, C.; Mocek, J.; Davezac, N.; Gabillet, J.; Guerrera, I.C.; Brouillard, F.; Tondelier, D.; Sermet-Gaudelus, I.; Danel, C.; et al. Coronin-1 is associated with neutrophil survival and is cleaved during apoptosis: Potential implication in neutrophils from cystic fibrosis patients. *J. Immunol.* **2009**, *182*, 7254–7263. [CrossRef] [PubMed]
15. Jayachandran, R.; Sundaramurthy, V.; Combaluzier, B.; Mueller, P.; Korf, H.; Huygen, K.; Miyazaki, T.; Albrecht, I.; Massner, J.; Pieters, J. Survival of mycobacteria in macrophages is mediated by coronin 1-dependent activation of calcineurin. *Cell* **2007**, *130*, 37–50. [CrossRef] [PubMed]
16. Mueller, P.; Massner, J.; Jayachandran, R.; Combaluzier, B.; Albrecht, I.; Gatfield, J.; Blum, C.; Ceredig, R.; Rodewald, H.R.; Rolink, A.G.; et al. Regulation of T cell survival through coronin-1-mediated generation of inositol-1,4,5-trisphosphate and calcium mobilization after T cell receptor triggering. *Nat. Immunol.* **2008**, *9*, 424–431. [CrossRef] [PubMed]
17. Föger, N.; Rangell, L.; Danilenko, D.M.; Chan, A.C. Requirement for coronin 1 in T lymphocyte trafficking and cellular homeostasis. *Science* **2006**, *313*, 839–842. [CrossRef]
18. Shiow, L.R.; Roadcap, D.W.; Paris, K.; Watson, S.R.; Grigorova, I.L.; Lebet, T.; An, J.; Xu, Y.; Jenne, C.N.; Föger, N.; et al. The actin regulator coronin 1A is mutant in a thymic egress-deficient mouse strain and in a patient with severe combined immunodeficiency. *Nat. Immunol.* **2008**, *9*, 1307–1315. [CrossRef]
19. Haraldsson, M.K.; Louis-Dit-Sully, C.A.; Lawson, B.R.; Gascoigne, N.R.J.; Argyrios, N.; Kono, D.H. The lupus-related Lmb3 locus contains a disease-suppressing Coronin-1A gene mutation. *Immunity* **2009**, *28*, 40–51. [CrossRef]
20. Stocker, T.; Pircher, J.; Skenderi, A.; Ehrlich, A.; Eberle, C.; Megens, R.; Petzold, T.; Zhang, Z.; Walzog, B.; Müller-Taubenberger, A.; et al. The actin regulator coronin-1A modulates platelet shape change and consolidates arterial thrombosis. *Thromb. Haemost.* **2018**, *118*, 2098–2111. [CrossRef]
21. Michelson, A.; Barnard, M.; Hechtman, H.; MacGregor, H.; Connolly, R.; Loscalzo, J.; Valeri, C. In vivo tracking of platelets: Circulating degranulated platelets rapidly lose surface P-selectin but continue to circulate and function. *Proc. Natl. Acad. Sci. USA* **1996**, *93*, 11877–11882. [CrossRef]
22. Piguet, P.F.; Vesin, C.; Ryser, J.E.; Senaldi, G.; Grau, G.E.; Tacchini-Cottier, F. An effector role for platelets in systemic and local lipopolysaccharide—Induced toxicity in mice, mediated by a CD11a- and CD54-dependent interaction with endothelium. *Infect. Immun.* **1993**, *61*, 4182–4187. [CrossRef] [PubMed]
23. Guo, J.; Piguet, P.F. Stimulation of thrombocytopoiesis decreases platelet beta2 but not beta1 or beta3 integrins. *Br. J. Haematol.* **1998**, *100*, 712–719. [CrossRef] [PubMed]
24. Piguet, P.F.; Vesin, C.; Rochat, A. Beta2 integrin modulates platelet caspase activation and life span in mice. *Eur. J. Cell Biol.* **2001**, *80*, 171–177. [CrossRef] [PubMed]
25. Zeiler, M.; Moser, M.; Mann, M. Copy number analysis of the murine platelet proteome spanning the complete abundance range. *Mol. Cell. Proteom.* **2014**, *13*, 3435–3445. [CrossRef]
26. Philippeaux, M.M.; Vesin, C.; Tacchini-Cottier, F.; Piguet, P.F. Activated human platelets express β2 integrin. *Eur. J. Haematol.* **2009**, *56*, 130–137. [CrossRef]
27. Zhu, X.; Subbaraman, R.; Sano, H.; Jacobs, B.; Sano, A.; Boetticher, E.; Muñoz, N.M.; Leff, A.R. A surrogate method for assessment of β2-integrin-dependent adhesion of human eosinophils to ICAM-1. *J. Immunol. Methods* **2000**, *240*, 157–164. [CrossRef]

28. Liu, J.; Muñoz, N.M.; Meliton, A.Y.; Zhu, X.; Lambertino, A.T.; Xu, C.; Myo, S.; Myou, S.; Boetticher, E.; Johnson, M.; et al. β2-Integrin adhesion caused by eotaxin but not IL-5 is blocked by PDE-4 inhibition and β2-adrenoceptor activation in human eosinophils. *Pulm. Pharmacol. Ther.* **2004**, *17*, 73–79. [CrossRef]
29. Jayachandran, R.; Gumienny, A.; Bolinger, B.; Ruehl, S.; Lang, M.J.; Fucile, G.; Mazumder, S.; Tchang, V.; Woischnig, A.-K.; Stiess, M.; et al. Disruption of coronin 1 signaling in T cells promotes allograft tolerance while maintaining anti-pathogen immunity. *Immunity* **2019**, *50*, 152–165. [CrossRef]
30. Falet, H. Anatomy of the platelet cytoskeleton. In *Platelets in Thrombotic and Non-Thrombotic Disorders*; Graesele, P., Kleiman, N.S., Lopez, J.A., Page, C.P., Eds.; Springer: Berlin, Germany, 2017; pp. 139–156.
31. Jarvis, G.E. Platelet aggregation: Turbidimetric measurements. *Methods Mol. Biol.* **2004**, *272*, 65–76.
32. Jarvis, G.E. Platelet aggregation in whole blood: Impedance and particle counting methods. *Methods Mol. Biol.* **2004**, *272*, 77–87.
33. Fagerholm, S.C.; Guenther, C.; Asens, M.L.; Savinko, T.; Uotila, L.M. Beta2-Integrins and interacting proteins in leukocyte trafficking, immune supression, and immunodeficiency disease. *Front. Immunol.* **2019**, *10*, 1–10. [CrossRef] [PubMed]
34. Cerletti, C.; de Gaetano, G.; Lorenzet, R. Platelet-leukocyte interactions: Multiple links between inflammation, blood coagulation and vascular risk. *Mediterr. J. Hematol. Infect. Dis.* **2010**, *2*, e2010023. [CrossRef] [PubMed]
35. Liu, X.; Bosedasgupta, S.; Jayachandran, R.; Studer, V.; Rühl, S.; Stiess, M.; Pieters, J. Activation of the cAMP/protein kinase A signalling pathway by coronin 1 is regulated by cyclin-dependent kinase 5 activity. *FEBS Lett.* **2016**, *590*, 279–287. [CrossRef] [PubMed]
36. Vinet, A.F.; Fiedler, T.; Studer, V.; Froquet, R.; Dardel, A.; Cosson, P.; Pieters, J. Initiation of multicellular differentiation in *Dictyostelium discoideum* is regulated by coronin A. *Mol. Biol. Cell* **2014**, *25*, 688–701. [CrossRef] [PubMed]
37. Li, X.; Zhong, K.; Yin, Z.; Hu, J.; Wang, W.; Li, L.; Zhang, H.; Zheng, X.; Wang, P.; Zhang, Z. The seven transmembrane domain protein MoRgs7 functions in surface perception and undergoes coronin MoCrn1-dependent endocytosis in complex with Gα subunit MoMagA to promote cAMP signaling and appressorium formation in *Magnaporthe oryzae*. *PLoS Pathog.* **2019**, *15*, e1007382. [CrossRef] [PubMed]
38. Rondina, M.T.; Weyrich, A.S. Targeting phosphodiesterases in anti-platelet therapy. *Handb. Exp. Pharmacol.* **2012**, *210*, 225–238.
39. Burkhart, J.M.; Vaudel, M.; Gambaryan, S.; Radau, S.; Walter, U.; Martens, L.; Geiger, J.; Sickmann, A.; Zahedi, R.P. The first comprehensive and quantitative analysis of human platelet protein composition allows the comparative analysis of structural and functional pathways. *Blood* **2012**, *120*, e73–e82. [CrossRef]
40. Thomas, S.G.; Poulter, N.S.; Bem, D.; Finney, B.; Machesky, L.M.; Watson, S.P. The actin binding proteins cortactin and HS1 are dispensable for platelet actin nodule and megakaryocyte podosome formation. *Platelets* **2017**, *28*, 372–379. [CrossRef]
41. Zuidscherwoude, M.; Green, H.L.H.; Thomas, S.G. Formin proteins in megakaryocytes and platelets: Regulation of actin and microtubule dynamics. *Platelets* **2019**, *30*, 23–30. [CrossRef]
42. Spoerl, Z.; Stumpf, M.; Noegel, A.A.; Hasse, A. Oligomerization, F-actin interaction, and membrane association of the ubiquitous mammalian coronin 3 are mediated by its carboxyl terminus. *J. Biol. Chem.* **2002**, *277*, 48858–48867. [CrossRef]

© 2020 by the authors. Licensee MDPI, Basel, Switzerland. This article is an open access article distributed under the terms and conditions of the Creative Commons Attribution (CC BY) license (http://creativecommons.org/licenses/by/4.0/).

Article

Arf6 Can Trigger Wave Regulatory Complex-Dependent Actin Assembly Independent of Arno

Vikash Singh, Anthony C. Davidson, Peter J. Hume and Vassilis Koronakis *

Department of Pathology, Tennis Court Road University of Cambridge, Cambridge CB2 1QP, UK; Vs399@cam.ac.uk (V.S.); acd49@cam.ac.uk (A.C.D.); pjh53@cam.ac.uk (P.J.H.)
* Correspondence: vk103@cam.ac.uk; Tel.: +44-01223-33715

Received: 21 February 2020; Accepted: 31 March 2020; Published: 2 April 2020

Abstract: The small GTPase ADP-ribosylation factor 6 (Arf6) anchors at the plasma membrane to orchestrate key functions, such as membrane trafficking and regulating cortical actin cytoskeleton rearrangement. A number of studies have identified key players that interact with Arf6 to regulate actin dynamics in diverse cell processes, yet it is still unknown whether Arf6 can directly signal to the wave regulatory complex to mediate actin assembly. By reconstituting actin dynamics on supported lipid bilayers, we found that Arf6 in co-ordination with Rac1(Ras-related C3 botulinum toxin substrate 1) can directly trigger actin polymerization by recruiting wave regulatory complex components. Interestingly, we demonstrated that Arf6 triggers actin assembly at the membrane directly without recruiting the Arf guanine nucleotide exchange factor (GEF) ARNO (ARF nucleotide-binding site opener), which is able to activate Arf1 to enable WRC-dependent actin assembly. Furthermore, using labelled *E. coli*, we demonstrated that actin assembly by Arf6 also contributes towards efficient phagocytosis in THP-1 macrophages. Taken together, this study reveals a mechanism for Arf6-driven actin polymerization.

Keywords: Arf GTPases; actin cytoskeleton; wave regulatory complex; phagocytosis; macrophages; host–pathogen interplay

1. Introduction

The ADP-ribosylation factor (Arf) protein family is involved in a plethora of cellular functions, including endocytosis, vesicle trafficking, phagocytosis, and cytoskeleton remodeling [1]. The involvement of Arf GTPases in a wide array of cellular functions is partly attributed to their diverse localization within the cell. For instance, Arfs from class I (Arf1 and 3) and class II (Arf4 and 5) are predominantly localized around the Golgi apparatus and hence play a key role in vesicle, lipid, and organelle trafficking [2]. On the other hand, the only class III Arf (Arf6) is found exclusively at the plasma membrane and on incoming endosomes/macropinosomes and plays a vital role in endocytosis, exocytosis, receptor, and endosome recycling to the cell surface and has a clear role in cortical cytoskeleton rearrangement [3,4].

Like other GTPases, Arf GTPases are also under tight spatial and temporal regulation by their guanine nucleotide-exchange factors (GEFs) and GTPase-activating proteins (GAPs), which catalyze GTP binding and hydrolysis, respectively [5]. Arf GTPases can also coordinate with other GTPases, such as Rho and Rab, to further orchestrate precise functions within the cell [6,7].

Among the various tasks performed by Arf GTPases, we are interested in understanding how Arf GTPases regulate the actin cytoskeleton at the plasma membrane. Previously, we established that Rac1 co-ordinates with Arf1 at the plasma membrane to regulate actin rearrangement via the direct recruitment of the wave regulatory complex (WRC) [8] and this drives the formation of lamellipodia in cells [9,10]. The cooperative recruitment and activation of WRC at the membrane is not restricted to Arf1, as the related Arf5, and Arl1, a distant member of the Arf GTPase family, could also achieve similar activity. Furthermore, in contrast to other Arf family members, which directly bind and activate the WRC, Arf6 signaled to WRC indirectly by recruiting the Arf1 GEF ARNO to the plasma membrane [11–13]. Since Arf6 is implicated in numerous actin remodeling processes, we sought to find out whether Arf6 exclusively operates via this ARF1/ARNO pathway, or whether it can also directly activate WRC signaling.

2. Results

2.1. Constitutively Active Arf6 and Rac1 Cooperate in Triggering Actin Assembly In Vitro

In order to examine whether Arf6 could trigger actin assembly independently of ARNO, we looked at the ability of a constitutively active mutant of Arf6, i.e., Arf6Q67L, hereafter "Arf6QL", to drive the formation of actin comets in porcine brain extract, as previously described [12–14]. We used porcine brain extract as we have shown previously that it lacks any detectable members of the ARNO family [12]. For this purpose, Arf6QL and Arf1QL, both alone and in combination with Rac1QL, were anchored to silica beads coated with a phospholipid bilayer composed of equal amounts of phosphatidylinositol and phosphatidylcholine (PC:PI). Control PC:PI lipid bilayers containing only Arf6QL, Arf1QL, or Rac1QL alone did not induce actin comet tail formation (Figure 1A,B). However, bilayers containing a combination of Arf6QL and Rac1QL were able to drive actin motility by forming actin comets similar to those achieved by Arf1QL and Rac1QL (Figure 1C). Inhibition of N-WASP-dependent activity by the addition of purified N-WASP ΔVCA had no effect on Arf6-driven actin assembly (Figure 1D), but actin motility was abolished by the addition of the Rac1 inhibitor EHT 1864, indicating that actin motility is Rac1 dependent (Figure 1D).

Rac1 is known to function upstream of WRC, but as we have previously shown [8], this requires a cooperating Arf GTPase. We therefore next sought to address whether Arf6 can cooperate with Rac1 to recruit WRC to the membrane. The anchored bilayers were incubated in porcine brain extract and the recruited proteins were subsequently analyzed by Coomassie Blue-stained SDS-PAGE, with their recruitment subsequently confirmed using immuno-blotting with the specific antibodies. As previously found [8], Rac1QL alone could not recruit any WRC components (Figure S1). However, as seen in Figure 2A,B, lipid bilayers containing both Arf6QL and Rac1QL recruited WRC components Cyfip, Nap1, WAVE1, and Abi1 similarly to Arf1QL:Rac1QL-containing bilayers. Densitometric quantification further revealed that Arf6QL:Rac1QL lipid bilayers had recruited approximately 50% less Cyfip, Nap1, and Wave than Arf1QL:Rac1QL bilayers (Figure 2C). Arf6 (WT) or a combination of Arf6WT: Rac1QL on PCPI monolayers failed to recruit WRC as shown in Figure S1. These differences in the WRC component recruitment are not due differences in the concentration of the GTPases present on the bilayers (Figure 2D). No recruited Arf1 was detected on theArf6QL:Rac1QL bilayers (using an antibody previously shown to detect porcine Arf1 [13]), suggesting that the WRC recruitment to these beads was achieved independently of Arf1. Taken together, these in vitro results suggest that Arf6 can recruit WRC independently of the ARNO-Arf1 signaling pathway. Lipid bilayers containing Arf6QL alone (control) did not recruit any detectable WRC components, thus this recruitment of WRC by Arf6 requires cooperation with Rac1.

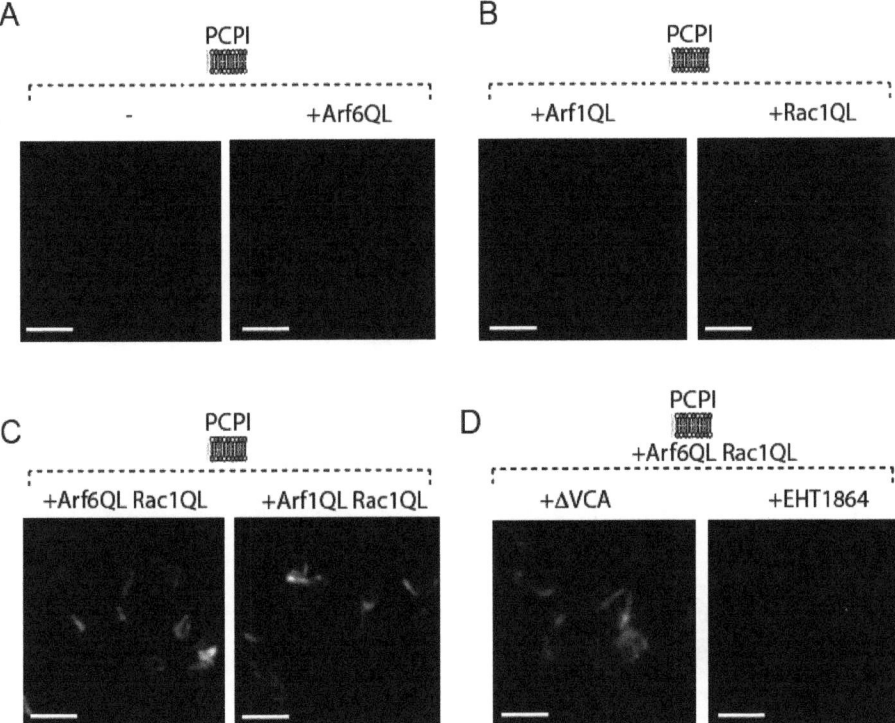

Figure 1. Fluorescence microscopy of rhodamine-actin assembly on the control, Arf6QL (**A**), Arf1QL, or Rac1QL (**B**) anchored PCPI membrane platforms in extract. (**C**) Fluorescence microscopy of rhodamine-actin assembly on Arf6QL:Rac1QL or Arf1QL:Rac1QL anchored PCPI membrane platforms in extract. (**D**) Rhodamine-actin assembly on Arf6QL: Rac1QL anchored PCPI membrane platforms in extract containing an inhibitor of Rac1 (EHT 1864) or N-WASP (n-waspΔvca) (scale bars: 10 μm).

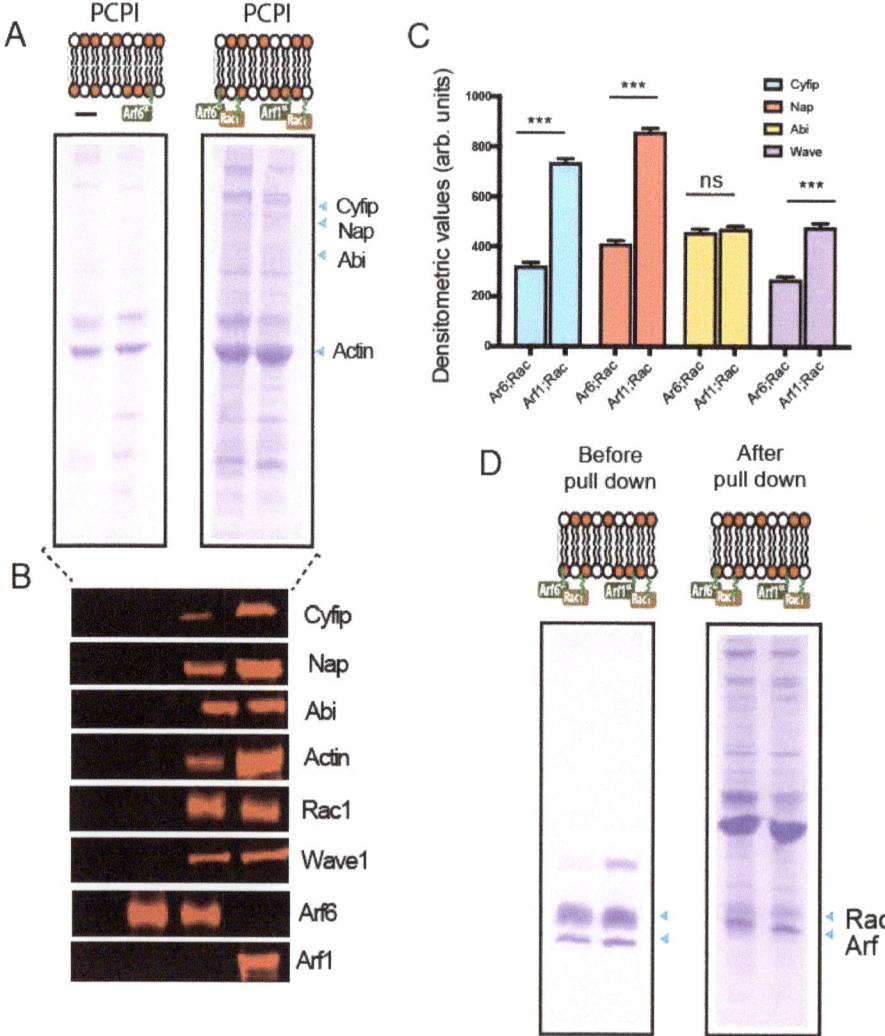

Figure 2. Coomassie blue staining depicting recruited protein from porcine brain extract on control (-), Arf6QL, Arf6QL; Rac1QL and Arf1QL; and Rac1QL anchored PCPI lipid bilayers (**A**). (**B**) Immunoblotting of samples from (**A**) with indicated antibodies. (**C**) Densitometric quantification of the wave-regulatory complex component bands recruited on Arf6QL; Rac1QL and Arf1QL; and Rac1QL anchored PCPI lipid bilayers as described in (**A**). (**D**) Coomassie blue staining depicting recruited proteins before and after incubating the Arf6QL; Rac1QL and Arf1QL; Rac1QL anchored PCPI lipid bilayers in brain extract. *** $p < 0.001$ (one-way ANOVA followed by a post hoc Dunnett comparison) relative to the control.

2.2. Arf6 Regulates Phagocytosis in Differentiated THP-1 Human Macrophages

Various cellular functions are controlled by Arf family proteins, for example, they play a vital role in innate immunity by regulating phagocytosis. Both Arf1 and Arf6 are recognized as regulators of phagocytosis [15,16]. Previously, a number of studies [17–19] have indicated the involvement of both Arf6 and Arf1 in controlling phagocytosis by regulating actin assembly. Furthermore, we have previously shown [20] that Arf1 achieves this by directly mediating WRC actin polymerization, which was crucial for the phagocytosis of pathogenic *E. coli*. Thus, we next examined whether the Arf6- and Rac1-dependent WRC recruitment had any effect on phagocytosis.

For this purpose, we depleted Arf1, Arf6, or control ArpC4 (Arp2/3 component) in PMA differentiated THP-1 human macrophages using siRNA and measured the percentage of internalized phRodo-labelled *E. coli*. As shown in Figure 3A,B, the percentage of internalized bacteria decreased significantly (80%) in control ArpC4-depleted cells. The Arf1-depleted THP-1 macrophages had 45% less internalized bacteria, whereas Arf6 knockdown cells exhibited 65% less internalized bacteria. As Arf6 knockdown had a small, though significant, additional effect on the phagocytosis of labelled *E. coli* compared to Arf1 knockdown, this suggests that the role of Arf6 in promoting actin assembly is not simply to activate Arf1, and that parallel direct recruitment of WRC by Arf6 may also be important. Arf6 knockdown presumably inhibits both of these pathways. Consistent with this, Arf1 and Arf6 double knockdown exhibited phagocytosis comparable to Arf6 knockdown alone (Figure 3B and Figure S2A). The level of phagocytosis in Arf1–Arf6 double knockdown cells was comparable to that in Hem1 knockdown cells (Figure 3D), suggesting that the two Arf pathways are the major activators of WRC. As phagocytosis was reduced further in ArpC4 knockdown cells (Figure 3B), there must also be additional pathways operating to activate Arp2/3 independently of WRC.

It is reported that Arf3 can compensate for the loss of Arf1 [21,22] in regulating actin cytoskeleton dynamics. Hence, we decided to knockdown both Arf3 and Arf1 and then measure the phagocytosis of labelled *E. coli*. As shown in Figure 3C,D, the loss of Arf3 does not appear to further reduce phagocytosis, again suggesting that Arf6 is directly regulating actin assembly without the need for Arf1 or Arf3.

GEFs play an imperative role in activating small GTPases by stimulating GDP dissociation, which allows GTP binding. ARNO is reported to function as a GEF for both Arf1 and to a lesser extent Arf6 [23,24]; importantly, Arf6 activation can signal to Arf1 via recruiting and activating ARNO. Once activated, both Arf1 and Arf6 are capable of recruiting and activating more ARNO, leading to further activation of Arf1. We next examined whether the lower levels of phagocytosis observed in Arf6-depleted cells compared to Arf1 + Arf3-depleted cells were due to there being an Arf6-dependent but Arf1/3 + ARNO-independent mechanism.

For this purpose, the ARNO inhibitor SecinH3 was added to control, Arf1-depleted, or Arf6-depleted THP-1 differentiated cells and the ability to phagocytose labeled bacteria was assessed. As shown in Figure 3C,D, ARNO inhibition in control cells resulted in a decrease in phagocytosis similar to that observed upon Arf1 knockdown (Figure 3B). Furthermore, the depletion of Arf1 in SecinH3-treated cells had no significant additional effect to that of cells treated with SecinH3 alone, suggesting that both Arno and Arf1 drive phagocytosis via the same pathway. However, the depletion of Arf6 in SecinH3-treated cells resulted in a further 18% drop in phagocytosis, suggesting that Arf6 can also regulate phagocytosis independently of ARNO. To confirm whether this significant ARNO-independent activity of Arf6 was WRC dependent, we depleted Hem-1 (Nap1 equivalent in macrophages) in THP-1 cells. The Hem-1 knockdown resulted in levels of phagocytosis similar to those observed in Arf6-depleted cells (Figure 3B,D). Taken together, these results, and those above, indicate that Arf6 can signal to the actin cytoskeleton via the recruitment of WRC to regulate phagocytosis independently of ARNO and Arf1.

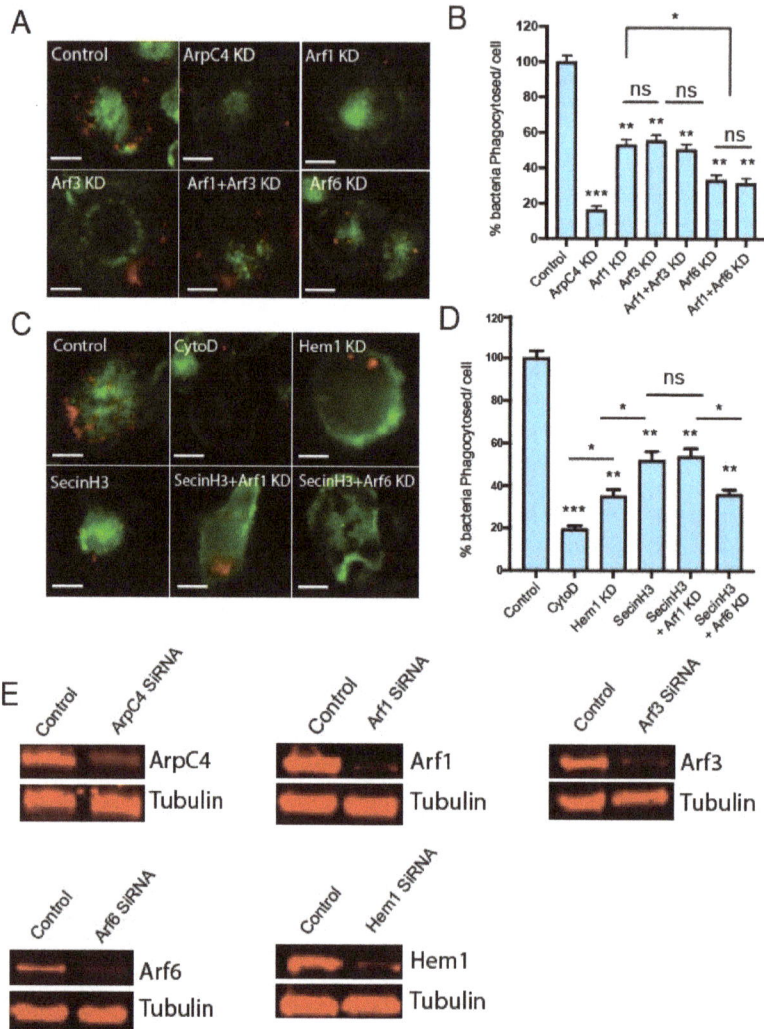

Figure 3. (**A**) Microscopy images depicting phagocytosis of labelled *E. coli* particles by PMA differentiated THP-1 macrophages (control) or upon silencing ArcC4, Arf1, Arf3, Arf1& Arf3, or Arf6 using siRNA. Internalized bacteria are shown in red while actin is stained using phalloidin (green). (**B**) Quantification of the phagocytosed *E. coli* by THP-1 macrophages as described in (**A**). Phagocytosis of phRodo-conjugated *E. coli* particles in THP-1 macrophages (control) or upon using actin inhibitor (CytoD), Arno inhibitor (SecinH3), or silencing of Arf1 or Arf6 in Arno-inhibited cells. Scale bar 10 μm (**D**). Represents quantification of the percentage of internalized bacteria under conditions as described for (**C**). (**E**) Immunoblot confirming the silencing of the mentioned proteins using SiRNA. Each bar represents the average of results from 3 separate experiments, and error bars represent SD, *** $p < 0.001$; ** $p < 0.01$; * $p < 0.05$; ns, not significant (one-way ANOVA followed by a post hoc Dunnett comparison) relative to the equivalent strain on WT THP-1 control cells. Lines indicate significance between pairs of conditions determined by Student's *t* test.

2.3. Arf6-Mediated Internalization of Salmonella is ARNO Dependent

To further confirm this, we examined *Salmonella* invasion in WT Hap, ΔArf6, and ΔNap1 Hap cells. *Salmonella* is a Gram-negative pathogen that uses its type-3 secretion system to force its entry into non-phagocytic cells by generating membrane ruffles that lead to macropinosome formation [25]. In order to generate these ruffles, *Salmonella* exploits WRC-mediated actin assembly. *Salmonella* activates Arf6 using host GEFs [12] and also generates $PI(3,4,5)P_3$ in the plasma membrane. This leads to the activation of ARNO, which results in recruitment of Arf1 at the plasma membrane. Activated Arf1 and Rac1 can then trigger WRC-dependent membrane ruffles, leading to the internalization of *Salmonella* into the host cell.

We endeavored to use this system to examine whether the ARNO-independent ability of Arf6 to drive WRC-mediated actin polymerization also contributes to *Salmonella*'s invasion of non-phagocytic cells. Previously, all the work demonstrating the significance of Arf6 [12,13] in *Salmonella* invasion was based on siRNA-mediated silencing and drug-mediated inhibition.

To more efficiently assess the role of Arf6, here, we used WT, ΔArf6, and ΔNap1 Hap1 cells, and performed a gentamycin protection assay measuring invasion at different times post infection (see methods). As can be seen in Figure 4A, in WT cells, the number of intracellular bacteria continued to increase until 60 min post-invasion. In both ΔArf6 and ΔNap1 cells, at all time points, there were significantly less internalized bacteria. Importantly, the inhibition of ARNO in WT cells impeded ī invasion to a similar level as that observed for Δ Arf6 cells. Subsequently, inhibiting ARNO (with secinH3) or Rac1 (using EHT 1864) did not further impede *Salmonella* invasion in ΔArf6 or ΔNap1 cells (Figure 4B), suggesting that in *Salmonella* entry, Arf6 acts exclusively via ARNO, signaling to WRC by recruiting ARf1 and that the Arf1/ARNO-independent pathway described above has a very small role in *Salmonella*'s invasion.

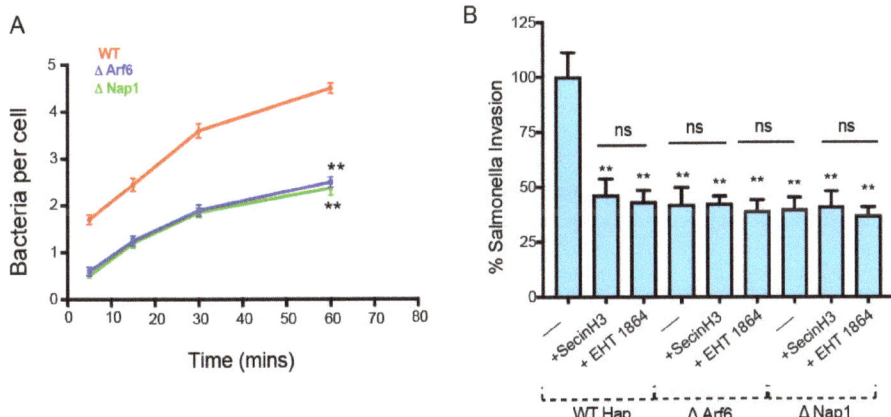

Figure 4. (**A**) Time course of *Salmonella* invasion in WT Haps, ΔArf6 (blue), or ΔNap1 (green) Hap cells with *Salmonella* bacteria carrying pM975 that express GFP inside pathogen-containing vacuoles. (**B**) Salmonella invasion in WT Haps, ΔArf6 Hap cells, or ΔNap1 Hap cells treated with Arno inhibitor (SecinH3) or Rac1 inhibitor (EHT 1864) after infecting *Salmonella* for 15 min. Each bar represents the average of the results from 3 separate experiments, and error bars represent SD. ** $p < 0.01$ (one-way ANOVA followed by a post hoc Dunnett comparison) relative to the control.

3. Discussion

Arf6 is the most predominant member of the Arf GTPase family that is extensively present at the plasma membrane. The role of Arf6 has long been established in regulating the actin cytoskeleton at the plasma membrane [26]. However, the molecular mechanism of how Arf6 directs actin polymerization is less well understood, majorly due to its diverse activities. Previously, our work has established that activation of Arf6, via receptor signaling, such as during EGF stimulation or indirectly by lipid modification by *Salmonella*, brings about actin polymerization at the leading edge of the cell by activating the Arf1 GEF ARNO. ARNO, upon activation, recruits Arf1 to the membrane, which can then co-ordinate with Rac1 to mediate actin polymerization via recruitment and activation of the WRC.

Here, by reconstituting actin assembly on lipid bilayers, we uncovered a potential direct role of Arf6 and demonstrated that it can assemble actin via the WRC, independently of ARNO. However, Arf6 has a poorer ability to drive WRC recruitment when compared to Arf1, as evident from the densitometric quantification of the recruited WRC components (Figure 2C). Hence, it is likely that Arf6 drives low levels of WRC activation, perhaps imitating the conditions in a resting unstimulated cell that would exhibit low-level actin remodeling for generic processes. When the cell is under an external stimulus, such as EGF, which results in acute PIP3 generation or during *Salmonella* entry [12,13], it switches to a more specific, stronger, short, and efficient means to bring about actin remodeling governed by Arf1 (Figure 5).

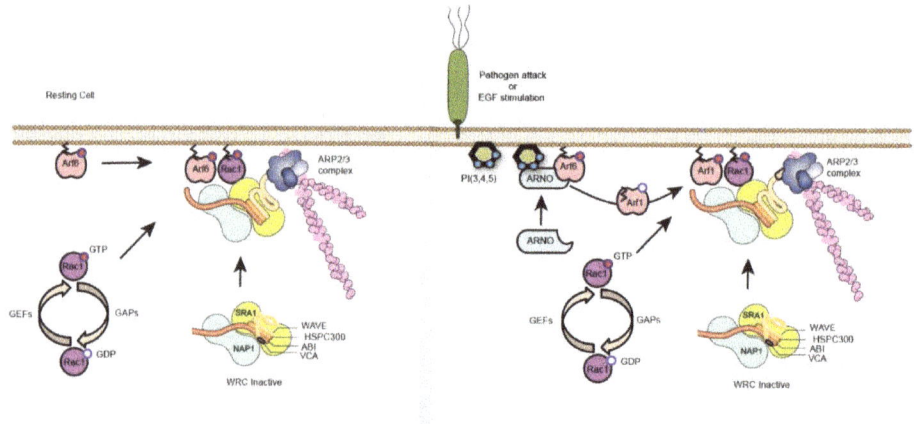

Figure 5. Schematic representation describing signaling to actin cytoskeleton by Arf6 GTPase. When the cell is in a resting state, Arf6 can mediate actin cytoskeleton rearrangement (via the wave regulatory complex) in co-operation with Rac1 to facilitate generic processes. This actin assembly is independent of ARNO and Arf1 interaction. On the contrary, when the cell is under an external stimulus, such as EGF, which results in acute PIP3 generation or during *Salmonella* entry, it switches to a more specific, stronger, short, and efficient means to bring about actin remodeling governed by ARNO/Arf1 recruitment, as described previously [12,13].

Both Arf1 and Arf6 are known for their role in phagocytosis [27,28]. Our results further illustrate that Arf6 by itself can contribute to phagocytosis by triggering WRC-mediated actin assembly without the involvement of Arf1/ARNO. Our results further provide a plausible explanation for the unexplained activation and localization of Arf6 at the tips of the pseudopods as previously reported by numerous studies [27–29]. Furthermore, these results are consistent with the reported role for Arf6 in Rac1 activation and lamellipodia formation [11,29].

Many pathogens have developed strategies to target Arf6 in order to facilitate their internationalization [1]. Whilst Arf6 can drive WRC activation independently of Arf1/ARNO as observed in vitro, the large reorganizations driven by *Salmonella* may well make use of the Arf6→ARNO→Arf1 pathway, which itself is amplified by positive feedback (as Arf1 can then recruit more ARNO), leading to a huge increase in WRC recruitment and activation required for this process [12,13]. The relatively low concentrated and scattered Arf6 in the cell alone is not sufficient, but the combination of Arf6 and PIP3 generation recruits ARNO, which activates Arf1, which in turn can recruit and activate more Arf1 at the plasma membrane [13]. This in turn results in far more efficient WRC recruitment and actin polymerization than that achieved by Arf6 alone. In a resting cell, Arf6 alone should not activate WRC unnecessarily, the requirement for a second downstream protein (Arf1 is required for acute short-lived but dramatic levels of actin turnover, driven by external stimuli, such as the enormous membrane ruffles generated by *Salmonella* during its invasion or when the cells are stimulated with epidermal growth factor (EGF)).

Our previous study [12] did not identify a direct role for Arf6 in actin remodeling, but it is worth noting that Arf6 loaded with the non-hydrolysable GTP analogue GTPγS was used to mimic active Arf6. In our original study, the loading of Arf6 with GTPγS was not quantified and thus may have been incomplete. In addition, the usage of GTPγS may not accurately recreate how Arf6 functions in a cell. The use of non-hydrolysable GTP analogues is reported to affect the activation state of other cellular proteins [30] and also GTPγS may not truly mimic the GTP-bound conformation of GTPase [31]. However, with the use of the constitutively active Arf6QL, we were able to uncover a potential direct role for Arf6 in regulating the actin cytoskeleton via the scar wave complex.

Unlike other Arfs, it seemed peculiar that Arf6 uniquely could not drive WRC activation; however, with better tools, we have now unraveled that Arf6 is directly capable of triggering actin polymerization independent of ARNO. Consistently, this ARNO-independent pathway of remodeling actin by Arf6 is distinct and may be utilized by the cell for more generic processes. Nonetheless, this study adds a new aspect of WRC control by defining how the Arf6 network further provides specificity in the regulation of WRC and highlights the elaborate spatiotemporal small GTPase control mechanisms that underlie actin polymerization specifically at the membrane.

4. Materials and Methods

4.1. Bacterial Strains

Salmonella SL1344 (gift from Jean Guard-Petter, Department of Agriculture, Athens, GA).

4.2. Antibodies

The following antibodies were purchased from Abcam (Rac1, ab33186; Arf6, ab81650; Arf1, ab58578; ArpC4, ab and tubulin, ab7291),Sigma (Abi1, A5106; actin, A2066; Cyfip, P0092; and Nap1, N3788) or were raised against recombinant peptides in rabbits by Diagnostics Scotland (WAVE1; amino acids 180–241).

4.3. Plasmids

The following plasmids were generated by Invitrogen Gateway methodology: pET20b-Arf6 encoding the Arf family *N*-myristoyltransferase site as previously described for Arf1, pET20b-Arf1, pET15b-Rac1 [8]. pM975-GFP from Wolf-Dietrich Hardt (EidgenössicheTechnische Hochschule, Zurich). GST- and His-tagged proteins were expressed in *Escherichia coli* Rosetta (Novagen, Merckmillipore, UK) at 18 °C overnight before affinity purification [12].

4.4. Cell Culture and Transfection

The human monocyte-like cell line THP1s (kind gift from Prof. Gordon Dougan) were cultured (37 °C 5% CO_2) in RPMI-1640 supplemented with 10% heat-inactivated fetal calf serum (FCS), 200mg/mL–1 streptomycin, and 100 U mL–1penicillin. THP1s were differentiated into mature macrophage-like cells by stimulation with 100 ng/mL Phorbol 12-myristate 13-acae-tate (PMA) for 2 days and then cultured for an additional day without PMA before phagocytosis assays.

For RNAi, siRNA from Qiagen against Arf1 (Hs_Arf1_1 sequence ACGTGGAAACCGTGGA GTACA, Hs_ARF1_11 sequence AGGGAAGACCACGATCCTCTA), Arf3 (Hs_ARF3_3 sequence CAGGGCTGACTGGGTATTCTA, Hs_ARF3_5 sequence CACCTATATGACCAATCCCTA); ARF6 (Hs_ARF6_5 sequence CAACGTGGAGACGGTGACTTA, Hs_ARF6_7 sequence AAGACCAGTATAG

TAAACTTA); ArpC4 (Hs_ARPC4_1 sequence CTGATAGGACCTTGATATATA, Hs_ARPC4_6 sequence CAGCATTAAAGCTGGCGCTTA); Hem1 (Hs_HEM1_1 sequence CAGGCATATACTAGTGTCTCA, Hs_HEM1_2 sequence TTCACTGAGATTATTCCTATA), or All Stars negative control siRNA (Qiagen) were combined for each individual gene (unless otherwise stated), and were introduced into differentiated THP1 cells with Oligofectamine transfection reagent (Invitrogen) according to the manufacturer's instructions. The transfection mixture was replaced after 24 h with complete growth medium and cells cultured for an additional day before phagocytosis assay.

WT Hap1 (C631) and verified-knockout ΔArf6 (HZGHC003403c006), ΔNap1 (HZGHC003401c004) cell lines were purchased from Horizon Discovery. Hap1 cells were maintained in Iscove's modified Dulbecco's medium (IMDM) supplemented with 10% FBS and 100 U/mL penicillin-streptomycin.

4.5. Salmonella Invasion of Non-Phagocytic Host Cells

Wild-type *Salmonella enterica* serovar Typhimurium SL1344 were used to assay the invasion into non-phagocytic cells as previously described [12,13]. *Salmonella* encoding pM975 that expresses GFP via the SPI2 promoter when bacteria are within *Salmonella*-containing vacuoles (SCVs) [32] were used to infect WT HAP, Arf6, and Nap1 knockout cells (15 min unless otherwise stated), and then the number of fluorescent bacteria were counted per cell using fluorescence microscopy. When appropriate, WT HAP cells were pre-treated with the following small molecular inhibitors for 30 min prior to *Salmonella* infection, 10 μM EHT 1864 (Rac1), 5 μM CytoD (actin), and 25 μM SecinH3 (ARNO).

4.6. Phagocytosis Assay

Differentiated Human THP-1 cells were incubated at 37 °C with pHrodo *E. coli* bioparticles (ThermoFisher Scientific, P35361) for 60 min as per the manufacturer's instructions, followed by fixing the cell using 4% paraformaldehyde (PFA) and actin was stained using AlexaFlour-488 phalloidin. Wherever indicated, cytochalasin (CytoD) or SecinH3 were added to the THP-1 cells 30 min prior to incubating the cells with pHrodo *E. coli* bioparticles. The amount of phagocytosis was assessed by counting the internalized bacteria (red) in a minimum of 50 cells per condition.

4.7. Actin-Based Motility and In Vitro Pull Downs

A 60-μL motility-mix (extract) was prepared on ice in the following order: 40 μL brain extract, 3 μL 20× energy mix (300 mM creatine phosphate, 40 mM MgCl2, 40 mM ATP), 3 μL G-actin/rhodamine actin, 6 μL 10× salt buffer (600 mM KCL, 200 mM 3-phosphoglycerate), 6 μL 50 mM BAPTA (Merck) and 1 μL 300 mM DTT (Merck), and, when appropriate, 1 μL 30 mM GTPγS (Roche). Actin-dependent motility assays were initiated by adding 0.1 vol phospholipids-coated beads to 10 μL motility mix, then 1μL was applied to a microscope slide and sealed under a glass coverslip with Vaseline:lanolin:paraffin (1:1:1), before viewing immediately under a fluorescence microscope (Leica DM IRBE) at Room Temperature. Digital images were captured (CCD camera, Hamamatsu) and analyzed (Volocity, Improvision), then figures assembled using Adobe Photoshop and Illustrator CS3.

The preparation of porcine brain extracts was as previously described [8,14]. Briefly, 40 fresh porcine brains were homogenized by 3 × 30 sec bursts of a Waring blender at 4 °C in an equal volume of extraction buffer (20 mM Hepes pH 7.4, 100 mM KCl 5 mM MgCl2, 0.5 mM EDTA, 1 mM EGTA with 0.5 mM ATP, 0.1 mM GTP, 1 mM DTT) supplemented with 10 μg/mL leupeptin, 10 μg/mL pepstatin, 10 μg/mL chymostatin, and Complete EDTA-free protease inhibitors (Roche). Homogenate was centrifuged (8000× *g*, 30 min, 4 °C) and the supernatant filtered through cheesecloth. Filtrate was clarified (12,000× *g*, 4 °C, 40 min), concentrated five-fold, and aliquots stored at −70 °C. Prior to use, thawed brain extract was clarified (100,000× *g*, 15 min).

For pull-down experiments, silica microspheres were coated with a bilayer composed of equal concentrations of phosphatidyl choline and phosphatidyl inositol (PC:PI). Indicated proteins (Arf6, Arf1, and Rac1) were anchored to these bilayers by incubating 15 μL of lipid-coated microspheres in 500 μL of HKS buffer containing approximately 20 μM of the protein to be anchored at room temperature for 1 h. The PC:PI bilayers were then washed by repeated (5×) low-speed centrifugation (1000× g) followed by resuspension in HKS buffer supplemented with 1 mM $MgCl_2$ (HKSM). Micropsheres were finally resuspended in 15 µL of HKSM, and incubated with clarified porcine brain extract for 15min at room temperature. The bilayers were then washed 3 times with HKSM (as above), before the final resuspension in SDS-Urea, and recruited proteins were analyzed by SDS-PAGE, and where indicated by Western blotting.

4.8. Immunoblotting and Densitometric Quantification

Briefly, samples were run on a 4–12% Bis-Tris Protein Gel and then transferred to a nitrocellulose membrane using the iBlot2 (ThermoScientific). The membrane was then blocked using the Odyssey blocking buffer for 1 h and incubated overnight with primary antibody at 4 °C with rotation, washed with PBST, incubated with secondary antibody, washed with PBST, and imaged. All images were obtained using a LiCOR Odyssey imager and quantified with LiCOR software. For quantification, an area was drawn around the band of interest and the respective fluorescence was recorded and normalized to the fluorescence of the corresponding control band. Secondary antibodies were used at 1:5000 dilution and purchased from LiCOR Biosciences. The specific antibodies used were as follows: IRDye® 800CW Goat anti Mouse IgG (925-32210), IRDye® 680LT Goat anti Mouse IgG (925-68020), IRDye® 800CW Goat anti Rabbit IgG (925-32211), and IRDye® 680LT Goat anti Rabbit IgG (925-68021).

5. Conclusions

We have previously reported that Arf6 can promote actin assembly by triggering recruitment and activation of ARNO, which in turn activates Arf1 to cooperate with Rac1 in activating WRC. Here, we have shown that in addition to this indirect pathway, Arf6 is able to directly recruit and activate WRC. This direct activity also requires cooperating Rac1, but is independent of the ARNO/Arf1 pathway. Both Arf6 pathways operate in THP-1 macrophages to drive the actin rearrangements necessary for phagocytosis. This work thus represents a new aspect of WRC control by a complex network of cooperating GTPases.

Supplementary Materials: Supplementary materials can be found at http://www.mdpi.com/1422-0067/21/7/2457/s1. Figure S1. Coomassie blue staining depicting recruited protein from porcine brain extract on control (-) and Rac1QL anchored PCPI lipid bilayers(A). (B) Immunoblotting of samples from (A) with indicated antibodies. (C) Coomassie blue staining depicting recruited protein from porcine brain extract on control Arf6WT (alone), and a combination of Arf6WT; Rac1QL anchored PCPI lipid bilayers. (D) Immunoblotting of samples from (C) depicting recruited proteins as indicated antibodies. Figure S2. (A) Microscopy images depicting phagocytosis of labelled E. coli particles by PMA differentiated THP-1 macrophages (control) or upon silencing Arf1 + Arf6 using siRNA. Internalized bacteria are shown in red while actin is stained using phalloidin (green). (B) Immunoblot confirming the silencing of the mentioned proteins using SiRNA. (C) Summarized Densitometric quantification depicting the efficiency of the silenced proteins as determined and normalized to control cell.

Author Contributions: Conceptualization, V.S., A.C.D.; P.J.H.; methodology, V.S., A.C.D., P.J.H.; investigation, V.S.; resources, V.S.; A.C.D.; data curation, V.S.; A.C.D. and P.J.H.; writing—original draft preparation, V.S.; writing—review and editing, V.S.; A.C.D. and P.J.H.; supervision, P.J.H. and V.K.; All authors have read and agreed to the published version of the manuscript.

Funding: This work was funded by a Wellcome Trust Senior Investigator Award (101828/Z/13/Z), by the Medical Research Council (grant MR/L008122/1), and by the Cambridge Isaac Newton Trust.

Acknowledgments: We would like to thank Rachael Stone for helping with the Arf6QL protein purification.

Conflicts of Interest: The authors declare no conflict of interest.

References

1. Van Acker, T.; Tavernier, J.; Peelman, F. The Small GTPase Arf6: An Overview of Its Mechanisms of Action and of Its Role in Host-Pathogen Interactions and Innate Immunity. *Int. J. Mol. Sci.* **2019**, *20*, 2209. [CrossRef]
2. D'Souza-Schorey, C.; Chavrier, P. ARF proteins: Roles in membrane traffic and beyond. *Nat. Rev. Mol. Cell Biol.* **2006**, *7*, 347–358. [CrossRef] [PubMed]

3. Donaldson, J.G. Multiple Roles for Arf6: Sorting, Structuring, and Signaling at the Plasma Membrane. *J. Biol. Chem.* **2003**, *278*, 41573–41576. [CrossRef] [PubMed]
4. Gaschet, J.; Hsu, V.W. Distribution of ARF6 between membrane and cytosol is regulated by its GTPase cycle. *J. Biol. Chem.* **1999**, *274*, 20040–20045. [CrossRef] [PubMed]
5. Goldberg, J. Structural basis for activation of ARF GTPase: Mechanisms of guanine nucleotide exchange and GTP-myristoyl switching. *Cell* **1998**, *95*, 237–248. [CrossRef]
6. Singh, V.; Davidson, A.C.; Hume, P.J.; Humphreys, D.; Koronakis, V. Arf GTPase interplay with Rho GTPases in regulation of the actin cytoskeleton. *Small GTPases* **2019**, *10*, 411–418. [CrossRef]
7. Shi, A.; Grant, B.D. Interactions between Rab and Arf GTPases regulate endosomal phosphatidylinositol-4,5-bisphosphate during endocytic recycling. *Small GTPases* **2013**, *4*, 106–109. [CrossRef]
8. Koronakis, V.; Hume, P.J.; Humphreys, D.; Liu, T.; Horning, O.; Jensen, O.N.; McGhie, E.J. WAVE regulatory complex activation by cooperating GTPases Arf and Rac1. *Proc. Natl. Acad. Sci. USA* **2011**, *108*, 14449–14454. [CrossRef]
9. Myers, K.R.; Casanova, J.E. Regulation of actin cytoskeleton dynamics by Arf-family GTPases. *Trends Cell Biol.* **2008**, *18*, 184–192. [CrossRef]
10. Humphreys, D.; Liu, T.; Davidson, A.C.; Hume, P.J.; Koronakis, V. The Drosophila Arf1 homologue Arf79F is essential for lamellipodium formation. *J. Cell Sci.* **2012**, *125 (Pt 23)*, 5630–5635. [CrossRef]
11. Boshans, R.L.; Szanto, S.; van Aelst, L.; D'Souza-Schorey, C. ADP-ribosylation factor 6 regulates actin cytoskeleton remodeling in coordination with Rac1 and RhoA. *Mol. Cell. Biol.* **2000**, *20*, 3685–3694. [CrossRef] [PubMed]
12. Humphreys, D.; Davidson, A.C.; Hume, P.J.; Makin, L.E.; Koronakis, V. Arf6 coordinates actin assembly through the WAVE complex, a mechanism usurped by *Salmonella* to invade host cells. *Proc. Natl. Acad. Sci. USA* **2013**, *110*, 16880–16885. [CrossRef] [PubMed]
13. Humphreys, D.; Davidson, A.; Hume, P.J.; Koronakis, V. Salmonella virulence effector SopE and Host GEF ARNO cooperate to recruit and activate WAVE to trigger bacterial invasion. *Cell Host Microbe* **2012**, *11*, 129–139. [CrossRef]
14. Hume, P.J.; Humphreys, D.; Koronakis, V. WAVE regulatory complex activation. *Methods Enzymol.* **2014**, *540*, 363–379. [CrossRef] [PubMed]
15. Uchida, H.; Kondo, A.; Yoshimura, Y.; Mazaki, Y.; Sabe, H. PAG3/Papalpha/KIAA0400, a GTPase-activating protein for ADP-ribosylation factor (ARF), regulates ARF6 in Fcγ receptor-mediated phagocytosis of macrophages. *J. Exp. Med.* **2001**, *193*, 955–966. [CrossRef] [PubMed]
16. Someya, A.; Moss, J.; Nagaoka, I. The guanine nucleotide exchange protein for ADP-ribosylation factor 6, ARF-GEP100/BRAG2, regulates phagocytosis of monocytic phagocytes in an ARF6-dependent process. *J. Biol. Chem.* **2010**, *285*, 30698–30707. [CrossRef] [PubMed]
17. Rougerie, P.; Miskolci, V.; Cox, D. Generation of membrane structures during phagocytosis and chemotaxis of macrophages: Role and regulation of the actin cytoskeleton. *Immunol. Rev.* **2013**, *256*, 222–239. [CrossRef]
18. Zhang, Q.; Cox, D.; Tseng, C.C.; Donaldson, J.G.; Greenberg, S. A requirement for ARF6 in Fcγ receptor-mediated phagocytosis in macrophages. *J. Biol. Chem.* **1998**, *273*, 19977–19981. [CrossRef]
19. Stuart, L.M.; Boulais, J.; Charriere, G.M.; Hennessy, E.J.; Brunet, S.; Jutras, I.; Goyette, G.; Rondeau, C.; Letarte, S.; Huang, H.; et al. A systems biology analysis of the Drosophila phagosome. *Nature* **2007**, *445*, 95–101. [CrossRef]
20. Humphreys, D.; Singh, V.; Koronakis, V. Inhibition of WAVE Regulatory Complex Activation by a Bacterial Virulence Effector Counteracts Pathogen Phagocytosis. *Cell Rep.* **2016**, *17*, 697–707. [CrossRef]
21. Volpicelli-Daley, L.A.; Li, Y.; Zhang, C.J.; Kahn, R.A. Isoform-selective effects of the depletion of ADP-ribosylation factors 1-5 on membrane traffic. *Mol. Biol. Cell* **2005**, *16*, 4495–4508. [CrossRef] [PubMed]
22. Kondo, Y.; Hanai, A.; Nakai, W.; Katoh, Y.; Nakayama, K.; Shin, H.W. Arf1 and Arf3 are required for the integrity of recycling endosomes and the recycling pathway. *Cell Struct. Funct.* **2012**, *37*, 141–154. [CrossRef] [PubMed]
23. Santy, L.C.; Ravichandran, K.S.; Casanova, J.E. The DOCK180/Elmo complex couples ARNO-mediated Arf6 activation to the downstream activation of Rac1. *Curr. Biol.* **2005**, *15*, 1749–1754. [CrossRef]

24. Garza-Mayers, A.C.; Miller, K.A.; Russo, B.C.; Nagda, D.V.; Goldberg, M.B. Shigella flexneri regulation of ARF6 activation during bacterial entry via an IpgD-mediated positive feedback loop. *MBio* **2015**, *6*, e02584. [CrossRef] [PubMed]
25. Hume, P.J.; Singh, V.; Davidson, A.C.; Koronakis, V. Swiss Army Pathogen: The *Salmonella* Entry Toolkit. *Front. Cell Infect. Microbiol.* **2017**, *7*, 348. [CrossRef] [PubMed]
26. Schafer, D.A.; D'Souza-Schorey, C.; Cooper, J.A. Actin assembly at membranes controlled by ARF6. *Traffic* **2000**, *1*, 892–903. [CrossRef]
27. Beemiller, P.; Hoppe, A.D.; Swanson, J.A. A phosphatidylinositol-3-kinase-dependent signal transition regulates ARF1 and ARF6 during Fcγ receptor-mediated phagocytosis. *PLoS Biol.* **2006**, *4*, e162. [CrossRef]
28. Egami, Y.; Fujii, M.; Kawai, K.; Ishikawa, Y.; Fukuda, M.; Araki, N. Activation-inactivation cycling of Rab35 and ARF6 is required for phagocytosis of zymosan in RAW264 macrophages. *J. Immunol. Res.* **2015**, *2015*, 429439. [CrossRef]
29. Radhakrishna, H.; Alawar, O.S.; Khachikian, Z.; Donaldson, J.G. ARF6 requirement for Rac ruffling suggests a role for membrane trafficking in cortical actin rearrangements. *J. Cell Sci.* **1999**, *112*, 855–866.
30. Shumilina, E.V.; khaitlina, S.Y.; Morachevskaya, E.A.; Negulyaev, Y.A. Non-hydrolyzable analog of GTP induces activity of Na+ channels via disassembly of cortical actin cytoskeleton. *FEBS Lett.* **2003**, *547*, 27–31. [CrossRef]
31. Wiegandt, D.; Vieweg, S.; Hofmann, F.; Koch, D.; Li, F.; Wu, Y.; Itzen, A.; Muller, M.P.; Goody, R.S. Locking GTPases covalently in their functional states. *Nat. Commun.* **2015**, *6*, 7773. [CrossRef] [PubMed]
32. Schlumberger, M.C.; Kappeli, R.; Wetter, M.; Muller, A.; Misselwitz, B.; Dilling, S.; Kremer, M.; Hardt, W. Two newly identified SipA domains (F1, F2) steer effector protein localization and contribute to Salmonella host cell manipulation. *Mol. Microbiol.* **2007**, *65*, 741–760. [CrossRef] [PubMed]

© 2020 by the authors. Licensee MDPI, Basel, Switzerland. This article is an open access article distributed under the terms and conditions of the Creative Commons Attribution (CC BY) license (http://creativecommons.org/licenses/by/4.0/).

Article

Arabidopsis Class II Formins AtFH13 and AtFH14 Can Form Heterodimers but Exhibit Distinct Patterns of Cellular Localization

Eva Kollárová, Anežka Baquero Forero, Lenka Stillerová, Sylva Přerostová and Fatima Cvrčková *

Department of Experimental Plant Biology, Faculty of Sciences, Charles University, Viničná 5, CZ 128 44 Prague, Czech Republic; eva.slikova@natur.cuni.cz (E.K.); anezka.houskova@natur.cuni.cz (A.B.F.); Lstillerova@seznam.cz (L.S.); prerostova@ueb.cas.cz (S.P.)
* Correspondence: fatima.cvrckova@natur.cuni.cz

Received: 2 December 2019; Accepted: 3 January 2020; Published: 5 January 2020

Abstract: Formins are evolutionarily conserved multi-domain proteins participating in the control of both actin and microtubule dynamics. Angiosperm formins form two evolutionarily distinct families, Class I and Class II, with class-specific domain layouts. The model plant *Arabidopsis thaliana* has 21 formin-encoding loci, including 10 Class II members. In this study, we analyze the subcellular localization of two *A. thaliana* Class II formins exhibiting typical domain organization, the so far uncharacterized formin AtFH13 (At5g58160) and its distant homolog AtFH14 (At1g31810), previously reported to bind microtubules. Fluorescent protein-tagged full length formins and their individual domains were transiently expressed in *Nicotiana benthamiana* leaves under the control of a constitutive promoter and their subcellular localization (including co-localization with cytoskeletal structures and the endoplasmic reticulum) was examined using confocal microscopy. While the two formins exhibit distinct and only partially overlapping localization patterns, they both associate with microtubules via the conserved formin homology 2 (FH2) domain and with the periphery of the endoplasmic reticulum, at least in part via the N-terminal PTEN (Phosphatase and Tensin)-like domain. Surprisingly, FH2 domains of AtFH13 and AtFH14 can form heterodimers in the yeast two-hybrid assay—a first case of potentially biologically relevant formin heterodimerization mediated solely by the FH2 domain.

Keywords: AtFH13; AtFH14; At5g58160; At1g31810; FH2 domain; class II formin; confocal laser scanning microscopy; PTEN-like domain

1. Introduction

The eukaryotic cytoskeleton represents a dynamic network of protein filaments and tubules that has been extensively studied in a variety of model systems. In addition to basic cellular functions, including nuclear division, cytokinesis, organelle positioning, membrane trafficking and cell expansion, the cytoskeleton plays an important role in processes such as polar cell growth, cell division plane positioning, or cell to cell communication, which are essential for proper morphogenesis and thus the development of multicellular bodies in both metazoans [1] and plants [2]. This dynamic network relies on a large ensemble of cytoskeleton-associated proteins controlling organization, remodeling, and crosstalk of cytoskeletal systems, as well as coordination of the cytoskeleton with cell membranes and organelles. While the cellular structures and organismal body plans vary greatly among the eukaryotes, many molecular and cellular mechanisms involving the two basic cytoskeletal systems—the actin microfilaments and microtubules—are conserved among divergent groups such as, e.g., plants and metazoans, which are connected only through the last eukaryotic common ancestor [3,4].

Formins, or FH2 proteins, members of an evolutionarily ancient and widely expressed family of eukaryotic cytoskeletal organizers, are a prominent example of such conserved components of the cell's morphogenetic machinery. They are generally characterized by the presence of the conserved formin homology 2 (FH2) domain whose dimer can nucleate and cap actin filaments. In addition to other variable regulatory domains, the FH2 domain is usually accompanied by the proline-rich formin homology 1 (FH1) domain that can bind actin-profilin complexes, providing a substrate for formin-driven assembly of unbranched actin filaments [5]. However, formins also associate with microtubules via the FH2 domain or via other specialized domains, affecting microtubule dynamics [6–9]. Thus, the physiological functions of formins related to morphogenesis, cell division, cytokinesis, cell polarity, and cell to cell trafficking are likely to be tightly coupled with their roles in cytoskeleton reorganization. All eukaryotic genomes examined, including fungi, metazoans, and plants, encode numerous formin paralogs [10–12], most of them experimentally hitherto uncharacterized. Formins are well established to form FH2 domain dimers [13], suggesting that heterodimerization among formin paralogs might further contribute to their functional diversification. However, there is so far very little evidence of formin heterodimerization, with the only documented cases involving closely related metazoan formins [14–16].

The *Arabidopsis thaliana* genome harbors 21 formin-encoding loci that can be divided into two phylogenetically distinct classes present in all angiosperms, referred to as Class I and Class II, based on their sequence similarity and domain organization [11,17,18]. Most Class I formins contain a N-terminal transmembrane domain that enables them to anchor cytoskeletal structures to membranes, while many Class II formins contain a N-terminal PTEN (Phosphatase and Tensin)-like domain (homologous to members of the widespread phosphatase and tensin homolog protein family) instead. The PTEN-like domain was predicted to associate with membranes [18]. This has later been confirmed in the case of the *Physcomitrella patens* For2A formin whose PTEN-like domain interacts with PI(3,5)P$_2$; this interaction is necessary for formin localization to the tip of apically growing cells [19]. Similarly, the PTEN-like domain of the rice Class II formin FH5 encoded by rice morphology determinant (RMD) gene determines tip localization of this protein in pollen tubes [20] and interestingly also mediates its anchorage to the outer surface of chloroplasts [21]. In Arabidopsis, the only typical Class II formin containing a PTEN-like domain experimentally characterized so far is AtFH14 (At1g31810), whose dimer can bind to actin barbed ends [22] and which associates with microtubules in mitotic BY-2 cells, participating in the control of mitosis and cytokinesis [23].

In this study, we compare the intracellular localization of AtFH14 with that of a hitherto uncharacterized *A. thaliana* Class II formin, AtFH13 (At5g58160) that shares overall domain structure composition with AtFH14, using *in planta* transient heterologous expression in native Australian tobacco (*Nicotiana benthamiana*) leaves. While both formins can associate with microtubules (MTs) and the endoplasmic reticulum (ER), they exhibit distinct subcellular localization patterns. In addition, we demonstrate that although these formins show different pattern of localization, their FH2 domains are capable of heterodimerization—a first such observation in plant formins.

2. Results

2.1. Construction of Fluorescent Protein-Tagged AtFH13 and AtFH14 Derivatives

Both AtFH13 (At5g58160) and AtFH14 (At1g31810) are typical Class II formins [18], containing a N-terminal PTEN-related domain, followed by a C2 domain (related to the Ca^{2+}-binding domain from protein kinase C), the proline-rich FH1 domain and the C-terminal FH2 domain. Although they share the overall domain structure, AtFH13 and AtFH14 are only 52% identical in their PTEN-like and C2 domains and 60% identical in their FH2 domains, and represent branches of the Class II formin clade that are clearly separate at least within Brassicaceae (Figure 1a), and possibly within angiosperms [11]. As usual for Class II formins, both proteins are encoded by genes of a complex exon–intron structure (Figure 1b).

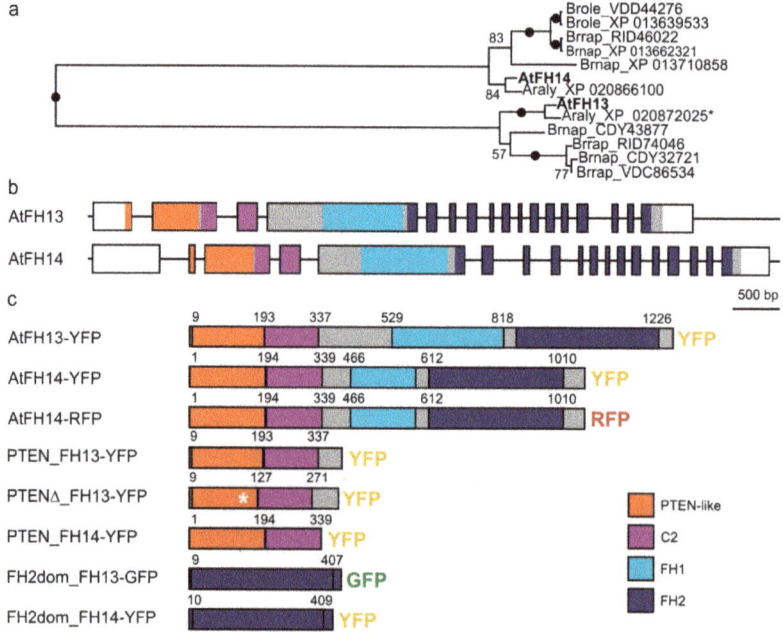

Figure 1. Evolutionary relationships of Arabidopsis Class II formins (AtFH13 and AtFH14), structure of the AtFH13 and AtFH14 genes, and constructs generated in this study. (**a**) Maximum likelihood phylogenetic tree of joined PTEN, C2 and FH2 domains of AtFH13, AtFH14 and their homologs from *Arabidopsis lyrata* (Araly), *Brassica napus* (Brnap), *B. rapa* (Brrap), and *B. oleracea* (Brole). Sequences are identified by their GenBank accession numbers; the asterisk marks a predicted protein sequence modified to include missing conserved exons (see Methods). Numbers denote bootstrap support (out of 100 replicates), branches with 100% support are marked by dots. (**b**) Exon–intron structure of both genes with non-coding (UTR) parts of exons represented by white boxes and coding exons shown either in grey or in color (for known domains). (**c**) Schematic representation of tagged protein constructs used in this study. Asterisk indicates the position of the 21 amino acids deletion in the mutant PTENΔ domain, starting from position 106 of the standard AtFH13 sequence. The numbers indicate amino acid positions related to the N-end. Domain abbreviations are defined in the text.

To examine in vivo subcellular localization of these formins, we constructed vectors expressing full length AtFH13 (Uniprot Acc. No. Q9LVN1) or a previously characterized variant of AtFH14 [23] tagged by C-terminal yellow fluorescent protein (YFP) or red fluorescent protein (RFP) fusion under the control of the constitutive ubiquitin 10 promoter (UBQ10) and used these constructs for transient transformation of *N. benthamiana* leaves. To examine contribution of individual domains of both proteins to their localization, we also prepared constructs expressing YFP-tagged N-terminal fragments of either protein containing the PTEN-like and C2 domains, as well as green fluorescent protein (GFP) and YFP-tagged isolated FH2 domains of AtFH13 and AtFH14. In addition, we also generated an YFP fusion of the N-terminal (PTEN-like and C2 domain-containing) fragment from a fortuitously cloned AtFH13 variant missing 21 amino acids within the PTEN-like domain (denoted PTENΔ) as a putative inactive variant of the PTEN domain (Figure 1c).

2.2. Both AtFH13 and AtFH14 Associate with Microtubules and the ER in Tobacco Epidermis

As usual in transient *N. benthamiana* leaf epidermis transformation, individual transformed cells exhibited variable signal intensity. Confocal imaging of YFP-tagged AtFH13 in cells with a relatively weak signal showed fibers, suggesting cytoskeletal association, as well as punctate structures of varying size in the cortical cytoplasm (Figure 2a), whereas cells which overexpressed the construct exhibited large, brightly fluorescent aggregates (Figure 2b). In case of AtFH14-YFP the cortical signal was mainly of fibrous character (Figure 2c), suggesting association with microtubules, consistent with previously observed localization of this protein's FH1-FH2 domain construct to the preprophase band, mitotic spindle, and phragmoplast in tobacco BY2 cells [23].

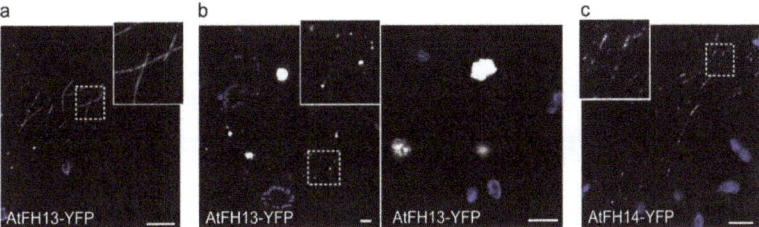

Figure 2. Localization of AtFH13-YFP and AtFH14-YFP in transiently transformed tobacco epidermis. (**a**) AtFH13-YFP localizes to fibrous structures at low expression levels. (**b**) Aggregates of overexpressed AtFH13-YFP. (**c**) Fibrous localization of AtFH14-YFP. All images are maximum intensity projections of confocal Z-stacks. YFP fluorescence is shown in grayscale, chlorophyll autofluorescence in blue. Scale bars are 10 µm, insets show the indicated details magnified 2.5 times compared to the main image. Construct abbreviations are defined in the text and in Figure 1.

To establish the relationship between the formin-labeled fibers and cytoskeletal structures, we co-expressed both tagged formins with the actin and microtubule markers LifeAct-RFP or KMD (kinesin motor domain)-RFP, respectively. Neither AtFH13 nor AtFH14 showed association with actin fibers (Figure 3) but both proteins differed in their localization patterns. AtFH13 exhibited patchy distribution following the microtubules, whereas AtFH14 decorated the microtubule network more homogeneously (Figure 4, Video S1, Video S2). An obvious, though statistically non-significant, trend towards co-localization with microtubules rather than microfilaments was detected also by quantitative image analysis (Figure S1).

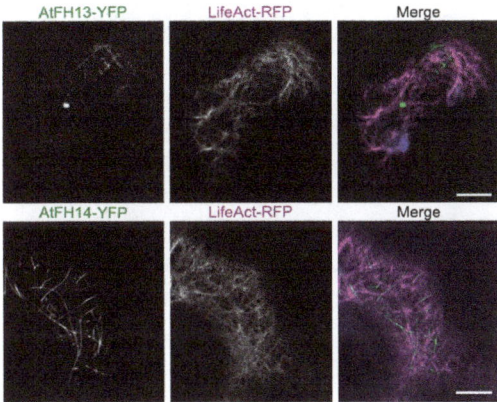

Figure 3. Single optical sections of epidermal cells co-expressing AtFH13-YFP or AtFH14-YFP with the actin marker LifeAct-RFP. Images from single channels are shown in grayscale, merged images display an overlay of three channels represented by green (YFP), magenta (RFP), and blue (chloroplast autofluorescence). Scale bars are 10 µm. Construct abbreviations are defined in the text and in Figure 1.

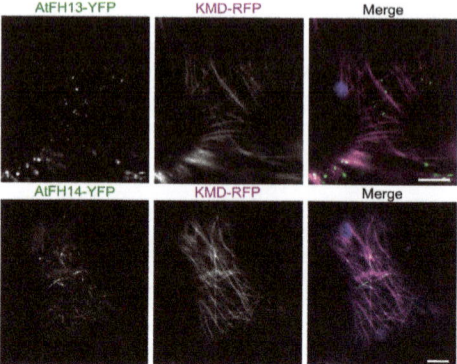

Figure 4. Single optical sections of epidermal cells co-expressing AtFH13-YFP or AtFH14-YFP with the microtubule marker KMD-RFP. Images from single channels are shown in grayscale, merged images display an overlay of three channels represented by green (YFP), magenta (RFP), and blue (chloroplast autofluorescence). Scale bar is 10 µm. Construct abbreviations are defined in the text and in Figure 1.

To confirm functional connection between the AtFH13-labelled foci and the microtubule cytoskeleton, we examined the effect of the microtubule depolymerizing drug oryzalin on localization of AtFH13-YFP co-expressed with KMD-RFP. The oryzalin treatment caused partial redistribution of the cytoplasmic dots and increased their mobility, which may be due to cytoplasmic streaming, not affected by microtubule disruption (Figure 5). This suggests that integrity of microtubules is essential to maintain subcellular distribution of AtFH13.

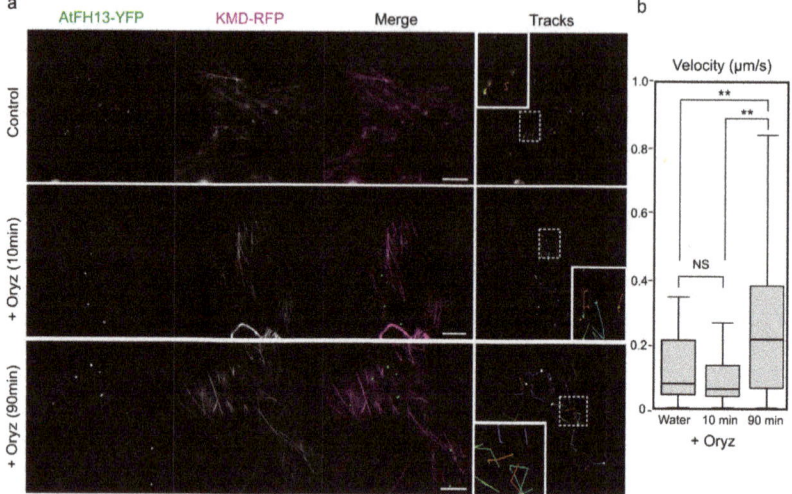

Figure 5. Prolonged oryzalin treatment increases mobility of AtFH13-YFP dots. (**a**) Epidermal cells co-expressing AtFH13-YFP and KMD-RFP mock-treated with water (control) and after 10 µM oryzalin treatment (+Oryz) for 10 min (no microtubule disruption observed) or 90 min (microtubules largely depolymerized). Single channel images are shown in grayscale, merged images display an overlay of the YFP channel in green with RFP in magenta. Particle tracks show trajectories of individual AtFH13-YFP dots over 50 s with insets magnified 2.5 times compared to the main image. (**b**) Box plot of mean velocities of AtFH13-YFP particles without, after 10 and after 90 min of oryzalin treatment. Scale bar is 10 µm. Two asterisks indicate a statistically significant difference (ANOVA, $p < 0.001$), NS: Non-significant. Construct abbreviations are defined in the text and in Figure 1.

Since both AtFH13 and AtFH14 contain the PTEN-like domain known to associate with membranes [20,24], we hypothesized that the fluorescent cytoplasmic punctate structures not associated with the cytoskeleton may be related to compartments of the endomembrane system. We thus examined co-localization of AtFH13-YFP and AtFH14-YFP with the ER marker ER-rk (ER red kanamycin). Remarkably, the two proteins exhibited obviously different localization patterns, with AtFH13-labelled structures peripherally associated with the ER network with minimal co-localization, while AtFH14-labeled structures exhibited noticeable overlap with the ER, which, however, could have been due to microtubule-ER co-alignment (Figure 6, Video S3, Video S4). Better co-localization of AtFH14 with the ER was supported also by a statistically non-significant trend detected quantitatively (Figure S1).

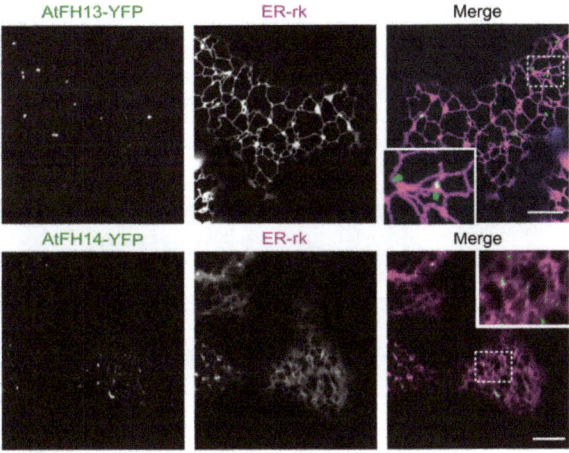

Figure 6. Single optical sections of epidermal cells co-expressing AtFH13-YFP and AtFH14-YFP with the endoplasmic reticulum marker ER-rk. Images from single channels are shown in grayscale, merged images display an overlay of three channels represented by green (YFP), magenta (RFP), and blue (chloroplast autofluorescence). Scale bars are 10 μm, insets are magnified 2.5 times compared to the main image. Construct abbreviations are defined in the text and in Figure 1.

2.3. Isolated FH2 Domains of AtFH13 and AtFH14 Exhibit Distinct Patterns of Microtubule Association

Next, we compared the localization of isolated FH2 domains of AtFH13 and AtFH14 in transiently transformed epidermal cells. GFP-tagged FH2 domain of AtFH13 was predominantly cytoplasmic with a few small punctate structures; in contrast, YFP-tagged FH2 domain of AtFH14 exhibited prominent fibrous distribution similar to the full-length protein (Figure 7a). Interestingly, upon co-expression with LifeAct-RFP, the GFP-tagged FH2 domain of AtFH13 organized into fibrous structures partially co-localizing with the actin marker, whereas the YFP-tagged FH2 domain of AtFH14 retained fibrous organization clearly distinct from the actin meshwork, similar to the full-length protein (Figure 7b). The microtubular nature of the network decorated by the YFP-tagged FH2 domain of AtFH14 was also confirmed by co-expression with the microtubule marker KMD-RFP. Remarkably, also the rare cytoplasmic dots labeled by GFP-tagged FH2 domain of AtFH13 were microtubule-associated (Figure 7c, Video S2).

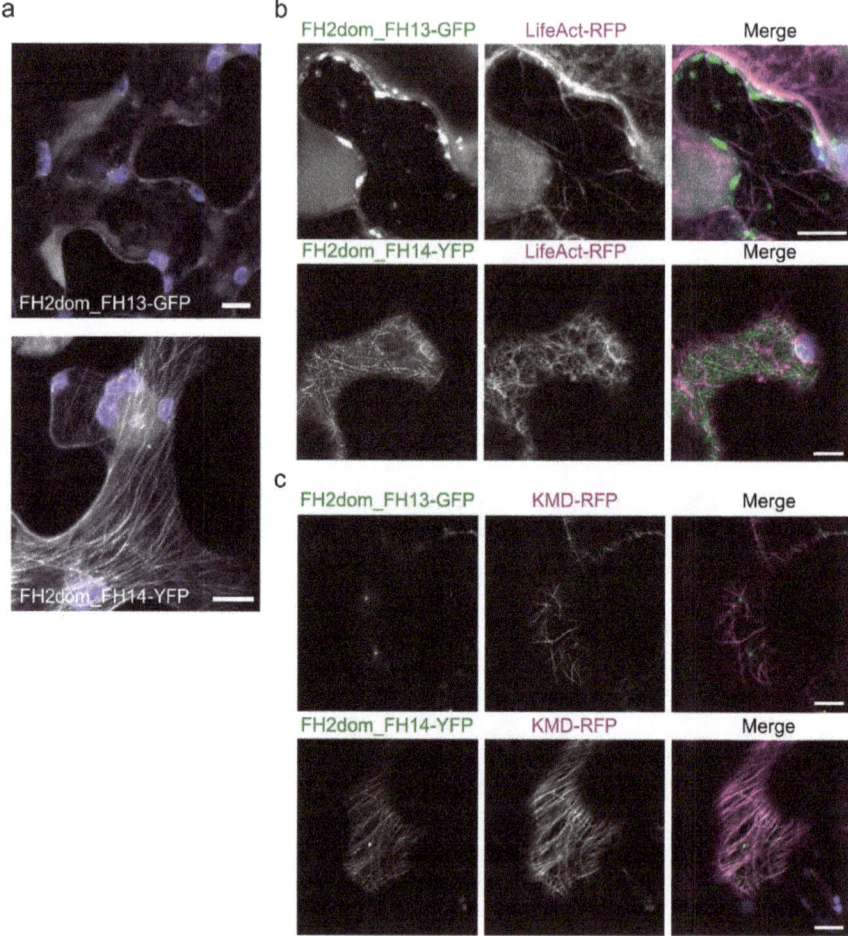

Figure 7. Localization of isolated FH2 domains of AtFH13 and AtFH14 in tobacco epidermal cells. (**a**) Maximal intensity projections of Z-stacks of optical sections from cells expressing FH2dom_AtFH13-GFP and FH2dom_AtFH14-YFP constructs, respectively. Green (GFP) or yellow (YFP) fluorescence shown in grayscale, chlorophyll autofluorescence in blue. (**b**) Single optical sections of cells co-expressing FH2dom_AtFH13-GFP or FH2dom_AtFH14-YFP with LifeAct-RFP. (**c**) Single optical sections of cells co-expressing FH2dom_AtFH13-GFP and FH2dom_AtFH14-YFP with KMD-RFP. In (**b**) and (**c**), images from single channels are shown in grayscale, merged images display an overlay of three channels represented by green (GFP or YFP), magenta (RFP), and blue (chloroplast autofluorescence. Scale bars are 10 µm. Construct abbreviations are defined in the text and in Figure 1.

2.4. The PTEN-Like Domain Mediates Association of AtFH13 and AtFH14 to the ER

To assess the possible contribution of the membrane-binding PTEN-like domain to the observed peripheral association of AtFH13 and AtFH14 with the ER, we examined the subcellular localization of YFP-tagged PTEN-like domains of AtFH13 and AtFH14. For control, we also included a fortuitously cloned deletion derivative of the PTEN-like domain from AtFH13, PTENΔ_FH13, presumed to be inactive due to disruption of a structurally important segment of the protein. The YFP-tagged PTEN domain of AtFH13 accumulated in structures reminiscent of the ER in the cytoplasm, whereas an analogous derivative of AtFH14 decorated small cytoplasmic punctate structures while exhibiting also strong accumulation in the nucleus and weak cytoplasmic background. In contrast, the deletion derivative of the AtFH13 PTEN-like domain showed only diffuse cytoplasmic distribution, consistent with the deletion altering the protein´s folding and consequently its subcellular localization (Figure 8a). Co-expression with the ER marker ER-rk shows that the YFP-tagged PTEN-like domain of AtFH13 peripherally associates with the ER, while its deletion derivative is cytoplasmic. The particles decorated by YFP-tagged PTEN-like domain of AtFH14 associated, and even partly overlapped, with the ER (Figure 8b).

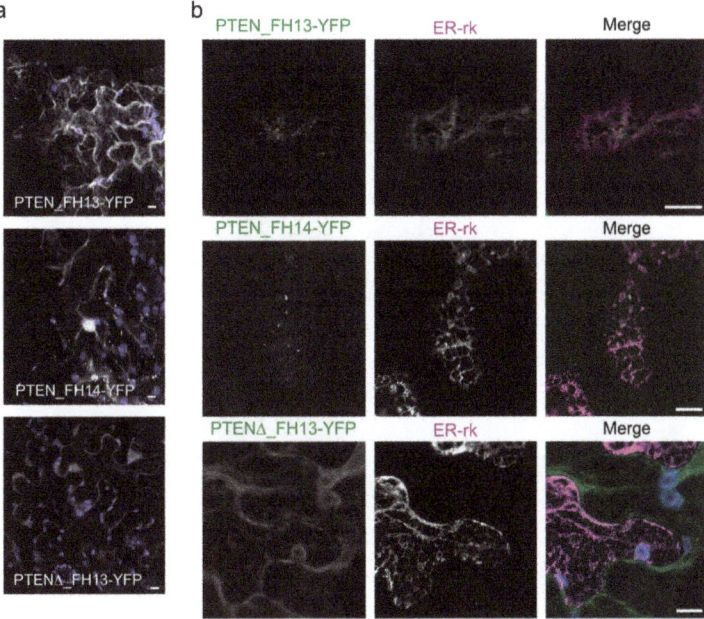

Figure 8. Localization of isolated PTEN domains of AtFH13 and AtFH14 in tobacco epidermal cells. (**a**) Maximal intensity projections of Z-stacks of optical sections from epidermal cells expressing PTEN_AtFH13-YFP, PTEN_AtFH14-YFP, or PTENΔ_AtFH13-YFP. Yellow (YFP) signal shown in grayscale, chlorophyll autofluorescence in blue. (**b**) Single optical sections of cells co-expressing PTEN_AtFH13-YFP, PTEN_AtFH14-YFP, or PTENΔ_AtFH13-YFP with ER-rk. Images from single channels are shown in grayscale, merged images display an overlay of three channels represented by green (YFP), magenta (RFP), and blue (chloroplast autofluorescence). Scale bars are 10 μm. Construct abbreviations are defined in the text and in Figure 1.

2.5. FH2 Domains of AtFH14 and AtFH13 Can Heterodimerize

It is well established that formins can form homodimers through their FH2 domains [13] but there is so far only limited evidence regarding their possible heterodimerization (see Introduction). The distinct localization patterns of AtFH13 and AtFH14 suggested that if heterodimerization between these formins does take place, it is unlikely to affect a substantial part of the cellular pool of both proteins. We nevertheless tested whether FH2 domains of AtFH13 and AtFH14 can heterodimerize by means of the yeast two hybrid assay.

In the yeast two hybrid system, an isolated FH2 domain of AtFH13 can form homodimers as expected. However, in our two-hybrid setup the FH2 domain of AtFH13 also readily formed heterodimers with the FH2 domain of AtFH14 (Figure 9). This is, at the first glance, surprising, given the distinct cellular localization of these two proteins. Nevertheless, when co-expressed, full length YFP-tagged AtFH13 and RFP-tagged AtFH14 exhibited partial co-localization in tobacco epidermal cells, in spite of their different overall localization patterns (Figure 10a). In particular, the puncta of AtFH13 often co-localized with fibers decorated by AtFH14, as documented by pixel intensity analyzes of the florescence signal (Figure 10b), consistent with partial co-localization which may involve FH2 domain-mediated heterodimerization. Quantitative estimate of co-localization of the two formins yielded values significantly higher than those obtained for either formin with the non-colocalizing actin marker LifeAct and also higher than those observed for microtubules and the ER, although in these cases the differences mostly were not statistically significant (Figure S1).

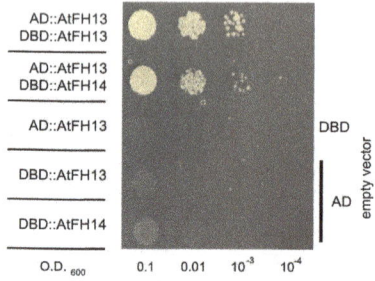

Figure 9. Yeast two-hybrid assays documenting heterodimerization of FH2 domains of AtFH13 and AtFH14. Serial dilution series of yeast transformants harboring the indicated constructs starting with $OD_{600} = 0.1$ were plated on media without adenine (-Ade), histidine (-His), leucine (-Leu), and tryptophan (-Trp) as indicated. DBD: DNA binding domain; AD: activation domain.

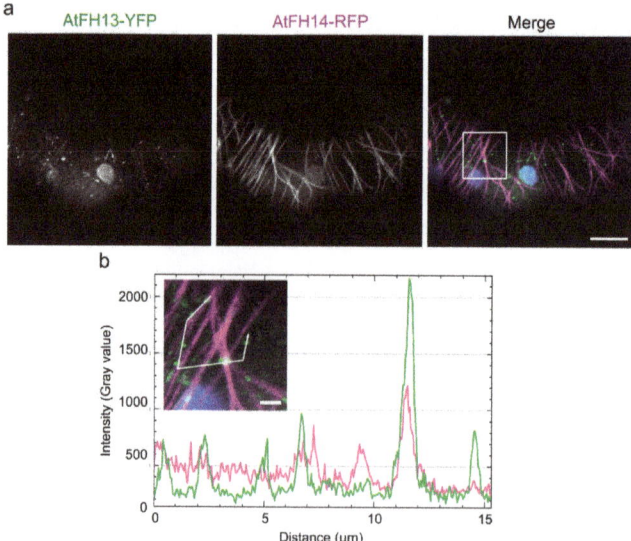

Figure 10. Co-localization of AtFH13 and AtFH14 in transiently transformed tobacco epidermal cells. (**a**) Single optical section of a cell co-expressing AtFH13-YFP and AtFH14-RFP. Images from single channels are shown in grayscale, merged images display an overlay of three channels represented by green (YFP), magenta (RFP), and blue (chloroplast autofluorescence). Note that one of the chloroplasts is probably damaged, exhibiting yellow autofluorescence. Scale bar is 10 µm. (**b**) Measurement of fluorescence intensity profiles of AtFH13-YFP and AtFH14-RFP along the indicated broken line in arrow direction. Scale bar is 1 µm. Construct abbreviations are defined in the text and in Figure 1.

3. Discussion

Formins are an evolutionarily ancient family of proteins [10–12] that can nucleate, cap, and bundle actin filaments, but also bind microtubules, resulting in an ability to modulate actin and microtubule organization and dynamics [7,9], well documented also in plants [8,25–27]. Angiosperm plants possess two clades of formins whose members often exhibit characteristic domain organization. While Class I formins are usually, though not always, integral membrane proteins [11,28], their Class II counterparts often harbor an N-terminal domain related to the mammalian PTEN (phosphatase and tensin homolog) lipid/protein phosphatase that was proposed to mediate membrane binding [11] and later shown to be responsible for membrane association and specific intracellular localization of Class II formins in the moss *Physcomitrella patens* [19,24] and in rice [21]. Thus, typical formins of both Class I and Class II are capable of membrane localization (reviewed in References [25,29]). In case of Class I formins, plasmalemma or endomembrane localization, as well as cytoskeletal, especially microtubule, association is well documented (e.g., References [6,30–35]). Much less is, however, known about in vivo intracellular localization of Class II formins.

In this study, we use a heterologous *in planta* expression system to document in vivo subcellular localization of fluorescent protein-tagged derivatives of two Arabidopsis Class II formins, the hitherto uncharacterized AtFH13 and somewhat characterized AtFH14 (see below). Although protein tagging by fusion with a relatively large polypeptide such as the YFP or GFP might influence its intracellular localization, such fusion proteins, including a plant formin [35], have been documented to be functional. Unfortunately, we could not confirm our observations by an independent method since we were unable to obtain specific antibodies suitable for immunostaining of formins, including AtFH13 [36]. Another theoretical alternative, expression of small tags visualized using cell permeable ligands, is well established in mammalian cell biology but practically restricted to the study of extracytoplasmic proteins in plants [37], with the exception of a single report with no published experimental follow-up [38].

Both formins studied exhibit the typical domain composition involving a N-terminal PTEN-like domain followed by a repetitive, proline-rich FH1 domain believed to mediate profilin-actin binding and by the C-terminal conserved dimerizing and actin-nucleating FH2 domain [11,18]. Despite the shared overall domain organization, the two proteins only exhibit limited similarity. Phylogenetic analysis of combined PTEN-like and FH2 domain sequences documented that they represent two distinct branches of the Class II formin clade that have separated no later than in basal Brassicaceae, and possibly earlier. While inner topology of the Class II formin clade in previous single-domain phylogenetic analyzes [11,18] had poor statistical support, it was nevertheless consistent with possible divergence of the clades represented by AtFH13 and AtFH14 as early as in ancestral angiosperms. The two proteins thus may exhibit different behavior and function, including distinct patterns of intracellular localization, which, indeed, turns out to be the case.

Formins, as known actin nucleators, can be expected to associate with actin filaments. Indeed, although AtFH14 was shown to preferentially bind microtubules, some co-localization of AtFH14 and actin was reported upon heterologous expression in mitotic tobacco BY2 cells [23]. In addition, using single molecule total internal reflection fluorescence microscopy, the FH1 and FH2 domains of AtFH14 were documented to directly bind the microfilament barbed end and act as a processive formin in vitro [22]. Nevertheless, in our *in planta* system co-expression of YFP-tagged full length AtFH14 or AtFH13 with actin markers did not reveal co-localization with the actin cytoskeleton, consistent with the strength, or indeed presence, of formin-microfilament association depending on the cellular context and molecular environment. Similar behavior has also been observed for the Class I formin AtFH4 which can bind and nucleate actin in vitro but does not decorate actin filaments in vivo [6]. Surprisingly, while the isolated FH2 domain of AtFH14 behaved comparably to the full length protein in co-expression with an actin marker, i.e., did not co-localize with actin fibers, the FH2 domain of AtFH13 showed a limited overlap with the actin network, suggesting differences in microfilament affinity between the two Class II formins.

Several Class II plant formins including AtFH14 were shown to interact with microtubules, sometimes more readily than with actin filaments [21,23,39,40]. We confirmed that YFP-tagged AtFH14 can co-localize with microtubules *in vivo*, and documented microtubule co-localization also for AtFH13. Remarkably, AtFH13 decorated the microtubules in a noticeably more discontinuous manner than AtFH14, often forming distinct particles along the microtubules. Mobility of these particles was greatly increased by the MT-depolymerizing drug oryzalin, proving that their organization is microtubule-dependent. Moreover, while the isolated FH2 domain of AtFH14 readily localized to the microtubules, the FH2 domain of AtFH13 was predominantly cytoplasmic, clearly indicating functional differences between the two FH2 domains. Analogously, metazoan formins have been documented to vary in their abilities to interact with microtubules and showed distinct localization patterns [41,42].

Interestingly, while both AtFH13 and AtFH14 associated with microtubules, only in the later the FH2 domain, which is also responsible for actin binding, was sufficient for microtubule interaction. Direct binding of FH2 domains to microtubules, or at least a specific contribution of FH2 to microtubule interaction, has been reported for several metazoan formins [41,43–45], as well as for plant formins of both Class I, such as rice FH15 [46], and Class II, namely rice FH5 [39]. In other cases, notably including Arabidopsis Class I formins AtFH4 and AtFH8 [6], the formin-microtubule interaction is mediated by specific domains or motifs outside FH2. In some cases, the association between formins and microtubules may be indirect, i.e., mediated by microtubule-binding proteins [47]. Together with previous reports, the observed different localization of isolated FH2 domains derived from AtFH13 and AtFH14 documents that while microtubule association appears to be a general feature of formins, its mechanism can vary substantially even within a relatively narrow clade such as the plant Class II formins.

In addition to microtubule co-localization, we observed partial overlap of AtFH14-labeled structures with the endoplasmic reticulum, as well as peripheral association of AtFH13-labelled particles with the ER network. Some derivatives of the Arabidopsis Class I formins, AtFH4 [6] and AtFH5 [25], as well as the non-conventional Class II formin AtFH16, which lacks the PTEN domain, can localize to the ER or to ER-like intracellular structures [25], and the transmembrane Class I formin AtFH1 was recently documented to pass through several endomembrane system compartments in differentiating root tissues [35]. The metazoan INF2 formin also associates with the ER periphery [48]. Our observations, which are consistent with the PTEN-like domains of AtFH13 and AtFH14 mediating peripheral association with the ER, are thus not surprising. However, the membrane nature and compartment identity of the observed formin-positive intracytoplasmic structures remains to be established.

Crystallographic studies revealed that the functional form of formin's FH2 domain is an antiparallel dimer capable of actin nucleation or barbed end capping [13]. In the case of AtFH14, dimerization and binding of the dimer to the barbed end of microfilaments has been documented for the FH1-FH2 domain combination, as well as for the isolated FH2 domain [22]. Using the yeast two hybrid assay, we confirmed that also the FH2 domain of AtFH13 can form homodimers. Although the localization patterns of AtFH13 and AtFH14 or their FH2 domains in transiently transformed *N. benthamiana* epidermis differed noticeably, we nevertheless observed co-localization of a fraction of the signal, and documented that the FH2 domains of these two Class II formins can form heterodimers in the yeast two-hybrid assay.

Heterodimerization of paralogous formins might theoretically provide a possible mechanism of formin activity regulation, as well as means of increasing functional diversity of fomins [18]; however, heterodimer formation may be, and often probably is, limited by structural divergence among formin paralogs [49]. Very few experimental studies address this topic, and so far, the only evidence of FH2 domain-mediated formin heterodimerization comes from metazoa. The FH2 domains of mammalian FRL2 and FRL3 are namely able to form hetero-oligomers when combined with N-terminal dimerization domains [16]. Other reported cases of mammalian Diaphanous-like formin heterodimerization [14,15] involve interactions mediated by domains other than FH2. Our finding that isolated FH2 domains of

AtFH13 and AtFH14 can form heterodimers thus presents, to our knowledge, the first documented case of exclusively FH2-mediated formin heterodimerization that might further increase the functional diversity of/not only plant formins.

4. Materials and Methods

For phylogenetic analysis, closest AtFH13 and AtFH14 relatives were identified by BLAST in the RefSeq section of GenBank after taxonomic restriction of the database to *Arabidopsis lyrata* or to *Brassica* sp. Only full-length sequences that retrieved the original query upon reverse BLAST search as the best match were considered. Missing exons in the predicted sequence of the AtFH13 ortholog from *A. lyrata* were added manually based on an experimentally sequenced partial cDNA (GenBank EFH40810). Multiple sequence alignment was constructed, manually cleaned of unreliably aligned segments and used to compute a maximum likelihood phylogenetic tree as described previously [50] except that Gamma distribution was used to model evolutionary rate differences among sites.

Constructs for transient expression *in planta* were prepared as follows. Vectors were generated using conventional restriction cloning or the Gateway (GW) cloning system (Invitrogen™, Carlsbad, CA, USA). For construction of entry clones, desired inserts were amplified using a PCR reaction with specific primer sets and templates summarized in Table S1. To amplify the AtFH14, the previously published cDNA clone [23] has been used as a template. Enzymatic cloning was used to create all entry clones derived from the GW-compatible donor vector pENTRA™1A (Invitrogen™) for further construction of expression vectors. Amplified inserts were digested with appropriate restriction enzymes (Fermentas, Vilnius, Lithuania; Table S1) and further ligated into the corresponding sites of the donor vector using T4 DNA ligase (Fermentas). For PCR-based entry clone construction, the pDONR™221 (Invitrogen™) vector was used. The PCR products were amplified by specific primers (Table S1) and transferred into donor vector by BP Clonase II (Invitrogen™). To obtain expression vectors, the resulting entry clones and binary destination vector pUBC-YFP-DEST/-RFP-DEST/-GFP-DES [51], which allows C-terminal fusion of the desired protein with YFP/RFP/GFP tag were recombined using LR Clonase II (Invitrogen™). The combinations of entry clones and destination vectors are summarized in Table S2. All newly cloned inserts where verified by sequencing.

For Agrobacterium-mediated transient transformation, expression vectors were transferred by electroporation into *Agrobacterium tumefaciens* strain GV3101 or C58C1. Plasmids carrying fluorescent protein markers, namely LifeAct-RFP [52], KMD-RFP [6], ER-rk [53], were used in co-localization experiments. Leaves from 2–3 weeks old *Nicotiana benthamiana* (native Australian tobacco) plants grown on peat pellets (Jiffy, Zwijndrecht, Netherlands) under the 8 h dark/16 h light photoperiod at 22 °C were used. Overnight *Agrobacterium* cultures were diluted in the infiltration medium (50 mM MES pH 5,6; 2 mM Na_3PO_4; 20 mM $MgSO_4$; 0.5% glucose, 100 μM acetosyringone). A mixture containing equal amounts of bacterial suspensions carrying each construct of interest, corresponding to final optical density (OD_{600}) 0.07 each, together with an amount of a suspension of agrobacteria expressing the viral silencing suppressor p19 [54] corresponding to OD_{600} 0.05, was infiltrated into the abaxial side of *N. benthamiana* leaves using a syringe. The fluorescent proteins were observed 48 h post-infiltration using confocal microscopy.

For the oryzalin treatment, oryzalin (Sigma-Aldrich, St Louis, MO, USA) was dissolved in dimethylsulfoxide (DMSO) as a 20 mM stock solution, stored at −20 °C, and later diluted to 10 μM final concentration with distilled water. Before observation, discs cut from infiltrated tobacco leaves were incubated in oryzalin solution (or in water for controls).

Microscopic images were obtained using the Zeiss LSM880 (Carl Zeiss, Oberkochen, Germany) confocal laser scanning microscope with a C-Apochromat 40×/1.2 W Korr FCS M27 objective. Fluorophores were excited with the 488 nm argon laser (YFP, GFP) and 561 nm laser (RFP), respectively, and detected by sensitive 32-chanell Gallium arsenide phosphide (GaAsP) spectral detector. All images were processed and analyzed using the Fiji software. All typical localization patterns that have been regularly observed in two to four independent biological replicates, each involving several leaves, are shown. Particle tracking was performed using the Fiji plugin TrackMate [55] v3.8.0 with the following settings: LoG detector, 2 μm diameter; no median filter; no sub-pixel localization; thresholding values were set to discard false-positive spots while the correctly identified spots were retained; simple LAP tracker; linking max distance = 5 μm; gap closing max distance = 5 μm; gap-closing max frame gap = 2. Data from at least five cells for each treatment with more than 200 trajectories detected were proceeded into a boxplot using the BoxPlotR tool [56]. To quantify protein co-localization, Pearson's coefficient was calculated from at least five images per construct pair using the Fiji/ImageJ plugin JACoP (Just Another Colocalisation Plugin) with manual threshold correction [57]. For statistic evaluation, an online tool [58] was used to perform one-way ANOVA with post-hoc Tukey HSD (honestly significant difference) test.

Constructs for the yeast two-hybrid assay were prepared by restriction cloning from previously prepared AtFH13 and AtFH14 clones (see above) and introduced into vectors pGADT7 (containing the GAL4 activation domain) and pGBKT7 (containing the GAL4 DNA binding domain). Primers, templates, and destination vectors are summarized in Table S3. The yeast two-hybrid assay was performed using Matchmaker GAL4 two-hybrid system 3 (Clontech, Mountain View, CA, USA) as described previously [50].

Supplementary Materials: The following are available online at http://www.mdpi.com/1422-0067/21/1/348/s1. Figure S1: Quantification of fluorescent protein-tagged AtFH13 and AtFH14 derivatives co-localization with other markers examined in this study; Table S1: Summarization of cloning methods used for each entry clone; Table S2: The combinations of entry clones and destination vectors used for generation of expression vectors used in this study; Table S3: Summarization of primers, templates, destination vectors, and restriction enzymes used to generate expression vectors for the yeast two-hybrid assay; Video S1: Co-expression of AtFH13-YFP with the microtubule marker KMD-RFP; Video S2: Co-expression of AtFH14-YFP with the microtubule marker KMD-RFP; Video S3: Co-expression of AtFH13-YFP with the endoplasmic reticulum marker ER-rk; Video S4: Co-expression of AtFH14-YFP with the endoplasmic reticulum marker ER-rk.

Author Contributions: Conceptualization, F.C. and E.K.; Methodology, E.K. and F.C.; Investigation, E.K., A.B.F., L.S., and S.P.; Writing—original draft preparation, E.K.; Writing—review and editing, F.C.; Visualization, E.K. and F.C. All authors have read and agreed to the published version of the manuscript.

Funding: This research was funded by the Ministry of Education of the Czech Republic, project NPUI LO1417.

Acknowledgments: We thank Viktor Žárský, Denisa Oulehlová, and Michal Hála for helpful discussion, Denisa Oulehlová, Haiyun Ren, and Jan Petrášek for sharing clones and vectors, Marta Čadyová for technical assistance and Ondřej Šebesta from the Faculty of Science Microscopy Facility for help with confocal microscopy.

Conflicts of Interest: The authors declare no conflict of interest. The funders had no role in the design of the study; in the collection, analyses, or interpretation of data; in the writing of the manuscript, or in the decision to publish the results.

Abbreviations

AD	activation domain
DBD	DNA-binding domain
ER	endoplasmic reticulum
ER-rk	ER red kanamycin
FH1	formin homology 1 domain
FH2	formin homology 2 domain
GFP	green fluorescent protein
GW	Gateway^{TM} cloning system
KMD	kinesin motor domain
MT	microtubule
PTEN	phosphatase and tensin homolog domain
RFP	red florescent protein
UBQ	ubiquitin
YFP	yellow florescent protein

References

1. Hasley, A.; Chavez, S.; Danilchik, M.; Wühr, M.; Pelegri, F. Vertebrate embryonic cleavage pattern Determination. *Adv. Exp. Med. Biol.* **2017**, *953*, 117–171. [PubMed]
2. Eng, R.C.; Sampathkumar, A. Getting into shape: The mechanics behind plant morphogenesis. *Curr. Opin. Plant Biol.* **2018**, *46*, 25–31. [CrossRef]
3. Vaškovičová, K.; Žárský, V.; Rösel, D.; Nikolič, M.; Buccione, R.; Cvrčková, F.; Brábek, J. Invasive cells in animals and plants: Searching for LECA machineries in later eukaryotic life. *Biol. Direct* **2013**, *8*, 8. [CrossRef] [PubMed]
4. Thomas, C.; Staiger, C.J. A dynamic interplay between membranes and the cytoskeleton critical for cell development and signaling. *Front. Plant Sci.* **2014**, *5*, 325. [CrossRef]
5. Paul, A.S.; Pollard, T.D. Review of the mechanism of processive actin filament elongation by formins. *Cell Motil. Cytoskelet.* **2009**, *66*, 606–617. [CrossRef] [PubMed]
6. Deeks, M.J.; Fendrych, M.; Smertenko, A.; Bell, K.S.; Oparka, K.; Cvrčková, F.; Žárský, V.; Hussey, P.J. The plant formin AtFH4 interacts with both actin and microtubules, and contains a newly identified microtubule-binding domain. *J. Cell Sci.* **2010**, *123*, 1209–1215. [CrossRef]
7. Bartolini, F.; Gundersen, G.G. Formins and microtubules. *Biochim. Biophys. Acta* **2010**, *1803*, 164–173. [CrossRef] [PubMed]
8. Wang, J.; Xue, X.; Ren, H. New insights into the role of plant formins: Regulating the organization of the actin and microtubule cytoskeleton. *Protoplasma* **2012**, *249*, S101–S107. [CrossRef] [PubMed]
9. Fernández-Barrera, J.; Alonso, M.A. Coordination of microtubule acetylation and the actin cytoskeleton by formins. *Cell. Mol. Life Sci.* **2018**, *75*, 3181–3191. [CrossRef]
10. Rivero, F.; Muramoto, T.; Meyer, A.-K.; Urushihara, H.; Uyeda, T.Q.P.; Kitayama, C. A comparative sequence analysis reveals a common GBD/FH3-FH1-FH2-DAD architecture in formins from Dictyostelium, fungi and metazoa. *BMC Genom.* **2005**, *6*, 28. [CrossRef]
11. Grunt, M.; Žárský, V.; Cvrčková, F. Roots of angiosperm formins: The evolutionary history of plant FH2 domain-containing proteins. *BMC Evol. Biol.* **2008**, *8*, 115. [CrossRef] [PubMed]
12. Pruyne, D. Probing the origins of metazoan formin diversity: Evidence for evolutionary relationships between metazoan and non-metazoan formin subtypes. *PLoS ONE* **2017**, *12*, e0186081. [CrossRef] [PubMed]
13. Xu, Y.; Moseley, J.B.; Sagot, I.; Poy, F.; Pellman, D.; Goode, B.L.; Eck, M.J. Crystal structures of a Formin Homology-2 domain reveal a tethered dimer architecture. *Cell* **2004**, *116*, 711–723. [CrossRef]
14. Copeland, S.J.; Green, B.J.; Burchat, S.; Papalia, G.A.; Banner, D.; Copeland, J.W. The diaphanous inhibitory domain/diaphanous autoregulatory domain interaction is able to mediate heterodimerization between mDia1 and mDia2. *J. Biol. Chem.* **2007**, *282*, 30120–30130. [CrossRef]
15. Gavard, J.; Patel, V.; Gutkind, J.S. Angiopoietin-1 prevents VEGF-induced endothelial permeability by sequestering Src through mDia. *Dev. Cell* **2008**, *14*, 25–36. [CrossRef] [PubMed]
16. Vaillant, D.C.; Copeland, S.J.; Davis, C.; Thurston, S.F.; Abdennur, N.; Copeland, J.W. Interaction of the N- and C-terminal autoregulatory domains of FRL2 does not inhibit FRL2 activity. *J. Biol. Chem.* **2008**, *283*, 33750–33762. [CrossRef]
17. Deeks, M.J.; Hussey, P.J.; Davies, B. Formins: Intermediates in signal-transduction cascades that affect cytoskeletal reorganization. *Trends Plant Sci.* **2002**, *7*, 492–498. [CrossRef]
18. Cvrčková, F.; Novotný, M.; Pícková, D.; Žárský, V. Formin homology 2 domains occur in multiple contexts in angiosperms. *BMC Genom.* **2004**, *5*, 44. [CrossRef]
19. Vidali, L.; van Gisbergen, P.A.C.; Guérin, C.; Franco, P.; Li, M.; Burkart, G.M.; Augustine, R.C.; Blanchoin, L.; Bezanilla, M. Rapid formin-mediated actin-filament elongation is essential for polarized plant cell growth. *Proc. Natl. Acad. Sci. USA* **2009**, *106*, 13341–13346. [CrossRef]
20. Li, G.; Yang, X.; Zhang, X.; Song, Y.; Liang, W.; Zhang, D. Rice Morphology Determinant-mediated actin filament organization contributes to pollen tube growth. *Plant Physiol.* **2018**, *177*, 255–270. [CrossRef]
21. Zhang, Z.; Zhang, Y.; Tan, H.; Wang, Y.; Li, G.; Liang, W.; Yuan, Z.; Hu, J.; Ren, H.; Zhang, D. Rice morphology determinant encodes the type II formin FH5 and regulates rice morphogenesis. *Plant Cell* **2011**, *23*, 681–700. [CrossRef] [PubMed]
22. Zhang, S.; Liu, C.; Wang, J.; Ren, Z.; Staiger, C.J.; Ren, H. A processive Arabidopsis formin modulates actin filament dynamics in association with profilin. *Mol. Plant* **2016**, *9*, 900–910. [CrossRef] [PubMed]

23. Li, Y.; Shen, Y.; Cai, C.; Zhong, C.; Zhu, L.; Yuan, M.; Ren, H. The type II Arabidopsis Formin14 interacts with microtubules and microfilaments to regulate cell division. *Plant Cell* **2010**, *22*, 2710–2726. [CrossRef] [PubMed]
24. Van Gisbergen, P.A.C.; Li, M.; Wu, S.-Z.; Bezanilla, M. Class II formin targeting to the cell cortex by binding PI(3,5)P$_2$ is essential for polarized growth. *J. Cell Biol.* **2012**, *198*, 235–250. [CrossRef]
25. Cvrčková, F.; Oulehlová, D.; Žárský, V. Formins: Linking cytoskeleton and endomembranes in plant cells. *Int. J. Mol. Sci.* **2014**, *16*, 1–18. [CrossRef]
26. Rosero, A.; Žárský, V.; Cvrčková, F. AtFH1 formin mutation affects actin filament and microtubule dynamics in Arabidopsis thaliana. *J. Exp. Bot.* **2013**, *64*, 585–597. [CrossRef]
27. Rosero, A.; Oulehlová, D.; Stillerová, L.; Schiebertová, P.; Grunt, M.; Žárský, V.; Cvrčková, F. Arabidopsis FH1 formin affects cotyledon pavement cell shape by nodulating cytoskeleton dynamics. *Plant Cell Physiol.* **2016**, *57*, 488–504. [CrossRef]
28. Cvrčková, F. Are plant formins integral membrane proteins? *Genome Biol.* **2000**, *1*, RESEARCH001. [CrossRef]
29. Cvrčková, F. Formins and membranes: Anchoring cortical actin to the cell wall and beyond. *Front. Plant Sci.* **2013**, *4*, 436. [CrossRef]
30. Favery, B.; Chelysheva, L.A.; Lebris, M.; Jammes, F.; Marmagne, A.; De Almeida-Engler, J.; Lecomte, P.; Vaury, C.; Arkowitz, R.A.; Abad, P. Arabidopsis formin AtFH6 is a plasma membrane-associated protein upregulated in giant cells induced by parasitic nematodes. *Plant Cell* **2004**, *16*, 2529–2540. [CrossRef]
31. Ingouff, M.; Fitz Gerald, J.N.; Guérin, C.; Robert, H.; Sørensen, M.B.; Van Damme, D.; Geelen, D.; Blanchoin, L.; Berger, F. Plant formin AtFH5 is an evolutionarily conserved actin nucleator involved in cytokinesis. *Nat. Cell Biol.* **2005**, *7*, 374–380. [CrossRef] [PubMed]
32. Deeks, M.J.; Cvrčková, F.; Machesky, L.M.; Mikitová, V.; Ketelaar, T.; Žárský, V.; Davies, B.; Hussey, P.J. Arabidopsis group Ie formins localize to specific cell membrane domains, interact with actin-binding proteins and cause defects in cell expansion upon aberrant expression. *New Phytol.* **2005**, *168*, 529–540. [CrossRef] [PubMed]
33. Martinière, A.; Gayral, P.; Hawes, C.; Runions, J. Building bridges: formin1 of Arabidopsis forms a connection between the cell wall and the actin cytoskeleton. *Plant J.* **2011**, *66*, 354–365. [CrossRef] [PubMed]
34. Lan, Y.; Liu, X.; Fu, Y.; Huang, S. Arabidopsis class I formins control membrane-originated actin polymerization at pollen tube tips. *PLoS Genet.* **2018**, *14*, e1007789. [CrossRef] [PubMed]
35. Oulehlová, D.; Kollárová, E.; Cifrová, P.; Pejchar, P.; Žárský, V.; Cvrčková, F. Arabidopsis class I formin FH1 relocates between membrane compartments during root cell ontogeny and associates with plasmodesmata. *Plant Cell Physiol.* **2019**, *60*, 1855–1870. [CrossRef] [PubMed]
36. Oulehlová, D.; Hála, M.; Potocký, M.; Žárský, V.; Cvrčková, F. Plant antigens cross-react with rat polyclonal antibodies against KLH-conjugated peptides. *Cell Biol. Int.* **2009**, *31*, 113–118. [CrossRef] [PubMed]
37. Sharma, I.; Russinova, E. Probing plant receptor kinase functions with labeled ligands. *Plant Cell Physiol.* **2018**, *59*, 1520–1527. [CrossRef]
38. Lang, C.; Schulze, J.; Mendel, R.R.; Hänsch, R. HaloTag: A new versatile reporter gene system in plant cells. *J. Exp. Bot.* **2006**, *57*, 2985–2992. [CrossRef]
39. Yang, W.; Ren, S.; Zhang, X.; Gao, M.; Ye, S.; Qi, Y.; Zheng, Y.; Wang, J.; Zeng, L.; Li, Q.; et al. Bent uppermost internode1 encodes the class II formin FH5 crucial for actin organization and rice development. *Plant Cell* **2011**, *23*, 661–680. [CrossRef]
40. Wang, J.; Zhang, Y.; Wu, J.; Meng, L.; Ren, H. AtFH16, an Arabidopsis type II formin, binds and bundles both microfilaments and microtubules, and preferentially binds to microtubules. *J. Integr. Plant Biol.* **2013**, *55*, 1002–1015. [CrossRef]
41. Gaillard, J.; Ramabhadran, V.; Neumanne, E.; Gurel, P.; Blanchoin, L.; Vantard, M.; Higgs, H.N. Differential interactions of the formins INF2, mDia1, and mDia2 with microtubules. *Mol. Biol. Cell* **2011**, *22*, 4575–4587. [CrossRef] [PubMed]
42. Courtemanche, N. Mechanisms of formin-mediated actin assembly and dynamics. *Biophys. Rev.* **2018**, *10*, 1553–1569. [CrossRef] [PubMed]
43. Bartolini, F.; Moseley, J.B.; Schmoranzer, J.; Cassimeris, L.; Goode, B.L.; Gundersen, G.G. The formin mDia2 stabilizes microtubules independently of its actin nucleation activity. *J. Cell Biol.* **2008**, *181*, 523–536. [CrossRef] [PubMed]

44. Roth-Johnson, E.A.; Vizcarra, C.L.; Bois, J.S.; Quinlan, M.E. Interaction between microtubules and the Drosophila formin Cappuccino and its effect on actin assembly. *J. Biol. Chem.* **2014**, *289*, 4395–4404. [CrossRef] [PubMed]
45. Foldi, I.; Szikora, S.; Mihály, J. Formin' bridges between microtubules and actin filaments in axonal growth cones. *Neural Regen. Res.* **2017**, *12*, 1971–1973. [PubMed]
46. Sun, T.; Li, S.; Ren, H. OsFH15, a class I formin, interacts with microfilaments and microtubules to regulate grain size via affecting cell expansion in rice. *Sci. Rep.* **2017**, *7*, 6538. [CrossRef] [PubMed]
47. DeWard, A.D.; Alberts, A.S. Microtubule stabilization: Formins assert their independence. *Curr. Biol.* **2008**, *18*, R605–R608. [CrossRef]
48. Chhabra, E.S.; Ramabhadran, V.; Gerber, S.A.; Higgs, H.N. INF2 is an endoplasmic reticulum-associated formin protein. *J. Cell. Sci.* **2009**, *122*, 1430–1440. [CrossRef]
49. Nezami, A.; Poy, F.; Toms, A.; Zheng, W.; Eck, M.J. Crystal structure of a complex between amino and carboxy terminal fragments of mDia1: Insights into autoinhibition of diaphanous-related formins. *PLoS ONE* **2010**, *5*, e12992. [CrossRef]
50. Baquero Forero, A.; Cvrčková, F. SH3Ps-evolution and diversity of a family of proteins engaged in plant cytokinesis. *Int. J. Mol. Sci.* **2019**, *20*, 5623. [CrossRef]
51. Grefen, C.; Donald, N.; Hashimoto, K.; Kudla, J.; Schumacher, K.; Blatt, M.R. A ubiquitin-10 promoter-based vector set for fluorescent protein tagging facilitates temporal stability and native protein distribution in transient and stable expression studies. *Plant J.* **2010**, *64*, 355–365. [CrossRef] [PubMed]
52. Fendrych, M.; Synek, L.; Pečenková, T.; Drdová, E.J.; Sekereš, J.; de Rycke, R.; Nowack, M.K.; Žárský, V. Visualization of the exocyst complex dynamics at the plasma membrane of *Arabidopsis thaliana*. *Mol. Biol. Cell* **2013**, *24*, 510–520. [CrossRef] [PubMed]
53. Nelson, B.K.; Cai, X.; Nebenführ, A. A multicolored set of in vivo organelle markers for co-localization studies in Arabidopsis and other plants. *Plant J.* **2007**, *51*, 1126–1136. [CrossRef] [PubMed]
54. Voinnet, O.; Rivas, S.; Mestre, P.; Baulcombe, D. An enhanced transient expression system in plants based on suppression of gene silencing by the p19 protein of tomato bushy stunt virus. *Plant J.* **2003**, *33*, 949–956, Retraction published in: *Plant J.* **2015**, *84*, 846. [CrossRef] [PubMed]
55. Tinevez, J.-Y.; Perry, N.; Schindelin, J.; Hoopes, G.M.; Reynolds, G.D.; Laplantine, E.; Bednarek, S.Y.; Shorte, S.L.; Eliceiri, K.W. TrackMate: An open and extensible platform for single-particle tracking. *Methods* **2017**, *115*, 80–90. [CrossRef] [PubMed]
56. Spitzer, M.; Wildenhain, J.; Rappsilber, J.; Tyers, M. BoxPlotR: A web tool for generation of box plots. *Nat. Methods* **2014**, *11*, 121–122. [CrossRef] [PubMed]
57. Bolte, S.; Cordelieres, F.P. A guided tour into subcellular colocalization analysis in light microscopy. *J. Microsc.* **2006**, *224*, 213–232. [CrossRef]
58. Online Web Statistical Calculators. Available online: https://astatsa.com (accessed on 30 November 2019).

© 2020 by the authors. Licensee MDPI, Basel, Switzerland. This article is an open access article distributed under the terms and conditions of the Creative Commons Attribution (CC BY) license (http://creativecommons.org/licenses/by/4.0/).

 International Journal of
Molecular Sciences

Article

Oligomerization Affects the Ability of Human Cyclase-Associated Proteins 1 and 2 to Promote Actin Severing by Cofilins

Vedud Purde [1,2], Florian Busch [1,3], Elena Kudryashova [1], Vicki H. Wysocki [1,2,3,4] and Dmitri S. Kudryashov [1,2,*]

1. Department of Chemistry and Biochemistry, The Ohio State University, Columbus, OH 43210, USA; purde.1@osu.edu (V.P.); busch.151@osu.edu (F.B.); kudryashova.1@osu.edu (E.K.); wysocki.11@osu.edu (V.H.W.)
2. The Ohio State Biochemistry Program, The Ohio State University, Columbus, OH 43210, USA
3. Resource for Native MS-Guided Structural Biology, The Ohio State University, Columbus, OH 43210, USA
4. Campus Chemical Instrument Center, Mass Spectrometry and Proteomics, The Ohio State University, Columbus, OH 43210, USA
* Correspondence: kudryashov.1@osu.edu; Tel.: +1-614-292-4848

Received: 18 October 2019; Accepted: 8 November 2019; Published: 12 November 2019

Abstract: Actin-depolymerizing factor (ADF)/cofilins accelerate actin turnover by severing aged actin filaments and promoting the dissociation of actin subunits. In the cell, ADF/cofilins are assisted by other proteins, among which cyclase-associated proteins 1 and 2 (CAP1,2) are particularly important. The N-terminal half of CAP has been shown to promote actin filament dynamics by enhancing ADF-/cofilin-mediated actin severing, while the central and C-terminal domains are involved in recharging the depolymerized ADP–G-actin/cofilin complexes with ATP and profilin. We analyzed the ability of the N-terminal fragments of human CAP1 and CAP2 to assist human isoforms of "muscle" (CFL2) and "non-muscle" (CFL1) cofilins in accelerating actin dynamics. By conducting bulk actin depolymerization assays and monitoring single-filament severing by total internal reflection fluorescence (TIRF) microscopy, we found that the N-terminal domains of both isoforms enhanced cofilin-mediated severing and depolymerization at similar rates. According to our analytical sedimentation and native mass spectrometry data, the N-terminal recombinant fragments of both human CAP isoforms form tetramers. Replacement of the original oligomerization domain of CAPs with artificial coiled-coil sequences of known oligomerization patterns showed that the activity of the proteins is directly proportional to the stoichiometry of their oligomerization; i.e., tetramers and trimers are more potent than dimers, which are more effective than monomers. Along with higher binding affinities of the higher-order oligomers to actin, this observation suggests that the mechanism of actin severing and depolymerization involves simultaneous or consequent and coordinated binding of more than one N-CAP domain to F-actin/cofilin complexes.

Keywords: cyclase-associated proteins; oligomerization; coiled coils; actin severing; actin depolymerization; α-barrels

---

1. Introduction

A dynamic balance between the polymerization of actin filaments at the barbed ends and depolymerization at the pointed ends is crucial for numerous cellular processes including cell adhesion, cell motility, cytokinesis, and morphogenesis [1–4]. To ensure a prompt response to cell needs and environmental challenges, fast rates of actin polymerization are underpinned by several mechanisms conferring high local concentrations of monomeric actin at polymerization sites [5,6]. To support these fast polymerization rates, prompt replenishing of the G-actin pool should be balanced by equally fast depolymerization of old filaments [2,3]. Fast depolymerization is achieved (1) via a severing-mediated increase in the number of the depolymerizing filament ends and (2) by promoting dissociation of actin subunits from the ends. Actin-depolymerizing factor (ADF)/cofilin family proteins play a key role in both mechanisms [3,7,8].

There are three isoforms of ADF/cofilins in mammals. Actin depolymerizing factor (ADF) is highly expressed in epithelial and endothelial cells; non-muscle cofilin 1 (CFL1) is expressed ubiquitously; and cofilin 2 (CFL2) is found predominantly, but not exclusively, in muscle tissues [8]. By selectively recognizing conformational changes in actin following ATP hydrolysis and P_i release [9,10], ADF/cofilins favorably decorate aged regions of F-actin, ensuring their prevalent disassembly. Binding of ADF/cofilins changes the filament twist [11] and reshapes the contacts between actin subunits [12–17], which results in severance of the borders between cofilin-decorated and non-decorated filament regions [18]. Since every severing event produces not only a new pointed end prone to depolymerization, but also a barbed end capable of elongation, the severing ability of cofilin was questioned initially as non-physiological [19–21]. This issue was resolved, however, by the discovery of proteins that assist ADF/cofilins in the controlled acceleration of actin depolymerization: actin-interacting protein 1 (Aip1), coronin, twinfilin, and cyclase-associated protein (Srv2/CAP) [22–26]. Working together, these proteins enable the mechanisms of selective nucleotide-dependent depolymerization of aged filaments [25,27], prevent polymerization at the barbed ends formed as a result of severing [26], accelerate dissociation of actin subunits from severed filaments [24], and stimulate dissociation of ADP from the depolymerized G-actin and its recharging by ATP [28]. Of these proteins, cyclase-associated proteins Srv2/CAPs [22,29,30] are of particular interest due to their multi-faceted contribution to the depolymerization process. The binding of Srv2/CAPs to cofilin-decorated actin via its N-terminal helical folded domain (HFD) potentiates cofilin-mediated severing [31] and, with the assistance of twinfilin, promotes disassembly of F-actin 3- and 17-fold at the barbed and pointed ends, respectively [32]. The central region of the protein contains a WH2 domain sandwiched between two proline-rich motifs (P_1 and P_2), which is followed by a C-terminal β-strand folded domain (Figure 1a) [33,34]. Binding of WH2 and β-strand domains to the ADP–G-actin/cofilin complex produced as a result of depolymerization promotes dissociation of both cofilin and ADP, and facilitates the recharging of the monomers with ATP [28]. The newly formed ATP–G-actin monomer is either released or actively transferred to profilin, thus completing the goal of regenerating the pool of polymerization-competent actin [28,35–37].

The full-length Srv2/CAP protein (FL CAP) has been found by gel filtration chromatography and analytical ultracentrifugation to form hexamers, while the N-terminal half (N-CAP) has presented a six-bladed shuriken shape in negative staining electron microscopy reconstructions [29,31,38]. The oligomerization is functionally important, and the N-terminal α-helix is thought to be critical for the formation of the hexamers, as its removal reduced the ability of the HFD domain to potentiate severing and depolymerization [31,38]. However, the role of Srv2/CAP oligomerization in this potentiation is not understood. Furthermore, the structure of the Srv2/CAP oligomerization is unclear. Approximately 98% of known coiled coils form low-order oligomers from dimers to tetramers, while the occurrence of natural coiled elements of a higher order is very rare [39] for reasons that are not well understood.

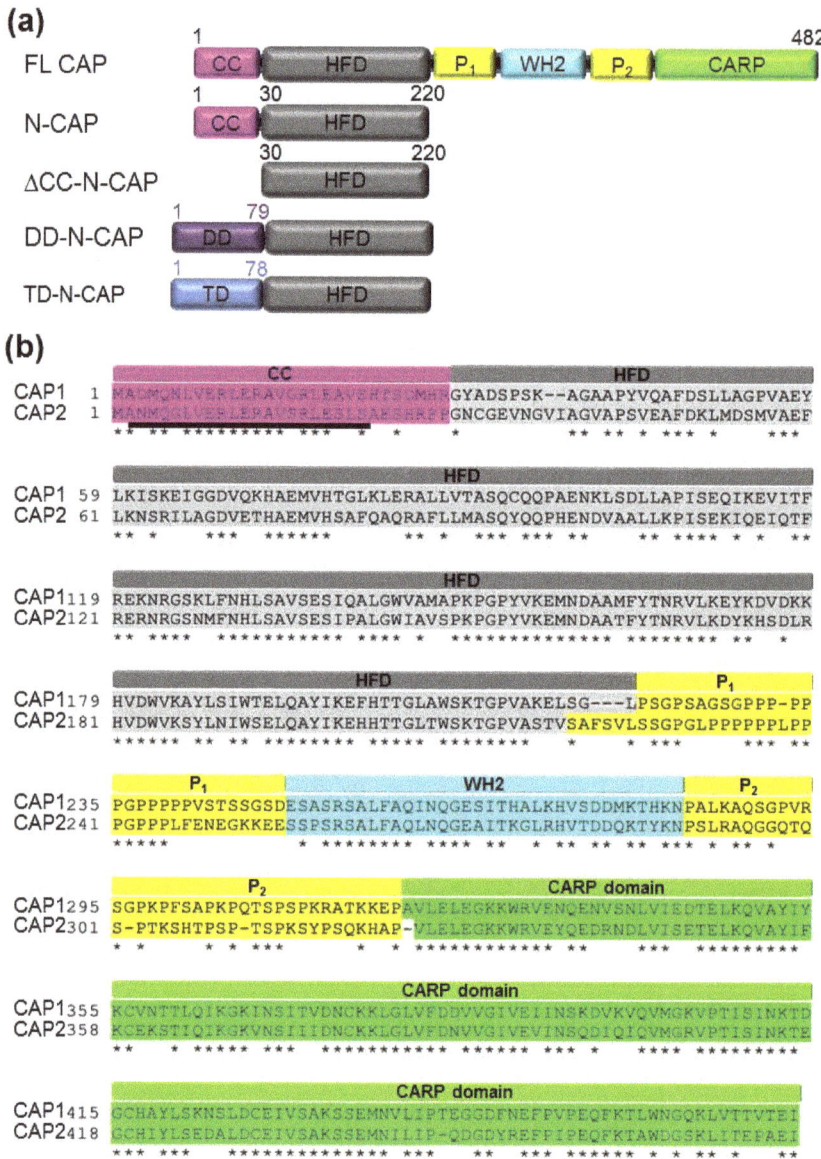

Figure 1. Domain structure of CAP proteins: (**a**) schematic representation of the full-length (FL) CAP domain structure and the truncation constructs used in this study. Amino acid numbering is shown for CAP1 only. CC: coiled coil domain (pink); HFD: helical folded domain (grey); P_1 and P_2: polyproline-rich regions (yellow); WH2: Wiskott-Aldrich syndrome protein (WASP)-homology 2 domain (cyan); CARP: C-terminal β-sheet domain (green); DD: dimerization domain; TD: trimerization domain. (**b**) Protein sequence alignment of human CAP1 and CAP2. Asterisks represent identical residues (64.1% identity) between the two isoforms. The underlined sequence with a predicted high helical propensity was used in CCBUILDER 2.0 for modeling coiled-coil oligomerization (see Figure 3). Domain abbreviation and coloring as in (a).

The two human isoforms of cyclase associated protein (CAP1 and CAP2) share 64%/75% sequence identity/similarity with each other (Figure 1b) and ~35%/50% identity/similarity with yeast Srv2 [40]. CAP1 is expressed ubiquitously, whereas CAP2 is produced predominantly in striated muscles, but is also found at lower levels in brain, testes, lung, liver, and skin tissues [40,41]. Accordingly, the most notable outcome of *CAP2* gene ablation in mouse is cardiomyopathy [42]. At the cellular level, CAP1 is found exclusively in the cytoplasm, while CAP2 shows dual localization in the cytoplasm and the nucleus [41].

Both human CAP isoforms have recently emerged as markers of invasive tumors. Thus, CAP1 has been implicated in breast, lung, pancreatic, and ovarian cancers along with glioma, hepatocellular, and head and neck squamous cell carcinomas [43–46]. CAP2 is overproduced in hepatocellular carcinoma, malignant melanoma, breast, and gastric cancers [47,48], and its expression is associated with poor clinical outcomes [44,49]. CAP2 is also overproduced by bladder, colon, kidney, and thyroid tumors [41], making it a common marker for various tumors. However, the exact role of CAPs in tumorigenesis is ambiguous, and anti-oncogenic action of the CAP1 and 2 isoforms for other tumors has also been proposed [45]. Interestingly, cancer-linked mutations in both CAP1 and CAP2 are enriched in the N-terminal domain, with the hot spot mutations in CAP1 are located at the Arg-29 residue [44], which is not conserved between the two isoforms. While mouse CAP1 (95% identical to its human counterpart) has been partially characterized [38], the mammalian CAP2 isoform or its fragments have never been biochemically explored. Therefore, the goals of the current study were (1) to evaluate and compare the abilities of the N-terminal fragments of the two human CAP isoforms to potentiate actin severing and depolymerization by human cofilins and (2) to explore how the oligomerization stoichiometry affects these abilities.

By conducting bulk actin depolymerization assays and monitoring single-filament severing using total internal reflection fluorescence (TIRF) microscopy, we found that the N-terminal domain of both isoforms (N-CAP) enhanced cofilin-mediated severing and depolymerization to a similar degree. Surprisingly, our analytical sedimentation and native mass spectrometry data showed that the N-terminal fragments of both human CAP isoforms formed tetramers rather than the hexamers reported for full-length Srv2 and mouse CAP1 [31,38]. By replacing the original oligomerization domains of CAPs with artificial coiled-coil sequences of known oligomerization patterns [50], we found that the activity of the proteins correlates with the stoichiometry of their oligomerization; i.e., tetramers and trimers more potently promoted severing and depolymerization of actin by both cofilin isoforms than dimers and monomers.

2. Results

2.1. Analytical Ultracentrifugation and Native Mass Spectrometry Reveaedl that the N-terminal Domains of Recombinant Human CAPs Form Tetramers

The N-terminal domain of N-Srv2 and mammalian CAPs contains a coiled-coil region followed by the helical folded domain (HFD) (Figure 1a). HFD interacts with F-actin and enhances cofilin-mediated F-actin disassembly [31,38], while the coiled coil contributes to oligomerization [38]. Specifically, the HFD of Srv2 has been proposed to form hexamers with radial symmetry and a shuriken-(star)-like appearance [31]. The extreme N-terminus of Srv2/CAP has been predicted to form a coiled-coil helix, but whether it can dictate the quaternary structure of CAPs is unclear, as coiled coils with a number of chains higher than four are rare in nature. We employed sedimentation velocity analytical ultracentrifugation (SV-AUC) to determine the oligomerization states of the recombinant constructs of human CAP1 and CAP2 HFDs with the original N-terminal coiled coils (N-CAP1 and N-CAP2) and 6xHis-tags placed at the C-termini to avoid interference with oligomerization. Raw AUC data (Supplementary Figure S1a) were analyzed using the SEDFIT analysis tool [51], and the molecular weights (MW) of the analyzed constructs were calculated based on their sedimentation coefficient values (Figure 2a and Table 1).

Surprisingly, the AUC-determined MWs of both N-CAP1 and N-CAP2 constructs with the original N-terminal coiled coils matched those of tetramers, and not hexamers as has been suggested for yeast Srv2 and mouse CAP1 counterparts [31,38]. To validate the unexpected results, we analyzed the oligomerization state of each construct using native mass spectrometry (native MS; Figure 2b and Supplementary Figure S1b) and confirmed that the main oligomeric state of both N-CAP1 and N-CAP2 constructs was tetrameric, with a minor presence of the lower oligomerization state species (Figure 2 and Table 1). Varying the buffer (Tris(hydroxymethyl)aminomethane (Tris)–HCl vs. phosphate-buffered saline (PBS)) and salt (30 mM NaCl vs. 135 mM KHPO$_4$) composition did not affect the oligomerization state of the proteins.

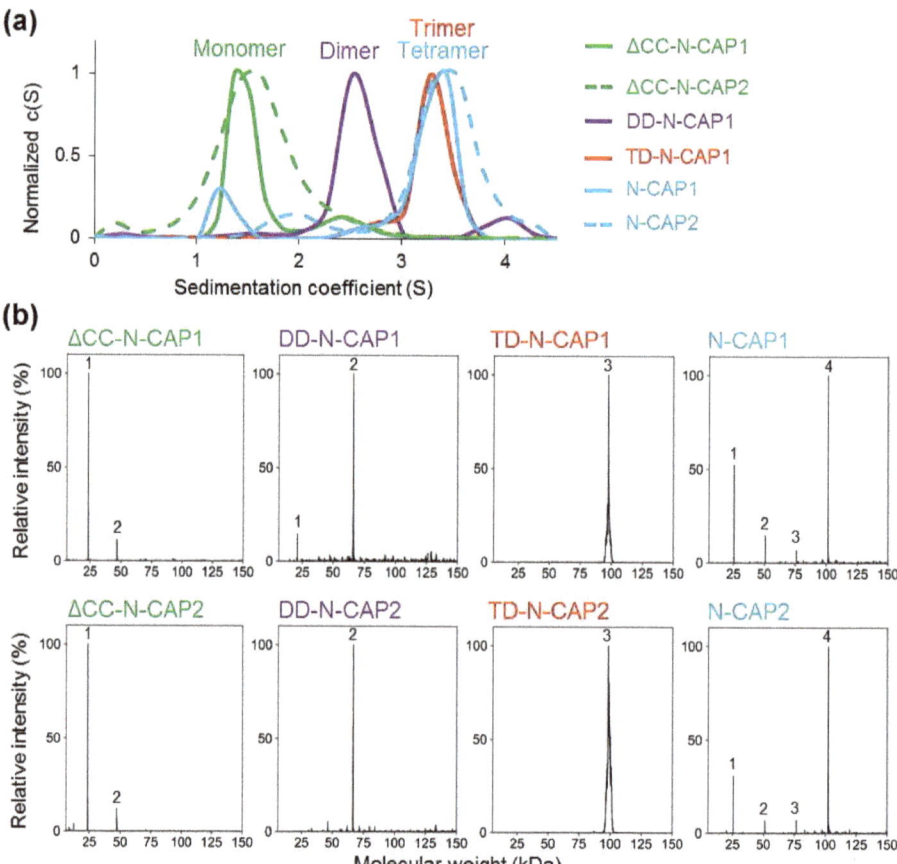

Figure 2. Oligomerization state of the N-terminal constructs of human CAPs: (**a**) sedimentation velocity analytical ultracentrifugation (SV-AUC) data were analyzed using SEDFIT software with a continuous sedimentation coefficient distribution model c(S). (**b**) Raw m/z data obtained by native mass spectrometry (MS) were deconvoluted with UniDec 4.0 and the molecular weights of the constructs were calculated. Numbers in the graphs indicate the oligomeric state of the CAP constructs present in solution: 1—monomer, 2—dimer, 3—trimer, 4—tetramer.

Table 1. Molecular weights (MW) of recombinant CAP constructs determined by SV-AUC and native MS. (f) —frictional ratio; (S)—sedimentation coefficient.

	AUC (f)	AUC (S)	AUC MW (kDa)	MS MW (kDa)	Theoretical MW (kDa)
ΔCC-N-CAP1	1.97	1.36	25.30	23.25	23.48 (monomer)
DD-N-CAP1	2.02	2.53	60.40	66.03	66.10 (dimer)
TD-N-CAP1	1.97	3.33	92.3	97.60	97.46 (trimer)
N-CAP1	2.17	3.43	92.1	101.46	101.46 (tetramer)
ΔCC-N-CAP2	2.07	1.62	23.50	23.69	23.69 (monomer)
DD-N-CAP2	1.44	3.78	66.8	66.63	66.52 (dimer)
TD-N-CAP2	1.41	4.84	92.3	98.36	98.03 (trimer)
N-CAP2	2.17	3.54	102.00	102.34	102.34 (tetramer)

In an attempt to resolve the difference between the reported data and our results, we computationally analyzed the ability of the CAP N-terminal helix to form coiled coils using CCBUILDER 2.0 software [52], which was created for prediction and de novo design of high-order α-helical oligomers. CCBUILDER predicted that the N-terminal helices of CAP1 and CAP2 (Figure 3) were substantially more likely to form hexamers than tetramers, as reflected in over two-fold higher negative energy amplitudes for interface packing, expressed in the BUDE (Bristol University Docking Engine) force field format [52] (Table 2). It is worth noticing, however, that heptamers of both CAP1 and CAP2 showed even lower BUDE energies, pointing to potential limitations of the approach.

Table 2. Predicted energies of N-terminal coiled-coil (CC) oligomers of human CAPs: provided values are means and standard deviations of 10 best scores of twenty Bristol University Docking Engine (BUDE) energies calculated by CCBUILDER 2.0.

	Oligomeric State	BUDE Energy
N-terminal CC of CAP1	Tetramer	−398.7 ± 13.7
	Pentamer	−671.4 ± 10.2
	Hexamer	−857.1 ± 73.0
	Heptamer	−916.4 ± 42.5
N-terminal CC of CAP2	Tetramer	−388.5 ± 15.4
	Pentamer	−601.2 ± 15.4
	Hexamer	−831.6 ± 65.1
	Heptamer	−878.6 ± 23.2

Analysis of the best-scored model structures showed that the tetramers followed a classical Type I "hxxhhxx" pattern (where "h" is a hydrophobic residue), with two "h" residues stabilizing neighboring partners and one shared between the hydrophobic coil-stabilizing seams (Figure 3b,c,g,h,l). Oligomers formed by five or more α-helices are called α-barrels as they have an internal cavity or channel. With the increased diameter, hexameric coiled-coil α-barrels require more hydrophobic residues to stabilize the core, and their residues more commonly follow Type II "hhxxhhx" heptad patterns [39]. The CCBUILDER model predicted that the role of the fourth hydrophobic residue in the heptad, in this case, would be played by the Cβ-Cγ atoms of the Arg residues, the charged side chains of which would contribute to stabilization of the helical barrels via salt bridge interactions (yellow dashed lines in Figure 3f) with Asp and Glu residues of the neighboring helices.

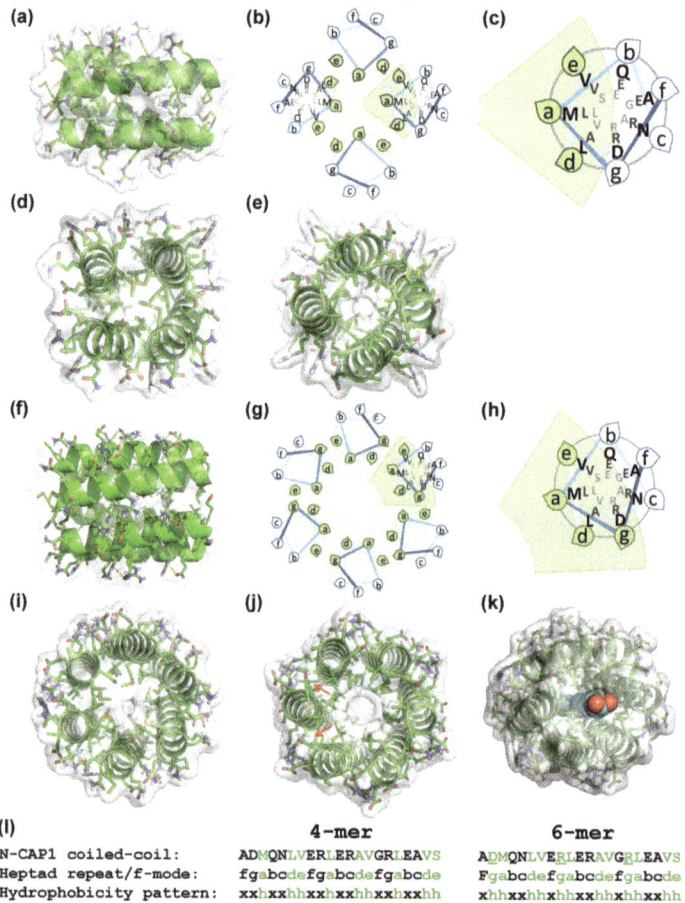

Figure 3. Modeling oligomerization of the N-terminal constructs of human CAPs: the CCBUILDER 2.0 web application (http://coiledcoils.chm.bris.ac.uk/ccbuilder2) was used to model the oligomeric coiled-coil structures of CAP1 (shown) and CAP2 N-terminal helices. The most stable tetramers (**a–e**) and hexamers (**f–j**) with the lowest energies are shown from three perspectives: side view (**a,f**), and along the helices from the N- (**d,i**), and C-termini (**e,j**). Schematic helical wheel diagrams show the relative orientations of the helices in the oligomers (**b,g**) and an enlarged view of the highlighted helices (c,h). Green shaded areas represent the hydrophobic cores of the oligomers. Up to four salt bridges between Arg and Asp/Glu residues contribute to the stabilization of each pair of neighboring helices (shown as yellow dotted lines in (**f**)). (**k**) Hypothetical stabilization of the hexamer structure by a fatty acid (blue) occupying the central, hydrophobic channel of Srv2/CAP (see Section 3). A low-energy CAP2 structure generated in CCBUILDER is shown. (l) The N-terminal coiled coil sequence of CAP1 (top row), with the conventional "a–f" designation of the heptad amino acids (middle row) and the hydrophobicity pattern (bottom row) characteristic of trimeric/tetrameric (xxhxxhh) and higher-order (xhhxxhh) coiled-coil structures. Residues involved in stabilization of the hydrophobic cores are designated by "h" in the hydrophobicity pattern and by green color elsewhere. Notice that in the hexamer, Cβ-Cγ atoms of the underlined Arg residues (l) contribute to the stabilization of the hydrophobic core (red arrows in (j)).

2.2. Construction of Monomers, Dimers, and Trimers of Human CAP1 and CAP2 HFD Domains

To address the role of the stoichiometry of N-CAP oligomers in their ability to enhance cofilin-mediated actin depolymerization, we generated recombinant constructs of the HFD domains with the native coiled-coil helices either deleted (ΔCC-N-CAPs) or replaced by helical elements with known dimer (DD-N-CAPs) and trimer (TD-N-CAPs) oligomerization stoichiometry [50] (Figure 1a).

The MW of the ΔCC-N-CAP1 and ΔCC-N-CAP2 constructs, experimentally determined by AUC, matched the theoretical MW of the respective constructs with 93 and 99% accuracy (Figure 2a and Table 1). The constructs designed as dimers and trimers matched the theoretical expectations with 95% and 94% accuracy, respectively (Figure 2a and Table 1). Furthermore, native MS confirmed the desired oligomerization states of the constructs with high accuracy, while also showing the presence of a minor population of dimers of both ΔCC-N-CAPs (Figure 2b and Table 1). The latter observation tentatively suggests that HFD may contribute to the interactions between the subunits and stabilization of the native quaternary state of the CAP oligomers.

2.3. Oligomerization through the N-terminal Coiled-Coil Region Affects the Binding Affinity of N-CAPs to F-actin

Assuming that the ability of Srv2 to enhance cofilin-mediated actin severing requires binding to F-actin via the N-terminal HFD domain [31], we tested the effects of oligomerization on the ability of human CAP1 and CAP2 HFD constructs to bind F-actin. By titrating a fixed concentration of N-CAP1 construct with various concentrations of F-actin, we found that the affinity of N-CAP1 constructs steadily increased from monomers to tetramers (Figure 4b–e,g). This higher affinity suggested that more than one subunit of the oligomers can contribute to the interaction with F-actin in natural and artificially designed N-CAP oligomers. Although N-CAP2 constructs showed a similar trend, the affinity for the monomeric form was lower, while the affinity of the tetrameric constructs was higher than that of N-CAP1 constructs (Figure 4f,g).

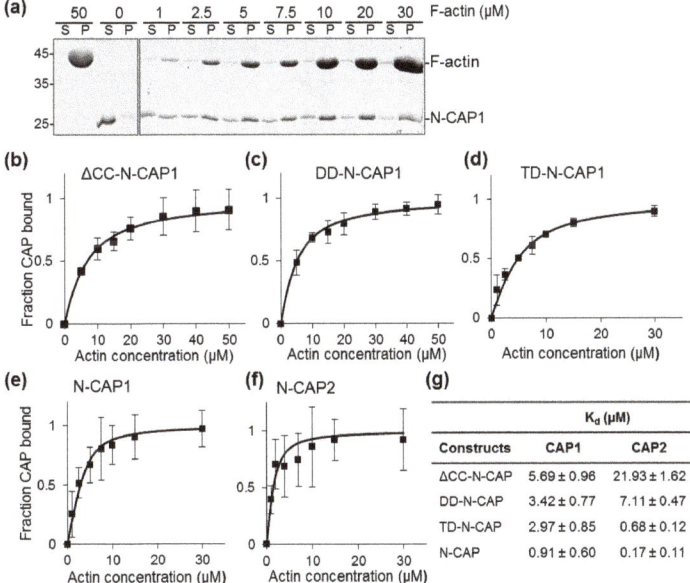

Figure 4. Comparison of F-actin binding affinities of human N-CAP constructs. CAP binding affinities to F-actin were analyzed by high-speed (300,000× g) cosedimentation: (**a**) representative sodium dodecyl sulfate polyacrylamide gel electrophoresis (SDS-PAGE) of supernatant (S) and pellet (P) fractions of N-CAP1 co-pelleted with F-actin (uncropped version of the gel is shown in Supplementary Figure S2); (**b–f**) binding curves with error bars representing the standard errors of the mean of three independent repetitions; (**g**) K_d values determined by fitting the experimental data to the binding isotherm equation defined in the Section 4.4.

2.4. Higher Affinities of the N-CAP Constructs to F-actin Correlated with Higher Potentiation of F-actin Depolymerization by Cofilins

To assess whether the increased affinity correlated with an improved function, we tested the ability of the N-CAP constructs to enhance cofilin-mediated F-actin disassembly in bulk pyrene–actin depolymerization assays. Given that the isoforms of human cofilins differ in their properties [8,53] and both CFL2 and CAP2 are predominantly found in muscle tissues, while both CFL1 and CAP1 are expressed ubiquitously, our goal was to establish whether the pairs of cofilin and CAP isoforms showed functional selectivity. We observed that in both the absence and presence of N-CAP constructs, depolymerizetion by CFL1 was overall more effective than that by CFL2 (Figure 5, Table 3, and Supplementary Figure S3). The concentrations of all the oligomers in these experiments were normalized to their monomeric forms, so the actual concentrations of the protein complexes were inversely proportional to their oligomerization state. Nonetheless, both N-CAP1 and N-CAP2 constructs potentiated actin depolymerization progressively better with increasing level of oligomerization, suggesting that proximity of the subunits in the oligomers positively contributed to the mechanism of severing and depolymerization. However, while dimers were notably more active than monomers, trimers were nearly as effective as tetramers, with the only exception being the N-CAP2 dimer, which was as effective as the CAP2 trimers and tetramers while cooperating with CFL2 (Supplementary Figure S3d and Table 3). There was no difference between the functional efficiencies of the CAP isoforms when assisting CFL1, and CAP2 was only marginally more effective than CAP1 when potentiating depolymerization by CFL2 (Figure 5, Supplementary Figure S3, and Table 3).

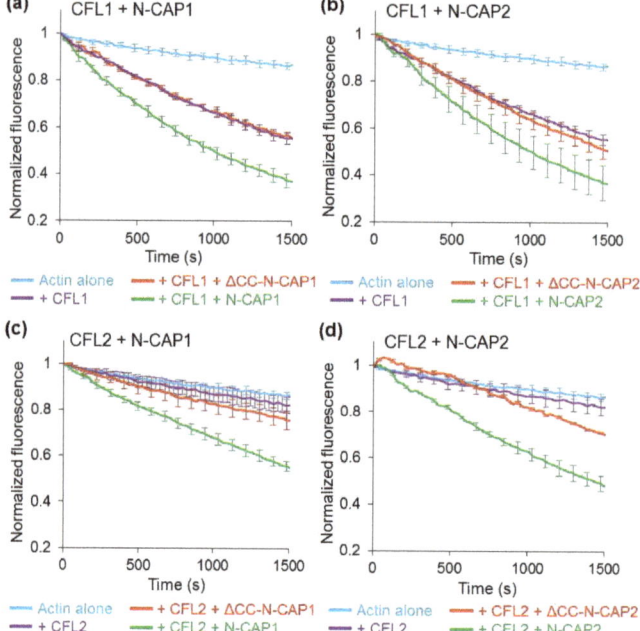

Figure 5. Effects of N-CAP oligomers on cofilin-mediated F-actin disassembly in bulk pyrene–actin depolymerization assays. Depolymerization of preassembled pyrene-labeled (10%) F-actin from the pointed ends (in the presence of CapZ) was initiated by the addition of a G-actin-sequestering drug latrunculin B along with 250 nM CFL1 (**a,b**) or CFL2 (**c,d**) in the absence of presence of 750 nM N-CAP1 (a,c) or N-CAP2 (b,d) oligomeric constructs. For clarity, error bars representing standard errors of the mean from three (CFL1) and four (CFL2) independent experiments are shown for every third data point.

Table 3. Effects of N-CAP constructs on cofilin-mediated F-actin depolymerization rates. Initial rates of F-actin depolymerization were measured from the slopes of the pyrene–actin depolymerization curves during the first 500 seconds. Rates are expressed in nM/min as mean values with standard errors from three (CFL1) and four (CFL2) experiments.

	No CAPs	CAP1	CAP2
Actin alone	12.9 ± 2.8		
CFL1	44.0 ± 4.0		
CFL1 + ΔCC-N-CAP		44 ± 4.0	48 ± 6.9
CFL1 + DD-N-CAP		52 ± 4.0	52 ± 4.0
CFL1 + TD-N-CAP		60 ± 12	64 ± 4.0
CFL1 + N-CAP		72 ± 3.4	68 ± 14
CFL2	15.8 ± 6.5		
CFL2 + ΔCC-N-CAP		24 ± 4.0	12 ± 1.4
CFL2 + DD-N-CAP		30 ± 0.8	48 ± 3.5
CFL2 + TD-N-CAP		42 ± 6.0	44 ± 4.0
CFL2 + N-CAP		39 ± 4.0	48 ± 2.1

To separate the effects due to the potentiation of severing from those caused by the acceleration of depolymerization from the ends, we analyzed severing effects at the single-filament level using TIRF microscopy. To this end, fluorescently labeled G-actin (33% Alexa 488-labeled actin and 1% biotinylated actin) was polymerized in a chamber and tethered to coverslips through biotin–streptavidin interactions. Filaments were grown up to 15 µm average size before either cofilin alone (50 nM) or with N-CAP constructs (750 nM) was flowed into the chamber to induce severing. The applied proteins did not contain G-actin to preclude polymerization from the severed ends of the filaments. As compared to CFL1 alone, we observed over 2.5-fold potentiation of severing in the presence of either N-CAP1 or N-CAP2 (Figure 6a–c) with no substantial difference between the two isoforms. Again, the severing events directly correlated with the N-CAP oligomeric stoichiometry, with the tetrameric constructs being notably more effective than their lower oligomerization state counterparts (Figure 6b,c). Interestingly, despite slower depolymerization in bulk assays, CFL2 severed filaments more effectively than CFL1 (Figure 6b–d). The severing was so fast and effective that both N-CAP constructs contributed only marginally to the severing activity of CFL2 (Figure 6d). These observations are in line with a higher severing efficiency of CFL2 towards skeletal actin [53,54] and its lower depolymerization capacity [8,54] reflected in its inability to increase the critical concentration for polymerization [17].

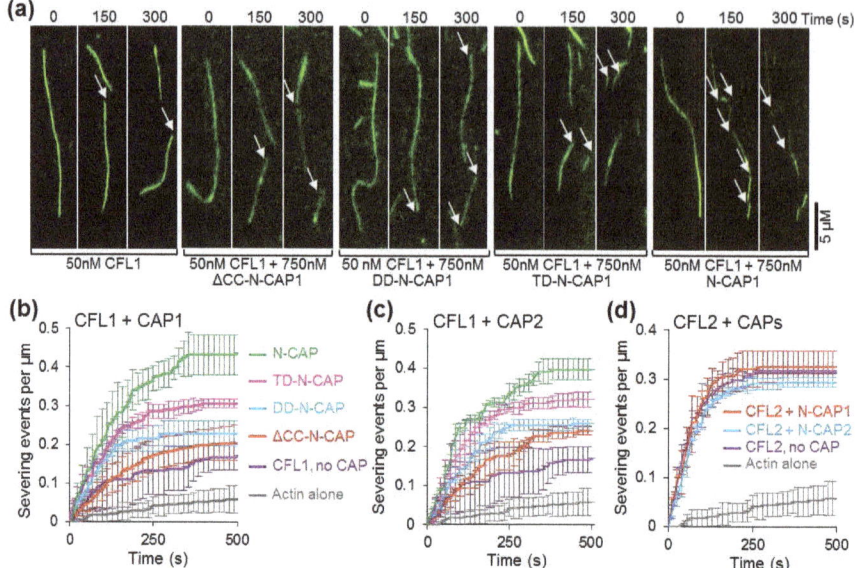

Figure 6. Effects of N-CAP constructs on cofilin-mediated F-actin severing observed by total internal reflection fluorescence (TIRF) microscopy. (**a**) Representative time-lapse images of Alexa 488-labeled F-actin severing upon addition of CFL1 in the absence or presence of N-CAP1 constructs. Arrows indicate severing events. (**b–d**) Analysis of severing activities: each data point represents the mean value of the number of severing events per micron of filament from three independent experiments (10–15 filaments per experiment). For clarity, error bars representing standard deviations of the mean are shown for every third data point.

2.5. Interaction of CAP1 with CFL1 and CFL2 Isoforms in Cells

To check whether the isoforms of cofilin and CAP made preferential interactions in cells, we employed the rolling cycle amplification-based proximity ligation assay (Duolink In Situ PLA), allowing detection of interacting partners when they are located within ~40 nm from each other (see Section 4.8). By testing various commercial antibodies, we were able to identify specific antibodies that recognized both recombinant proteins and a single band of the correct size for CFL1, CFL2, and CAP1 (Figure 7a and Supplementary Figures S4 and S5). Unfortunately, CAP2 antibodies either detected a single band but of the wrong size in cell extracts and failed to detect recombinant CAP2 (Santa Cruz Biotechnology #sc-377471; Supplementary Figure S6a), or detected the recombinant protein and the right size protein in cell lysates, but only as a weak-intensity band among about a dozen higher-intensity bands (Abnova #H00010486-M01; Supplementary Figure S6b), which made the antibody inappropriate for immunocytochemistry (and PLA, in particular). To explore the localization of CAP1 with both isoforms of cofilin, we screened over two dozen cell lines (Supplementary Figure S5) and selected Hs 578T breast epithelial carcinoma cells, which express high levels of both CFL1 and CFL2 (Figure 7a). Despite CFL1 being the more prevalent of the two isoforms in these cells, the CFL2/CAP1 pair generated notably more proximity ligation events (Figure 7b), suggesting their more prominent co-localization and interaction. Notice, however, that more prominent localization may result from the low depolymerization ability of CFL2 (Table 3), leading to more prolonged, but functionally less effective, cooperation.

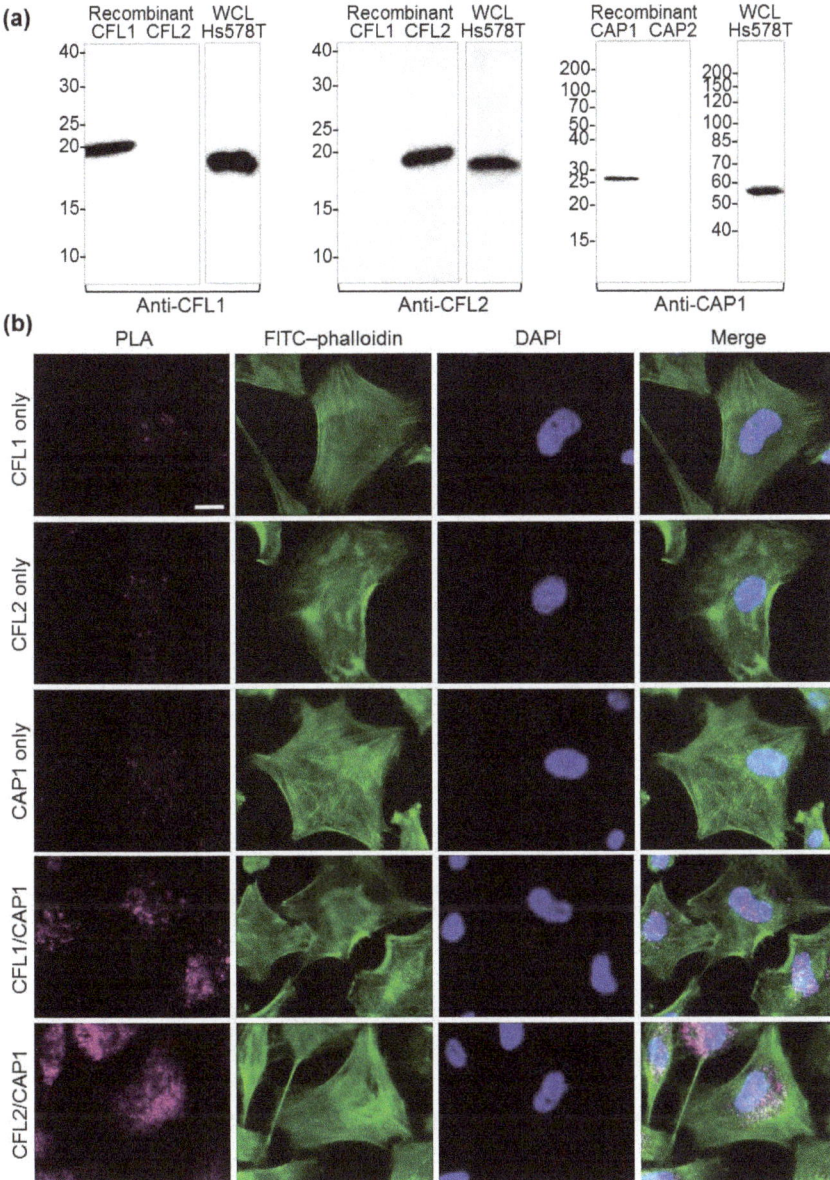

Figure 7. Interaction of native CAP1 with cofilin isoforms in cells: (**a**) western blot analysis demonstrates the specificity of the primary isoform-specific anti-CFL1, anti-CFL2, and anti-CAP1 antibodies used in immunofluorescence proximity ligation assay (PLA) experiments. WCL: whole-cell lysate. Uncropped versions of the blots are shown in Supplementary Figures S4 and S5. (**b**) Duolink in situ PLA assay performed on Hs 578T cells, as described in the Section 4.8. Cells stained using a single primary antibody (CFL1 only, CFL2 only, and CAP1 only) are shown as negative controls. PLA signal (magenta) using pairs of CFL1/CAP1 and CFL2/CAP1 antibodies represents CAP1/cofilin interaction events. Cells were counter-stained with fluorescein isothiocyanate (FITC)–phalloidin for F-actin (green) and nuclear 4′,6-diamidino-2-phenylindole (DAPI, blue). Scale bar is 20 μm.

3. Discussion

Elevated expression of both human CAP isoforms (CAP1 and CAP2) is linked to poor prognosis in multiple different metastatic cancers [45,55]. High invasiveness of cancer cells overexpressing CAP1 and CAP2 correlates with their roles in stimulating the turnover of actin filaments, which is essential for cell motility via promoted formation and turnover of pro-migratory structures such as filopodia and lamellipodia, leading to metastasis [45]. Since cancer-related mutations in both CAP1 and CAP2 are more prevalent in the N-terminal part of the CAP isoforms, we compared the biochemical properties of the N-terminal region of both isoforms, which are recognized to bind F-actin and enhance cofilin-mediated filament disassembly.

Although native Srv2 and CAP1 are reported to form hexamers, as originally revealed by gel filtration and reconstruction of negatively stained electron microscopy images [29,38] and recently confirmed by AUC [56], we found that the recombinant N-terminal regions of human CAP1 and CAP2, that contained the coiled-coil helix and HFD domains both formed tetramers. A computational approach also suggested that for the N-terminal coiled coils of both CAPs, hexamers are more enthalpically favorable than tetramers (Figure 3 and Table 2). Several factors could contribute to this discrepancy. First, in contrast to tetramers, hexamers were predicted to have an internal channel inlaid by Leu side chains and large enough to incorporate water (Figure 3g,j). Such channels would be entropically disfavored unless occupied by nonpolar molecules, e.g., fatty acids. Such complexes with fatty acids, trans-retinol, and Vitamin D have been demonstrated for the pentameric α-barrel of cartilage oligomeric matrix protein (COMP) [57,58]. We noticed that saturated fatty acids in their extended conformation fit well in the central cavity of the CAP1/2 models (Figure 3k). Therefore, it is appealing to consider that, under physiological conditions of the cell, CAP hexamers are stabilized by such nonpolar molecules. Next, the hexamers of N-CAPs can be disfavored by the lower configurational entropy of HFD domains, which were expected to be constrained in hexamers but loosely placed in lower-level oligomers. In the full-length proteins, the equilibrium can be shifted towards the hexameric state by the enthalpic forces of the pairwise association between the CAP C-terminal domains [28]. One could also speculate that the formation of hexamers or other high-order constitutive oligomers in the cell occurs co-translationally on polyribosomes, as has been shown for the assembly of vault particles [59]. This mechanism could also reduce non-productive or mistargeted interactions with other proteins and disfavor the formation of hetero-hexamers of CAP1 and CAP2. In the absence of such mechanisms, the proposed hetero-hexamers are likely upon simultaneous expression of the CAP1 and CAP2 isoforms, due to the high conservation of the coiled-coil domains varying by only four conserved amino acids (Figure 1b).

Although the stoichiometry of the recombinant CAP constructs did not match the reported stoichiometry of the native complexes, the recombinant tetramers were nearly 2-fold more effective in assisting cofilin in actin severing and depolymerization as compared to the HFD monomers (Figures 5 and 6). Furthermore, oligomers of a reduced stoichiometry (i.e., trimers and dimers) also assisted cofilin better than monomers (Supplementary Figure S3). Along with the higher binding affinities of the oligomers to actin (Figure 4g), this observation suggested that the mechanism of actin severing and depolymerization involves simultaneous or consequent and coordinated binding of more than one HFD domain to F-actin–cofilin complexes. It is tempting to speculate that the CAP oligomers may roll at the depolymerizing end of F-actin as a wheel, with "spokes" of HFD binding to both terminal and adjacent actin subunits, separating the former from the latter and from the filament body. Since severing occurs at the interface between the cofilin-decorated and cofilin-free filament regions characterized by different twists, the HFD domains of CAPs may serve to stabilize the original actin twist and provide a distinct border between the two states.

We did not observe a measurable difference between the CAP isoforms in their ability to assist in actin depolymerization (Figure 5). This observation correlates with the fact that both CAP1 and CAP2 are recognized as makers of tumorigenesis, promoting invasiveness and cytokinesis of tumor cells. While more effective in severing (Figure 6b–d), CFL2 was notably less potent than CFL1 in accelerating filament depolymerization in bulk assays (Table 3). This difference between CFL1 and CFL2 persisted in the presence of N-CAP1/2 constructs, and may account for the more abundant association of CFL2 with CAP1 observed in cells (Figure 7b). Indeed, slower dissociation of actin subunits from the filament ends might imply cycling via less productive but more extended in time association of CFL2 with CAPs. Since our experiments did not reveal a substantial functional difference between N-CAP1 and N-CAP2, it is conceivable that the difference between the isoforms is limited to their differential regulation and/or their C-terminal regions involved in recharging the depolymerized ADP–G-actin/cofilin complexes with ATP and profilin [28].

In addition to promoting actin turnover, ADF/cofilins mediate active, importin-mediated transport of actin to the nucleus [60–62], stabilize thick parallel F-actin bundles (rods) both in the cytoplasm and the nucleus under stress conditions [63–65], regulate mitochondrial dynamics [66–68], and, controversially, contribute to apoptosis upon translocation to mitochondria [69–73]. These activities appear to correlate with a low energy state or a decreased cell's reducing power, and are designed to either compensate for these deficiencies by reducing actin treadmilling and the resulted energy consumption (e.g., by sequestering ADP–actin in actin rods) or exacerbating the condition by promoting cell death (e.g., apoptosis). Nuclear transport of actin may serve similar compensatory purposes, as high levels of nuclear actin inhibit transcription by RNA polymerase II [74]. Whether or not CAP1/CAP2 assist cofilin in these processes is unknown and remains to be established in future works.

4. Materials and Methods

4.1. Molecular Cloning

N-CAP constructs with the original coiled-coil N-terminal helices were cloned with C-terminal 6×His-tags into pET21b plasmid (Novagen, Madison, WI, USA.). Dimerization 2L4HC2_23 (5J0K), and trimerization 2L6HC3_6 (5J0J) domain sequences [50] were amplified from plasmids generously provided by Dr David Baker (University of Washington, Seattle, WA, USA). Coiled-coil regions of N-CAPs (a.a. 1–29) were deleted (to generate ΔCC-N-CAP constructs) or replaced with either dimerization or trimerization domains (to generate DD-N-CAPs and TD-N-CAPs, respectively) and cloned with N-terminal 6×His-tags into pColdI vector (Clontech, San Francisco, CA, USA) using NEBuilder HiFi DNA Assembly Master Mix (New England BioLabs, Ipswich, MA, USA). Sequences of all constructs were verified by Sanger DNA sequencing (Genomics Shared Resource, OSU Comprehensive Cancer Center, Columbus, OH, USA).

4.2. Protein Purification

Actin was purified from skeletal muscle acetone powder: either of rabbit origin purchased from Pel-Freez Biologicals (Rogers, AR, USA) or of chicken origin prepared in-house from flash-frozen chicken breast (Trader Joe's, Columbus, OH, USA), as previously described [75–77].

CAP1 and CAP2 constructs (N-CAP, Δ-CC-CAP, DD-N-CAP, TD-N-CAP) were expressed in BL21-CodonPlus(DE3)pLysS *Escherichia coli* (Agilent Technologies, Santa Clara, CA, USA) grown in nutrient-rich bacterial growth medium (1.25% tryptone, 2.5% yeast extract, 125 mM NaCl, 0.4% glycerol, 50 mM tris(hydroxymethyl)aminomethane hydrochloride (Tris-HCl), pH 8.2). After reaching an OD_{600} of 1–1.2, the cells were incubated for 30 minutes on ice, and expression was induced by the addition of 1 mM isopropyl-β-D-thiogalactoside (IPTG), after which the cells were grown at 15 °C for 15–20 hours. CAP constructs were purified by immobilized metal affinity chromatography on TALON metal affinity resin (Clontech, San Francisco, CA, USA) and eluted in buffer containing 50 mM sodium phosphate, pH 7.4, 300 mM NaCl, 0.1 mM phenylmethylsulfonyl fluoride (PMSF), 2 mM benzamidine–HCl, and 250 mM imidazole. Purified constructs were dialyzed against buffer containing 20 mM Tris-HCl, pH 8.0, 30 mM NaCl, 2 mM dithiothreitol (DTT), and 0.1 mM PMSF. CAP constructs were aliquoted, flash-frozen in liquid nitrogen, and stored at −80 °C.

Tagless full-length human cofilins were expressed in *E. coli* BL21-CodonPlus(DE3) cells. Transformed bacterial cells were grown at 37 °C in 4 L of a rich medium as described above. Cultures were induced with 1 mM IPTG, and incubated at 37 °C in a shaking incubator for 4 h. Cells were pelleted at 4 °C and resuspended in ice-cold buffer containing 10 mM piperazine-N,N'-bis(2-ethanesulfonic acid) (PIPES), pH 6.8, 0.5 mM ethylenediaminetetraacetic acid (EDTA), 10 mM 2-mercaptoethanol, 1 mM PMSF, 5 mM benzamidine, SIGMAFAST protease inhibitor cocktail (EDTA-free), and lysed using a French press. Cofilins were purified by sequential anion and cation exchange chromatography as described previously [78]. Briefly, cell lysates were passed through DE52 (DEAE cellulose) and suplphopropyl (SP)-sepharose (Sigma-Aldrich, St. Louis, MO, USA) columns connected in tandem (in this order), followed by elution from the SP-sepharose column with a gradient of 50 to 500 mM NaCl in buffer containing 10 mM PIPES, pH 6.8, 0.5 mM EDTA, 10 mM 2-mercaptoethanol, and 0.5 mM PMSF. The recombinant cofilins were purified to >95% homogeneity by size-exclusion fast protein liquid chromatography (FPLC) in 10 mM 3-(N-morpholino)propanesulfonic acid (MOPS), pH 7.0, 25 mM NaCl, 1 mM DTT, and 0.1 mM PMSF.

Concentrations of all proteins were determined based on their absorption: for actin in the presence of 0.5 M NaOH, an A (1%) at 290 nm of 11.5 cm^{-1} was assumed; extinction coefficients at 280 nm for CFL1 and CFL2 as well as for all N-CAP1 and N-CAP2 constructs were predicted based on their sequences using the Expasy ProtParam online resource, Switzerland [79].

4.3. Sedimentation Velocity Analytical Ultracentrifugation (SV-AUC)

Sedimentation velocity experiments were performed in a ProteomeLab XL-I analytical ultracentrifugation system (Beckman Coulter, Chaska, MN, USA). Briefly, 50 µM protein samples were loaded into AUC cell assemblies at a 12 mm path length. To achieve chemical and thermal equilibrium, the An-50 TI rotor with loaded samples was allowed to equilibrate in the centrifuge at 20 °C for ~4 h. The rotor was spun at 50,000 rpm and absorption at 280 nm data were collected for up to 6 hours and a total of 42 scans. SEDFIT software (http://sedfitsedphat.nibib.nih.gov, version 16-1c, USA) was used to perform the data analysis with a continuous sedimentation coefficient distribution model $c(S)$. Values for 20 mM Tris-HCl, pH 8.0 buffer viscosity (0.010102 poise), density (1.02 g/mL), and partial specific volume (0.73 mL/g) were used, and confidence level was set to 0.68.

4.4. F-Actin Binding Cosedimentation Assays

Ca^{2+} in the nucleotide cleft of G-actin was switched to Mg^{2+} by adding MgCl$_2$ and ethylene glycol-bis(β-aminoethyl ether)-N,N,N',N'-tetraacetic acid (EGTA) to 0.1 and 0.5 mM, respectively, and incbuating on ice for 10 min. G-actin was then polymerized by supplementing MgCl$_2$ and KCl to 2 and 50 mM, respectively, in 20 mM Tris-HCl, pH 7.5, and incubating at room temperature for 1 h. CAPs were used at a final concentration of 5 µM, while actin concentration varied from 2.5 to 50 µM. Reaction mixtures were incubated either 1 h at room temperature or overnight at 4 °C. Reactions were centrifuged at 300,000× *g* at 25 °C for 30 min using a TLA-100 rotor in an Optima TLX ultracentrifuge (Beckman Coulter, Chaska, MN, USA). Supernatants and pellets were carefully separated, balanced by volume, and analyzed by SDS-PAGE. Gels were stained with Coomassie Brilliant Blue and quantified using ImageJ software version 2.0.0.-rc-69/1.52p, USA [80]. Binding efficiency expressed as an equilibrium dissociation constant (K_d) was quantified by fitting the data to the binding isotherm equation

$$\Delta F / \Delta F_{max} = (P + A + K_d - \sqrt{((P + A + K_d)^2 - 4PA}) / 2P$$

where A is the concentration of F-actin and P is the concentration of N-CAP constructs.

4.5. TIRF Microscopy

TIRF microscopy was conducted as described previously [76,77]. Briefly, immediately before the experiment, flow chambers were functionalized by incubation with 0.1 mg/ml streptavidin in phosphate-buffered saline (PBS) and blocked for 3 min in blocking buffer (1% (w/v) bovine serum albumin (BSA) in 50 mM Tris-HCl, pH 7.5, 150 mM NaCl), followed by successive washes with the blocking buffer and 1× TIRF buffer (10 mM imidazole, pH 7.0, 50 mM KCl, 50 mM DTT, 1 mM $MgCl_2$, 1 mM EGTA, 0.2 mM ATP, 50 μM $CaCl_2$, 15 mM glucose, 20 μg/mL catalase, 100 μg/mL glucose oxidase, 15% glycerol, and 0.5% methylcellulose-400cP (Sigma Aldrich, St. Louis, MO, USA)). Skeletal actin (33% Alexa 488-labeled, 1% biotinylated; 1.5 μM final concentration) was incubated with an exchange buffer (50 μM $MgCl_2$, 0.1 mM EGTA) for 2 min in order to switch from Ca^{2+}- to Mg^{2+}-ATP actin. Polymerization of actin was initiated by the addition of an equal volume of 2× TIRF buffer and the mixture was transferred to the flow chamber within 15 seconds. Filaments were grown to ~15 μM average length. Free actin monomers were then removed by flushing the desired concentrations of proteins in 1× TIRF buffer. Images were collected every 5 s with a Nikon Eclipse Ti-E microscope, through-the-objective TIRF illumination system, 100× oil objective, and a DS-QiMc camera (Nikon Instruments Inc., Melville, NY, USA). Data were analyzed using ImageJ software: severing events per μM of filament length were calculated by measuring the filament length in the frame prior to the flow of proteins and then manually counting the number of severing events.

4.6. Bulk F-actin Disassembly Assays

The final concentration of 2 μM, 10% pyrene-labeled F-actin was mixed with 4 μM latrunculin B and 100 nM CapZ in F-buffer (20 nM Tris-HCl, pH 7.5, 50 mM KCl, 0.2 mM ATP, 1 mM $MgCl_2$, 0.5 mM EGTA, and 1 mM DTT). At time zero, disassembly was induced by addition of either F-buffer alone, F-buffer containing cofilin, or F-buffer containing cofilin and CAPs. A decrease in fluorescence was monitored for 3000 s at 25 °C at excitation wavelength (λ_{ex} = 365 nm) and emission wavelength (λ_{em} = 407 nm) using an Infinite M1000 Pro plate reader (Tecan, Baldwin Park, CA, USA).

4.7. Native Mass Spectrometry (Native MS)

Sample purity and integrity were analyzed by online buffer exchange MS using an UltiMate™ 3000 RSLC coupled to an Exactive Plus EMR Orbitrap instrument (Thermo Fisher Scientific, Grand Island, NY, USA) modified to incorporate a quadrupole mass filter and allow for surface-induced dissociation [81]. Between 100 and 300 pmole protein (referring to monomer) were injected and online buffer exchanged to 200 mM ammonium acetate, pH 6.8 by a self-packed buffer exchange column [81,82] (P6 polyacrylamide gel; Bio-Rad Laboratories, Hercules, CA, USA) at a flow rate of 100 μL per min. Mass spectra were recorded for 1000–14000 m/z at 8750 resolution, as defined at 200 m/z. The injection time was set to 200 ms. Voltages applied to the transfer optics were optimized to allow ion transmission while minimizing unintentional ion activation. Mass spectra were deconvoluted with UniDec version 4.0.0 beta, England [83].

4.8. Cell Culture, Western Blotting, and In Situ Proximity Ligation Assay (PLA)

To reveal the native interactions of CAP proteins with cofilin isoforms in cells, we utilized a Duolink in situ proximity ligation immunofluorescence assay (Sigma-Aldrich, St. Louis, MO, USA). In this assay, two primary antibodies raised in different species are used to detect two unique protein targets (e.g., CFL and CAP) followed by binding of the corresponding species-specific secondary antibodies conjugated with oligonucleotides (PLA probes). If the target proteins interact with each other, the corresponding PLA probes will be close to each other (<40 nm) and hybridizing connector oligos will join the PLA probes, consequently amplifying the localized signal up to 1000-fold by rolling-circle amplification. The signal is then visualized as discrete spots by microscope imaging.

Primary antibodies for Duolink PLA were evaluated by western blot analysis using whole-cell lysates (WCL; 50 µg per lane) and purified recombinant proteins (50 ng per lane). Rabbit anti-CFL1 (#5175 Cell Signaling Technology, Danvers, MA, USA) and rabbit anti-CFL2 (#AP20625c Abgent, San Diego, CA, USA) specifically recognized the respective recombinant proteins, as well as native proteins in WCLs at 1:1000 dilution (Figure 7a and Supplementary Figure S5), while rabbit anti-CFL2 (#GTX100213 GeneTex, Irvine, CA, USA) failed to recognize recombinant CFL2. Mouse anti-CAP1 (#SAB1406999 Sigma-Aldrich, St. Louis, MO, USA) specifically recognized recombinant ΔCC-N-CAP1 and native CAP1 in WCLs at 1:500 (Figure 7a and Supplementary Figure S4). However, mouse anti-CAP2 antibody (#sc-377471 Santa Cruz Biotechnology, Dallas, TX, USA) raised against amino acids 77–121 of human CAP2 failed to recognize the recombinant human ΔCC-N-CAP2 (a.a. 30–220) (Supplementary Figure S6a). Furthermore, while mouse anti-CAP2 (#H00010486-M01 Abnova, Taiwan) raised against GST-tagged full-length recombinant human CAP2 specifically recognized the recombinant human ΔCC-N-CAP2, staining of WCLs produced multiple major non-specific bands (Supplementary Figure S6b), which made this antibody unusable for PLA applications. Secondary antibodies used for western blotting analysis were anti-mouse and anti-rabbit IgG conjugated to horseradish peroxidase (#A4416 and #A0545 Sigma-Aldrich, St. Louis, MO, USA), both at 1:10,000. The signal was detected using a WesternBright Sirius chemiluminescent HRP substrate (#K-12043 Advansta, San Jose, CA, USA) in an OmegaLum G Aplegen imager (Gel Company, San Francisco, CA, USA).

The following cell lines were grown according to ATCC recommendations: HeLa, CaCo 2, LS 174T, HEK 293, HT 1080, U2OS, PANC 1, MDA-MB-231, -436, -468, SKBR 3, MCF 7, Raw 264.7, 3T3, CHOK1, CCL-39, MDCK, IEC-18, WI-38, and Hs 578T. Anti-cofilin western blots revealed that among these cell lines, Hs 578T demonstrated prominent expression of both CFL1 and CFL2 isoforms (Supplementary Figure S5). Anti-CAP1 western blotting confirmed similarly high levels of CAP1 expression in Hs 578T, compared to HeLa, HT 1080, U2OS, MDA-MB-436, SKBR 3, and WI-38 (Supplementary Figure S4). Therefore, the Hs 578T cell line was selected for the PLA studies.

For the Duolink in situ PLA, Hs 578T cells were plated at 30%–50% confluency on µ-slide ibiTreat 8 well plates (#80826 ibidi, Germany), allowed to attach overnight, and fixed/permeabilized for 15 min in PBS containing 4% formaldehyde and 0.1% Triton X-100. Duolink in situ PLA staining was performed according to the manufacturer's user guide. Pairs of primary antibodies were used as followed: rabbit anti-CAP1 (#SAB1406999 Sigma-Aldrich, St. Louis, MO, USA) 1:100 with either mouse anti-CFL1 (#5175, 1:400; Cell Signaling Technology, Danvers, MA, USA) or mouse anti-CFL2 (#AP20625c, 1:100; Abgent, San Diego, CA, USA). For the negative control staining, only one primary antibody was used, followed by both PLA probes and rolling cycle amplification. Cells were counter-stained with FITC–phalloidin (Thermo Fisher Scientific; 30 nM final concentration in PBS), mounted in Duolink in situ mounting medium with DAPI (Sigma-Aldrich), and imaged using Eclipse Ti-E microscope with a 60× oil objective and DS-QiMc camera (Nikon Instruments Inc., Melville, NY, USA).

Supplementary Materials: Supplementary materials can be found at http://www.mdpi.com/1422-0067/20/22/5647/s1.

Author Contributions: Conceptualization, D.S.K.; methodology, V.P., F.B., V.H.W., E.K., D.S.K.; data analysis, V.P., F.B.; investigation, V.P., F.B., E.K.; writing—original draft preparation, D.S.K., V.P.; writing—review and editing, V.P., E.K., D.S.K., F.B., V.H.W.; visualization, V.P., E.K., D.S.K.; supervision, D.S.K., V.H.W.; funding acquisition, D.S.K., V.H.W.

Funding: This work was supported by the National Institute of General Medical Sciences of the NIH under award number R01 GM114666, to D.S.K. Native MS measurements were funded by the National Institutes of Health, grant number P41 GM128577, to V.H.W.

Acknowledgments: We thank David Baker (University of Washington, Seattle) for providing plasmids encoding dimerization "2L4HC2_23 (5J0K)" and trimerization "2L6HC3_6 (5J0J)" domain sequences.

Conflicts of Interest: The authors declare no conflict of interest. The funders had no role in the design of the study; in the collection, analyses, or interpretation of data; in the writing of the manuscript, or in the decision to publish the results.

Abbreviations

ADF	Actin depolymerizing factor
CFL	Cofilin
Srv2/CAP	Cyclase-associated protein
CC	Coiled-coil
HFD	Helical folded domain
P_1 and P_2	Proline-rich regions 1 and 2
WH2	Wiskott–Aldrich homology 2
CARP	C-terminal β-sheet domain
Aip1	Actin interacting protein 1
ATP	Adenosine triphosphate
ADP	Adenosine diphosphate
P_i	Inorganic phosphate
SV-AUC	Sedimentation velocity analytical ultracentrifugation
c(S)	Sedimentation coefficient distribution
TIRF	Total internal reflection fluorescence
MS	Mass spectrometry
PLA	Proximity ligation assay
SDS-PAGE	Sodium dodecyl sulfate polyacrylamide gel electrophoresis
BUDE	Bristol University Docking Engine
COMP	Cartridge oligomeric matrix protein
WCL	Whole-cell lysate
FITC	Fluorescein isothiocyanate
DAPI	4′,6-Diamidino-2-phenylindole
IPTG	Isopropyl-β-D-thiogalactoside
PMSF	Phenylmethylsulfonyl fluoride
DTT	Dithiothreitol
TRIS	Tris(hydroxymethyl)aminomethane
PIPES	Piperazine-N,N′-bis(2-ethanesulfonic acid)
EDTA	Ethylenediaminetetraacetic acid
MOPS	3-(N-morpholino)propanesulfonic acid
EGTA	Ethylene glycol-bis(β-aminoethyl ether)-N,N,N′,N′-tetraacetic acid
PBS	Phosphate-buffered saline
GST	Glutathione S-transferase
BSA	Bovine serum albumin
HRP	Horseradish peroxidase

References

1. Pantaloni, D.; Carlier, M.F.; Korn, E.D. The Interaction between ATP-Actin and ADP-Actin. A Tentative Model for Actin Polymerization. *J. Biol. Chem.* **1985**, *260*, 6572–6578. [PubMed]
2. Pollard, T.D. Rate Constants for the Reactions of ATP- and ADP-Actin with the Ends of Actin Filaments. *J. Cell Biol.* **1986**, *103*, 2747–2754. [CrossRef] [PubMed]
3. Chen, H.; Berstain, B.; Bamburg, J. Regulating Actin-Filament Dynamics in Vivo. *Trends Biochem. Sci.* **2000**, *25*, 19–23. [CrossRef]
4. Pollard, T.D.; Cooper, J.A. Actin, a Central Player in Cell Shape and Movement. *Science* **2009**, *326*, 1208–1212. [CrossRef] [PubMed]
5. Dominguez, R. Actin Filament Nucleation and Elongation Factors–Structure–Function Relationships. *Crit. Rev. Biochem. Mol. Biol.* **2009**, *44*, 351–366. [CrossRef] [PubMed]
6. Courtemanche, N. Mechanisms of Formin-Mediated Actin Assembly and Dynamics. *Biophys. Rev.* **2018**, *10*, 1553–1569. [CrossRef] [PubMed]
7. Ono, S. Mechanism of Depolymerization and Severing of Actin Filaments and Its Significance in Cytoskeletal Dynamics. *Int. Rev. Cytol.* **2007**, *258*, 1–82. [PubMed]

8. Vartiainen, M.K. The Three Mouse Actin-Depolymerizing Factor/Cofilins Evolved to Fulfill Cell-Type-Specific Requirements for Actin Dynamics. *Mol. Biol. Cell* **2002**, *13*, 183–194. [CrossRef] [PubMed]
9. Kudryashov, D.S.; Grintsevich, E.E.; Rubenstein, P.A.; Reisler, E. A Nucleotide State-Sensing Region on Actin. *J. Biol. Chem.* **2010**, *285*, 25591–25601. [CrossRef] [PubMed]
10. Kudryashov, D.S.; Reisler, E. ATP and ADP Actin States. *Biopolymers* **2013**, *99*, 245–256. [CrossRef] [PubMed]
11. McGough, A.; Pope, B.; Chiu, W.; Weeds, A. Cofilin Changes the Twist of F-Actin: Implications for Actin Filament Dynamics and Cellular Function. *J. Cell Biol.* **1997**, *138*, 771–781. [CrossRef] [PubMed]
12. Galkin, V.E.; Orlova, A.; Kudryashov, D.S.; Solodukhin, A.; Reisler, E.; Schroder, G.F.; Egelman, E.H. Remodeling of Actin Filaments by ADF/Cofilin Proteins. *Proc. Natl. Acad. Sci. USA* **2011**, *108*, 20568–20572. [CrossRef] [PubMed]
13. Tanaka, K.; Takeda, S.; Mitsuoka, K.; Oda, T.; Kimura-Sakiyama, C.; Maéda, Y.; Narita, A. Structural Basis for Cofilin Binding and Actin Filament Disassembly. *Nat. Commun.* **2018**, *9*, 1860. [CrossRef] [PubMed]
14. Muhlrad, A.; Kudryashov, D.; Michael Peyser, Y.; Bobkov, A.A.; Almo, S.C.; Reisler, E. Cofilin Induced Conformational Changes in F-Actin Expose Subdomain 2 to Proteolysis. *J. Mol. Biol.* **2004**, *342*, 1559–1567. [CrossRef] [PubMed]
15. Bobkov, A.A.; Muhlrad, A.; Shvetsov, A.; Benchaar, S.; Scoville, D.; Almo, S.C.; Reisler, E. Cofilin (ADF) Affects Lateral Contacts in F-Actin. *J. Mol. Biol.* **2004**, *337*, 93–104. [CrossRef] [PubMed]
16. Bobkov, A.A.; Muhlrad, A.; Kokabi, K.; Vorobiev, S.; Almo, S.C.; Reisler, E. Structural Effects of Cofilin on Longitudinal Contacts in F-Actin. *J. Mol. Biol.* **2002**, *323*, 739–750. [CrossRef]
17. Kudryashov, D.S.; Galkin, V.E.; Orlova, A.; Phan, M.; Egelman, E.H.; Reisler, E. Cofilin Cross-Bridges Adjacent Actin Protomers and Replaces Part of the Longitudinal F-Actin Interface. *J. Mol. Biol.* **2006**, *358*, 785–797. [CrossRef] [PubMed]
18. Huehn, A.; Cao, W.; Elam, W.A.; Liu, X.; De La Cruz, E.M.; Sindelar, C.V. The Actin Filament Twist Changes Abruptly at Boundaries between Bare and Cofilin-Decorated Segments. *J. Biol. Chem.* **2018**, *293*, 5377–5383. [CrossRef] [PubMed]
19. Carlier, M.F.; Laurent, V.; Santolini, J.; Melki, R.; Didry, D.; Xia, G.X.; Hong, Y.; Chua, N.H.; Pantaloni, D. Actin Depolymerizing Factor (ADF/Cofilin) Enhances the Rate of Filament Turnover: Implication in Actin-Based Motility. *J. Cell Biol.* **1997**, *136*, 1307–1322. [CrossRef] [PubMed]
20. Ressad, F.; Didry, D.; Xia, G.; Hong, Y.; Chua, N.; Pantaloni, D.; Carlier, M. Kinetic Analysis of the Interaction of Actin-Depolymerizing Factor (ADF)/Cofilin with G- and F-Actins. *J. Biol. Chem.* **1998**, *273*, 20894–20902. [CrossRef] [PubMed]
21. Ressad, F.; Didry, D.; Egile, C.; Pantaloni, D.; Carlier, M.F. Control of Actin Filament Length and Turnover by Actin Depolymerizing Factor (ADF/Cofilin) in the Presence of Capping Proteins and ARP2/3 Complex. *J. Biol. Chem.* **1999**, *274*, 20970–20976. [CrossRef] [PubMed]
22. Balcer, H.I.; Goodman, A.L.; Rodal, A.A.; Smith, E.; Kugler, J.; Heuser, J.E.; Goode, B.L. Coordinated Regulation of Actin Filament Turnover by a High-Molecular-Weight Srv2/CAP Complex, Cofilin, Profilin, and Aip1. *Curr. Biol.* **2003**, *13*, 2159–2169. [CrossRef] [PubMed]
23. Mattila, P.K.; Quintero-Monzon, O.; Kugler, J.; Moseley, J.B.; Almo, S.C.; Lappalainen, P.; Goode, B.L. A High-Affinity Interaction with ADP-Actin Monomers Underlies the Mechanism and in Vivo Function of Srv2/Cyclase-Associated Protein. *Mol. Biol. Cell* **2004**, *15*, 5158–5171. [CrossRef] [PubMed]
24. Okada, K.; Ravi, H.; Smith, E.M.; Goode, B.L. Aip1 and Cofilin Promote Rapid Turnover of Yeast Actin Patches and Cables: A Coordinated Mechanism for Severing and Capping Filaments. *Mol. Biol. Cell* **2006**, *17*, 2855–2868. [CrossRef] [PubMed]
25. Gandhi, M.; Achard, V.; Blanchoin, L.; Goode, B.L. Coronin Switches Roles in Actin Disassembly Depending on the Nucleotide State of Actin. *Mol. Cell* **2009**, *34*, 364–374. [CrossRef] [PubMed]
26. Jansen, S.; Collins, A.; Chin, S.M.; Ydenberg, C.A.; Gelles, J.; Goode, B.L. Single-Molecule Imaging of a Three-Component Ordered Actin Disassembly Mechanism. *Nat. Commun.* **2015**, *6*, 7202. [CrossRef] [PubMed]
27. Ge, P.; Durer, Z.A.O.; Kudryashov, D.; Zhou, Z.H.; Reisler, E. Cryo-EM Reveals Different Coronin Binding Modes for ADP–and ADP–BeFx Actin Filaments. *Nat. Struct. Mol. Biol.* **2014**, *21*, 1075–1081. [CrossRef] [PubMed]
28. Kotila, T.; Kogan, K.; Enkavi, G.; Guo, S.; Vattulainen, I.; Goode, B.L.; Lappalainen, P. Structural Basis of Actin Monomer Re-Charging by Cyclase-Associated Protein. *Nat. Commun.* **2018**, *9*, 1892. [CrossRef] [PubMed]

29. Quintero-Monzon, O.; Jonasson, E.M.; Bertling, E.; Talarico, L.; Chaudhry, F.; Sihvo, M.; Lappalainen, P.; Goode, B.L. Reconstitution and Dissection of the 600-KDa Srv2/CAP Complex. *J. Biol. Chem.* **2009**, *284*, 10923–10934. [CrossRef] [PubMed]
30. Ono, S. The Role of Cyclase-Associated Protein in Regulating Actin Filament Dynamics—More than a Monomer-Sequestration Factor. *J. Cell Sci.* **2013**, *126*, 3249–3258. [CrossRef] [PubMed]
31. Chaudhry, F.; Breitsprecher, D.; Little, K.; Sharov, G.; Sokolova, O.; Goode, B.L. Srv2/Cyclase-Associated Protein Forms Hexameric Shurikens That Directly Catalyze Actin Filament Severing by Cofilin. *Mol. Biol. Cell* **2013**, *24*, 31–41. [CrossRef] [PubMed]
32. Johnston, A.B.; Collins, A.; Goode, B.L. High-Speed Depolymerization at Actin Filament Ends Jointly Catalysed by Twinfilin and Srv2/CAP. *Nat. Cell Biol.* **2015**, *17*, 1504–1511. [CrossRef] [PubMed]
33. Kakurina, G.V.; Kolegova, E.S.; Kondakova, I.V. Adenylyl Cyclase-Associated Protein 1: Structure, Regulation, and Participation in Cellular Processes. *Biochemistry* **2018**, *83*, 45–53. [CrossRef] [PubMed]
34. Chaudhry, F.; Little, K.; Talarico, L.; Quintero-Monzon, O.; Goode, B.L. A Central Role for the WH2 Domain of Srv2/CAP in Recharging Actin Monomers to Drive Actin Turnover in Vitro and in Vivo. *Cytoskeleton* **2010**, *67*, 120–133. [CrossRef] [PubMed]
35. Witke, W. The Role of Profilin Complexes in Cell Motility and Other Cellular Processes. *Trends Cell Biol.* **2004**, *14*, 461–469. [CrossRef] [PubMed]
36. Goode, B.L.; Eck, M.J. Mechanism and Function of Formins in the Control of Actin Assembly. *Annu. Rev. Biochem.* **2007**, *76*, 593–627. [CrossRef] [PubMed]
37. Pantaloni, D.; Carlier, M.F. How Profilin Promotes Actin Filament Assembly in the Presence of Thymosin B4. *Cell* **1993**, *75*, 1007–1014. [CrossRef]
38. Jansen, S.; Collins, A.; Golden, L.; Sokolova, O.; Goode, B.L. Structure and Mechanism of Mouse Cyclase-Associated Protein (CAP1) in Regulating Actin Dynamics. *J. Biol. Chem.* **2014**, *289*, 30732–30742. [CrossRef] [PubMed]
39. Woolfson, D.N.; Bartlett, G.J.; Bruning, M.; Thomson, A.R. New Currency for Old Rope: From Coiled-Coil Assemblies to α-Helical Barrels. *Curr. Opin. Struct. Biol.* **2012**, *22*, 432–441. [CrossRef] [PubMed]
40. Yu, G.; Swiston, J.; Young, D. Comparison of Human CAP and CAP2, Homologs of the Yeast Adenylyl Cyclase-Associated Proteins. *J. Cell Sci.* **1994**, *107*, 1671–1678. [PubMed]
41. Peche, V.; Shekar, S.; Leichter, M.; Korte, H.; Schröder, R.; Schleicher, M.; Holak, T.A.; Clemen, C.S.; Ramanath-Y., B.; Pfitzer, G.; et al. CAP2, Cyclase-Associated Protein 2, Is a Dual Compartment Protein. *Cell. Mol. Life Sci.* **2007**, *64*, 2702–2715. [CrossRef] [PubMed]
42. Peche, V.S.; Holak, T.A.; Burgute, B.D.; Kosmas, K.; Kale, S.P.; Wunderlich, F.T.; Elhamine, F.; Stehle, R.; Pfitzer, G.; Nohroudi, K.; et al. Ablation of Cyclase-Associated Protein 2 (CAP2) Leads to Cardiomyopathy. *Cell. Mol. Life Sci.* **2013**, *70*, 527–543. [CrossRef] [PubMed]
43. Tan, M.; Song, X.; Zhang, G.; Peng, A.; Li, X.; Li, M.; Liu, Y.; Wang, C. Overexpression of Adenylate Cyclase-Associated Protein 1 Is Associated with Metastasis of Lung Cancer. *Oncol. Rep.* **2013**, *30*, 1639–1644. [CrossRef] [PubMed]
44. Xie, S.; Shen, C.; Tan, M.; Li, M.; Song, X.; Wang, C. Systematic Analysis of Gene Expression Alterations and Clinical Outcomes of Adenylate Cyclase-Associated Protein in Cancer. *Oncotarget* **2017**, *8*, 27216–27239. [PubMed]
45. Hasan, R.; Zhou, G.L. The Cytoskeletal Protein Cyclase-Associated Protein 1 (CAP1) in Breast Cancer: Context-Dependent Roles in Both the Invasiveness and Proliferation of Cancer Cells and Underlying Cell Signals. *Int. J. Mol. Sci.* **2019**, *20*, 2653. [CrossRef] [PubMed]
46. Hua, M.; Yan, S.; Deng, Y.; Xi, Q.; Liu, R.; Yang, S.; Liu, J.; Tang, C.; Wang, Y.; Zhing, J. CAP1 Is Overexpressed in Human Epithelial Ovarian Cancer and Promotes Cell Proliferation. *Int. J. Mol. Med.* **2015**, *35*, 941–949. [CrossRef] [PubMed]
47. Li, L.; Fu, L.Q.; Wang, H.J.; Wang, Y.Y. CAP2 Is a Valuable Biomarker for Diagnosis and Prognostic in Patients with Gastric Cancer. *Pathol. Oncol. Res.* **2018**, *37*, 17784–17792. [CrossRef] [PubMed]
48. Xu, L.; Peng, S.; Huang, Q.; Liu, Y.; Jiang, H.; Li, X.; Wang, J. Expression Status of Cyclase-Associated Protein 2 as a Prognostic Marker for Human Breast Cancer. *Oncol. Rep.* **2016**, *36*, 1981–1988. [CrossRef] [PubMed]
49. Masugi, Y.; Tanese, K.; Emoto, K.; Yamazaki, K.; Effendi, K.; Funakoshi, T.; Mori, M.; Sakamoto, M. Overexpression of Adenylate Cyclase-Associated Protein 2 Is a Novel Prognostic Marker in Malignant Melanoma. *Pathol. Int.* **2015**, *65*, 627–634. [CrossRef] [PubMed]

50. Boyken, S.E.; Chen, Z.; Groves, B.; Langan, R.A.; Oberdorfer, G.; Ford, A.; Gilmore, J.M.; Xu, C.; DiMaio, F.; Pereira, J.H.; et al. De Novo Design of Protein Homo-Oligomers with Modular Hydrogen-Bond Network-Mediated Specificity. *Science* **2016**, *352*, 680–687. [CrossRef] [PubMed]
51. Zhao, H.; Brautigam, C.A.; Ghirlando, R.; Schuck, P. Overview of Current Methods in Sedimentation Velocity and Sedimentation Equilibrium Analytical Ultracentrifugation. *Curr. Protoc. Protein Sci.* **2013**, *71*. [CrossRef] [PubMed]
52. Wood, C.W.; Woolfson, D.N. CCBuilder 2.0: Powerful and Accessible Coiled-Coil Modeling. *Protein Sci.* **2018**, *27*, 103–111. [CrossRef] [PubMed]
53. Paavilainen, V.O.; Oksanen, E.; Goldman, A.; Lappalainen, P. Structure of the Actin-Depolymerizing Factor Homology Domain in Complex with Actin. *J. Cell Biol.* **2008**, *182*, 51–59. [CrossRef] [PubMed]
54. Chin, S.M.; Jansen, S.; Goode, B.L. TIRF Microscopy Analysis of Human Cof1, Cof2, and ADF Effects on Actin Filament Severing and Turnover. *J. Mol. Biol.* **2016**, *428*, 1604–1616. [CrossRef] [PubMed]
55. Effendi, K.; Yamazaki, K.; Mori, T.; Masugi, Y.; Makino, S.; Sakamoto, M. Involvement of Hepatocellular Carcinoma Biomarker, Cyclase-Associated Protein 2 in Zebrafish Body Development and Cancer Progression. *Exp. Cell Res.* **2013**, *319*, 35–44. [CrossRef] [PubMed]
56. Mu, A.; Fung, T.S.; Kettenbach, A.N.; Chakrabarti, R.; Higgs, H.N. A Complex Containing Lysine-Acetylated Actin Inhibits the Formin INF2. *Nat. Cell Biol.* **2019**, *21*, 592–602.
57. Guo, Y.; Bozic, D.; Malashkevich, V.N.; Kammerer, R.A.; Schulthess, T.; Engel, J. All-Trans Retinol, Vitamin D and Other Hydrophobic Compounds Bind in the Axial Pore of the Five-Stranded Coiled-Coil Domain of Cartilage Oligomeric Matrix Protein. *EMBO J.* **1998**, *17*, 5265–5272. [CrossRef] [PubMed]
58. MacFarlane, A.A.; Orriss, G.; Okun, N.; Meier, M.; Klonisch, T.; Khajehpour, M.; Stetefeld, J. The Pentameric Channel of COMPcc in Complex with Different Fatty Acids. *PLoS ONE* **2012**, *7*, e48130. [CrossRef] [PubMed]
59. Mrazek, J.; Toso, D.; Ryazantsev, S.; Zhang, X.; Zhou, Z.H.; Fernandez, B.C.; Kickhoefer, V.A.; Rome, L.H. Polyribosomes Are Molecular 3D Nanoprinters That Orchestrate the Assembly of Vault Particles. *ACS Nano* **2014**, *8*, 11552–11559. [CrossRef] [PubMed]
60. Dopie, J.; Skarp, K.P.; Rajakylä, E.K.; Tanhuanpää, K.; Vartiainen, M.K. Active Maintenance of Nuclear Actin by Importin 9 Supports Transcription. *Proc. Natl. Acad. Sci. USA* **2012**, *109*, E544–E552. [CrossRef] [PubMed]
61. Pendleton, A.; Pope, B.; Weeds, A.; Koffer, A. Latrunculin B or ATP Depletion Induces Cofilin-Dependent Translocation of Actin into Nuclei of Mast Cells. *J. Biol. Chem.* **2003**, *278*, 14394–14400. [CrossRef] [PubMed]
62. Munsie, L.N.; Desmond, C.R.; Truant, R. Cofilin Nuclear-Cytoplasmic Shuttling Affects Cofilin-Actin Rod Formation during Stress. *J. Cell Sci.* **2012**, *125*, 3977–3988. [CrossRef] [PubMed]
63. Minamide, L.S.; Striegl, A.M.; Boyle, J.A.; Meberg, P.J.; Bamburg, J.R. Neurodegenerative Stimuli Induce Persistent ADF/Cofilin-Actin Rods That Disrupt Distal Neurite Function. *Nat. Cell Biol.* **2000**, *2*, 628–636. [CrossRef] [PubMed]
64. Ono, S.; Abe, H.; Nagaoka, R.; Obinata, T. Colocalization of ADF and Cofilin in Intranuclear Actin Rods of Cultured Muscle Cells. *J. Muscle Res. Cell Motil.* **1993**, *14*, 195–204. [CrossRef] [PubMed]
65. Bamburg, J.R.; Bernstein, B.W. Actin Dynamics and Cofilin-Actin Rods in Alzheimer Disease. *Cytoskeleton* **2016**, *73*, 477–497. [CrossRef] [PubMed]
66. Beck, H.; Flynn, K.; Lindenberg, K.S.; Schwarz, H.; Bradke, F.; Di Giovanni, S.; Knöll, B. Serum Response Factor (SRF)-Cofilin-Actin Signaling Axis Modulates Mitochondrial Dynamics. *Proc. Natl. Acad. Sci. USA* **2012**, *109*, E2523–E2532. [CrossRef] [PubMed]
67. Li, G.B.; Zhang, H.W.; Fu, R.Q.; Hu, X.Y.; Liu, L.; Li, Y.N.; Liu, Y.X.; Liu, X.; Hu, J.J.; Deng, Q.; et al. Mitochondrial Fission and Mitophagy Depend on Cofilin-Mediated Actin Depolymerization Activity at the Mitochondrial Fission Site. *Oncogene* **2018**, *37*, 1485–1502. [CrossRef] [PubMed]
68. Li, G.; Zhou, J.; Budhraja, A.; Hu, X.; Chen, Y.; Cheng, Q.; Liu, L.; Zhou, T.; Li, P.; Liu, E.; et al. Mitochondrial Translocation and Interaction of Cofilin and Drp1 Are Required for Erucin-Induced Mitochondrial Fission and Apoptosis. *Oncotarget* **2015**, *6*, 1834–1849. [CrossRef] [PubMed]
69. Chua, B.T.; Volbracht, C.; Tan, K.O.; Li, R.; Yu, V.C.; Li, P. Mitochondrial Translocation of Cofilin Is an Early Step in Apoptosis Induction. *Nat. Cell Biol.* **2003**, *5*, 1083–1089. [CrossRef] [PubMed]
70. Rehklau, K.; Gurniak, C.B.; Conrad, M.; Friauf, E.; Ott, M.; Rust, M.B. ADF/Cofilin Proteins Translocate to Mitochondria during Apoptosis but Are Not Generally Required for Cell Death Signaling. *Cell Death Differ.* **2012**, *19*, 958–967. [CrossRef] [PubMed]

71. Hoffmann, L.; Rust, M.B.; Culmsee, C. Actin(g) on Mitochondria—A Role for Cofilin1 in Neuronal Cell Death Pathways. *Biol. Chem.* **2019**, *400*, 1089–1097. [CrossRef] [PubMed]
72. Li, G.; Cheng, Q.; Liu, L.; Zhou, T.; Shan, C.; Hu, X.; Zhou, J.; Liu, E.; Li, P.; Gao, N. Mitochondrial Translocation of Cofilin Is Required for Allyl Isothiocyanate-Mediated Cell Death via ROCK1/PTEN/PI3K Signaling Pathway. *Cell Commun. Signal.* **2013**, *11*, 50. [CrossRef] [PubMed]
73. Zhang, Y.; Fu, R.; Liu, Y.; Li, J.; Zhang, H.; Hu, X.; Chen, Y.; Liu, X.; Li, Y.; Li, P.; et al. Dephosphorylation and Mitochondrial Translocation of Cofilin Sensitizes Human Leukemia Cells to Cerulenin-Induced Apoptosis via the ROCK1/Akt/JNK Signaling Pathway. *Oncotarget* **2016**, *7*, 20655–20668. [CrossRef] [PubMed]
74. Serebryannyy, L.A.; Parilla, M.; Annibale, P.; Cruz, C.M.; Laster, K.; Gratton, E.; Kudryashov, D.; Kosak, S.T.; Gottardi, C.J.; de Lanerolle, P. Persistent Nuclear Actin Filaments Inhibit Transcription by RNA Polymerase II. *J. Cell Sci.* **2016**, *129*, 3412–3425. [CrossRef] [PubMed]
75. Spudich, J.A.; Watt, S. The Regulation of Rabbit Skeletal Muscle Contraction. I. Biochemical Studies of the Interaction of the Tropomyosin-Troponin Complex with Actin and the Proteolytic Fragments of Myosin. *J. Biol. Chem.* **1971**, *246*, 4866–4871. [PubMed]
76. Heisler, D.B.; Kudryashova, E.; Grinevich, D.O.; Suarez, C.; Winkelman, J.D.; Birukov, K.G.; Kotha, S.R.; Parinandi, N.L.; Vavylonis, D.; Kovar, D.R.; et al. ACD Toxin–Produced Actin Oligomers Poison Formin-Controlled Actin Polymerization. *Science* **2015**, *349*, 535–539. [CrossRef] [PubMed]
77. Kudryashova, E.; Heisler, D.B.; Williams, B.; Harker, A.J.; Shafer, K.; Quinlan, M.E.; Kovar, D.R.; Vavylonis, D.; Kudryashov, D.S. Actin Cross-Linking Toxin Is a Universal Inhibitor of Tandem-Organized and Oligomeric G-Actin Binding Proteins. *Curr. Biol.* **2018**, *28*, 1536–1547. [CrossRef] [PubMed]
78. Yehl, J.; Kudryashova, E.; Reisler, E.; Kudryashov, D.; Polenova, T. Structural Analysis of Human Cofilin 2/Filamentous Actin Assemblies: Atomic-Resolution Insights from Magic Angle Spinning NMR Spectroscopy. *Sci. Rep.* **2017**, *7*, 44506. [CrossRef] [PubMed]
79. Gasteiger, E. ExPASy: The Proteomics Server for in-Depth Protein Knowledge and Analysis. *Nucleic Acids Res.* **2003**, *31*, 3784–3788. [CrossRef] [PubMed]
80. Schneider, C.A.; Rasband, W.S.; Eliceiri, K.W. NIH Image to ImageJ: 25 Years of Image Analysis. *Nat. Methods* **2012**, *9*, 671–675. [CrossRef] [PubMed]
81. Vanaernum, Z.L.; Gilbert, J.D.; Belov, M.E.; Makarov, A.A.; Horning, S.R.; Wysocki, V.H. Surface-Induced Dissociation of Noncovalent Protein Complexes in an Extended Mass Range Orbitrap Mass Spectrometer. *Anal. Chem.* **2019**, *91*, 3611–3618. [CrossRef] [PubMed]
82. VanAernum, Z.; Buch, F.; Jones, B.J.; Jia, M.; Chen, Z.; Boyken, S.E.; Sahasrabuddhe, A.; Baker, D.; Wysocki, V. Rapid Online Buffer Exchange: A Method for Screening of Proteins, Protein Complexes, and Cell Lysates by Native Mass Spectrometry. *ChemRxiv* **2019**. [CrossRef]
83. Marty, M.T.; Baldwin, A.J.; Marklund, E.G.; Hochberg, G.K.A.; Benesch, J.L.P.; Robinson, C.V. Bayesian Deconvolution of Mass and Ion Mobility Spectra: From Binary Interactions to Polydisperse Ensembles. *Anal. Chem.* **2015**, *87*, 4370–4376. [CrossRef] [PubMed]

© 2019 by the authors. Licensee MDPI, Basel, Switzerland. This article is an open access article distributed under the terms and conditions of the Creative Commons Attribution (CC BY) license (http://creativecommons.org/licenses/by/4.0/).

Review

ERM Proteins at the Crossroad of Leukocyte Polarization, Migration and Intercellular Adhesion

Almudena García-Ortiz † and Juan Manuel Serrador *

Interactions with the Environment Program, Immune System Development and Function Unit, Centro de Biología Molecular "Severo Ochoa" (CBMSO). CSIC-UAM, 28049 Madrid, Spain; almudenagor@gmail.com
* Correspondence: jmserrador@cbm.csic.es; Tel.: +34-911964547
† Current Address: Departamento de Hematología Traslacional, Servicio de Hematología, Hospital Universitario 12 de Octubre, 28041 Madrid, Spain.

Received: 18 January 2020; Accepted: 19 February 2020; Published: 22 February 2020

Abstract: Ezrin, radixin and moesin proteins (ERMs) are plasma membrane (PM) organizers that link the actin cytoskeleton to the cytoplasmic tail of transmembrane proteins, many of which are adhesion receptors, in order to regulate the formation of F-actin-based structures (e.g., microspikes and microvilli). ERMs also effect transmission of signals from the PM into the cell, an action mainly exerted through the compartmentalized activation of the small Rho GTPases Rho, Rac and Cdc42. Ezrin and moesin are the ERMs more highly expressed in leukocytes, and although they do not always share functions, both are mainly regulated through phosphatidylinositol 4,5-bisphosphate (PIP_2) binding to the N-terminal band 4.1 protein-ERM (FERM) domain and phosphorylation of a conserved Thr in the C-terminal ERM association domain (C-ERMAD), exerting their functions through a wide assortment of mechanisms. In this review we will discuss some of these mechanisms, focusing on how they regulate polarization and migration in leukocytes, and formation of actin-based cellular structures like the phagocytic cup-endosome and the immune synapse in macrophages/neutrophils and lymphocytes, respectively, which represent essential aspects of the effector immune response.

Keywords: ezrin; moesin; actin; leukocytes; polarization; immune synapse

1. Introduction

The plasma membrane (PM)-associated cytoskeleton, namely the cell cortex, is a dense network of microfilaments and motor proteins of the myosin II family that coordinately produces tension under the PM of cells. Such PM-associated tension controls cell shape, polarization, motility and cell–cell interactions, among other important cellular functions. In addition to actin and myosin II, the cell cortex contains roughly one hundred actin binding proteins (ABPs) that are involved in organization of the actin meshworks and are important for the generation and regulation of tension near the PM [1]. Ezrin, radixin and moesin proteins (ERMs) and merlin are among the ABPs that regulate organization of actin filaments (F-actin) under the PM (reviewed in [2]). ERMs localize to PM protrusions (e.g., microvilli, filopodia, retraction fibers and pseudopods), cell–cell junctions and the cleavage furrow of dividing cells. The ERM-related protein merlin, the neurofibromatosis type 2 (NF2) tumor suppressor gene product, is also associated with cell–cell junctions. However, its functions in leukocytes will not be addressed in this review since their study has been mainly restricted to cancer cells and cells of the nervous system, in which merlin regulates signaling pathways associated not only with the PM but also with the cytoplasmic and nuclear compartments (reviewed in [3]). Ezrin (named after Ezra Cornell University, where it was isolated) was originally identified as a component of microvilli in chicken intestinal epithelial cells, while radixin (from the Latin radix, which means root) and moesin (membrane-organizing extension spike protein) were isolated from the adherens junctions of rat liver hepatocytes and smooth muscle cells of the bovine uterus, respectively [4–6]. By anchoring F-actin to the cytoplasmic tail of transmembrane proteins, ERMs can regulate cortex tension and stiffness throughout the PM and PM-associated domains of polarized cells, taking part in the formation of complex tissue-associated structures including the brush border of intestinal villi [7]; organization of photoreceptors in the retina [8,9]; and formation of tubules in blood vessels, the excretory intracellular canal of *Caenorhabditis*, and terminal cells of the *Drosophila* tracheal system [10–12]. Although cultured cells express ERMs to a greater or lesser extent, the expression of particular ERM members is strictly regulated in certain tissues: endothelial cells mainly express moesin, ezrin is expressed in intestinal epithelial cells but is absent in hepatocytes, whereas the opposite holds true for radixin. Moesin is the most abundant ERM in leukocytes, whereas ezrin is less expressed and radixin is nearly absent [13–16]. In this review, we describe the intrinsic features that enable ERMs to work as efficient PM-cytoskeleton cross-linkers, and offer a perspective on the functional role of ERMs in leukocyte polarization, migration and intercellular adhesion, focusing on the phagocytic cup and the immune synapse (IS) as paradigmatic PM-associated actin-based structures for the function of leukocytes in the immune system.

2. ERM Tools for Plasma Membrane-to-Cytoskeleton Bridging

Given the high degree of homology shared among the three ERMs (73% amino acid identity) and the expression of more than one in many cell types, overlapping or even compensatory functions have been proposed. This suggests that they work in a similar way, a view that has been confirmed at structural level except for some cases in which specific activities have been assigned to individual ERMs. ERMs bear two well-defined functional domains connected through a long α-helix region: the N-terminal FERM (band 4.1 protein-ERM) domain and the C-terminal ERM association domain (C-ERMAD, 50% sequence homology among ezrin, radixin and moesin). The FERM domain is composed of three subdomains (F1, a ubiquitin-like domain; F2, with four α-helices; and F3, a pleckstrin homology domain) and shows over 75% sequence homology [3] (Figure 1). The presence of the FERM domain is critical for the function that ERMs exert as linkers of the PM and the actin cell cortex.

Biochemical studies and structural analyses of protein complexes with the cytoplasmic tail of adhesion molecules ICAM-2, PSGL-1, CD43 and CD44 [17–21] have shown that ERMs can directly bind to these adhesion receptors through a juxtamembrane cytoplasmic region containing a positively charged cluster and a contiguous nonpolar amino acid motif (R/K)-(aa$_2$/aa$_3$)-(Y/L)-aa-(L/V/I) (where aa represents any amino acid), a finding that can be extended to other known ERM-binding proteins (e.g., ICAM-1 [22], ICAM-3 [23], VCAM-1 [24] and N-CAM-L1 [25]). Such binding to ERMs takes place in a groove formed between a β-strand and an α-helix of the FERM F3 subdomain. In addition to this consensus motif, specific Ser in the cytoplasmic tail of adhesion molecules can regulate their binding to ERMs through phosphorylation-dependent mechanisms. Interactions between Ser and the FERM domain have been reported in ICAM-3, PSGL-1, N-CAM-L1 and L-selectin; whereas phosphomimetic mutations of key Ser residues susceptible to phosphorylation by PKC in the cytoplasmic tail of ICAM-3 (Ser6), CD43 (Ser76), CD44 (Ser2) and L-selectin (Ser9) interfere with their binding to the FERM domain, likely by reducing the net positive charge of their FERM-binding motifs [17,26–30]. The FERM domain can also bind indirectly to ion transporters and other transmembrane receptors (e.g., the β_2-adrenergic receptor, Na^+/H^+ exchangers [NHE3], and the cystic fibrosis transmembrane conductance regulator, CFTR) through two PDZ domains in the scaffolding ERM-binding phosphoprotein 50 (EBP50, also called NHERF1) and NHE3 kinase A regulatory proteins (E3KARP, also called NHERF-2) (reviewed in [31]). Crystal structures of the EBP50 and E3KARP C-terminal peptides bound to radixin have identified a consensus amino acid sequence that can bind to a region of the FERM domain that, despite barely overlapping with the binding site to adhesion molecules, can interfere with their binding by transmitting conformational changes in the F3 subdomain [32].

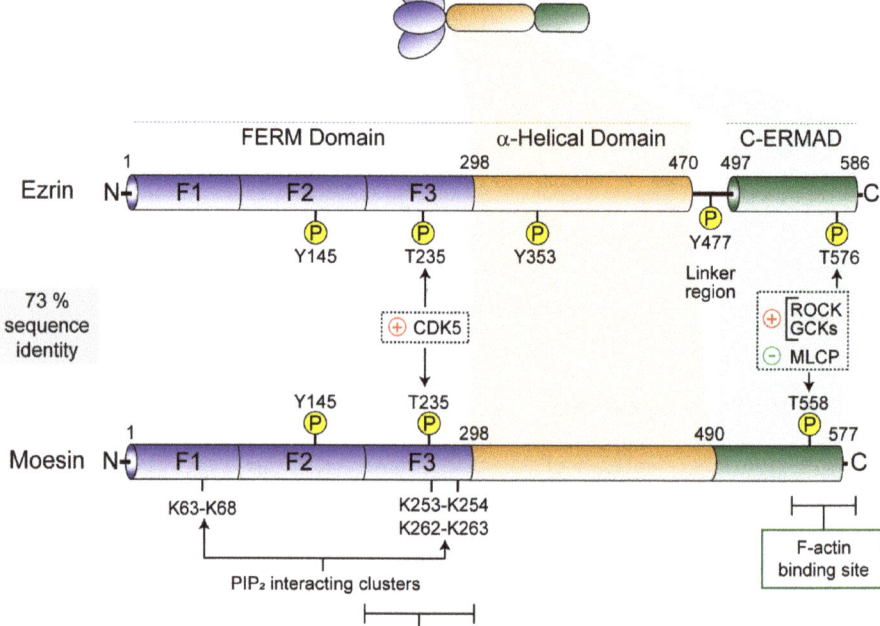

Figure 1. Schematic comparison of the conserved domain structure of human ezrin and moesin showing their sequence identity. The three subdomains (F1–F3) of the N-terminal band 4.1 protein ezrin, radixin and moesin (FERM) domain, the α-helical region, and the C-terminal ERM association domain (C-ERMAD) are depicted. Note that ezrin bears a linker region containing a regulatory Tyr (Y477) that is absent in moesin. The binding sites for PIP$_2$, adhesion receptors and the PDZ domain-containing proteins EBP50 and E3KARP in the FERM domain, and for the F-actin binding site in the C-ERMAD, are shown together with the regulatory Tyr and Thr. Ser/Thr-specific ERM-associated kinases (CDK5, cyclin-dependent kinase 5; ROCK, Rho kinase; GCKs, germinal center kinases, e.g., LOK, lymphocyte-oriented kinase) and phosphatases (MLCP, myosin light chain phosphatase) are also depicted.

The FERM domain and the C-ERMAD can bind each other in a head-to-tail manner, leading to a closed/inactive conformation [33,34]. The release of the C-ERMAD from the FERM domain is necessary for the activation of ERMs, unmasking their F-actin- and PM binding sites. The C-ERMAD can also bind to F-actin after phosphorylation on a conserved Thr in ezrin, radixin and moesin (Thr576, Thr564 and Thr558, respectively) [35,36], which is an important feature for the fine regulation of the PM-to-actin cytoskeleton-linking activity of ERMs. Although phosphorylation of Thr in the F-actin binding site containing C-ERMAD is essential for the activation of ERMs, our current view of how ERMs bind to both PM and F-actin requires the participation of phosphatidylinositol 4,5-bisphosphate (PIP$_2$) [37–39]. Among the mechanisms by which PIP$_2$ may regulate activation of ERMs, it is worth noting recent studies suggesting that, during the interaction between CD44 and ERMs, two molecules of ERM and two molecules of CD44 are indirectly bound by PIP$_2$ forming a heterotetramer at the PM. PIP$_2$ can bind to two sites on the FERM domain (Lys63-Lys68 of the F1 subdomain, and clusters Lys253-Lys254 and Lys262-Lys263 of the subdomain F3) through a mechanism by which one molecule of PIP$_2$ sequentially binds the FERM subdomains, changing the conformational structure of ERMs in such a way that renders the F-actin binding site of the C-ERMAD more accessible for Thr phosphorylation [40,41]. Initial studies reported that the key Thr on the C-ERMAD was phosphorylated by Rho-kinase (ROCK) both in vitro and in vivo [36]; however, ROCK-independent mechanisms have also been described, suggesting that C-ERMAD may be phosphorylated by other kinases [42,43]. From then, the number of Ser/Thr kinases that are able to phosphorylate the conserved Thr on the C-ERMAD of ERMs has greatly increased. The PKC isoenzymes PKC-θ and PKC-α phosphorylate moesin and ezrin in vitro and associate with them in human T lymphocytes and breast carcinoma cells, respectively [44,45]. Moreover, recent attention has been given to germinal center kinases (GCK), a subfamily of the mammalian sterile 20-like kinases (Mst) including lymphocyte-oriented kinase (LOK), Mst4, SLK and Nck interacting kinase (NIK), as the main kinases that phosphorylate the regulatory Thr of ERMs during cell motility and division [46–51]. To this list of kinases, we must now add two sterile 20-like kinases identified in *Drosophila*, misshapen (an orthologue of NIK) and Slik/SLK [52–54].

ERMs are also regulated by phosphatases, as dephosphorylation of the key regulatory Thr of the C-ERMAD detaches ERMs from the cell cortex, adopting a closed/inactive conformation in the cytoplasm. In mammalian cells, several phosphatases can dephosphorylate the regulatory Thr of ERMs. Pioneering studies have reported the association of moesin and ezrin with myosin light chain phosphatase (MLCP) and their coordinated regulation by ROCK-mediated phosphorylation and MLCP-mediated dephosphorylation downstream of the activity of the GTPase Rho [55]. Moreover, protein phosphatase 1 (PP1, the catalytic domain of MLCP) and 2C (PP2C) can dephosphorylate moesin both in vitro and in vivo from human platelets and at the cortex poles of anaphase cells, respectively [56,57]. More recently, involvement of the tumor suppressor PTEN phosphatase in the dephosphorylation of moesin has also been described in chemoattractant-treated neutrophils [58].

Although the regulatory Thr in the C-ERMAD is the most recognized target for phosphorylation-mediated ERM activation, there are other important targets of phosphorylation: Thr235 in the interface between the FERM and the C-ERMAD, the ezrin-specific Tyr353 and 477, and Tyr145 (conserved in all three ERM members) [59–61] (Figure 1). Some evidence suggests that, at least in ezrin, Thr235 is phosphorylated by cyclin-dependent kinase 5 (CDK5) and cooperates with Thr576 for its full activation and the cell morphology changes induced in osteosarcoma cells during senescence [62]. On the other hand, ezrin Tyr145, 353 and 477 can be phosphorylated by Src kinases and the intrinsic Tyr kinase activity of the growth factor receptors for EGF, HGF and PDGF. Tyr145 and 477 seem to play a role in cell adhesion and migration, whereas ezrin Tyr353 has been linked to reorganization of the actin cytoskeleton and activation of B cells in response to tetraspanin CD81- and B-cell receptor (BCR)-mediated stimulation [61,63–66]. However, the importance of these posttranslational modifications on the activation and function of ezrin has been much less studied than the effects of the regulatory Thr of the C-ERMAD. Therefore, although promising, extensive work is required to draw a clear view of the relationship between these posttranslational modifications and how they regulate ERM functions, paying particular attention to the possibility that they may explain some of the specific cellular functions described for ezrin in leukocytes.

3. ERMs in Leukocyte Polarization and Migration

In leukocytes, polarization and migration are interconnected processes regulated by ERMs and their interaction with guanine nucleotide exchange factors (GEFs) and Rho GDP-dissociation inhibitors (RhoGDI) of the small Rho GTPases Rho, Rac and Cdc42 in the two major PM-associated compartments of polarized cells, the leading edge and the uropod [67,68]. These two cell poles are characterized by their respective clustering of chemoattractant receptors and enrichment of adhesion molecules on ERM-organized microspikes and microvilli [69,70]. Leukocytes egress from hematopoietic niches and lymphoid organs to patrol the organism following endothelial cell-presented adhesion receptors and chemoattractant trails that permit their exit from blood and lymphatic vessels and arrival to target tissues (e.g., secondary lymphoid organs and inflammatory foci) [71]. Leukocytes responding to chemoattractants convert mechanical forces into directional locomotion as a result of their marked front-to-rear polarity. Hence, Rac-dependent actin polymerization at the leading edge and retraction at the trailing edge by RhoA/ROCK/phosphorylated myosin light chain (MLC)-stimulated actomyosin contraction near the uropod are coordinately regulated by each other to maintain polarity and generate the main forces pushing leukocytes forward [72,73].

Studies with primary T and B lymphocytes, neutrophils and HL-60 human myeloid cells clearly show that a considerable proportion of ERMs are constitutively activated by phosphorylation and bound to the PM of resting leukocytes [16,74–76]. PM tension and cell symmetry can be maintained by the inactivation of small Rho GTPases as result of the binding of their corresponding GEFs (e.g., PDZRhoGEF, Vav1 and α-PIX) to activated ERMs. However, cell symmetry can be broken by chemoattractants, which induce leukocyte polarization through transient dephosphorylation of ERMs by the phosphatase activity of the PP1c subunit of MLCP that, in response to the G protein-coupled receptor (GPCR)-associated heterotrimeric protein $G\alpha_i$ and the hematopoietic cell-specific actin regulatory protein Hem-1, is recruited to the emerging leading edge [76]. Once there, MLCP dephosphorylates ERMs, in turn releasing GEFs to activate Rac and Cdc42, stimulating F-actin polymerization and subsequent PM protrusive activity at the cell front. Almost immediately, ERMs can be re-phosphorylated by LOK and/or RhoA-stimulated ROCK and redistributed to the cell rear, reinforcing PM tension and preventing the formation of secondary pseudopods (Figure 2). Adhesion to substratum-coated surfaces (e.g., fibrinogen, fibronectin and VCAM-1) via β1 integrins is also a prerequisite to break the symmetry of leukocytes in chemoattractant-induced polarization. Physical tension induced at the substratum-attached cell rear polarizes SRGAP-2 (a Bin-Amphiphysin-Rvs (BAR) domain containing Rac-1-GAP), which binds and deforms the PM, co-recruiting activated myosin II (pMLC) and the synthesis of PIP_2 by PIP5K [77–79]. Since PIP_2 is essential for the full activation of ERMs, it is feasible that its synthesis at the rear of adhered leukocytes may contribute to recruitment and retention of ERMs in the emerging uropod. In this regard, there is evidence indicating that ERMs play an important role in the organization of the trailing edge and the formation of the uropod. Phosphorylated ERMs organize at the cell rear with flotillin-containing lipid rafts, forming clusters that can activate RhoA by either sequestering its inhibitor Rho GDI or by binding to Dbl, a Rho GEF concentrated in the cell rear [43,80]. In this cell compartment, activation of Rho kinase (ROCK) by RhoA phosphorylates MLC to stimulate actomyosin contraction, which together with F-actin binding to ERMs and polymerization can form the uropod [43,73]. Thus, in polarized leukocytes, it seems that capping of ERMs at the cell rear not only establishes where the uropod should be formed, but it is actively involved in its formation by activating RhoA-ROCK and maintaining the PM tension required to impede the formation of pseudopods anywhere other than on the leading edge.

Although initial studies did not find any significant differences between the adhesion and migration capabilities of moesin-deficient and wild-type mouse fibroblasts [81], more recent studies have suggested a non-redundant role for ezrin and moesin in lymphocyte migration. The number of lymphocytes that egressed from primary and secondary lymphoid organs was impaired in the moesin-knockout mice, reducing the populations of T and B cells in the peripheral blood and lymph nodes [82]. Furthermore, siRNA silencing of moesin in T lymphocytes from conditional ezrin-deficient mice showed that ERMs are important for integrin β1-mediated adhesion to fibronectin and homing to lymphoid organs, but for chemotaxis solely when cells must pass through constricted spaces [79]. In this regard, an increasing body of evidence suggests that ERMs are involved in both intra-tissue leukocyte chemotaxis and in the main steps of the chemoattractant-stimulated leukocyte recruitment cascade: tethering, rolling, firm adhesion and transendothelial migration (TEM) [83].

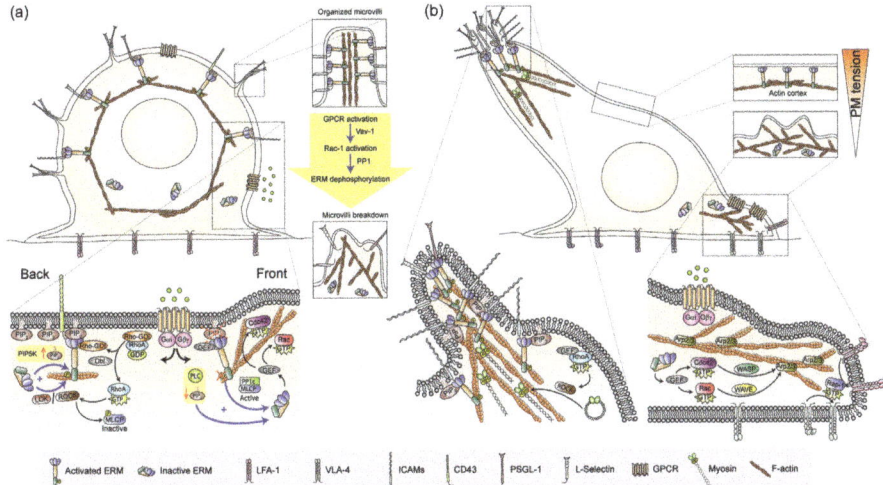

Figure 2. Compartmentalized activation of ERMs and small Rho GTPases in polarized motile leukocytes. (**a**) Initial symmetry breaking events in resting leukocytes stimulated with chemoattractants. In response to chemoattractants, G protein-coupled receptor (GPCR)-associated heterotrimeric proteins stimulate ERM activation at the cell rear through phosphorylation by the kinase activity of LOK and RhoA-stimulated ROCK, production of PIP$_2$ by PIP5K; and ERM inactivation at the cell front through PIP$_2$ hydrolysis by phospholipase C (PLC) and the phosphatase activity of the PP1 subunit of MLCP, which subsequently activates Rac and Cdc42 through the release of their corresponding guanine nucleotide exchange factors (GEFs). On the upper right, breakdown of microvilli by chemoattractant-stimulated Rac-1 activation is depicted. (**b**) Polarized motile leukocyte showing Rac- and Cdc42-stimulated actin polymerization by the Arp2/3 complex in the leading edge and RhoA-ROCK-mediated actomyosin contraction in the uropod, which provide the main forces required to move the cell forward. Rap-1-mediated β1 (e.g., VLA-4) and β2 (e.g., LFA-1) integrin activation in the leading edge and adhesion molecule-bearing microvilli on the uropod are also depicted. On the upper right, plasma membrane (PM) tension near the uropod and at the leading edge are compared.

3.1. Tethering and Rolling

Activated ERMs and ERM-binding adhesion receptors (e.g., PSGL-1, L-selectin, ICAMs, CD43 and CD44) work together in the organization of microvilli, F-actin-based finger-shaped PM protrusions that are important for tethering and rolling during the initial contacts of leukocytes with endothelial cells [84,85]. It has been proposed that in the bloodstream, PSGL-1 and L-selectin on the tips of leukocyte microvilli are the first adhesion molecules that establish contact with their counter-receptors (E/P-selectin and CD34, respectively) on activated endothelial cells of postcapillary venules. L-selectin also contacts the addressins MadCAM-1 and GlyCAM-1 on the high endothelial venules (HEVs) of lymphatics, or even with PSGL-1 during secondary contacts among bystander and adhered leukocytes. In this regard, there is much evidence indicating that moesin and ezrin regulate the tethering and rolling velocity of leukocytes both in vivo and in vitro. Deficiency of Rap-1, a small Rho GTPase that in its inactive form fosters ERM phosphorylation in resting leukocytes by activating LOK, disturbs rolling of naïve T lymphocytes on P-selectin and on addressins by inhibiting tethering [86]. The rolling velocity of neutrophils in cremaster muscle venules after trauma- or TNF-α-induced inflammation is detrimentally higher in moesin-deficient than control mice [87]. 32D myeloblast-like cells expressing an ERM-binding defective PSGL-1 mutant show increased rolling velocity and reduced tethering on L-, P- and E-selectin, whereas only tethering on PSGL-1 is affected in the corresponding ERM-binding defective L-selectin mutant [85,88]. Lastly, treatment of mouse splenic B lymphocytes with the phosphatase inhibitor calyculin A (which increases ERM phosphorylation) or overexpression of a phosphomimetic ezrin mutant impairs microvilli formation, chemotaxis and B cell migration to the spleen and lymph nodes [14]. This suggests that ERMs may support early steps in the leukocyte transmigration cascade, including secondary leukocyte tethering on leukocytes adhered to endothelial cells, through the formation of PSGL-1- and L-selectin-bearing microvilli.

3.2. Firm Adhesion

Rolling leukocytes can detect chemoattractants bound to proteoglycans on endothelial cells and spread on them upon binding to G protein-coupled receptors (GPCRs), which can trigger breakdown of microvilli through the transient dephosphorylation of ERMs by Rac-1- and Rap-1-mediated PP2A/PP1 activation and LOK inhibition, respectively [86,89]. This early morphological alteration may help cells to reduce PM tension and increase the surface of contact at the leukocyte–endothelial cell interface. In addition, chemoattractants induce clustering and conformational activation of the β1 integrin VLA-4 and β2 integrins LFA-1 (in lymphocytes and monocytes) and Mac-1 (in monocytes and neutrophils) and their respective interaction with VCAM-1 and ICAM-1. The latter are organized in the apically localized tetraspanin microdomains of endothelial cells, adhesion platforms that can be linked to the actin cytoskeleton through the binding of CD81 and its partners EWI-2 and EWI-F to ERMs [90,91], giving rise to leukocyte crawling and subsequent firm adhesion and arrest. However, endothelial cells also participate actively in this process. In response to the interaction of β2 integrins with ICAM-1, they form an F-actin-based docking structure which embraces the leukocyte with ICAM-1- and VCAM-1-rich PM projections containing PIP_2 and phosphorylated moesin and ezrin (among other ABPs) [24]. This actin cytoskeleton-integrated structure is thought to enable the dynamic transition between leukocyte firm adhesion and TEM. Although the interaction between VCAM-1 and moesin is involved in leukocyte adhesion to the endothelium, PM projections containing moesin and ICAM-1 facilitate TEM [24,92]. The contribution of ERMs to the formation of the docking structure and their role in leukocyte adhesion and TEM have been studied both in an experimental model of leukocyte chemotaxis in COS-7 cells expressing an ERM-binding defective mutant of ICAM-1, and also in human endothelial cells infected with *Neisseria meningitidis*, in which the pathogen competes with leukocytes for the recruitment of ERMs, ICAM-1 and VCAM-1 to their adhesion sites. Overexpression of ezrin and moesin in *N. meningitidis*-infected endothelium rescued the formation of the docking structure, whereas expression of the ezrin FERM domain blocked its formation along with adhesion and diapedesis of leukocytes, similarly to the effect of expression of the ERM-binding defective ICAM-1 mutant in COS-7 cells [93,94], confirming that the complex organization of adhesion receptors and F-actin in the docking structure is regulated by ERMs.

3.3. Transendothelial Migration

To cross the endothelial cell monolayer and reach the abluminal side of the vessel, leukocytes must pass through narrowed spaces, overcoming the resistance offered by endothelial cells and their basement membrane. To achieve this, PM deformation, force-generated extension at the leading edge and retraction of the uropod are required [95]. Chemoattractant-induced ERM dephosphorylation regulates PM tension and deformation in neutrophils and T lymphocytes. The importance of the regulation of PM tension by ERMs during TEM has been studied in transgenic mice constitutively expressing the phosphomimetic ezrin mutant T567E, whose T lymphocytes present increased PM tension, defective migration in vitro and impaired homing to lymph nodes [96]. Although these lymphocytes showed no defective tethering and rolling on HEVs (perhaps because, as noticed by the authors, these cells did not present alterations in the length and number of microvilli, possibly as a consequence of the low levels of ezrin T567E expressed in cells), their passage to the parenchyma was seriously impaired, as most of them either did not cross the endothelial monolayer or took much longer to cross it than wild type lymphocytes. Accordingly, neutrophils from PTEN knockout mice, a phosphatase that can dephosphorylate moesin on Thr558, are defective in chemotaxis and recruitment to the peritoneum in a thioglycolate-induced mouse model of acute peritonitis [58]. Furthermore, treatment of T lymphocytes with glucocorticoids, which increase ERM phosphorylation through gene expression-independent mechanisms, reduce transmigration in vitro [97]. On the other hand, leukocytes crawling on endothelial cells under shear stress can ventrally extend short, Cdc42-dependent exploratory filopodia and Wiskott–Aldrich syndrome protein (WASP)-regulated invasive podosomes for transcellular and paracellular TEM, respectively. These are reorganized into a main pseudopod when an appropriate place for diapedesis, the passage of leukocytes through capillary endothelial cells, is found [98,99]. WASP is considered an effector of many of the mechanisms by which Cdc42 promotes actin polymerization in leukocytes. Further studies have shown that a deficiency of Cdc42 interacting protein 4 (CIP4), which interacts with both Cdc42 and WASP, impairs interaction of T lymphocytes with immobilized ICAM-1 and VCAM-1 under shear flow and their transmigration across TNF-α-activated endothelial cells [100]. Knowing whether regulation of ERM dephosphorylation can facilitate this passage by fostering PM deformation and activating Cdc42-dependent actin polymerization for formation of the invasive pseudopod would thus provide new insights on how TEM is regulated. Interestingly, recent findings from the study of inflammatory monocytes suggest that ERMs can work differentially in TEM, since ezrin binds first to L-selectin and promotes formation of a main pseudopod, whereas moesin interacts later on, fostering shedding of the L-selectin ectodomain and restricting the appearance of additional pseudopods that would disturb directional transmigration [30,101].

On the other hand, taking advantage of the fact that moesin is the sole ERM member expressed in *Drosophila*, an important role for moesin in control of the leading pseudopod has been also associated with persistence and directionality of collective cell migration in vivo, a process by which groups of cells coordinately move through tissues [102]. In border cells of the egg chamber in the *Drosophila* ovary, which form a small cluster that migrates directionally by means of a pseudopod strictly formed at the front of the cluster's leader cell, silencing of moesin or the moesin kinase misshapen promotes formation of protrusions in non-leader cells and disturbs polarized migration of cell clusters [52,103], suggesting that localization of moesin at cell–cell contacts can foster formation of the leading pseudopod by increasing cortical membrane stiffness. In the immune system, most leukocytes move as solitary cells, but in certain lymphoid malignancies they can also move as aggregates in tissues [104]. This opens the interesting question of whether ERMs may also regulate the collective migration of leukocytes by preventing formation of additional pseudopods in non-leader cells.

4. ERMs and Intercellular Adhesion: the Phagocytic Cup and the Immune Synapse as Paradigms of Ezrin- and Moesin-Mediated PM Organization in Leukocytes

In order to exert cytokine secretion-independent defensive functions, macrophages, DCs, neutrophils and other phagocytic leukocytes must establish contact with microorganisms, tumoral and apoptotic cells in a pathophysiological context using PM-associated receptors (e.g., high-affinity IgG-binding receptors (FcγRs), complement receptor (CR)3, toll-like receptor (TLR)-4, the receptor of apoptotic cells stabilin-2, or the scavenger receptors CD36 and Dectin-1). This contact permits their engulfment via the phagocytic cup and internalization within the phagosome, an intracellular vesicle formed by invagination of the PM. The phagosome is responsible for degradation and, in the case of professional antigen-presenting cells (APCs), ultimate processing of proteins to initiate adaptive immune responses by presenting antigen to the antigen-specific T cell receptor (TCR) at the immune synapse (IS), a PM-associated intercellular compartment that regulates the activation of T lymphocytes by APCs, but that is also found in NK cells and B lymphocytes [105–107].

Although the phagocytic cup and IS are quite different structures in regard to the identity of the PM-associated receptors and cell types involved in their formation, they still share some important features [108]. Among them, it is worth noting the involvement of the actin cytoskeleton and ERMs in the dynamics of receptors driving their functional organization (Figure 3).

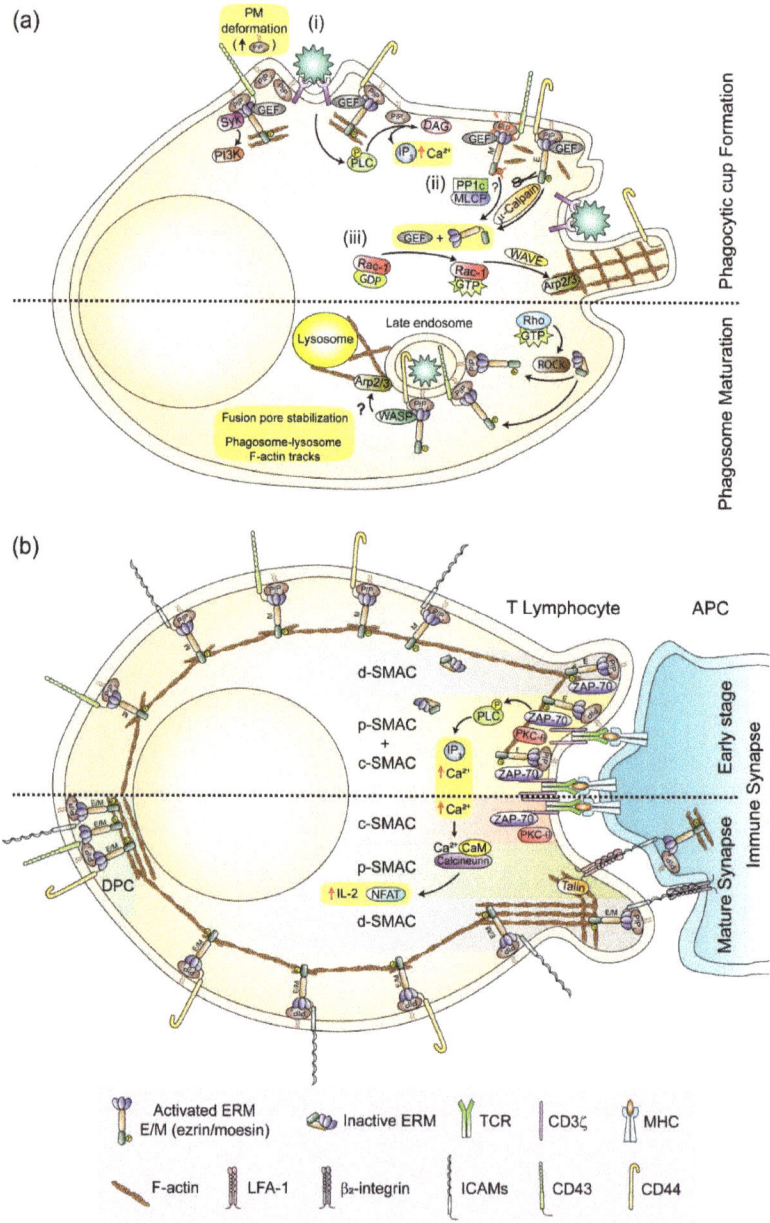

Figure 3. Organization of leukocyte interactions by ERMs. (**a**) The phagocytic cup and the phagosome. Upper part, three-step phagocytic cup formation in direct receptor-mediated phagocytosis (e.g., scavenger and apoptotic cell recognition receptors): (i) receptor-mediated cell binding and recruitment of PIP_2 and ERMs to deformed PM. PLC and Syk signaling from the phagocytic receptor and the ERM-bearing immunoreceptor tyrosine-based activation motif (ITAM) are respectively depicted; (ii) activation of PLC

from the phagocytic receptor reduces PM PIP$_2$ and increases intracellular Ca^{2+} levels, which may release ERMs from the PM and activate Rac-1 through the specific cleavage of ezrin by calpain and dephosphorylation of ERMs by myosin light chain phosphatase (MLCP); (iii) activation of Rac-1 by GEFs released from inactivated/closed ERMs may stimulate actin polymerization by the WAVE-Arp2/3 complex, giving rise to the phagocytic cup. Lower part, phagosome maturation: In late phagosomes, ERMs return to their intracellular side through a Rho-dependent mechanism, promoting phagolysosome formation through WASP-ARP2/3-mediated actin polymerization. (**b**) The T cell immune synapse. Upper part, early immune synapse (IS) formation: ERMs are transiently inactivated and ezrin interacts with ZAP-70 along the IS fostering intracellular fluxes of Ca^{2+}, whereas moesin and the adhesion molecules CD43, CD44 and ICAMs are excluded from the IS. Lower part, mature IS: Ezrin and moesin are both excluded from the IS and preferentially localized together with adhesion molecules and F-actin at the d-SMAC and the distal pole complex (DPC), promoting the activation of TCR-associated proximal and distal elements (PLC-γ activation/Ca^{2+} mobilization and NFAT dephosphorylation/IL-2 production, respectively).

4.1. The phagocytic Cup and the Phagosome

ERMs and small Rho-GTPases are involved in the early and late steps of phagosome formation. Phagocytic leukocytes expand their PM surface upon receptor-mediated contact with target cells, forming the phagocytic cup. ERMs control lateral diffusion of receptors on PM by forming a fence-and-picket structure (also termed "corral") of F-actin bundles indirectly bound to transmembrane proteins (e.g., CD44) through ERMs [109]. This constraint is less strict in the leading edge of phagocytic leukocytes, where receptors can diffuse early upon ligand recognition to form clusters that bind multivalent targets, and may be subsequently confined at the emerging phagocytic cup by ERM-sustained corrals. However, considering that ERMs can also work as protein-associated modules to transmit signaling through the binding of the Src kinase Syk to the immunoreceptor tyrosine-based activation motif (ITAM) of their FERM domain [110], a provocative study has suggested that ERMs may stimulate phagocytosis through receptor-independent mechanisms. In this way, they would act as phylogenetically conserved mechanotransducers that activate PI3K in response to the deformations of the PM, which accumulate PIP$_2$ at the contact sites with foreign cells and particles, recruiting ERMs to the emerging phagocytic cup [111]. Nevertheless, while ERMs can regulate phagocytosis through a signaling pathway similar to that triggered by the ITAMs of FcγRs, the absence of moesin does not disturb internalization of IgG-coated particles, suggesting that ERMs are not essential for opsonization-mediated internalization. On the contrary, ERMs seem to improve phagocytosis mediated by those mechanisms in which neither opsonization nor ITAM-bearing receptors are involved, such as the clearance of microorganisms and apoptotic cells by direct binding of scavenger receptors and receptors of phosphatidyl serine (PS), respectively [111]. Although attractive, the mechanotransduction model by which the PIP$_2$–ERM–ITAM–Syk axis may collaborate with receptor-mediated opsonisation-independent phagocytosis leads to some unanswered questions. For instance, since ezrin and moesin are already activated at the PM of resting leukocytes, connecting the PM to the actin cortex in flat areas as well as in microvilli and other small protrusions on which the phagocytic cup will be formed, the positioning of ERMs at the PM seems not to be sufficient for phagocytic cup formation, suggesting that additional factors must cooperate with them to trigger phagocytosis. Moreover, some evidence suggests that ERMs detach from the PM early during the phagocytic process, only returning once the phagosome has been formed [112]. In this regard, a specific role for ezrin in rearrangements of the actin cytoskeleton leading to the growth of PM projections surrounding the phagocytic cup of neutrophils has been proposed on the basis that ezrin is the sole ERM regulated by the proteolytic activity of μ-calpain [113]. Ligand binding to phagocytic receptors activates phospholipase (PL)C/D to produce IP$_3$, which can trigger an increase of intracellular Ca^{2+}, inducing the translocation of μ-calpain from the cytosol to the emerging phagocytic cup where it is activated to cleave ezrin and break the linkage between the PM and the actin cortex (reviewed in [114]). The specific role of ezrin in the formation of the phagocytic cup is supported by a recent report showing that ezrin (but not moesin) promotes the formation of functional phagocytic cup-like invasive structures during infection of mammalian cells by extracellular amastigotes of *Trypanosoma cruzi* [115]. Detachment of ezrin from the PM may relax it but also release Rac-1-GEFs and thereby stimulate activation of Rac-1 and its effector WASP-family verprolin-homologous protein (WAVE), which can trigger localized actin polymerization by the Arp2/3 complex, pushing the PM away to form the phagocytic cup [116]. On the other hand, in using FRET-based biosensors to visualize the activity of small Rho-GTPase in real-time imaging, recent studies on the clearance of apoptotic thymocytes through stabilin-2, one of the PS receptors, have shown that RhoA is transiently activated immediately before phagocytic cup closure and internalization. Furthermore, a constitutively activated form of RhoA inhibits phagocytosis through ROCK, suggesting that besides Rac-1, RhoA is also important for early phagocytic events [117]. In macrophages, ROCK regulates maturation of phagosomes generated from the clearance of apoptotic cells [118], a function that would take place through the re-phosphorylation of ERMs and their binding to PIP_2 and/or some of their abundantly expressed PM partners (e.g., CD43 and CD44) on the cytoplasmic side of late phagosomes. At this site, ERMs can interact with WASP via the FERM domain, and induce de novo F-actin polymerization through Arp2/3, enabling phagolysosome formation by transmission of lysosomal content to late phagosomes [112] (Figure 3a). Nevertheless, how ERMs can facilitate phagosome–lysosome fusion is still an open question. Although some authors assign a direct action to ERMs in the stabilization of the fusion pores that connect the two organelles during the acidification of the phagosome [119], the possibility that localization of ERMs on phagosomes may still provide F-actin tracks for phagosome–lysosome approach and membrane fusion seems not be entirely excluded.

4.2. The Immune Synapse

To carry out their functions during adaptive immunity, lymphocytes must first be activated in the IS. The prototypical IS of T cells is organized in a central area or supramolecular activation cluster (c-SMAC) in which the TCR-CD3 signaling complex is concentrated, a ring-shaped peripheral (p)-SMAC where the β2 integrin LFA-1 is clustered with talin and β-actin, and a more distal lamellipodium-like (d)-SMAC characterized by a ring of F-actin [120]. Among its functions, the IS modulates T cell activation, attenuating or sustaining signaling by degradation of TCR-ligand complexes at the c-SMAC or stabilization of signaling microclusters at the p-SMAC, respectively [121,122].

Several research groups simultaneously reported the localization of ezrin and moesin at the T cell IS, and they explored their possible functions in its organization with a focus on their phosphorylation and co-distribution with the large ERM-binding glycoprotein CD43 (Figure 3b). They found that ERMs were transiently dephosphorylated in response to TCR-triggered stimuli and localized with F-actin at the d-SMAC, or beyond it, in the distal pole complex (DPC) when T cells adopt a striking polarized morphology, excluding CD43 from the IS [123–125]. This ERM function was also observed in the killer cell immunoglobulin-like receptor (KIR)- and NKG2A receptor-triggered inhibitory IS of NK cells, but only partially in the NK cell activating IS in which ezrin co-localizes with granules of perforin, whereas moesin can be redistributed towards the DPC [15,126,127]. While exclusion of CD43 from the IS was initially regarded as a mechanism to facilitate T cell activation by preventing steric interference of TCR binding to major histocompatibility complex (MHC)-loaded antigenic peptides, further studies with hyperproliferative CD43-deficient T cells expressing CD43 protein constructs have shown that regulation of T cell activation by CD43 is not exerted by its extracellular domain, but rather by signaling from its cytoplasmic tail independently of where it was localized on the PM [128]. Later, interesting nonredundant roles for moesin and ezrin in organization and function of the IS of T cells were proposed in light of results showing that ezrin was specifically localized in the IS associated with ZAP-70, a key adaptor-kinase for TCR-triggered signal transmission to proximal elements of activation (e.g., the adaptor protein linker for activation of T cells LAT), whereas moesin excludes CD43 from the IS [129]. Moreover, by expressing phosphomimetic- and phosphorylation-defective ERM mutants in Jurkat cells, the study showed that phosphorylation of ezrin on Thr567 and moesin on Thr558 are both important for IS formation, but that only phosphorylation of ezrin was involved in the recruitment of ZAP-70 to the IS and TCR-triggered Ca^{2+} mobilization. However, in using T lymphocytes from ezrin-knockout mice, another study has shown that ezrin can exert a slight impact on the organization of the IS, and although ezrin (but not moesin) is recruited to the IS at early times, both are subsequently recruited to the DPC, working together to promote activation of TCR proximal (PLC-γ and ZAP-70 phosphorylation; and Ca^{2+} intracellular flux) and distal (NF-AT transcriptional activity and IL-2 production) elements [130]. The apparent discrepancies between these two studies have been mainly assigned to differences in the cellular models used (human Jurkat T cells vs. primary mouse T lymphocytes) [131], off-target effects of the exogenously expressed ERM mutants, possible cross-reactivity of the Abs used, and residual endogenous expression of moesin in ezrin knockout mouse T cells interfered with moesin-specific siRNAs. Regardless, there seems little doubt that ERMs are involved in organization of the IS, a role also supported by the finding that efficient formation of antigen-specific T cell–APC conjugates requires Vav1-Rac-1-regulated ERM dephosphorylation and subsequent PM detachment from the actin cortex at the T cell–APC contact interface [132]. This may promote conjugate formation by increasing PM flexibility and the avidity of LFA-1 for ICAM-1 on the APC, since the transient detachment of LFA-1 from depolymerized F-actin can increase its lateral mobility to form small aggregates and clusters at the p-SMAC [133]. However, ERMs can also regulate the interaction between LFA-1 and ICAM-1 from the APC side of the IS. Anchoring of ICAM-1 to the actin cortex can provide the resistance required to stretch the F-actin-associated β2 chain of LFA-1, thus inducing the conformational changes required to increase the affinity of LFA-1 for ICAM-1 [134]. This function is also observed in the activating IS of NK cells, in which ERMs tether ICAM-1 to the actin cortex of NK-sensitive target cells, facilitating the polarized secretion of cytolytic granules through the interaction between ICAM-1 and LFA-1 [135].

ERMs have been also involved in the formation of the B cell IS. In resting B lymphocytes, BCRs are included in lipid rafts, and their diffusion is constrained by a ERM-sustained F-actin corral [136,137]. However, like in TCR-stimulated T cells, ERMs are transiently dephosphorylated in B lymphocytes upon BCR-mediated stimulation, inducing actin depolymerization and the dissociation of PM-associated BCR-containing lipid rafts from the actin cortex, which facilitates their coalescence at the IS [138]. BCR triggering induces B cell spread on APCs, giving rise to the rapid ERM-mediated disorganization of the corral's F-actin, which can permit signaling by lateral diffusion of small, BCR-containing microclusters to form larger, more stable ones that, after ERM re-phosphorylation, re-attach to the PM and are redistributed towards the c-SMAC for internalization by the combined actions of the centripetal retrograde flow of actin and actomyosin contraction (reviewed in [137]). The importance of ERMs for organization of signaling microclusters at the IS of B cells has been demonstrated by overexpression of the FERM domain of ezrin, which disturbed the coalescence of BCR microclusters to the c-SMAC, similar to that previously described for TCR microclusters in the T cell IS [123]. Furthermore, overexpression of the phosphomimetic ezrin T567D mutant or a lack of both ezrin and moesin also impaired the coalescence of BCR microclusters to the c-SMAC [75], suggesting that coordinated ERM binding to and detachment from the PM and F-actin are required for actin cytoskeleton-regulated BCR signaling from microclusters. More recently, a differential function for ERMs has been established between naïve and germinal center (GC) B cells, since the latter form ezrin and moesin-containing podosome-like projections with BCR clusters at their tips, which facilitate antigen extraction from the PM of APCs through "pulling" forces that preferentially engage high-affinity antigens, perhaps to establish thresholds for antigen processing that would permit the spatial-temporal regulation of antigen presentation to follicular T cells [139].

5. ERMs and Immune Regulation

A large body of evidence indicates that ERMs are important for the function of lymphocytes. X-linked moesin-associated immune deficiency (X-MAID), a human genetic disorder caused by the missense mutation R171W in the moesin gene, is characterized by extensive lymphopenia, resembling the phenotype observed in moesin-knockout mice, with low proliferation of T lymphocyte in which the proportion of naïve CD4$^+$ and senescent and exhausted CD8$^+$ T cell subtypes is unusually high [82,140,141]. On the other hand, overexpression of a phosphomimetic moesin mutant attenuates spontaneous autoimmunity in Rap-1-deficient mice by reducing the number of inflammatory T lymphocytes recruited to the colon, whereas T lymphocytes infiltrating inflamed kidneys from systemic lupus erythematosus (SLE) patients show high levels of ERM phosphorylation [86,142]. However, ERMs do not only regulate autoimmunity through cell migration but also by promoting the production of CD4$^+$ and CD8$^+$ regulatory T cells (Tregs). CD4$^+$ Treg production is stimulated by ERM binding and stabilization of the TGF-βR-I and -II on the PM as a positive TGF-β-dependent feedback mechanism that increases the expression of moesin in cells, whereas CD8$^+$ Tregs are incremented by fostering the IL-15 signaling pathway that maintains their homeostasis [143,144]. Moreover, ezrin also regulates B cell-mediated autoimmunity, since conditional deletion of ezrin attenuates lupus in mice deficient for the Src kinase Lyn by reducing B cell activation, leukocyte infiltration and IgG deposition in the kidney glomeruli [145]. Regardless, no important defects in the homeostasis of B lymphocytes have been observed in ezrin-deficient mice except for an increase in IL-10-producing Bregs, which make conditional ezrin-deficient mice more prone to trigger pro-inflammatory responses to sublethal doses of LPS in vivo by increasing IL-10 production via TLR4, an effect also observed in B cells treated with an ezrin-specific inhibitor that dephosphorylates ezrin on Thr567 [146,147]. Altogether, these reports suggest that the precise threshold of ERM phosphorylation on the regulatory Thr of the C-ERMAD is critical to prevent defective lymphocyte functions.

6. Conclusions and Future Perspectives

A large body of evidence now indicates that ERMs are multifunctional proteins that, through complex regulation by kinases and phosphatases, organize the PM and actin cortex and transmit information between the external milieu and the cell in a bidirectional way, thus linking PM structure to function. In leukocytes this feature is paradigmatic, since ERMs contribute to the process of leukocyte polarization and migration through the control of PM tension and the formation of uropods and microvilli, and they regulate intracellular signaling in lymphocyte activation by locally organizing signaling receptors at the IS.

Although it has been well established that many ERM functions are exerted through activation of small Rho GTPases upon closing/inactivation, it remains unclear how RhoA is activated in the uropods of migrating leukocytes while ERMs can maintain a GEF-sequestering open/active conformation at the PM. In this regard, the mechanism by which ERMs and their partners are relocated to the cell rear during leukocyte polarization (the uropod of motile leukocytes and the DPC of the IS) is still not well understood. Whether the redistribution of ERM-adhesion receptor complexes towards the cell rear is the result of centripetal retrograde flow of the actin cortex, consecutive cycles of ERM phosphorylation/re-phosphorylation, or simply the reassembly of ERMs and their adhesion receptor partners at the actin cortex of the cell rear, remains to be investigated in depth.

Another open question is whether microtubules (MTs), whose centrosome-organized network is packed in the uropod of polarized motile leukocytes, play any role in the function of ERMs. Although pioneering studies on the organization of the erythrocyte marginal band and the association of ezrin with cytoskeletal components in insect cells have suggested some interaction between ERMs and MTs [148,149], it was not until recent studies, which showed an important role for MTs in the mechanisms by which Dmoesin controls the stability and orientation of microtubules in the mitotic spindle of *Drosophila* cells, that the interaction between ERMs and MTs at the cell cortex was reported [53,150]. In this regard, some findings support the notion that ERMs can also interact with MTs in leukocytes. MT disruption by nocodazole breaks polarity and induces clustering of ERMs in T lymphocytes, probably through the release of a Rho-GEF from MTs and the subsequent activation of RhoA-ROCK [151]. Also, the persistence of directional migration of neutrophil-like cells toward chemoattractant gradients has been proposed to be dependent on the polarization of ERMs at the cell rear and their regulation by MTs, since their disruption reduced persistent directional migration through moesin mislocalization [152]. Moreover, ezrin can regulate the organization and function of TCR microclusters in the T cell IS by binding to the PDZ domain containing scaffold protein Dlg1, which may facilitate the interaction of MTs with the cell cortex and, therefore, the organization and function of the MT network in the IS [153].

Further studies will be required to explore the significance of the possible interaction between ERMs and MTs in leukocytes and the impact that ERMs might exert on their function as pivotal linkers of the PM to both the actin and tubulin cytoskeletons.

Author Contributions: A.G.-O. and J.M.S. contributed to the conceptualization and writing of this review. All authors have read and agreed to the published version of the manuscript.

Funding: This research was funded by Ministerio de Ciencia Innovación y Universidades (MICIU)/FEDER, grant number RTI2018-100815-B-100 (J.M.S.). We acknowledge support of the publication fee by the Spanish Research Council (CSIC) Open Access Publication Support Initiative.

Acknowledgments: We thank F. Sánchez-Madrid, N. Martín-Cófreces, F. Carrasco and A. Borroto for critical readings and N. Beach for editorial assistance.

Conflicts of Interest: The authors declare no conflicts of interest.

References

1. Chugh, P.; Clark, A.G.; Smith, M.B.; Cassani, D.A.; Dierkes, K.; Ragab, A.; Paluch, E.K. Actin cortex architecture regulates cell surface tension. *Nat. Cell Biol.* **2017**, *19*, 689–697. [CrossRef]
2. Fehon, R.G.; McClatchey, A.I.; Bretscher, A. Organizing the cell cortex: The role of ERM proteins. *Nat. Rev. Mol. Cell Biol.* **2010**, *11*, 276–287. [CrossRef]
3. Michie, K.A.; Bermeister, A.; Robertson, N.O.; Goodchild, S.C.; Curmi, P.M. Two Sides of the Coin: Ezrin/Radixin/Moesin and Merlin Control Membrane Structure and Contact Inhibition. *Int. J. Mol. Sci.* **2019**, *20*, 1996. [CrossRef] [PubMed]
4. Lankes, W.; Griesmacher, A.; Grünwald, J.; Schwartz-Albiez, R.; Keller, R. A heparin-binding protein involved in inhibition of smooth-muscle cell proliferation. *Biochem. J.* **1988**, *251*, 831–842. [CrossRef] [PubMed]
5. Bretscher, A. Purification of an 80,000-dalton protein that is a component of the isolated microvillus cytoskeleton, and its localization in nonmuscle cells. *J. Cell Biol.* **1983**, *97*, 425–432. [CrossRef] [PubMed]
6. Tsukita, S.; Hieda, Y.; Tsukita, S. A new 82-kD barbed end-capping protein (radixin) localized in the cell-to-cell adherens junction: Purification and characterization. *J. Cell Biol.* **1989**, *108*, 2369–2382. [CrossRef]
7. Crawley, S.W.; Mooseker, M.S.; Tyska, M.J. Shaping the intestinal brush border. *J. Cell Biol.* **2014**, *207*, 441–451. [CrossRef]
8. Chorna-Ornan, I.; Tzarfaty, V.; Ankri-Eliahoo, G.; Joel-Almagor, T.; Meyer, N.E.; Huber, A.; Minke, B. Light-regulated interaction of Dmoesin with TRP and TRPL channels is required for maintenance of photoreceptors. *J. Cell Biol.* **2005**, *171*, 143–152. [CrossRef]
9. Bonilha, V.L.; Finnemann, S.C.; Rodriguez-Boulan, E. Ezrin promotes morphogenesis of apical microvilli and basal infoldings in retinal pigment epithelium. *J. Cell Biol.* **1999**, *147*, 1533–1548. [CrossRef]
10. Wang, Y.; Kaiser, M.S.; Larson, J.D.; Nasevicius, A.; Clark, K.J.; Wadman, S.A.; Essner, J.J. Moesin1 and Ve-cadherin are required in endothelial cells during in vivo *tubulogenesis*. *Development* **2010**, *137*, 3119–3128. [CrossRef]
11. JayaNandanan, N.; Mathew, R.; Leptin, M. Guidance of subcellular tubulogenesis by actin under the control of a synaptotagmin-like protein and Moesin. *Nat. Commun.* **2014**, *5*, 3036. [CrossRef] [PubMed]
12. Khan, L.A.; Jafari, G.; Zhang, N.; Membreno, E.; Yan, S.; Zhang, H.; Gobel, V. A tensile trilayered cytoskeletal endotube drives capillary-like lumenogenesis. *J. Cell Biol.* **2019**, *218*, 2403–2424. [CrossRef] [PubMed]
13. Serrador, J.M.; Nieto, M.; Alonso-Lebrero, J.L.; del Pozo, M.A.; Calvo, J.; Furthmayr, H.; Sánchez-Madrid, F. CD43 interacts with moesin and ezrin and regulates its redistribution to the uropods of T lymphocytes at the cell-cell contacts. *Blood* **1998**, *91*, 4632–4644. [CrossRef] [PubMed]
14. Parameswaran, N.; Matsui, K.; Gupta, N. Conformational switching in ezrin regulates morphological and cytoskeletal changes required for B cell chemotaxis. *J. Immunol.* **2011**, *186*, 4088–4097. [CrossRef] [PubMed]

15. Ramoni, C.; Luciani, F.; Spadaro, F.; Lugini, L.; Lozupone, F.; Fais, S. Differential expression and distribution of ezrin, radixin and moesin in human natural killer cells. *Eur. J. Immunol.* **2002**, *32*, 3059–3065. [CrossRef]
16. Yoshinaga-Ohara, N.; Takahashi, A.; Uchiyama, T.; Sasada, M. Spatiotemporal regulation of moesin phosphorylation and rear release by Rho and serine/threonine phosphatase during neutrophil migration. *Exp. Cell Res.* **2002**, *278*, 112–122. [CrossRef]
17. Mori, T.; Kitano, K.; Terawaki, S.I.; Maesaki, R.; Fukami, Y.; Hakoshima, T. Structural basis for CD44 recognition by ERM proteins. *J. Biol. Chem.* **2008**, *283*, 29602–29612. [CrossRef]
18. Takai, Y.; Kitano, K.; Terawaki, S.I.; Maesaki, R.; Hakoshima, T. Structural basis of the cytoplasmic tail of adhesion molecule CD43 and its binding to ERM proteins. *J. Mol. Biol.* **2008**, *381*, 634–644. [CrossRef]
19. Takai, Y.; Kitano, K.; Terawaki, S.I.; Maesaki, R.; Hakoshima, T. Structural basis of PSGL-1 binding to ERM proteins. *Genes Cells* **2007**, *12*, 1329–1338. [CrossRef]
20. Yonemura, S.; Hirao, M.; Doi, Y.; Takahashi, N.; Kondo, T.; Tsukita, S.; Tsukita, S. Ezrin/radixin/moesin (ERM) proteins bind to a positively charged amino acid cluster in the juxta-membrane cytoplasmic domain of CD44, CD43, and ICAM-2. *J. Cell Biol.* **1998**, *140*, 885–895. [CrossRef]
21. Alonso-Lebrero, J.L.; Serrador, J.M.; Domınguez-Jiménez, C.; Barreiro, O.; Luque, A.; Del Pozo, M.A.; Lozano, F. Polarization and interaction of adhesion molecules P-selectin glycoprotein ligand 1 and intercellular adhesion molecule 3 with moesin and ezrin in myeloid cells. *Blood* **2000**, *95*, 2413–2419. [CrossRef] [PubMed]
22. Heiska, L.; Alfthan, K.; Grönholm, M.; Vilja, P.; Vaheri, A.; Carpén, O. Association of ezrin with intercellular adhesion molecule-1 and -2 (ICAM-1 and ICAM-2) Regulation by phosphatidylinositol 4, 5-bisphosphate. *J. Biol. Chem.* **1998**, *273*, 21893–21900. [CrossRef] [PubMed]
23. Serrador, J.M.; Alonso-Lebrero, J.L.; Pozo, M.A.D.; Furthmayr, H.; Schwartz-Albiez, R.; Calvo, J.; Sánchez-Madrid, F. Moesin interacts with the cytoplasmic region of intercellular adhesion molecule-3 and is redistributed to the uropod of T lymphocytes during cell polarization. *J. Cell Biol.* **1997**, *138*, 1409–1423. [CrossRef] [PubMed]
24. Barreiro, O.; Yáñez-Mó, M.; Serrador, J.M.; Montoya, M.C.; Vicente-Manzanares, M.; Tejedor, R.; Sánchez-Madrid, F. Dynamic interaction of VCAM-1 and ICAM-1 with moesin and ezrin in a novel endothelial docking structure for adherent leukocytes. *J. Cell Biol.* **2002**, *157*, 1233–1245. [CrossRef] [PubMed]
25. Dickson, T.C.; Mintz, C.D.; Benson, D.L.; Salton, S.R. Functional binding interaction identified between the axonal CAM L1 and members of the ERM family. *J. Cell Biol.* **2002**, *157*, 1105–1112. [CrossRef] [PubMed]
26. Serrador, J.M.; Vicente-Manzanares, M.; Calvo, J.; Barreiro, O.; Montoya, M.C.; Schwartz-Albiez, R.; Sánchez-Madrid, F. A novel serine-rich motif in the intercellular adhesion molecule 3 is critical for its ezrin/radixin/moesin-directed subcellular targeting. *J. Biol. Chem.* **2002**, *277*, 10400–10409. [CrossRef]
27. Serrador, J.M.; Urzainqui, A.; Alonso-Lebrero, J.L.; Cabrero, J.R.; Montoya, M.C.; Vicente-Manzanares, M.; Sánchez-Madrid, F. A juxta-membrane amino acid sequence of P-selectin glycoprotein ligand-1 is involved in moesin binding and ezrin/radixin/moesin-directed targeting at the trailing edge of migrating lymphocytes. *Eur. J. Immunol.* **2002**, *32*, 1560–1566. [CrossRef]
28. Cheng, L.; Itoh, K.; Lemmon, V. L1-mediated branching is regulated by two ezrin-radixin-moesin (ERM)-binding sites, the RSLE region and a novel juxtamembrane ERM-binding region. *J. Neurosci.* **2005**, *25*, 395–403. [CrossRef]
29. Cannon, J.L.; Mody, P.D.; Blaine, K.M.; Chen, E.J.; Nelson, A.D.; Sayles, L.J.; Burkhardt, J.K. CD43 interaction with ezrin-radixin-moesin (ERM) proteins regulates T-cell trafficking and CD43 phosphorylation. *Mol. Biol. Cell* **2011**, *22*, 954–963. [CrossRef]
30. Newe, A.; Rzeniewicz, K.; König, M.; Schroer, C.F.; Joachim, J.; Rey-Gallardo, A.; Ivetic, A. Serine Phosphorylation of L-Selectin Regulates ERM Binding, Clustering, and Monocyte Protrusion in Transendothelial Migration. *Front. Immunol.* **2019**, *10*, 2227. [CrossRef]
31. Bretscher, A.; Chambers, D.; Nguyen, R.; Reczek, D. ERM-Merlin and EBP50 protein families in plasma membrane organization and function. *Annu. Rev. Cell Dev. Biol.* **2000**, *16*, 113–143. [CrossRef] [PubMed]
32. Terawaki, S.; Maesaki, R.; Hakoshima, T. Structural basis for NHERF recognition by ERM proteins. *Structure* **2006**, *14*, 777–789. [CrossRef] [PubMed]
33. Gary, R.; Bretscher, A. Ezrin self-association involves binding of an N-terminal domain to a normally masked C-terminal domain that includes the F-actin binding site. *Mol. Biol. Cell* **1995**, *6*, 1061–1075. [CrossRef] [PubMed]

34. Magendantz, M.; Henry, M.D.; Lander, A.; Solomon, F. Interdomain interactions of radixin in vitro. *J. Biol. Chem.* **1995**, *270*, 25324–25327. [CrossRef]
35. Nakamura, F.; Amieva, M.R.; Furthmayr, H. Phosphorylation of threonine 558 in the carboxyl-terminal actin-binding domain of moesin by thrombin activation of human platelets. *J. Biol. Chem.* **1995**, *270*, 31377–31385. [CrossRef] [PubMed]
36. Matsui, T.; Maeda, M.; Doi, Y.; Yonemura, S.; Amano, M.; Kaibuchi, K.; Tsukita, S. Rho-kinase phosphorylates COOH-terminal threonines of ezrin/radixin/moesin (ERM) proteins and regulates their head-to-tail association. *J. Cell Biol.* **1998**, *140*, 647–657. [CrossRef] [PubMed]
37. Tsukita, S.; Yonemura, S. Cortical actin organization: Lessons from ERM (ezrin/radixin/moesin) proteins. *J. Biol. Chem.* **1999**, *274*, 34507–34510. [CrossRef]
38. Fievet, B.T.; Gautreau, A.; Roy, C.; Del Maestro, L.; Mangeat, P.; Louvard, D.; Arpin, M. Phosphoinositide binding and phosphorylation act sequentially in the activation mechanism of ezrin. *J. Cell Biol.* **2004**, *164*, 653–659. [CrossRef]
39. Nakamura, F.; Huang, L.; Pestonjamasp, K.; Luna, E.J.; Furthmayr, H. Regulation of F-actin binding to platelet moesin in vitro by both phosphorylation of threonine 558 and polyphosphatidylinositides. *Mol. Biol. Cell* **1999**, *10*, 2669–2685. [CrossRef]
40. Ben-Aissa, K.; Patino-Lopez, G.; Belkina, N.V.; Maniti, O.; Rosales, T.; Hao, J.J.; Shaw, S. Activation of moesin, a protein that links actin cytoskeleton to the plasma membrane, occurs by phosphatidylinositol 4,5-bisphosphate (PIP2) binding sequentially to two sites and releasing an autoinhibitory linker. *J. Biol. Chem.* **2012**, *287*, 16311–16323. [CrossRef]
41. Chen, X.; Khajeh, J.A.; Ju, J.H.; Gupta, Y.K.; Stanley, C.B.; Do, C.; Bu, Z. Phosphatidylinositol 4,5-bisphosphate clusters the cell adhesion molecule CD44 and assembles a specific CD44-Ezrin heterocomplex, as revealed by small angle neutron scattering. *J. Biol. Chem.* **2015**, *290*, 6639–6652. [CrossRef] [PubMed]
42. Yonemura, S.; Matsui, T.; Tsukita, S.; Tsukita, S. Rho-dependent and -independent activation mechanisms of ezrin/radixin/moesin proteins: An essential role for polyphosphoinositides in vivo. *J. Cell Sci.* **2002**, *115*, 2569–2580. [PubMed]
43. Lee, J.H.; Katakai, T.; Hara, T.; Gonda, H.; Sugai, M.; Shimizu, A. Roles of p-ERM and Rho-ROCK signaling in lymphocyte polarity and uropod formation. *J. Cell Biol.* **2004**, *167*, 327–337. [CrossRef] [PubMed]
44. Pietromonaco, S.F.; Simons, P.C.; Altman, A.; Elias, L. Protein kinase C-theta phosphorylation of moesin in the actin-binding sequence. *J. Biol. Chem.* **1998**, *273*, 7594–7603. [CrossRef] [PubMed]
45. Ng, T.; Parsons, M.; Hughes, W.E.; Monypenny, J.; Zicha, D.; Gautreau, A.; Parker, P.J. Ezrin is a downstream effector of trafficking PKC-integrin complexes involved in the control of cell motility. *EMBO J.* **2001**, *20*, 2723–2741. [CrossRef]
46. Yin, H.; Shi, Z.; Jiao, S.; Chen, C.; Wang, W.; Greene, M.I.; Zhou, Z. Germinal center kinases in immune regulation. *Cell Mol. Immunol.* **2012**, *9*, 439–445. [CrossRef]
47. Belkina, N.V.; Liu, Y.; Hao, J.J.; Karasuyama, H.; Shaw, S. LOK is a major ERM kinase in resting lymphocytes and regulates cytoskeletal rearrangement through ERM phosphorylation. *Proc. Natl. Acad. Sci. USA* **2009**, *106*, 4707–4712. [CrossRef]
48. Kschonsak, Y.T.; Hoffmann, I. Activated ezrin controls MISP levels to ensure correct NuMA polarization and spindle orientation. *J. Cell Sci.* **2018**, *131*. [CrossRef]
49. Ten Klooster, J.P.; Jansen, M.; Yuan, J.; Oorschot, V.; Begthel, H.; Di Giacomo, V.; Clevers, H. Mst4 and Ezrin induce brush borders downstream of the Lkb1/Strad/Mo25 polarization complex. *Dev. Cell* **2009**, *16*, 551–562. [CrossRef]
50. Vitorino, P.; Yeung, S.; Crow, A.; Bakke, J.; Smyczek, T.; West, K.; Ndubaku, C. MAP4K4 regulates integrin-FERM binding to control endothelial cell motility. *Nature* **2015**, *519*, 425–430. [CrossRef]
51. Baumgartner, M.; Sillman, A.L.; Blackwood, E.M.; Srivastava, J.; Madson, N.; Schilling, J.W.; Barber, D.L. The Nck-interacting kinase phosphorylates ERM proteins for formation of lamellipodium by growth factors. *Proc. Natl. Acad. Sci. USA* **2006**, *103*, 13391–13396. [CrossRef] [PubMed]
52. Plutoni, C.; Keil, S.; Zeledon, C.; Delsin, L.E.A.; Decelle, B.; Roux, P.P.; Emery, G. Misshapen coordinates protrusion restriction and actomyosin contractility during collective cell migration. *Nat. Commun.* **2019**, *10*, 3940. [CrossRef] [PubMed]

53. Machicoane, M.; de Frutos, C.A.; Fink, J.; Rocancourt, M.; Lombardi, Y.; Garel, S.; Echard, A. SLK-dependent activation of ERMs controls LGN-NuMA localization and spindle orientation. *J. Cell Biol.* **2014**, *205*, 791–799. [CrossRef] [PubMed]
54. Carreno, S.; Kouranti, I.; Glusman, E.S.; Fuller, M.T.; Echard, A.; Payre, F. Moesin and its activating kinase Slik are required for cortical stability and microtubule organization in mitotic cells. *J. Cell Biol.* **2008**, *180*, 739–746. [CrossRef]
55. Fukata, Y.; Kimura, K.; Oshiro, N.; Saya, H.; Matsuura, Y.; Kaibuchi, K. Association of the myosin-binding subunit of myosin phosphatase and moesin: Dual regulation of moesin phosphorylation by Rho-associated kinase and myosin phosphatase. *J. Cell Biol.* **1998**, *141*, 409–418. [CrossRef]
56. Rodrigues, N.T.; Lekomtsev, S.; Jananji, S.; Kriston-Vizi, J.; Hickson, G.R.; Baum, B. Kinetochore-localized PP1-Sds22 couples chromosome segregation to polar relaxation. *Nature* **2015**, *524*, 489–492. [CrossRef]
57. Hishiya, A.; Ohnishi, M.; Tamura, S.; Nakamura, F. Protein phosphatase 2C inactivates F-actin binding of human platelet moesin. *J. Biol. Chem.* **1999**, *274*, 26705–26712. [CrossRef]
58. Li, Y.; Jin, Y.; Liu, B.; Lu, D.; Zhu, M.; Jin, Y.; Yin, Y. PTENalpha promotes neutrophil chemotaxis through regulation of cell deformability. *Blood* **2019**, *133*, 2079–2089. [CrossRef]
59. Yang, H.S.; Hinds, P.W. Increased ezrin expression and activation by CDK5 coincident with acquisition of the senescent phenotype. *Mol. Cell.* **2003**, *11*, 1163–1176. [CrossRef]
60. Krieg, J.; Hunter, T. Identification of the two major epidermal growth factor-induced tyrosine phosphorylation sites in the microvillar core protein ezrin. *J. Biol. Chem.* **1992**, *267*, 19258–19265.
61. Heiska, L.; Carpen, O. Src phosphorylates ezrin at tyrosine 477 and induces a phosphospecific association between ezrin and a kelch-repeat protein family member. *J. Biol. Chem.* **2005**, *280*, 10244–10252. [CrossRef] [PubMed]
62. Yang, H.S.; Hinds, P.W. Phosphorylation of ezrin by cyclin-dependent kinase 5 induces the release of Rho GDP dissociation inhibitor to inhibit Rac1 activity in senescent cells. *Cancer Res.* **2006**, *66*, 2708–2715. [CrossRef] [PubMed]
63. Srivastava, J.; Elliott, B.E.; Louvard, D.; Arpin, M. Src-dependent ezrin phosphorylation in adhesion-mediated signaling. *Mol. Biol. Cell* **2005**, *16*, 1481–1490. [CrossRef] [PubMed]
64. Elliott, B.E.; Meens, J.A.; SenGupta, S.K.; Louvard, D.; Arpin, M. The membrane cytoskeletal crosslinker ezrin is required for metastasis of breast carcinoma cells. *Breast Cancer Res.* **2005**, *7*, R365–R373. [CrossRef]
65. Parameswaran, N.; Enyindah-Asonye, G.; Bagheri, N.; Shah, N.B.; Gupta, N. Spatial coupling of JNK activation to the B cell antigen receptor by tyrosine-phosphorylated ezrin. *J. Immunol.* **2013**, *190*, 2017–2026. [CrossRef]
66. Coffey, G.P.; Rajapaksa, R.; Liu, R.; Sharpe, O.; Kuo, C.C.; Krauss, S.W.; Robinson, W.H. Engagement of CD81 induces ezrin tyrosine phosphorylation and its cellular redistribution with filamentous actin. *J. Cell Sci.* **2009**, *122*, 3137–3144. [CrossRef]
67. Sanchez-Madrid, F.; del Pozo, M.A. Leukocyte polarization in cell migration and immune interactions. *EMBO J.* **1999**, *18*, 501–511. [CrossRef]
68. Del Pozo, M.A.; Vicente-Manzanares, M.; Tejedor, R.; Serrador, J.M.; Sánchez-Madrid, F. Rho GTPases control migration and polarization of adhesion molecules and cytoskeletal ERM components in T lymphocytes. *Eur. J. Immunol.* **1999**, *29*, 3609–3620.
69. McFarland, W. Microspikes on the lymphocyte uropod. *Science* **1969**, *163*, 818–820. [CrossRef]
70. Bhalla, D.K.; Braun, J.; Karnovsky, M.J. Lymphocyte surface and cytoplasmic changes associated with translational motility and spontaneous capping of Ig. *J. Cell Sci.* **1979**, *39*, 137–147.
71. Zabel, B.A.; Rott, A.; Butcher, E.C. Leukocyte chemoattractant receptors in human disease pathogenesis. *Annu. Rev. Pathol.* **2015**, *10*, 51–81. [CrossRef] [PubMed]
72. Xu, J.; Wang, F.; Van Keymeulen, A.; Herzmark, P.; Straight, A.; Kelly, K.; Bourne, H.R. Divergent signals and cytoskeletal assemblies regulate self-organizing polarity in neutrophils. *Cell* **2003**, *114*, 201–214. [CrossRef]
73. Sanchez-Madrid, F.; Serrador, J.M. Bringing up the rear: Defining the roles of the uropod. *Nat. Rev. Mol. Cell Biol.* **2009**, *10*, 353–359. [CrossRef] [PubMed]
74. Hao, J.J.; Liu, Y.; Kruhlak, M.; Debell, K.E.; Rellahan, B.L.; Shaw, S. Phospholipase C-mediated hydrolysis of PIP2 releases ERM proteins from lymphocyte membrane. *J. Cell Biol.* **2009**, *184*, 451–462. [CrossRef]
75. Treanor, B.; Depoil, D.; Bruckbauer, A.; Batista, F.D. Dynamic cortical actin remodeling by ERM proteins controls BCR microcluster organization and integrity. *J. Exp. Med.* **2011**, *208*, 1055–1068. [CrossRef]

76. Liu, X.; Yang, T.; Suzuki, K.; Tsukita, S.; Ishii, M.; Zhou, S.; Oh, M.J. Moesin and myosin phosphatase confine neutrophil orientation in a chemotactic gradient. *J. Exp. Med.* **2015**, *212*, 267–280. [CrossRef]
77. Ren, C.; Yuan, Q.; Braun, M.; Zhang, X.; Petri, B.; Zhang, J.; Fan, R. Leukocyte Cytoskeleton Polarization Is Initiated by Plasma Membrane Curvature from Cell Attachment. *Dev. Cell* **2019**, *49*, 206–219. [CrossRef]
78. Lacalle, R.A.; Peregil, R.M.; Albar, J.P.; Merino, E.; Martínez-A, C.; Mérida, I.; Mañes, S. Type I phosphatidylinositol 4-phosphate 5-kinase controls neutrophil polarity and directional movement. *J. Cell Biol.* **2007**, *179*, 1539–1553. [CrossRef]
79. Chen, E.J.; Shaffer, M.H.; Williamson, E.K.; Huang, Y.; Burkhardt, J.K. Ezrin and moesin are required for efficient T cell adhesion and homing to lymphoid organs. *PLoS ONE* **2013**, *8*, e52368. [CrossRef]
80. Martinelli, S.; Chen, E.J.; Clarke, F.; Lyck, R.; Affentranger, S.; Burkhardt, J.K.; Niggli, V. Ezrin/Radixin/Moesin proteins and flotillins cooperate to promote uropod formation in T cells. *Front. Immunol.* **2013**, *4*, 84. [CrossRef]
81. Doi, Y.; Itoh, M.; Yonemura, S.; Ishihara, S.; Takano, H.; Noda, T.; Tsukita, S. Normal development of mice and unimpaired cell adhesion/cell motility/actin-based cytoskeleton without compensatory up-regulation of ezrin or radixin in moesin gene knockout. *J. Biol. Chem.* **1999**, *274*, 2315–2321. [CrossRef] [PubMed]
82. Hirata, T.; Nomachi, A.; Tohya, K.; Miyasaka, M.; Tsukita, S.; Watanabe, T.; Narumiya, S. Moesin-deficient mice reveal a non-redundant role for moesin in lymphocyte homeostasis. *Int. Immunol.* **2012**, *24*, 705–717. [CrossRef] [PubMed]
83. Sanz, M.J.; Kubes, P. Neutrophil-active chemokines in in vivo imaging of neutrophil trafficking. *Eur. J. Immunol.* **2012**, *42*, 278–283. [CrossRef] [PubMed]
84. Yonemura, S.; Tsukita, S.; Tsukita, S. Direct involvement of ezrin/radixin/moesin (ERM)-binding membrane proteins in the organization of microvilli in collaboration with activated ERM proteins. *J. Cell Biol.* **1999**, *145*, 1497–1509. [CrossRef]
85. Ivetič, A.; Florey, O.; Deka, J.; Haskard, D.O.; Ager, A.; Ridley, A.J. Mutagenesis of the ezrin-radixin-moesin binding domain of L-selectin tail affects shedding, microvillar positioning, and leukocyte tethering. *J. Biol. Chem.* **2004**, *279*, 33263–33272. [CrossRef]
86. Ishihara, S.; Nishikimi, A.; Umemoto, E.; Miyasaka, M.; Saegusa, M.; Katagiri, K. Dual functions of Rap1 are crucial for T-cell homeostasis and prevention of spontaneous colitis. *Nat. Commun.* **2015**, *6*, 8982. [CrossRef]
87. Matsumoto, M.; Hirata, T. Moesin regulates neutrophil rolling velocity in vivo. *Cell Immunol.* **2016**, *304–305*, 59–62. [CrossRef]
88. Spertini, C.; Baisse, B.; Spertini, O. Ezrin-radixin-moesin-binding sequence of PSGL-1 glycoprotein regulates leukocyte rolling on selectins and activation of extracellular signal-regulated kinases. *J. Biol. Chem.* **2012**, *287*, 10693–10702. [CrossRef]
89. Nijhara, R.; van Hennik, P.B.; Gignac, M.L.; Kruhlak, M.J.; Hordijk, P.L.; Delon, J.; Shaw, S. Rac1 mediates collapse of microvilli on chemokine-activated T lymphocytes. *J. Immunol.* **2004**, *173*, 4985–4993. [CrossRef]
90. Barreiro, O.; Yáñez-Mó, M.; Sala-Valdés, M.; Gutiérrez-López, M.D.; Ovalle, S.; Higginbottom, A.; Sánchez-Madrid, F. Endothelial tetraspanin microdomains regulate leukocyte firm adhesion during extravasation. *Blood* **2005**, *105*, 2852–2861. [CrossRef]
91. Sala-Valdés, M.; Ursa, Á.; Charrin, S.; Rubinstein, E.; Hemler, M.E.; Sánchez-Madrid, F.; Yáñez-Mó, M. EWI-2 and EWI-F link the tetraspanin web to the actin cytoskeleton through their direct association with ezrin-radixin-moesin proteins. *J. Biol. Chem.* **2006**, *281*, 19665–19675. [CrossRef] [PubMed]
92. Carman, C.V.; Jun, C.D.; Salas, A.; Springer, T.A. Endothelial cells proactively form microvilli-like membrane projections upon intercellular adhesion molecule 1 engagement of leukocyte LFA-1. *J. Immunol.* **2003**, *171*, 6135–6144. [CrossRef] [PubMed]
93. Doulet, N.; Donnadieu, E.; Laran-Chich, M.P.; Niedergang, F.; Nassif, X.; Couraud, P.O.; Bourdoulous, S. Neisseria meningitidis infection of human endothelial cells interferes with leukocyte transmigration by preventing the formation of endothelial docking structures. *J. Cell Biol.* **2006**, *173*, 627–637. [CrossRef] [PubMed]
94. Oh, H.M.; Lee, S.; Na, B.R.; Wee, H.; Kim, S.H.; Choi, S.C.; Jun, C.D. RKIKK motif in the intracellular domain is critical for spatial and dynamic organization of ICAM-1: Functional implication for the leukocyte adhesion and transmigration. *Mol. Biol. Cell* **2007**, *18*, 2322–2335. [CrossRef]
95. Filippi, M.D. Mechanism of Diapedesis: Importance of the Transcellular Route. *Adv. Immunol.* **2016**, *129*, 25–53.

96. Liu, Y.; Belkina, N.V.; Park, C.; Nambiar, R.; Loughhead, S.M.; Patino-Lopez, G.; von Andrian, U.H. Constitutively active ezrin increases membrane tension, slows migration, and impedes endothelial transmigration of lymphocytes in vivo in mice. *Blood* **2012**, *119*, 445–453. [CrossRef]
97. Müller, N.; Fischer, H.J.; Tischner, D.; van den Brandt, J.; Reichardt, H.M. Glucocorticoids induce effector T cell depolarization via ERM proteins, thereby impeding migration and APC conjugation. *J. Immunol.* **2013**, *190*, 4360–4370. [CrossRef]
98. Shulman, Z.; Shinder, V.; Klein, E.; Grabovsky, V.; Yeger, O.; Geron, E.; Laudanna, C. Lymphocyte crawling and transendothelial migration require chemokine triggering of high-affinity LFA-1 integrin. *Immunity* **2009**, *30*, 384–396. [CrossRef]
99. Carman, C.V.; Sage, P.T.; Sciuto, T.E.; Miguel, A.; Geha, R.S.; Ochs, H.D.; Springer, T.A. Transcellular diapedesis is initiated by invasive podosomes. *Immunity* **2007**, *26*, 784–797. [CrossRef]
100. Koduru, S.; Kumar, L.; Massaad, M.J.; Ramesh, N.; Le Bras, S.; Ozcan, E.; King, S. Cdc42 interacting protein 4 (CIP4) is essential for integrin-dependent T-cell trafficking. *Proc. Natl. Acad. Sci. USA* **2010**, *107*, 16252–16256. [CrossRef]
101. Rey-Gallardo, A.; Tomlins, H.; Joachim, J.; Rahman, I.; Kitscha, P.; Frudd, K.; Ivetic, A. Sequential binding of ezrin and moesin to L-selectin regulates monocyte protrusive behaviour during transendothelial migration. *J. Cell Sci.* **2018**, *131*. [CrossRef]
102. Ilina, O.; Friedl, P. Mechanisms of collective cell migration at a glance. *J. Cell Sci.* **2009**, *122*, 3203–3208. [CrossRef]
103. Ramel, D.; Wang, X.; Laflamme, C.; Montell, D.J.; Emery, G. Rab11 regulates cell-cell communication during collective cell movements. *Nat. Cell boil.* **2013**, *15*, 317–324. [CrossRef]
104. Malet-Engra, G.; Yu, W.; Oldani, A.; Rey-Barroso, J.; Gov, N.S.; Scita, G.; Dupré, L. Collective cell motility promotes chemotactic prowess and resistance to chemorepulsion. *Curr. Biol.* **2015**, *25*, 242–250. [CrossRef]
105. Grakoui, A.; Bromley, S.K.; Sumen, C.; Davis, M.M.; Shaw, A.S.; Allen, P.M.; Dustin, M.L. The immunological synapse: A molecular machine controlling T cell activation. *Science* **1999**, *285*, 221–227. [CrossRef]
106. Tolar, P. Cytoskeletal control of B cell responses to antigens. *Nat. Rev. Immunol.* **2017**, *17*, 621–634. [CrossRef]
107. Lagrue, K.; Carisey, A.; Oszmiana, A.; Kennedy, P.R.; Williamson, D.J.; Cartwright, A.; Davis, D.M. The central role of the cytoskeleton in mechanisms and functions of the NK cell immune synapse. *Immunol. Rev.* **2013**, *256*, 203–221. [CrossRef]
108. Niedergang, F.; Di Bartolo, V.; Alcover, A. Comparative Anatomy of Phagocytic and Immunological Synapses. *Front. Immunol.* **2016**, *7*, 18. [CrossRef]
109. Freeman, S.A.; Vega, A.; Riedl, M.; Collins, R.F.; Ostrowski, P.P.; Woods, E.C.; Mayor, S. Transmembrane Pickets Connect Cyto- and Pericellular Skeletons Forming Barriers to Receptor Engagement. *Cell* **2018**, *172*, 305–317. [CrossRef]
110. Urzainqui, A.; Serrador, J.M.; Viedma, F.; Yáñez-Mó, M.; Rodríguez, A.; Corbí, A.L.; Sánchez-Madrid, F. ITAM-based interaction of ERM proteins with Syk mediates signaling by the leukocyte adhesion receptor PSGL-1. *Immunity* **2002**, *17*, 401–412. [CrossRef]
111. Mu, L.; Tu, Z.; Miao, L.; Ruan, H.; Kang, N.; Hei, Y.; Du, Y. A phosphatidylinositol 4,5-bisphosphate redistribution-based sensing mechanism initiates a phagocytosis programing. *Nat. Commun.* **2018**, *9*, 4259. [CrossRef]
112. Defacque, H.; Egeberg, M.; Habermann, A.; Diakonova, M.; Roy, C.; Mangeat, P.; Griffiths, G. Involvement of ezrin/moesin in de novo actin assembly on phagosomal membranes. *EMBO J.* **2000**, *19*, 199–212. [CrossRef]
113. Shcherbina, A.; Bretscher, A.; Kenney, D.M.; Remold-O'Donnell, E. Moesin, the major ERM protein of lymphocytes and platelets, differs from ezrin in its insensitivity to calpain. *FEBS Lett.* **1999**, *443*, 31–36. [CrossRef]
114. Roberts, R.E.; Hallett, M.B. Neutrophil Cell Shape Change: Mechanism and Signalling during Cell Spreading and Phagocytosis. *Int. J. Mol. Sci.* **2019**, *20*, 1383. [CrossRef]
115. Ferreira, É.R.; Bonfim-Melo, A.; Cordero, E.M.; Mortara, R.A. ERM Proteins Play Distinct Roles in Cell Invasion by Extracellular Amastigotes of Trypanosoma cruzi. *Front. Microbiol.* **2017**, *8*, 2230. [CrossRef]
116. Nakaya, M.; Kitano, M.; Matsuda, M.; Nagata, S. Spatiotemporal activation of Rac1 for engulfment of apoptotic cells. *Proc. Natl. Acad. Sci. USA* **2008**, *105*, 9198–9203. [CrossRef]
117. Kim, S.Y.; Kim, S.; Bae, D.J.; Park, S.Y.; Lee, G.Y.; Park, G.M.; Kim, I.S. Coordinated balance of Rac1 and RhoA plays key roles in determining phagocytic appetite. *PLoS ONE* **2017**, *12*, e0174603. [CrossRef]

118. Erwig, L.P.; McPhilips, K.A.; Wynes, M.W.; Ivetic, A.; Ridley, A.J.; Henson, P.M. Differential regulation of phagosome maturation in macrophages and dendritic cells mediated by Rho GTPases and ezrin-radixin-moesin (ERM) proteins. *Proc. Natl. Acad. Sci. USA* **2006**, *103*, 12825–12830. [CrossRef]
119. Marion, S.; Hoffmann, E.; Holzer, D.; Le Clainche, C.; Martin, M.; Sachse, M.; Griffiths, G. Ezrin promotes actin assembly at the phagosome membrane and regulates phago-lysosomal fusion. *Traffic* **2011**, *12*, 421–437. [CrossRef]
120. Dustin, M.L. Hunter to gatherer and back: Immunological synapses and kinapses as variations on the theme of amoeboid locomotion. *Annu. Rev. Cell Dev. Biol.* **2008**, *24*, 577–596. [CrossRef]
121. Fooksman, D.R.; Vardhana, S.; Vasiliver-Shamis, G.; Liese, J.; Blair, D.A.; Waite, J.; Dustin, M.L. Functional anatomy of T cell activation and synapse formation. *Annu. Rev. Immunol.* **2009**, *28*, 79–105. [CrossRef]
122. Vardhana, S.; Choudhuri, K.; Varma, R.; Dustin, M.L. Essential role of ubiquitin and TSG101 protein in formation and function of the central supramolecular activation cluster. *Immunity* **2010**, *32*, 531–540. [CrossRef]
123. Roumier, A.; Olivo-Marin, J.C.; Arpin, M.; Michel, F.; Martin, M.; Mangeat, P.; Alcover, A. The membrane-microfilament linker ezrin is involved in the formation of the immunological synapse and in T cell activation. *Immunity* **2001**, *15*, 715–728. [CrossRef]
124. Delon, J.; Kaibuchi, K.; Germain, R.N. Exclusion of CD43 from the immunological synapse is mediated by phosphorylation-regulated relocation of the cytoskeletal adaptor moesin. *Immunity* **2001**, *15*, 691–701. [CrossRef]
125. Allenspach, E.J.; Cullinan, P.; Tong, J.; Tang, Q.; Tesciuba, A.G.; Cannon, J.L.; Sperling, A.I. ERM-dependent movement of CD43 defines a novel protein complex distal to the immunological synapse. *Immunity* **2001**, *15*, 739–750. [CrossRef]
126. McCann, F.E.; Vanherberghen, B.; Eleme, K.; Carlin, L.M.; Newsam, R.J.; Goulding, D.; Davis, D.M. The size of the synaptic cleft and distinct distributions of filamentous actin, ezrin, CD43, and CD45 at activating and inhibitory human NK cell immune synapses. *J. Immunol.* **2003**, *170*, 2862–2870. [CrossRef]
127. Masilamani, M.; Nguyen, C.; Kabat, J.; Borrego, F.; Coligan, J.E. CD94/NKG2A inhibits NK cell activation by disrupting the actin network at the immunological synapse. *J. Immunol.* **2006**, *177*, 3590–3596. [CrossRef]
128. Tong, J.; Allenspach, E.J.; Takahashi, S.M.; Mody, P.D.; Park, C.; Burkhardt, J.K.; Sperling, A.I. CD43 regulation of T cell activation is not through steric inhibition of T cell-APC interactions but through an intracellular mechanism. *J. Exp. Med.* **2004**, *199*, 1277–1283. [CrossRef]
129. Ilani, T.; Khanna, C.; Zhou, M.; Veenstra, T.D.; Bretscher, A. Immune synapse formation requires ZAP-70 recruitment by ezrin and CD43 removal by moesin. *J. Cell Biol.* **2007**, *179*, 733–746. [CrossRef]
130. Shaffer, M.H.; Dupree, R.S.; Zhu, P.; Saotome, I.; Schmidt, R.F.; McClatchey, A.I.; Burkhardt, J.K. Ezrin and moesin function together to promote T cell activation. *J. Immunol.* **2009**, *182*, 1021–1032. [CrossRef]
131. Blumenthal, D.; Burkhardt, J.K. Multiple actin networks coordinate mechanotransduction at the immunological synapse. *J. Cell Biol.* **2020**, *219*. [CrossRef] [PubMed]
132. Faure, S.; Salazar-Fontana, L.I.; Semichon, M.; Tybulewicz, V.L.; Bismuth, G.; Trautmann, A.; Delon, J. ERM proteins regulate cytoskeleton relaxation promoting T cell-APC conjugation. *Nat. Immunol.* **2004**, *5*, 272–279. [CrossRef] [PubMed]
133. Van Kooyk, Y.; van Vliet, S.J.; Figdor, C.G. The actin cytoskeleton regulates LFA-1 ligand binding through avidity rather than affinity changes. *J. Biol. Chem.* **1999**, *274*, 26869–26877. [CrossRef] [PubMed]
134. Comrie, W.A.; Li, S.; Boyle, S.; Burkhardt, J.K. The dendritic cell cytoskeleton promotes T cell adhesion and activation by constraining ICAM-1 mobility. *J. Cell Biol.* **2015**, *208*, 457–473. [CrossRef]
135. Gross, C.C.; Brzostowski, J.A.; Liu, D.; Long, E.O. Tethering of intercellular adhesion molecule on target cells is required for LFA-1-dependent NK cell adhesion and granule polarization. *J. Immunol.* **2010**, *185*, 2918–2926. [CrossRef]
136. Treanor, B.; Depoil, D.; Gonzalez-Granja, A.; Barral, P.; Weber, M.; Dushek, O.; Batista, F.D. The membrane skeleton controls diffusion dynamics and signaling through the B cell receptor. *Immunity* **2010**, *32*, 187–199. [CrossRef]
137. Parameswaran, N.; Gupta, N. Re-defining ERM function in lymphocyte activation and migration. *Immunol. Rev.* **2013**, *256*, 63–79. [CrossRef]

138. Gupta, N.; Wollscheid, B.; Watts, J.D.; Scheer, B.; Aebersold, R.; DeFranco, A.L. Quantitative proteomic analysis of B cell lipid rafts reveals that ezrin regulates antigen receptor-mediated lipid raft dynamics. *Nat. Immunol.* **2006**, *7*, 625–633. [CrossRef]
139. Kwak, K.; Quizon, N.; Sohn, H.; Saniee, A.; Manzella-Lapeira, J.; Holla, P.; Spillane, K.M. Intrinsic properties of human germinal center B cells set antigen affinity thresholds. *Sci. Immunol.* **2018**, *3*, eaau6598. [CrossRef]
140. Delmonte, O.M.; Biggs, C.M.; Hayward, A.; Comeau, A.M.; Kuehn, H.S.; Rosenzweig, S.D.; Notarangelo, L.D. First Case of X-Linked Moesin Deficiency Identified After Newborn Screening for SCID. *J. Clin. Immunol.* **2017**, *37*, 336–338. [CrossRef]
141. Lagresle-Peyrou, C.; Luce, S.; Ouchani, F.; Soheili, T.S.; Sadek, H.; Chouteau, M.; Lambert, N. X-linked primary immunodeficiency associated with hemizygous mutations in the moesin (MSN) gene. *J. Allergy Clin. Immunol.* **2016**, *138*, 1681–1689. [CrossRef] [PubMed]
142. Li, Y.; Harada, T.; Juang, Y.T.; Kyttaris, V.C.; Wang, Y.; Zidanic, M.; Tsokos, G.C. Phosphorylated ERM is responsible for increased T cell polarization, adhesion, and migration in patients with systemic lupus erythematosus. *J. Immunol.* **2007**, *178*, 1938–1947. [CrossRef] [PubMed]
143. Satooka, H.; Nagakubo, D.; Sato, T.; Hirata, T. The ERM Protein Moesin Regulates CD8(+) Regulatory T Cell Homeostasis and Self-Tolerance. *J. Immunol.* **2017**, *199*, 3418–3426. [CrossRef] [PubMed]
144. Ansa-Addo, E.A.; Zhang, Y.; Yang, Y.; Hussey, G.S.; Howley, B.V.; Salem, M.; Howe, P.H. Membrane-organizing protein moesin controls Treg differentiation and antitumor immunity via TGF-beta signaling. *J. Clin. Investig.* **2017**, *127*, 1321–1337. [CrossRef]
145. Pore, D.; Huang, E.; Dejanovic, D.; Parameswaran, N.; Cheung, M.B.; Gupta, N. Cutting Edge: Deletion of Ezrin in B Cells of Lyn-Deficient Mice Downregulates Lupus Pathology. *J. Immunol.* **2018**, *201*, 1353–1358. [CrossRef]
146. Pore, D.; Matsui, K.; Parameswaran, N.; Gupta, N. Cutting Edge: Ezrin Regulates Inflammation by Limiting B Cell IL-10 Production. *J. Immunol.* **2016**, *196*, 558–562. [CrossRef]
147. Pore, D.; Parameswaran, N.; Matsui, K.; Stone, M.B.; Saotome, I.; McClatchey, A.I.; Gupta, N. Ezrin tunes the magnitude of humoral immunity. *J. Immunol.* **2013**, *191*, 4048–4058. [CrossRef]
148. Birgbauer, E.; Solomon, F. A marginal band-associated protein has properties of both microtubule- and microfilament-associated proteins. *J. Cell Biol.* **1989**, *109*, 1609–1620. [CrossRef]
149. Martin, M.; Roy, C.; Montcourrier, P.; Sahuquet, A.; Mangeat, P. Three determinants in ezrin are responsible for cell extension activity. *Mol. Biol. Cell* **1997**, *8*, 1543–1557. [CrossRef]
150. Solinet, S.; Mahmud, K.; Stewman, S.F.; Ben El Kadhi, K.; Decelle, B.; Talje, L.; Carreno, S. The actin-binding ERM protein Moesin binds to and stabilizes microtubules at the cell cortex. *J. Cell Biol.* **2013**, *202*, 251–260. [CrossRef]
151. Takesono, A.; Heasman, S.J.; Wojciak-Stothard, B.; Garg, R.; Ridley, A.J. Microtubules regulate migratory polarity through Rho/ROCK signaling in T cells. *PLoS ONE* **2010**, *5*, e8774. [CrossRef] [PubMed]
152. Prentice-Mott, H.V.; Meroz, Y.; Carlson, A.; Levine, M.A.; Davidson, M.W.; Irimia, D.; Shah, J.V. Directional memory arises from long-lived cytoskeletal asymmetries in polarized chemotactic cells. *Proc. Natl. Acad. Sci. USA* **2016**, *113*, 1267–1272. [CrossRef] [PubMed]
153. Lasserre, R.; Charrin, S.; Cuche, C.; Danckaert, A.; Thoulouze, M.I.; De Chaumont, F.; Etienne-Manneville, S. Ezrin tunes T-cell activation by controlling Dlg1 and microtubule positioning at the immunological synapse. *EMBO J.* **2010**, *29*, 2301–2314. [CrossRef] [PubMed]

© 2020 by the authors. Licensee MDPI, Basel, Switzerland. This article is an open access article distributed under the terms and conditions of the Creative Commons Attribution (CC BY) license (http://creativecommons.org/licenses/by/4.0/).

Review

Roles of Actin in the Morphogenesis of the Early *Caenorhabditis elegans* Embryo

Dureen Samandar Eweis [1,2] and Julie Plastino [1,2,]*

[1] Laboratoire Physico-Chimie Curie, Institut Curie, PSL Research University, CNRS, 75005 Paris, France; dureen.samandar-eweis@curie.fr
[2] Sorbonne Université, 75005 Paris, France
* Correspondence: julie.plastino@curie.fr

Received: 6 April 2020; Accepted: 19 May 2020; Published: 21 May 2020

Abstract: The cell shape changes that ensure asymmetric cell divisions are crucial for correct development, as asymmetric divisions allow for the formation of different cell types and therefore different tissues. The first division of the *Caenorhabditis elegans* embryo has emerged as a powerful model for understanding asymmetric cell division. The dynamics of microtubules, polarity proteins, and the actin cytoskeleton are all key for this process. In this review, we highlight studies from the last five years revealing new insights about the role of actin dynamics in the first asymmetric cell division of the early *C. elegans* embryo. Recent results concerning the roles of actin and actin binding proteins in symmetry breaking, cortical flows, cortical integrity, and cleavage furrow formation are described.

Keywords: actin cytoskeleton; myosin; *C. elegans* embryo; asymmetric cell division

1. Introduction

Actin is one of the most abundant proteins in the cell, existing as globular monomers (G-actin) that polymerize into helical filaments (F-actin). F-actin is polar with a fast-growing, dynamic 'barbed end' and a slow-growing, less dynamic 'pointed end'. The dynamic assembly and disassembly of F-actin, as well as the myosin molecular motors that associate with actin, produce forces within the cell and between cells that drive cellular and tissue reorganization. Actin dynamics are controlled by actin-binding proteins, which variously activate or inhibit F-actin formation, stabilize/destabilize existing actin structures, or bind actin monomers [1]. One of the most important actin regulators is the Arp2/3 complex that nucleates the formation of new actin filaments as branches off the sides of existing filaments. Another important class of actin polymerization nucleators is the formin family of proteins. Formins create new unbranched filaments, and also associate with the barbed end of the actin filament, enhancing actin assembly. Once formed, F-actin is remodeled by actin bundling and cross-linking proteins—such as fascin, plastin/fimbrin, and filamin—which promote the formation of parallel and antiparallel bundles of F-actin or cross-linked arrays, respectively. Additionally, myosin remodels F-actin structures, using actin filaments as tracks, sliding antiparallel filaments in relation to each other to create contraction. G-actin binding proteins include profilin, which is key for actin dynamics in the cell, as it reduces spontaneous nucleation and also prevents pointed end polymerization thus allowing for controlled, directed actin assembly in vivo. Capping proteins, which bind barbed ends and prevent further polymerization, and ADF/cofilin proteins, which sever actin filaments, are also important for in vivo actin dynamics and function [2]. Many actin binding proteins, including the main polymerization nucleators and myosin, are regulated by the small GTPases Rho, Rac, and Cdc42, which are in turn controlled by guanine-nucleotide-exchange factors (GEFs) and GTPase-activating proteins (GAPs) downstream of extracellular signals.

The diversity of actin-binding proteins leads to a diversity of actin architectures in the cell, adapted to different functions [1]. The actomyosin cortex is a thin layer of cross-linked actin filaments interspersed by myosin, attached to the inner face of the cell membrane. In the moving cell, the cortex at the back of the cell contracts to squeeze the cell forward, while protrusive actin structures, lamellipodia and filopodia, form at the front. Blebs, another type of cell protrusion, occur when the cell membrane detaches from the cytoskeleton and balloons outward, initially devoid of actin, due to actomyosin contractility in the cortex.

Caenorhabditis elegans has been used as a model organism to investigate the regulation and dynamics of actin networks in developmental processes [3]. In particular, the first asymmetric division of the *C. elegans* single cell embryo has been studied extensively in order to understand symmetry breaking, polarity establishment, microtubule assembly, spindle positioning, and the cell shape changes that accompany asymmetric cell division [4,5]. Briefly, as concerns the actin cytoskeleton and related proteins, in the just-fertilized zygote, cortical ruffles are evident all around the circumference of the embryo due to the highly dynamic and contractile cortical actomyosin layer. Symmetry is broken when the sperm contents approach the future posterior pole, locally downregulating contractility there and initiating the retraction of the actomyosin cortex to the future anterior pole [6,7]. This flow of actomyosin density towards the anterior pole leads to an invagination at the boundary between high and low actomyosin activity similar to ruffles but much deeper, called the pseudocleavage furrow. Anterior-directed cortical flow is concomitant with the segregation of the polarity proteins PAR-3, PAR-6, and PKC-3 to the anterior of the embryo, while PAR-1 and PAR-2 are recruited to the posterior cortex [8].

The result of this polarization phase in the embryo is the formation of two cortical domains that have different actomyosin activity and different PAR protein occupancy. During this period, known as the maintenance phase, a complex network of reciprocally supportive and antagonistic interactions between PAR proteins exists reinforcing their localization at the poles of the embryo. This also includes effects on actomyosin wherein PAR-1 and PAR-2 at the posterior pole have a role in inhibiting the posterior localization of non-muscle myosin II (NMY-2), while anterior PARs—PAR-3 and PAR-6/PKC-3—lead to the accumulation of NMY-2 at the anterior, establishing and maintaining a more contractile anterior pole [9]. During the maintenance phase, in a series of steps that are largely microtubule dependent, the maternal pronucleus joins the paternal pronucleus at the future posterior side of the cell, the complex recenters and then forms the spindle, which is pulled posteriorly again during anaphase to give the final asymmetry in division. Importantly, the positioning of the cleavage furrow and site of division is highly dependent on the positioning of the spindle [10], and it is the cortical polarity of the embryo that controls the mitotic spindle shift since cortical pulling forces are more pronounced at the posterior pole [6,11].

Although the first cell division of the *C. elegans* embryo has been studied for decades, many open questions remain. Here we review results from the last five years that address some of the outstanding questions in the field and demonstrate the multitude of roles played by the actin cytoskeleton in the *C. elegans* embryo.

2. Role of Actin in the Just-Fertilized Embryo during Completion of Meiosis

As for many organisms, *C. elegans* oocytes complete meiosis upon fertilization. While meiotic divisions are occurring at what will become the anterior pole of the embryo, the sperm contents, including genetic material, are retained at the site of sperm entry at the posterior pole, despite cytoplasmic streaming in the embryo.

A recent finding reported that the actin cytoskeleton was what was confining sperm DNA to the posterior pole of the embryo, and keeping it from getting captured by the meiotic spindle [12]. In embryos where actin polymerization was reduced, either by interfering with the profilin/formin mode of actin assembly or via application of inhibitory drugs, sperm DNA was distributed throughout the embryo due to cytoplasmic flows. This study further showed that the sperm contents were not restricted to the posterior cortex by cytoplasmic actin via a sieving effect as could have been expected. Rather, the study suggested that it was the cortical actin pool that was important for sperm content confinement by an as-yet-unidentified mechanism. Although most studies on the one-cell embryo pay particular attention to actin structures and functions during and after polarity establishment, this work showed that F-actin was an important player prior to these events.

3. Actomyosin Dynamics in Symmetry Breaking

The morphological and biochemical changes defining the anteroposterior axis of the embryo occur downstream of sperm entry, which has been shown to break the symmetry of the embryo and define its future posterior pole [6,7,13]. At the moment of fertilization, actomyosin foci are present over the entire embryo surface, along with actomyosin-based cortical contractions. At the end of meiosis II and just as mitosis is beginning, the sperm centrosome moves close to the embryo cortex at the future posterior pole. As described in the introduction, this produces an immediate cessation of cortical activity and a flow of actomyosin foci away from this region. As reviewed recently [6], this local transformation in the actomyosin cytoskeleton is known to be due to the removal of the RhoGEF ECT-2 from the cortex.

Until recently, the molecular nature of the cue delivered by the sperm centrosome to downregulate posterior actomyosin activity and initiate cortical flow was unknown. Novel studies in the past year have identified the mitotic kinase Aurora A (AIR-1) as the previously unidentified centrosome component (Figure 1). One study showed that the phosphorylated, active form of AIR-1 was released from the centrosomes into the cytoplasm, driving the inhibition of posterior cortical actomyosin networks in the vicinity of the centrosomes [14]. This was discovered based on the finding that the GFP-tagged version of AIR-1 was not completely wild-type. GFP-labeled AIR-1 embryos performed AIR-1-dependent processes normally, including centrosome maturation, but failed to correctly clear actomyosin from the future posterior pole during symmetry breaking. It was hypothesized that this effect was due to a defect in diffusion of the GFP-labeled protein. This hypothesis was supported by experiments that manipulated the position of the centrosome, moving it closer and further away from the cortex, improving and exacerbating, respectively, the actomyosin clearing defect [14]. Another recent study came to a similar conclusion concerning the identity of the centrosome-derived cue [15]. Both studies showed that AIR-1's role in symmetry breaking was a result of its effect on ECT-2, altering its localization and perhaps its GEF activity by an unknown mechanism. AIR-1 could also be downregulating myosin activity via its phosphorylation of other RHO-1 pathway effectors that are upstream of ECT-2.

In addition to its centrosomal role in initiating cortical flow, non-centrosomal AIR-1 has been shown in recent studies to globally downregulate cortical actomyosin activity during polarity establishment. Embryos lacking AIR-1 showed hypercontractility of the cortex, and became bipolar, with NMY-2-poor/PAR-2-rich domains at both poles of the embryo, and weak cortical flows directed toward the embryo center from both sides [14–17]. One study further showed that this bipolarization could occur even in wild-type embryos when fertilized with acentrosomal sperm [16]. Putting all this together, it seemed that there was a basal activity of non-centrosomal AIR-1, which kept actomyosin downregulated globally and prevented spontaneous bipolarization events, while centrosomal AIR-1 downregulated actomyosin locally to initiate the polarizing cortical flows that establish embryo polarity. Why PAR-2 domains form in a bipolar manner was unclear, but it was proposed that curvature could be the determining factor [16]. This hypothesis was tested by placing *air-1* depleted embryos in triangular chambers. It was observed that PAR-2 domains emerged in regions with the highest curvature. PAR-2 accumulation at curved regions could be biochemically driven by lipid affinities or due to geometrical considerations, where the curved surface of the poles restricts diffusion out of the immediate vicinity [16]. An additional function for AIR-1 was recently observed in the steps leading up to fertilization and symmetry breaking in the oocyte, where AIR-1 was shown to play a role via another cell cycle kinase PLK-1 (polo-like kinase) in regulating anterior PAR loading and activation at the membrane, although the details of the actomyosin cortex are not addressed in this study [17]. This regulation prevents premature polarization of the embryo, and enforces the dependence on the centrosome cue.

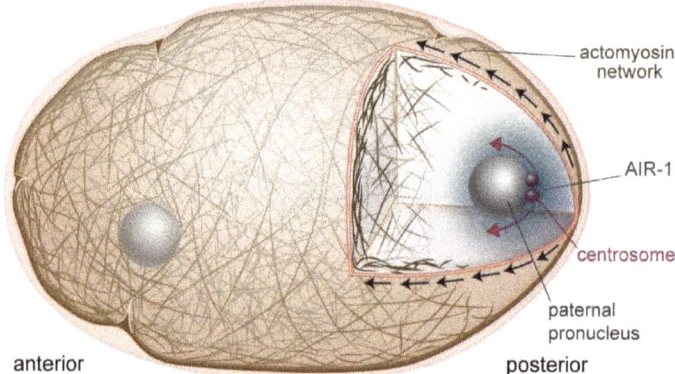

Figure 1. Symmetry breaking in the one-cell embryo. AIR-1 (blue cloud) is the cue that initiates polarization of the embryo. AIR-1 diffuses from the centrosome (red spheres) and downregulates actomyosin at the adjacent cortex. This causes a local weakening, and produces cortical flows (black arrows) directed away from this point, which also serve to separate the centrosomes (red arrows).

4. Cortical Flows during Polarity Establishment

Anterior-directed actomyosin contractility creates cortical flow, which sets up the polarity axis of the embryo as concerns both actin cytoskeleton and PAR proteins. Several recent studies have shed light on the precise details of how actomyosin-driven dynamics are coupled to PAR protein polarization. One study showed that cortical actomyosin tension facilitates the clustering of PAR-3 perhaps by inducing conformational changes that allow oligomerization [18]. Indeed, embryos lacking cortical tension molecules like NMY-2 did not exhibit PAR-3 clusters, but clusters could be rescued by artificial increases in cortical tension applied via osmotic shock for example. PAR-3 clustering induced clustering of PKC-3, and the clustering of both proteins was important for their proper transport to the anterior pole. A coincident study concurred, proposing that either clustering reduced the effective diffusion of membrane-associated anterior PARs, or the larger size of clusters allowed them to interact more effectively with the cortical actomyosin layer, both of which would favor advective transport by flows [19]. Another study came to a similar conclusion, showing via a single-cell extract technique that PAR-3 clusters were more efficiently transported by cortical flow due to the longer residence time of the PAR-3 oligomers at the cortex [20]. These and other findings concerning the roles of PLK-1 and CDC-42 in PAR protein clustering and cortical transport to the anterior pole are nicely reviewed in [21].

Another recent study showed the very clear link between cortical flows and PAR domain location. In this work, they used a novel focused-light-induced cytoplasmic streaming (FLUCS) system to induce controllable cytoplasmic flows in the embryo via temperature changes [22]. Cytoplasmic flows were shown to drive cortical flows, which were exactly mirrored by PAR protein domain relocation [22]. Compellingly, when the PAR-2 domain was moved to the anterior pole by flow, the embryo divided with an inverted size asymmetry (smaller anterior cell). All together, recent work confirms and extends the importance of actomyosin cortical flow for PAR domain establishment in the embryo.

The studies discussed above were principally about how flows driven by actomyosin contractility play a role in PAR protein localization, however the converse is also true: PAR proteins affect NMY-2 recruitment to the cortex and thus control flows. This complex interplay was investigated in a recent study where the kinetics of NMY-2 association and dissociation from the posterior and anterior cortex were measured [23]. It was observed that NMY-2 association with the cortex was identical for the posterior and anterior domains. Dissociation on the other hand was twice as high in the posterior domain as compared to the anterior region. This was shown to depend on PAR-6 specifically, where increases in PAR-6 were accompanied by decreases in the dissociation of NMY-2. It was proposed that this mechanochemical feedback between actomyosin localization and PAR proteins ensured robustness of embryo polarity.

A previously unidentified role for cortical flow was highlighted in a recent study demonstrating that flow contributed to centrosome separation [24]. The centrosome pair present on the paternal pronucleus must be separated to correctly form the bipolar mitotic spindle necessary for the first cell division. The authors showed that cortical dynein, imbedded in the actomyosin cortex, and also bound to microtubules emanating from the centrosomes, was the main player in this process. Cortical dynein was swept anteriorly by actomyosin flows, pulling with it the centrosomes. When cortical flow was impaired by depletion of NMY-2 or RHO-1, centrosome separation was retarded although there was no effect on cell cycle progression. Since AIR-1 on the centrosome is what is responsible for breaking the symmetry of the embryo and triggering cortical flow, centrosomes are perfectly positioned to harness flow for separation (Figure 1).

PAR proteins control flows, but actin-binding proteins are also known to affect flows, presumably by modifying the organization and mechanical properties of the cortex. This point was addressed by carefully characterizing flow velocities, flow pulsatility and myosin foci size and density in *C. elegans* embryos upon depletion of different actin-binding proteins and myosin regulators [25]. Many were found to affect the measured parameters in sometimes subtle ways. One general result that came out of this analysis was that bigger, more sparsely-distributed myosin foci were correlated with slower average flow velocities. However, the main point of this study was that data clustering revealed classes of proteins with sometimes dissimilar molecular activities that nevertheless affected cortical properties in similar ways. This pointed to a degeneracy in the molecular components needed to produce a given phenotype. A related paper specifically examined the origins of pulsatility, asking why this did not lead to instability and collapse of the system [26]. They showed evidence that Rho activity oscillated, thus damping down myosin foci contraction and preventing collapse of the cortex. Although not pertaining to the one-cell embryo, another study came to a similar conclusion about the presence of a Rho oscillator [27]. These studies all together illustrate that actomyosin contractility and resulting flows are robustly controlled in the *C. elegans* embryo.

5. Cortical Actin Architecture and Dynamics

Many proteins that affect flows in the polarizing embryo also have effects on the stability of the actomyosin cortex in the maintenance phase, for example, the actin filament bundling protein plastin (PLST-1) [28]. Deletion of *plst-1* led to smaller and more dispersed NMY-2 foci as well as weaker and less coherent directed cortical flows suggesting a need for PLST-1 for proper NMY-2 coalescence and resulting flows. Predictably, since PAR protein organization depends in part on cortical flows, there was also a defect in the polarity of PAR proteins in embryos lacking PLST-1. Via laser ablations and measurements of cortex recoil velocity, lack of PLST-1 was shown to decrease the tension of the anterior cortex of the embryo during the maintenance phase, indicating reduced cortical stiffness. Overall, the authors proposed that, by controlling actin network connectivity, PLST-1 regulated the mechanical properties of the cortex. Imaging techniques that allow visualization of nanoscale structures of the actin cortex in the presence and absence of PLST-1 would be very interesting to obtain in order to get a molecular picture of connectivity differences.

Another interesting study on cortical stability addressed the functions of cadherin (HMR-1) in the *C. elegans* embryo [29]. HMR-1 is normally involved in cell-cell adhesions, but these are not present in the one-cell embryo. It was further shown that HMR-1's ability to interact with the permeability barrier surrounding the plasma membrane of the embryo was not important for HMR-1's role at this stage, so it appeared that HMR-1 was playing an adhesion-independent role. The observation was that clusters of HMR-1 formed during the polarity establishment phase of the single cell embryo, and associated with cortical F-actin. Clusters were transported to the anterior pole by flows where, during the maintenance phase, they appeared to antagonize cortical NMY-2. Indeed, upon depletion of *hmr-1*, the level of anterior cortical NMY-2 significantly increased while there was no significant effect on the global concentration of NMY-2 or cortical F-actin. Non-junctional HMR-1 clusters appeared to control cortical NMY-2 via its upstream regulator, RHO-1: the amount of active cortical RHO-1 increased upon *hmr-1* depletion. Finally, in conditions of depleted *hmr-1*, cortical flows were accentuated and the actin cortex was observed to tear away from the cell membrane, probably due to the combined effect of increased flow/contractility and loss of stabilizing linkages conferred by HMR-1, which bridges the cell membrane to the actin network. Overall, this study brought to light a role for non-adhesive cadherin clusters in regulating actomyosin cortex stability and flows, and attachment to the cell membrane.

While much of the work reviewed here deals with the actomyosin cortex, which is attached to the cell membrane, not many studies have investigated the involvement of the cell membrane itself in the *C. elegans* embryo. A new study showed that the lipid phosphatidylinositol 4,5-bisphosphate (PIP2) was non-uniform in the polarized embryo. It was enriched at the anterior pole where it controlled actin dynamics and PAR protein recruitment [30]. However, this study conflicts with another recent work, which demonstrated that the increased formation of membrane folds at the anterior pole lead to an enrichment of lipids in general, not just PIP2 [31]. In fact, these membrane folds were shown to be filopodia, dependent on the formin CYK-1 and the Arp2/3 complex. The existence of PIP2 membrane microdomains in the *C. elegans* zygote therefore does not appear likely.

Overall, these recent studies detail the molecular regulation of the structure and dynamics of the actomyosin cortex during the post-polarization phase, expanding what is already known for symmetry breaking and polarity establishment.

6. Contractile Ring Formation and Positioning

Once polarity is triggered and established, the main steps preceding cytokinesis are contractile ring formation and spindle positioning, which is what defines the position of the cleavage furrow [10]. The ring is comprised of F-actin, myosin and accessory proteins and it accomplishes cytokinesis of the single cell embryo. Detailed, up-to-date reviews of the players needed for ensuring proper formation of the cytokinetic ring and furrow ingression have been published [6,32].

Insight as to how the contractile ring is constructed was provided by a recent study, where it was shown that F-actin alignment due to converging cortical flow at the cell equator was sufficient to drive ring formation (Figure 2) [33]. No localized actin polymerization was necessary, and in fact did not exist for ring formation during pseudocleavage, although such mechanisms exist for contractile ring formation during cleavage. Reducing flow rates by perturbing myosin machinery had predictable effects on the pseudocleavage furrow: slight reductions still permitted actin filaments to align to create a furrow while drastic reductions abolished furrow formation. A related work showed that mechanical compression of embryos inhibited longitudinal flows, but favoured rotational flows, which had the same end result of efficiently aligning filaments to create a contractile ring [34]. This study further showed that myosin flow characteristics were not quite the same as actin flows. In particular, NMY-2 flows were more long-ranged, a difference explained by the fact that actin was disassembled by compression as it flowed to the equator while myosin was not [34]. Cortical flows were also shown to be important for ring dynamics during constriction [35]. This study found that new cortex, including NMY-2, was being pulled into the ring due to myosin activity and polar relaxation. The added myosin in turn led to increased flow into the ring, and an exponential increase in the amount of ring components. This was counterbalanced by disassembly-coupled ring shortening. The end result was that, although NMY-2 levels and the levels of other ring components appeared to remain constant during ring constriction, the ring was in fact undergoing dramatic restructuring. These recent studies emphasize the robustness of cortical flows for creating and maintaining a proper furrow structure.

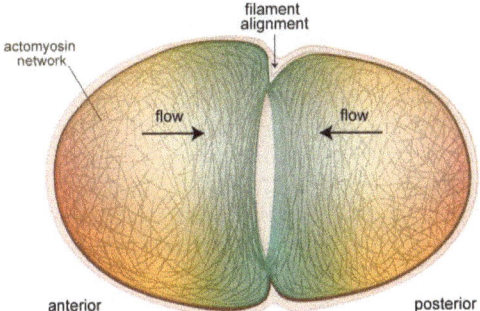

Figure 2. Cortical flows align filaments to form the contractile ring for cytokinesis of the one-cell embryo. The actomyosin cortex at the poles flows toward the equator of the embryo (black arrows), and these flows progressively transform the unorganized actin filament network at the poles (red shading) into the aligned structure of the cytokinetic ring (blue shading). In cases where actin-binding proteins are perturbed, thus altering cortical properties and flows, the formation and ingression of the contractile ring are impacted (see text).

One recent study revealed that both the Arp2/3 complex and the formin CYK-1 were important for the kinetics of cytokinesis even though the Arp2/3 complex did not localize to the cleavage furrow [36]. CYK-1 was shown to be enriched in the contractile ring and essential to elongate filaments to form it: *cyk-1* depletion led to a decrease in actin bundles at the equatorial cortex, and a furrow that initiated and ingressed more slowly. However, the Arp2/3 complex was also found to positively regulate the kinetics of contractile ring assembly and ring constriction, although it had the opposite effect on furrow initiation. Intriguingly, the slowing down of cytokinesis upon Arp2/3 complex inhibition was a result of an increase in CYK-1-mediated actin polymerization at the cell cortex and in the contractile ring. The conclusion was that Arp2/3 complex activity was required in the cortex to temper CYK-1-based polymerization, possibly via a competition for actin monomers as has been shown in other systems [37]. Too much cortical actin could be supposed to interfere with the cell shape changes accompanying cell division, while excessive actin in the furrow could inhibit contraction by impeding ring disassembly.

Proteins that affect cortical flows in the polarization phase and actomyosin cortex integrity in the maintenance phase also affect the speed of contractile ring formation and its contraction, due to the dependence of all these processes on similar parameters. The new studies concerning PLST-1 and HMR-1, mentioned previously in the context of cortex integrity, showed that these proteins also play roles in furrow formation. Lack of PLST-1 resulted in furrows that formed more slowly, although once formed, they ingressed at the normal speed, while lack of HMR-1 produced faster ingression than wild-type although with a tendency for the cortex to peel away from the cell membrane at the anterior pole [28,29].

As concerns furrow ingression, some studies have shown evidence that it is not myosin motor activity that drives contraction of the cytokinetic ring, but rather myosin's actin cross-linking activity coupled with depolymerization [38–40]. New studies in the *C. elegans* embryo have demonstrated unequivocally that NMY-2 motor activity was required throughout the formation and constriction of the actomyosin ring [41]. During ring assembly, the activity of NMY-2, rather than its F-actin cross-linking ability, were shown to control proper densification and alignment of actin in the cytokinetic ring, and timely deformation of the cell equator. In general, however, actin cross-linking activity is known to play a role in modulating contractility, and it was shown that for the *C. elegans* cytokinetic ring, an optimal degree of cross-linking existed, while too much or too little cross-linking activity was deleterious to contraction [42].

Contractile ring formation is important, but so is its correct positioning in the cell. Although positioning is mostly determined by astral microtubule dynamics, two recent papers show the importance of the downregulation of actomyosin contractility at the embryo cortex for correct placement of the furrow. One study brought to light the essential role of AIR-1 in clearing contractile ring proteins such as anillin, but also actin, from the poles of the embryo [43]. At the onset of anaphase, a conserved activator of AIR-1, TPXL-1, became localized to astral microtubules, and it was shown that its capacity to bind and activate AIR-1 was key for polar clearing. The hypothesis was that activated AIR-1 promoted clearing of contractile ring proteins from the poles of the embryo via phosphorylation of target proteins, but this mechanism has yet to be investigated in detail. Taken all together with results described previously in this review, this study provides another example of the AIR-1's varied roles in the regulation of the actin cytoskeleton during the development of the early *C. elegans* embryo.

The role of cortical dynamics in furrow positioning was also demonstrated in another recent study, which described a role for bleb formation in releasing cortical tension in mutant embryos that had excessive anterior myosin contractility [44]. In these embryos, the furrow initially formed in an anterior position, however DNA segregation defects were avoided by a posterior shift of the furrow concomitant with an anterior shift of the anterior nucleus. These shifts were shown to be produced by bleb formation near the anterior side of the furrow. The release of tension due to bleb formation allowed repositioning of the nascent cleavage furrow, and also created cytoplasmic flows that contributed to moving the nucleus posteriorly.

7. Conclusions

The *C. elegans* embryo is a powerful model system for understanding the many roles of the actomyosin cytoskeleton in asymmetric cell division. Results obtained using this model over the last five years have answered outstanding questions, as summarized in this review. In particular, the nature of the centrosomal cue for symmetry breaking has at last been identified. Moreover, other recent studies have delved deeper into the molecular mechanisms of how actomyosin dynamics drives morphological changes in the embryo. In the past, the microtubule cytoskeleton and PAR proteins have taken center stage in the *C. elegans* embryo. Our summary in this review shows the importance of the actin cytoskeleton in polarity establishment of the *C. elegans* zygote as well as the cellular shape changes that ensure the first asymmetric division of the embryo.

Funding: J.P. acknowledges financial support from the Fondation ARC (Grant PJA 20191209604) and the program "Investissements d'Avenir" launched by the French Government and implemented by ANR (ANR-10-LABX-0038 and ANR-10-IDEX-0001-02 PSL). D.S.E. received funding from the European Union's Horizon 2020 research and innovation program under the Marie Skłodowska-Curie grant agreement N° 666003.

Acknowledgments: We acknowledge Agnieszka Kawska at IlluScientia.com for graphical design of the figures.

Conflicts of Interest: The authors declare no conflict of interest.

References

1. Blanchoin, L.; Boujemaa-Paterski, R.; Sykes, C.; Plastino, J. Actin dynamics, architecture and mechanics in cell motility. *Physiol. Rev.* **2014**, *94*, 235–263. [CrossRef] [PubMed]
2. Plastino, J.; Blanchoin, L. Dynamic stability of the actin ecosystem. *J. Cell Sci.* **2018**, *132*, jcs219832. [CrossRef]
3. Agarwal, P.; Zaidel-Bar, R. Principles of actomyosin regulation in vivo. *Trends Cell Biol.* **2019**, *29*, 150–163. [CrossRef] [PubMed]
4. Cowan, C.R.; Hyman, A.A. Asymmetric cell division in *C. elegans*: Cortical polarity and spindle positioning. *Annu. Rev. Cell Dev. Biol.* **2004**, *20*, 427–453. [CrossRef]
5. Begasse, M.L.; Hyman, A.A. The first cell cycle of the *Caenorhabditis elegans* embryo: Spatial and temporal control of an asymmetric cell division. *Results Probl. Cell Differ.* **2011**, *53*, 109–133. [PubMed]
6. Pacquelet, A. Asymmetric cell division in the one-cell *C. elegans* embryo: Multiple steps to generate cell size asymmetry. *Results Probl. Cell Differ.* **2017**, *61*, 115–140. [PubMed]

7. Rose, L.; Gonczy, P. Polarity establishment, asymmetric division and segregation of fate determinants in early *C. elegans* embryos. *WormBook* **2014**. Available online: http://www.wormbook.org/chapters/www_asymcelldiv.2/asymcelldiv.2.html (accessed on 21 May 2020). [CrossRef]
8. Munro, E.; Nance, J.; Priess, J.R. Cortical flows powered by asymmetrical contraction transport PAR proteins to establish and maintain anterior-posterior polarity in the early *C. elegans* embryo. *Dev. Cell* **2004**, *7*, 413–424. [CrossRef]
9. Lang, C.F.; Munro, E. The PAR proteins: From molecular circuits to dynamic self-stabilizing cell polarity. *Development* **2017**, *144*, 3405–3416. [CrossRef]
10. Glotzer, M. Cleavage furrow positioning. *J. Cell Biol.* **2004**, *164*, 347–351. [CrossRef]
11. Gonczy, P. Mechanisms of asymmetric cell division: Flies and worms pave the way. *Nat. Rev. Mol. Cell Biol.* **2008**, *9*, 355–366. [CrossRef] [PubMed]
12. Panzica, M.T.; Marin, H.C.; Reymann, A.C.; McNally, F.J. F-actin prevents interaction between sperm DNA and the oocyte meiotic spindle in *C. elegans*. *J. Cell Biol.* **2017**, *216*, 2273–2282. [CrossRef] [PubMed]
13. Cowan, C.R.; Hyman, A.A. Acto-myosin reorganization and PAR polarity in *C. elegans*. *Development* **2007**, *134*, 1035–1043. [CrossRef] [PubMed]
14. Zhao, P.; Teng, X.; Tantirimudalige, S.N.; Nishikawa, M.; Wohland, T.; Toyama, Y.; Motegi, F. Aurora-A breaks symmetry in contractile actomyosin networks independently of its role in centrosome maturation. *Dev. Cell* **2019**, *48*, 631–645. [CrossRef] [PubMed]
15. Kapoor, S.; Kotak, S. Centrosome Aurora A regulates RhoGEF ECT-2 localisation and ensures a single PAR-2 polarity axis in *C. elegans* embryos. *Development* **2019**, *146*, dev174565. [CrossRef]
16. Klinkert, K.; Levernier, N.; Gross, P.; Gentili, C.; von Tobel, L.; Pierron, M.; Busso, C.; Herrman, S.; Grill, S.W.; Kruse, K.; et al. Aurora A depletion reveals centrosome-independent polarization mechanism in *Caenorhabditis elegans*. *eLife* **2019**, *8*, e44552. [CrossRef]
17. Reich, J.D.; Hubatsch, L.; Illukkumbura, R.; Peglion, F.; Bland, T.; Hirani, N.; Goehring, N.W. Regulated activation of the PAR polarity network ensures a timely and specific response to spatial cues. *Curr. Biol.* **2019**, *29*, 1911–1923. [CrossRef]
18. Wang, S.C.; Low, T.Y.F.; Nishimura, Y.; Gole, L.; Yu, W.; Motegi, F. Cortical forces and CDC-42 control clustering of PAR proteins for *Caenorhabditis elegans* embryonic polarization. *Nat. Cell Biol.* **2017**, *19*, 988–995. [CrossRef]
19. Rodriguez, J.; Peglion, F.; Martin, J.; Hubatsch, L.; Reich, J.; Hirani, N.; Gubieda, A.G.; Roffey, J.; Fernandes, A.R.; St Johnston, D.; et al. aPKC cycles between functionally distinct PAR protein assemblies to drive cell polarity. *Dev. Cell* **2017**, *42*, 400–415. [CrossRef]
20. Dickinson, D.J.; Schwager, F.; Pintard, L.; Gotta, M.; Goldstein, B. A single-cell biochemistry approach reveals PAR complex dynamics during cell polarization. *Dev. Cell* **2017**, *42*, 416–434. [CrossRef]
21. Munro, E. Protein clustering shapes polarity protein gradients. *Dev. Cell* **2017**, *42*, 309–311. [CrossRef] [PubMed]
22. Mittasch, M.; Gross, P.; Nestler, M.; Fritsch, A.W.; Iserman, C.; Kar, M.; Munder, M.; Voigt, A.; Alberti, S.; Grill, S.W.; et al. Non-invasive perturbations of intracellular flow reveal physical principles of cell organization. *Nat. Cell Biol.* **2018**, *20*, 344–351. [CrossRef] [PubMed]
23. Gross, P.; Kumar, K.V.; Goehring, N.W.; Bois, J.S.; Hoege, C.; Julicher, F.; Grill, S.W. Guiding self-organized pattern formation in cell polarity establishment. *Nat. Phys.* **2019**, *15*, 293–300. [CrossRef] [PubMed]
24. De Simone, A.; Nedelec, F.; Gonczy, P. Dynein transmits polarized actomyosin cortical flows to promote centrosome separation. *Cell Rep.* **2016**, *14*, 2250–2262. [CrossRef] [PubMed]
25. Naganathan, S.R.; Furthauer, S.; Rodriguez, J.; Fievet, B.T.; Julicher, F.; Ahringer, J.; Cannistraci, C.V.; Grill, S.W. Morphogenetic degeneracies in the actomyosin cortex. *eLife* **2018**, *7*, e37677. [CrossRef]
26. Nishikawa, M.; Naganathan, S.R.; Julicher, F.; Grill, S.W. Controlling contractile instabilities in the actomyosin cortex. *eLife* **2017**, *6*, e19595. [CrossRef]
27. Michaux, J.B.; Robin, F.B.; McFadden, W.M.; Munro, E.M. Excitable RhoA dynamics drive pulsed contractions in the early *C. elegans* embryo. *J. Cell Biol.* **2018**, *217*, 4230–4252. [CrossRef]
28. Ding, W.Y.; Ong, H.T.; Hara, Y.; Wongsantichon, J.; Toyama, Y.; Robinson, R.C.; Nedelec, F.; Zaidel-Bar, R. Plastin increases cortical connectivity to facilitate robust polarization and timely cytokinesis. *J. Cell Biol.* **2017**, *216*, 1371–1386. [CrossRef]

29. Padmanabhan, A.; Ong, H.T.; Zaidel-Bar, R. Non-junctional E-cadherin clusters regulate the actomyosin cortex in the *C. elegans* zygote. *Curr. Biol.* **2017**, *27*, 103–112. [CrossRef]
30. Scholze, M.J.; Barbieux, K.S.; De Simone, A.; Boumasmoud, M.; Suess, C.C.N.; Wang, R.; Gonczy, P. PI(4,5)P2 forms dynamic cortical structures and directs actin distribution as well as polarity in *Caenorhabditis elegans* embryos. *Development* **2018**, *145*, dev164988. [CrossRef]
31. Hirani, N.; Illukkumbura, R.; Bland, T.; Mathonnet, G.; Suhner, D.; Reymann, A.C.; Goehring, N.W. Anterior-enriched filopodia create the appearance of asymmetric membrane microdomains in polarizing *C. elegans* zygotes. *J. Cell Sci.* **2019**, *132*, jcs230714. [CrossRef] [PubMed]
32. Leite, J.; Osorio, D.S.; Sobral, A.F.; Silva, A.M.; Carvalho, A.X. Network contractility during cytokinesis-from molecular to global views. *Biomolecules* **2019**, *9*, 194. [CrossRef] [PubMed]
33. Reymann, A.C.; Staniscia, F.; Erzberger, A.; Salbreux, G.; Grill, S.W. Cortical flow aligns actin filaments to form a furrow. *eLife* **2016**, *5*, e17807. [CrossRef]
34. Singh, D.; Odedra, D.; Dutta, P.; Pohl, C. Mechanical stress induces a scalable switch in cortical flow polarization during cytokinesis. *J. Cell Sci.* **2019**, *132*, jcs231357. [CrossRef]
35. Khaliullin, R.N.; Green, R.A.; Shi, L.Z.; Gomez-Cavazos, J.S.; Berns, M.W.; Desai, A.; Oegema, K. A positive-feedback-based mechanism for constriction rate acceleration during cytokinesis in *Caenorhabditis elegans*. *eLife* **2018**, *7*, e36073. [CrossRef]
36. Chan, F.Y.; Silva, A.M.; Saramago, J.; Pereira-Sousa, J.; Brighton, H.E.; Pereira, M.; Oegema, K.; Gassmann, R.; Carvalho, A.X. The ARP2/3 complex prevents excessive formin activity during cytokinesis. *Mol. Biol. Cell* **2019**, *30*, 96–107. [CrossRef] [PubMed]
37. Burke, T.A.; Christensen, J.R.; Barone, E.; Suarez, C.; Sirotkin, V.; Kovar, D.R. Homeostatic actin cytoskeleton networks are regulated by assembly factor competition for monomers. *Curr. Biol.* **2014**, *24*, 579–585. [CrossRef]
38. Lord, M.; Laves, E.; Pollard, T.D. Cytokinesis depends on the motor domains of myosin-II in fission yeast but not in budding yeast. *Mol. Biol. Cell* **2005**, *16*, 5346–5355. [CrossRef]
39. Ma, X.; Kovacs, M.; Conti, M.A.; Wang, A.; Zhang, Y.; Sellers, J.R.; Adelstein, R.S. Nonmuscle myosin II exerts tension but does not translocate actin in vertebrate cytokinesis. *Proc. Natl. Acad. Sci. USA* **2012**, *109*, 4509–4514. [CrossRef]
40. Mendes Pinto, I.; Rubinstein, B.; Kucharavy, A.; Unruh, J.R.; Li, R. Actin depolymerization drives actomyosin ring contraction during budding yeast cytokinesis. *Dev. Cell* **2012**, *22*, 1247–1260. [CrossRef]
41. Osorio, D.S.; Chan, F.Y.; Saramago, J.; Leite, J.; Silva, A.M.; Sobral, A.F.; Gassmann, R.; Carvalho, A.X. Crosslinking activity of non-muscle myosin II is not sufficient for embryonic cytokinesis in *C. elegans*. *Development* **2019**, *146*, dev179150. [CrossRef] [PubMed]
42. Descovich, C.P.; Cortesa, D.B.; Ryan, S.; Nash, J.; Zhang, L.; Maddox, P.S.; Nedelec, F.; Maddox, A.S. Cross-linkers both drive and brake cytoskeletal remodeling and furrowing in cytokinesis. *Mol. Biol.Cell* **2018**, *29*, 622–631. [CrossRef] [PubMed]
43. Mangal, S.; Sacher, J.; Kim, T.; Osorio, D.S.; Motegi, F.; Carvalho, A.X.; Oegema, K.; Zanin, E. TPXL-1 activates Aurora A to clear contractile ring components from the polar cortex during cytokinesis. *J. Cell Biol.* **2018**, *217*, 837–848. [CrossRef] [PubMed]
44. Pacquelet, A.; Jousseaume, M.; Etienne, J.; Michaux, G. Simultaneous regulation of cytokinetic furrow and nucleus positions by cortical tension contributes to proper DNA segregation during late mitosis. *Curr. Biol.* **2019**, *29*, 3766–3777. [CrossRef] [PubMed]

© 2020 by the authors. Licensee MDPI, Basel, Switzerland. This article is an open access article distributed under the terms and conditions of the Creative Commons Attribution (CC BY) license (http://creativecommons.org/licenses/by/4.0/).

Review

Actin Mutations and Their Role in Disease

Francine Parker, Thomas G. Baboolal and Michelle Peckham *

School of Molecular and Cellular Biology, University of Leeds, Leeds LS2 9JT, UK; f.parker@leeds.ac.uk (F.P.); Thomas.baboolal@gmail.com (T.G.B.)
* Correspondence: m.peckham@leeds.ac.uk; Tel.: +44-(0)1133-434348

Received: 25 March 2020; Accepted: 7 May 2020; Published: 10 May 2020

Abstract: Actin is a widely expressed protein found in almost all eukaryotic cells. In humans, there are six different genes, which encode specific actin isoforms. Disease-causing mutations have been described for each of these, most of which are missense. Analysis of the position of the resulting mutated residues in the protein reveals mutational hotspots. Many of these occur in regions important for actin polymerization. We briefly discuss the challenges in characterizing the effects of these actin mutations, with a focus on cardiac actin mutations.

Keywords: actin; mutation; polymerization; myosin

1. Introduction

Actin is a globular protein (G-actin) that assembles into filaments (F-actin) and is important for cell movement, intracellular movement, muscle contraction and many other functions. There are six actin genes in the human genome. Three of these encode the α-actin isoforms found in cardiac, skeletal or smooth muscle (*ACTC1*, *ACTA1* and *ACTA2*, respectively). Two encode γ-actin, of which one is widely expressed (*ACTG1*) and the other is smooth muscle specific (*ACTG2*). The final gene encodes the widely expressed β-actin (*ACTB*). These actin isoforms are highly (>90%) conserved at the protein level. Actin is a promiscuous protein, interacting with many other proteins [1], and is also subject to many different post-translational modifications [2].

The structure of actin has been solved multiple times in its monomeric [3] and, more recently, in its filamentous form, the latter building on advances in cryo-electron microscopy [4–8]. In addition, there are structures of F-actin in the complex with a myosin motor domain, and/or with other F-actin filament binding proteins such as cofilin and tropomodulin [9,10]. These structures show that the actin monomer is divided into two halves (inner and outer domains) by a cleft that binds nucleotide and cation (Mg^{2+}). Each half is further divided into two domains, with one domain comprising subdomains 1 and 2, and the other subdomains 3 and 4. Subdomains 1 and 2 are found on the outer edge of the actin filament. In the actin filament, each monomer makes interactions longitudinally within a protofilament (along the actin filament), as well as laterally (between the two protofilaments), such that each actin monomer (subunit) interacts with its three surrounding subunits. The currently available structures of G- and F-actin provide a rich resource for understanding how myosin interacts with actin, how the actin monomer forms filaments and interacts with a variety of actin binding proteins and how disease-causing mutations in actin affect its biological function.

Disease-causing mutations have been reported for each of the six actin genes, demonstrating the importance of actin for normal cell behaviour and function in a variety of cell types. The majority of these (>90% for five out of the six actin genes) result in missense mutations in the protein (Supplementary Table S1), and typically, these mutations are dominant. The type of disease that mutations in a specific actin gene commonly cause reflect its expression pattern, as detailed below. Moreover, mutations occur throughout the entire sequence for each actin gene. It is therefore of interest to determine if there are any specific residues or common functional regions in the encoded proteins in which these missense mutations are found. The main aim of this article is to evaluate these missense mutations, determine if there are any mutational 'hotspots' and the potential consequences for actin function. We then go on to discuss how a range of approaches is needed to test experimentally what these consequences are, and how they might lead to disease, with a specific focus on cardiac actin mutations.

2. Disease-causing Mutations in the Six Actin Genes, an Overview

The gene with the highest number of reported mutations (over 220) is *ACTA1*, which encodes the isoform of α-actin almost exclusively expressed in skeletal muscle. Mutations in *ACTA1* are found throughout the sequence and >92% result in single amino acid substitutions in the protein (Figure 1 and Supplementary Table S1). They are a common cause of Nemaline Myopathy (reviewed in [11,12]), a non-progressive skeletal muscle disease that commonly has an early onset, with severe cases diagnosed at birth. This disease typically results in muscle weakness, particularly in the respiratory muscles, which can cause breathing difficulties, but also difficulties in swallowing in severe cases.

The second most commonly mutated gene is in *ACTA2* with over 80 mutations, of which 93% are missense (Figure 1 and Supplementary Table S1). *ACTA2* encodes an α-actin isoform that is highly expressed in specific smooth muscle cells associated with the vasculature. Perhaps not unsurprisingly, mutations in this gene are strongly associated with familial thoracic aortic aneurysms, and *ACTA2* is the most frequent gene mutated in this disorder [13]. In rarer cases, mutations in *ACTA2* cause cerebral arteriopathy, in which the most commonly affected residue is Arg179 [14].

Mutations in *ACTC1*, the gene that encodes the isoform of α-actin found predominantly in cardiac muscle, are the next most common, with over 70 mutations, of which over 90% are missense (Figure 1 and Supplementary Table S1). Mutations in this gene cause heart disease [15]. Over 50% of the known mutations in *ACTC1* cause hypertrophic cardiomyopathy (HCM) and about 20% cause dilated cardiomyopathy (DCM). The remainder cause left ventricular non-compaction or other heart defects. *ACTC* is one of eight disease genes that contain missense mutations causing HCM, and one of 20 disease genes causing DCM [16,17]. Mutations in *ACTC* are common in patients with apical HCM (50%) and patients are heterozygous for the mutant allele. HCM affects 1:500 to 1:250 people and is a common cause of premature death in young adults [18,19]. In HCM, typically the left ventricular wall and/or septum between left and right ventricles thickens, and the myocytes (muscle cells) become disorganised. DCM is less common, affecting around 1:3000 people [20]. In DCM, the left ventricular wall becomes thinner. Ventricular relaxation is impaired, restricting the atrial emptying into the ventricles, and resulting in dilation of the atria [21]. The high incidence of sudden death seen in DCM patients results from the impaired systolic function of the left ventricle, which can lead to related pathologies such as thromboembolic events and arrhythmias.

About 70 mutations have been described for *ACTB*, which encodes the widely expressed β-actin (Figure 1, Supplementary Table S1). However, in this case only ~50% are missense mutations. Consistent with its widespread expression, mutations result in a broad range of defects, including a specific facial appearance, intellectual disability, hearing loss, heart and renal defects, brain abnormalities, neuronal migration defects and muscle wasting, typical of a syndrome named Baraitser-Winter syndrome [22–24]. The remaining mutations in *ACTB* include deletions, premature stop codons and frameshift mutations, of which most (40%) are gross deletions, many of which lead to complete gene loss [25]. The resulting haploinsufficiency of β-actin is associated with widespread effects, including developmental delay, organ malformations, and growth retardation, although this phenotype is considered distinct from the symptoms associated with Baraitser-Winter syndrome [25]. Mutations in *ACTB* have also more recently been associated with bleeding disorders [26].

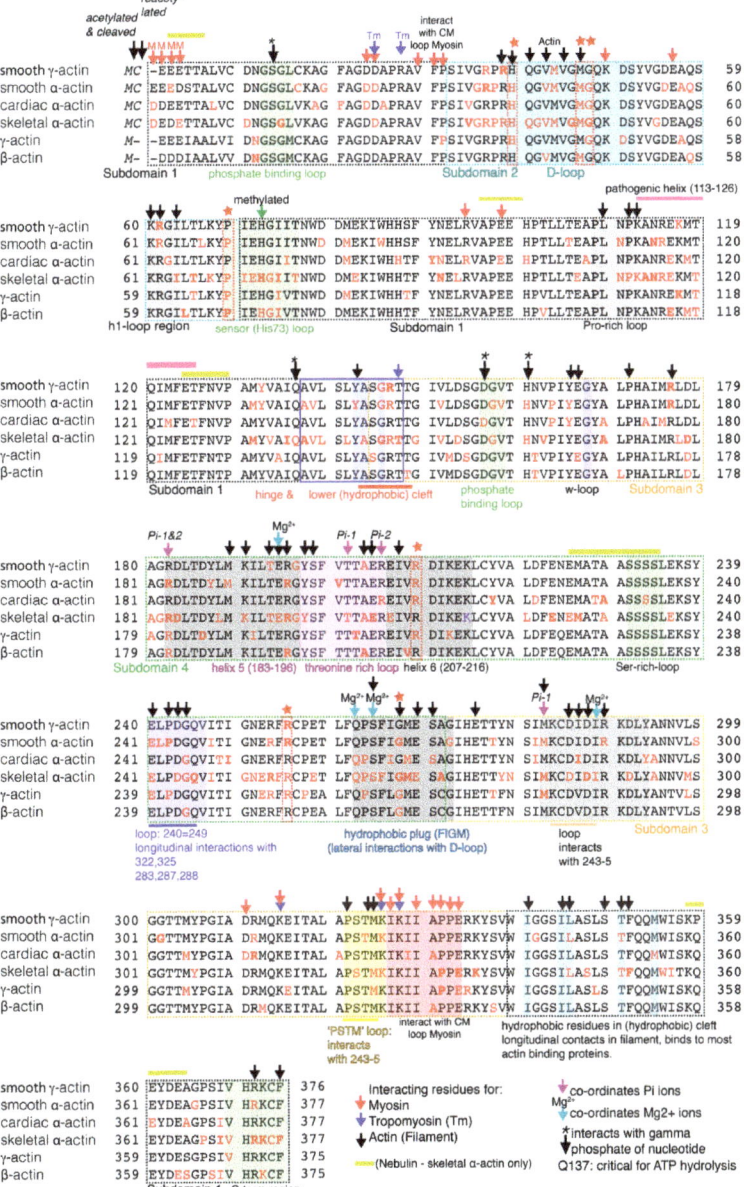

Figure 1. Annotated alignment of the six actin sequences for *Homo Sapiens* showing the positions of mutations. All six sequences were retrieved from UNIPROT (P63267: smooth γ-actin, P60709: β-actin: P68133: skeletal α-actin, P63261: γ-actin; P60382: cardiac α-actin, P62736: smooth α-actin). The first one–two residues are typically acetylated and cleaved, such that the third residue is the first residue in the expressed sequence. The positions of key interaction regions in the sequences is indicated together with positions of mutated residues (red font) in each sequence. Residues in which mutations are found in four out of the six isoforms are indicated by a red star. See text for more details. Numbering in the text refers to the residue number for α-actin isoforms. Mutations were retrieved from the Human Genome Mutation Database (HGMD).

A second widely expressed protein is γ-actin, encoded by the gene *ACTG1*. Over 50 mutations have been reported for *ACTG1*, of which all but one are missense mutations (Figure 1 and Supplementary Table S1). Just over half of these mutations cause deafness [27]. This is consistent with the expression and role of γ-actin in the sensory epithelial cells of the inner ear, in which γ-actin is an essential component of the stereocilia, along with β-actin. Distortion of the stereocilia is essential for the detection of sound. Although stereocilia form normally in a mouse γ-actin knockout model, maintenance of these structures is affected and the mice show a progressive loss of hearing [28]. This research also demonstrated that γ-actin (*ACTG1*) is not strictly required for development, possibly because levels of β-actin increased in the knockout mouse model, partly compensating for the loss of γ-actin. However, a high proportion of the remaining mutations in *ACTG1* cause Baraitser-Winter syndrome, consistent with the widespread expression of γ-actin in multiple tissues [29]. It is not clear why some mutations in *ACTG1* appear to have a more limited effect than others, as affected residues for both types of disease are distributed throughout the sequence (Supplementary Table S1).

Mutations in *ACTG2* are the least common, with just over 20 described. Almost all of these (96%) are missense (Figure 1 and Supplementary Table S1). The expression of this γ-actin isoform is restricted to smooth muscle cells in the gut, prostate, bladder and adrenal gland. Mutations in *ACTG2* cause Megacystis microcolon-Intestinal hypoperistalsis syndrome [30], also known as chronic intestinal pseudo-obstruction, visceral myopathy (or degenerative leiomyopathy) [31]. All of these are disorders of enteric smooth muscle function. They mainly affect the intestine leading to chronic intestinal obstruction, and can also affect the bladder. The smooth muscle cells in the smooth muscle layers that surround the gut epithelium are important for moving the contents of the gut along its length. The organisation of these smooth muscle cells is often disordered in this disease, likely leading to decreased smooth muscle contraction and the resultant intestinal obstruction [31,32].

3. Positional Analysis of Disease-Causing Missense Actin Mutations

An analysis of the missense mutations for each of the actin isoforms shows that, while they are found throughout the sequence, there are common mutational hotspots where the numbers of mutations tend to be higher than elsewhere (Figures 1 and 2a,b). The total number of missense mutations (sum total for mutations, Figure 2a) are somewhat dominated by the large number of mutations in skeletal α-actin. However, removing these mutations from the plot (Figure 2b) reveals similar mutational hotspots. This suggests there are key functional regions in all of the actin isoforms where disease-causing mutations are more likely to result in a phenotype.

A major hotspot for mutations is the DNAse-1 loop (or D-loop) in subdomain 2 (SD2). Its name is derived from its ability to bind to DNAse-1, which inhibits F-actin formation, and this interaction was instrumental in generating the first crystal structure for G-actin [33]. The D-loop is crucial for actin polymerisation and is also the target of many actin binding proteins [3]. These include tropomodulin, which caps the pointed end of actin filaments preventing polymerisation and depolymerisation [10], and cofilin, which severs actin filaments [34]. The D-loop is important for polymerisation as it is involved in both lateral (between the two protofilaments) and longitudinal contacts (along the protofilament) between actin monomers, and its structure is sensitive to the occupancy of the nucleotide binding site. The residues Ser14 and the methylated His73 detect the nucleotide state and transmit that information to the D-loop, which can then move its position [8].

In longitudinal contacts, the D-loop inserts into a cleft in the adjacent actin monomer (magenta, Figure 2). Residues in the D-loop (Gln41, Val43 and Val45) interact with residues in SD1 (Tyr143, Arg372 and Phe375) and SD3 (Leu110, Tyr169, Ala170 and Pro172) of the adjacent actin monomer [5,7]. Residues Arg39 and His40 in the proximal region of this loop are additionally involved in only one of only two lateral (inter-protofilament) contacts with adjacent actin monomers. This occurs through an interaction of these D-loop residues in one monomer with residues in the so-called hydrophobic plug of the second monomer (Arg39 interacts with Glu270, His40 interacts with Ser265 and Gly268) [4,6,7]. The distal part of the D-loop is found towards the outer edge of the filament and is involved in interactions with myosin. For example, Glu570 in non-muscle myosin 2C (NM2C) probably forms an electrostatic interaction with Lys49 in the D-loop [35]. A similar interaction with this residue has also been demonstrated for loop 3 of Myo6 [6].

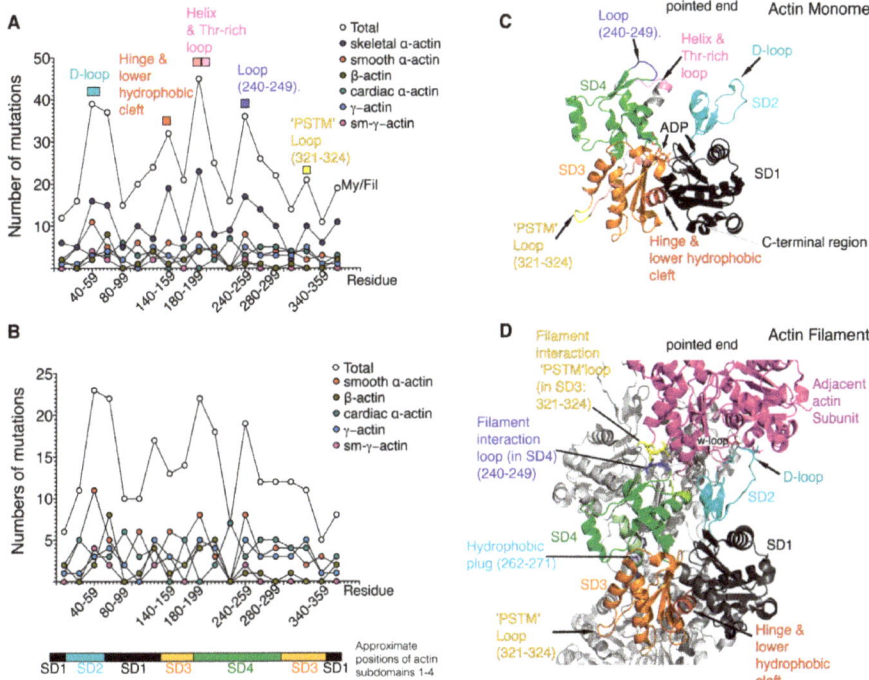

Figure 2. Analysis of actin mutations in all six *Homo sapiens* actin isoforms. Numbers of mutations in residues 1–19, 20–39 and so on, were summed and plotted as shown across the whole sequence for each individual isoform and for all isoforms (Total). (**A**) Panel A shows this analysis for each six actin isoforms, and the total for all six. The protein names are shown in the key. (**B**) Panel B is similar to A, except the mutations for ACTA1 were excluded. Mutations were taken from the HGMD (accessed November 2019, Supplementary Table S1). The approximate positions of the four subdomains found in the G-actin monomer (**C**) is additionally shown underneath panels A and B for reference. (**C**) The G-actin monomer is made up of four subdomains (denoted by black (subdomain-1, SD1), cyan (subdomain-2, SD2) orange (subdomain-3, SD3) and green (subdomain-4, SD4). Labels indicate regions of the structure that are hotspots for mutations. (**D**) The F-actin filament. One actin monomer is shown as in C, with the SD1-4 in black, cyan, orange and green, respectively. Positions where mutations are common are highlighted. The interaction between residues 243–245 in in SD4 (coloured blue), with residues 322–325 in SD3 (coloured yellow) in the adjacent actin subunit (coloured magenta) is indicated. Longitudinal contacts between the DNAse-1 loop (or D-loop; 38-52) and residues in the C-terminal region of the adjacent monomer (residues R372, F375), Pro-rich and w-loops are also shown. The helix coloured red (residues 135–146) in D is found in the lower hydrophobic cleft. The hydrophobic plug (residues 262–272, light blue) is important for lateral actin monomer contact between protofilaments. The regions of sequence for each subdomain have been defined as follows SD1 (residues 1–32, 70–144, and 338–372), SD2 (residues 33–69) SD3 (residues 145–180 and 270–337) SD4 (residues 181–269) [33].

A closer look at mutations in the D-loop reveals three specific residues (His40, Met47 and Gly48) that are mutated in four out of the six actin isoforms (indicated by the red stars in Figure 1). Gly48 is not apparently directly involved in actin or myosin binding, but substitution of this small amino acid residue is likely to have an overall effect on the flexibility of the D-loop, and thus, indirectly affect actin filament polymerization. His40 is critical for longitudinal contacts between monomers in the filament, as discussed above. Moreover, it has long been known that selective carboethoxylation of this residue inhibits actin polymerization [36]. Mutations in His40 are therefore likely to decrease the numbers of filaments. Its mutation to Tyr in skeletal α-actin, cardiac α-actin and γ-actin cause nemaline myopathy, hypertrophic cardiomyopathy and Baraitser-Winter syndrome, respectively [37–39]. Its mutation to Asn in smooth α-actin causes thoracic aortic aneurysms and dissections [40] (Supplementary Table S1).

Met47 is one of two residues (Met44 and Met47) in the D-loop that are oxidised by MICAL (molecule interacting with CasL). The resulting oxidation causes rapid and catastrophic depolymerisation of the actin filament [41,42] and the resulting monomers do not polymerise as efficiently as non-modified actin monomers. Thus, oxidation of Met44 and Met47 is an alternative strategy for regulating actin polymerisation, in addition to the action of actin severing proteins such as cofilin [43]. Mutation of Met47 to Leu abolishes a longitudinal actin-actin M37-O-T351 contact, and prevents this catastrophic filament disassembly [42]. Met47 is mutated in all three α-actin isoforms and in β-actin. Disease mutations (Supplementary Table S1) in this residue that prevent its oxidation might therefore be expected to stabilise actin filaments. In non-muscle cells, filament remodelling and turnover is important for cell motility, and thus, stabilization of β-actin filaments would be expected to affect the behaviour and migration of these cells. However, while mutation of Met47 to Thr in β-actin is known to cause Baraitser-Winter syndrome [24], this disease mutation has not yet been tested to determine if or how it affects actin filament dynamics in cells.

The effects of disease mutations in Met47 in muscle specific α-actin isoforms is less clear. Typically, actin in striated muscle tends to be stably incorporated into filaments. The rate of actin synthesis and turnover in cardiac and skeletal muscle is relatively slow (occurring over weeks [44]). Moreover, the roles of MICAL in striated muscle are also not well understood. In cardiac muscle, mutation of Met47 to Leu in cardiac α-actin causes hypertrophic cardiomyopathy [45]. However, while MICAL3 is expressed in the heart [46], if or how this plays a role in actin polymerization in the heart or in filament maintenance is unclear. Mutation of Met47 to Val in skeletal α-actin causes nemaline myopathy ([47], Supplementary Table S1). Although mutations in MICAL cause contractile muscle filaments to become disorganised in skeletal muscle in *Drosophila* [48], this pathway and its role in actin polymerization has hardly been explored in mammalian skeletal muscle. In smooth α-actin, mutation of Met47 causes thoracic aortic aneurysms [49]. Smooth muscle also expresses MICAL, but its role in actin organization in smooth muscle has not been investigated. In addition to the potential interaction with MICAL, and downstream effects on filament polymerization, it is also possible that a mutation in Met47 simply alters the structure of the D-loop and thus destabilises the actin filament. Clearly, effects of mutations in this residue need exploring further.

Two further mutational hotspots are found in the two loops that make a longitudinal interaction between two actin monomers along a protofilament (Figure 2a–d). One loop comprises residues 240-249 in SD4 and the other, residues 321–324 in SD3. The residues Pro243, Asp244 and Gly245 in the loop in SD4 (coloured blue in Figure 2a,c,d) interact with residues Pro322 and Met325 in a second loop in SD3 (coloured yellow in Figure 2a,c,d), and with residues Met283, Ile287 and Asp288 in an adjoining loop in SD3 (not highlighted in Figure 2) [7]. Almost all of these residues are mutated in one or more isoforms (Figure 1 and Supplementary Table S1). A mutation in the adjacent acidic residue Glu241, to a basic residue (Lys) in the SD4 loop, is found in smooth and skeletal α-actin, and in γ-actin, causing thoracic aortic aneurysms, nemaline myopathy and deafness, respectively [50–52] (Supplementary Table S1). This residue interacts with Ser323 and Thr324 in the adjacent actin subunit [7], and mutations are likely to destabilise the actin filament.

A fourth hotspot is in the Thr-rich loop and its flanking helices H5 and H6. The numbers of mutations in this region, across all actin isoforms, are particularly high. Interestingly, not only is this region important in actin-actin interactions in the filament, but specific residues co-ordinate phosphate and Mg^{2+} ions (Figure 1), and this co-ordination is also important for filament stability [7]. The structure of H5 and the Thr-rich loop changes between G- and F-actin [7] and the Thr-rich loop becomes involved in longitudinal contacts between actin monomers. Arg210 in H6 is mutated in four out of the six actin isoforms; to Cys in γ-actin (causing deafness) [53], to Asn in smooth γ-actin and smooth α-actin (causing visceral myopathy and thoracic artery disease, respectively) [54,55] and to His in cardiac α-actin (causing HCM and DCM) [56] (Supplementary Table S1). While this residue has not been identified as part of a specific interaction, it must play a critical role in actin-actin interaction in some way. From investigating the orientation of the side-chain rotamers for Arg210 and a glutamate residue downstream (Glu214) in the actin structure, it seems likely that these two residues could form an ionic interaction (i, i+4), which would help to stabilise this helix, as seen in single α-helical (SAH) domains [57].

Two further highly mutated residues include Arg256 found in a helix in subdomain 4 and Pro70 found at the start of the sensor (His73) loop at the boundary between SD2 and SD1. Arg256 is mutated in four out of six actin isoforms: to His or Cys in γ -smooth actin [30,58], to His in smooth α-actin [59], to His or Gly in skeletal α-actin [60] and to Trp in γ-actin [22]). These cause Megacystis microcolon-intestinal hypoperistalsis syndrome, or visceral myopathy (smooth γ-actin), aortic disease (smooth α-actin), nemaline myopathy (skeletal α-actin) and Baraitser-Winter Syndrome (γ-actin) (Supplementary Table S1). Arg256 is part of a pathogenic network, in which the so-called pathogenic helix (residues 113–126, Figure 1) and the C-terminal helix (residues 370–375) are interconnected. Arg256 modulates the lateral interaction of Lys113 with Glu195 in the adjacent monomer. Arg256 is also close to and likely to interact with Ile191 in H5. This network of interactions is thought to sense the binding of actin-binding proteins, and to communicate this to the rest of the actin molecule.

Pro70, found at the start of the sensor loop (His73), is mutated in every actin isoform except cardiac and γ-smooth actin. Mutation to Leu in β-actin causes Baraitser-Winter syndrome, and in γ-actin, causes Ocular coloboma [23,61]. A second mutation to Ala in β-actin also causes Baraitser-Winter syndrome [62]). Mutation to Arg in skeletal α-actin causes congenital myopathy [47] and to Gln in smooth α-actin causes thoracic aortic aneurisms [55]. The sensor loop that contains this residue is thought to function as a switch, linking changes in the nucleotide site to structural transitions in SD2, and in particular in the D-loop. Mutations would be expected to affect this sensor function and associated changes in structure in the D-loop, with downstream effects of filament stability.

Finally, mutations in Gly268 in the hydrophobic plug (FIGM) are found in four out of six isoforms. The hydrophobic plug is important for one of only two lateral interactions between actin monomers in the filament, in which it interacts with the D-loop. Gly268 interacts specifically with His40 as described above. This residue is mutated to Arg in both skeletal and smooth α-actin, causing nemaline myopathy [52,63] and aortic disease [64], respectively, and two further mutations (to Asp [65] or Cys [63]) in skeletal α-actin cause nemaline myopathy. The equivalent mutation in γ- and β-actin causes deafness [27] and Baraitser-Winter syndrome [66], respectively. Mutation of Gly268 is likely to weaken the lateral interaction with adjacent monomers, destabilizing the actin filament. Indeed, it is interesting that both His40 and Gly268 are mutated in multiple actin isoforms.

This analysis also shows some interesting differences in the positions of the mutations in each isoform. One of these is the lack of mutations in the region encompassing residues 220–230, just after the Thr-rich loop and H6, in isoforms other than skeletal and cardiac α-actin. This region has been identified as part of the actin binding interface for nebulin [67]. Nebulin is a large (~800 kDa) protein expressed in skeletal muscle that extends along the thin filament, binding to actin and tropomyosin and other sarcomeric proteins [68]. It contains a C-terminal SH3 domain, located in the Z-disc and 178 nebulin repeats. Recent data suggest that it stiffens the thin filament and contributes to thin filament activation in skeletal muscle [69]. Mutations in nebulin are a common cause of nemaline myopathy [12]. Therefore, in skeletal muscle, mutations in this region could disrupt the binding of nebulin to actin in the thin filament, weaken the thin filament and, thus, cause disease.

In cardiac muscle, nebulette and not nebulin is expressed. Nebulette is a member of the nebulin family of actin-binding proteins. Its N- and C-terminal regions are similar to those of nebulin, but it only has 32 nebulin repeats [68], is much smaller (~120 kDa) than nebulin and is not large enough to extend along the length of the thin filament. It is found in the Z-disc, and either projects a short distance along the filament from the Z-disc, interacts with desmin (as does nebulin) [70] and/or crosslinks actin filaments from adjacent sarcomeres within the Z-disc (reviewed in [71]). Mutations in nebulette have been linked to heart disease (HCM and DCM (reviewed in [71])). Assuming nebulette interacts with actin in a similar way to that of nebulin, mutations in actin in this region will also affect its ability to bind to cardiac α-actin, with potential downstream effects on actin filament stability.

A detailed knowledge of the structure of actin is important in predicting the precise effects of mutations. Indeed, the analysis presented here suggests that many mutations are likely to affect actin polymerization. However, some mutations will also affect the interaction of actin with its many other interacting proteins [1]. Other specific mutations can affect post-translational modifications [72] important for actin polymerization.

Three different post-translational modifications of actin have been linked to human cancers. First, Asp3 of β- (but not γ-) actin is arginylated by the enzyme ATE1 (arginyl-tRNA-protein transferase 1, which acts after the first two residues are removed [73]. A lack of arginylation reduces actin polymerization and its interaction with actin binding proteins [74]. Second is the specific acetylation of the N-terminal residues of β-actin and γ-actin by NAA80 (N-alpha acetyltransferase 80). A lack of NAA80 increases the numbers of actin filaments [75]. Third is the MICAL-mediated oxidation of Met44 and Met47 mentioned above. Levels of ATE1 are reduced in some human cancers and mutations in both NAA80 and ATE1 in cancer cell databases have also been reported. This suggests that both these enzymes could directly contribute to abnormal cell behaviour and metastasis through their effects on actin dynamics [2]. The depolymerization of F-actin initiated by MICAL is enhanced by the Abl non-receptor tyrosine kinase, and Abl is upregulated in several cancers (reviewed in [76]), suggesting a more indirect but important role of MICAL.

4. Mutations in Cardiac α-actin that Affect Myosin Interaction with Actin, Directly or Indirectly

Although the effects of mutations can be predicted, there remains a considerable challenge in experimentally determining their effects. This challenge is exemplified by considering a few examples of mutation in cardiac α-actin predicted to interact with myosin, and how they cause disease. These include mutations at the N-terminal region of actin, in the outer region of the D-loop and in subdomain 3 (between residues 311 and 335). The outer region of the D-loop is part of the 'Milligan' contact, in which myosin loop 3 (H551-G576) in the L50 domain interacts with actin SD1 and the D-loop of the adjacent actin subunit [77,78]. The precise mechanism by which myosin binds to actin can be somewhat variable between different myosin isoforms (e.g., compare NM2C and Myo6 [6,35]). However, each myosin primarily binds to residues in SD1, with some interactions in SD2 (such as the D-loop) of the adjacent actin, and SD3. Thus, mutations in residues within SD1 have the potential to disrupt myosin binding. For tropomyosin, key residues are D25 and the triad of residues, K326, K328 and R147, which interact with tropomyosin in the blocked (off) state of actin (reviewed in [79], and see Figure 1).

A relatively well understood mutation in cardiac α-actin is E99K (Glu99Lys), which causes HCM [80]. It is found in a region of actin thought to be involved in binding to the lower 50K domain of the myosin motor. E99K decreases the overall negative charge of the binding site within actin, and thus, weakens the acto-myosin interaction. Although one report suggested that E99K does not fold as well as wild type actin [81], a second report [82] showed that it does. Our lab has expressed this mutation as an eGFP-cardiac α-actin fusion construct in adult cardiomyocytes, using an adenoviral expression system [83] and found that it incorporates normally into the muscle sarcomere (Supplemental Figure S1). In contrast, a recent study using C-elegans [84] reported that this mutation does form some aggregates, but when expressed in mouse hearts, this is not the case [85]. However, levels of expression are likely to affect these data, with higher expression levels more likely to lead to aggregates. Moreover, eGFP fused to actin can affect filament incorporation. For example, we found that eGFP-fused to the C-terminus of cardiac α-actin did not work well compared to eGFP-fused to the N-terminus (unpublished observation). Other work has also reported that the position of tags on actin can affect its properties [86,87]. Experimentally, myosin has been shown to bind more weakly to purified E99K F-actin and in vitro motility assays show that myosin moves both actin filaments [82] and reconstituted actin filaments (containing tropomyosin and troponin) [88] more slowly than wild type. Thus, the experimental data suggest that this mutation does not strongly affect the ability of actin to incorporate into thin filaments; however, once incorporated, this mutation affects the force output by reducing strong binding of myosin to actin, leading to compensatory hypertrophy.

A331P (Ala331Pro), which causes DCM [80], is in a region of actin that interacts with the cardiomyopathy loop in myosin [4,35] and with the tropomyosin binding site in the absence of Ca^{2+} [89]. However, rather than interfering with myosin binding directly, this mutation seems to do this indirectly by modulating the behaviour of tropomyosin. Recent work, using recombinant A331P cardiac α-actin, expressed using the baculovirus/insect cell expression system, showed that it polymerises faster than wild type actin, but in reconstituted thin filaments, there is a weaker interaction between myosin and actin [90]. This mutation was suggested to affect the interaction of tropomyosin with nearby residues, and, in particular, residues D25, R28 and P33, which form a bulge that defines the position of tropomyosin on actin in the 'off' (or blocked) state. A331P could increase the potential for tropomyosin to inhibit the acto-myosin interaction, by reducing the likelihood of movement of tropomyosin from its position in the 'off' state, to the 'on' state, thus explaining the decreased myosin interaction with the reconstituted thin filament. A331P also decreased the binding affinity for the C0C2 subunits of myosin binding protein C (MYBPC) to actin [91]. C0–C1 interact with both actin and tropomyosin to regulate contraction. In agreement with these findings, in unpublished work from our lab we found that eGFP-A331P was able to incorporate into muscle sarcomere (Supplemental Figure S1) and while we did not find a significant effect on contraction (unpublished observation), this may be dependent on the levels of expression.

E361G (Glu361Gly), an HCM causing mutation [80], is close to the C-terminal myosin binding region of actin. However, it seems to exert its effect by indirectly affecting the binding of myosin to actin, through its effects on the Ca^{2+} sensitivity of the thin filament. E361G incorporates into thin filaments in the heart muscle in transgenic mice [92]. Our unpublished work also shows that eGFP-E361G incorporates normally into thin filaments in isolated cardiomyocytes, when expressed using an adenoviral system (Supplementary Figure S1) and we also found it did not affect cardiomyocyte contraction (unpublished data). Thin filaments isolated from the transgenic mice, with E361G expressed at ~50% of the total actin show normal myosin driven motility in in vitro motility assays [92] and the mice have a very mild phenotype. However, cardiac contraction in these mice is not sensitive to phosphorylation of troponin I. The N-terminal peptide sequence of cardiac troponin-I is phosphorylated by protein kinase A (PKA) in response to β1-adrenergic signalling (reviewed in [93]). Through its interaction with troponin C, this decreases the Ca^{2+} binding affinity of troponin C, thus increasing the rate of Ca^{2+} dissociation and allowing the rate of twitch relaxation to increase. In turn, this allows the heart rate to increase, and increase force output [92]. Thus, the E361G mutation in actin uncouples the normal relationship between β1 adrenergic signalling, Ca^{2+} sensitivity and troponin I phosphorylation.

Finally, the mutation Arg312His (R312H), which causes DCM [17], is also likely to indirectly affect myosin binding to actin, by affecting the actin–tropomyosin interaction [88]. The myosin-driven velocity of reconstituted actin R312H filaments in in vitro motility assays, is reduced at high Ca^{2+} but increased at low Ca^{2+} concentrations compared to wild type actin [88]. However, strong binding of myosin to the mutant actin was unaffected. Thus, the observed changes in motility and Ca^{2+} sensitivity could be explained by an interaction of this region of actin with tropomyosin. In contrast, work by others has shown that R312H actin is less stable, less well able to polymerise, releases phosphate from its nucleotide site at a faster rate compared to wild type actin [94] and has a decreased actin-activated myosin ATPase linked to its reduced stability [95]. Both studies used recombinant actin produced from insect cells. In our unpublished work (Supplementary Figure S1), we found that eGFP-R312H was able to incorporate into muscle sarcomeres in isolated adult rat cardiomyocytes and found no effect on contractility (unpublished observation). Given that the mutant actin appears to be able to incorporate into thin filaments in cells, a change to Ca^{2+} sensitivity, mediated through troponin/tropomyosin, may be more likely to account for the effects of this mutation.

5. Conclusions

The studies of a small number of mutations in cardiac α-actin discussed in the previous section demonstrate the challenges of trying to understand the effects of these mutations. Perhaps it is not surprising then that despite the large numbers of actin mutations reported, only a small number have been characterised in detail. A wide range of assays, both in vivo and in vitro, are generally needed to fully understand the effects of these mutations. Studies in vitro require purified actin and additional proteins (e.g., myosin, tropomyosin, troponin). This is not straightforward. Actin cannot be expressed and purified from *Escherichia. coli*, as it needs chaperones to fold correctly. However, it can be expressed and purified successfully using the Sf9/baculovirus system [82,86,96]. Studies in vivo require transgenic animals or human samples. The effects of mutations in intact skeletal fibres from humans or mouse models have also been characterised, as described above. Further examples include the demonstration that Phe352Ser increases contractile function [97], Asp286Gly prevents strong myosin binding [98] and His40Tyr inhibits co-polymerisation of wild type and mutant actin isoforms, and affects conformational changes in actin during contraction [99]. The use of model organisms, such as the indirect flight muscle of *Drosophila melanogaster*, has also proved useful [100]. This approach brings the advantage that the actin can be expressed on a null background, and can either be expressed as heterozygous (alongside a wild type copy of the gene) or homozygous, at levels typical of normal expression levels, and effects can be assessed in a mature muscle fibre. For example, this approach has shown that Asp292Val affects the regulation of contraction [100] through its interaction with tropomyosin [4].

Each of these systems has its own advantages and disadvantages. For example, in vitro expression systems do work but may only produce small amounts of actin [82,101]. The use of mouse models allows analysis of effects on adult skeletal fibres or the heart, but they are time consuming to make individual mutations and expensive to maintain mouse colonies. Human biopsies can be hard to obtain, and this research needs ethical approval. The use of *Drosophila Melanogaster* brings several advantages, however there are differences between the overall structure of its indirect flight muscle and the skeletal or cardiac muscles in humans.

In conclusion, disease-causing mutations are present in all six actin isoforms, and yet, the effects of many of these have not been well explored, particular those in smooth and non-muscle actin isoforms. While the effects of these mutations can be predicted from their positions in the structure, these predictions still need to be tested experimentally, to understand their potentially complex effects. The presence of mutations in similar regions of the sequence across the different actin isoforms, and their likely effects on filament stability, suggests it may well be worth exploring simple mammalian expression systems to analyse the effects of these mutations. CRISPR-based gene engineering approaches to directly edit the gene, while avoiding the use of tags, would allow analysis of actin dynamics in live cells, followed by purification of the expressed actin, to complement the many in vitro and in vivo assays already being used. The exploration of the role of PTMs in cancer biology is underexplored. It will be exciting to see the outcome of this type of research in the future.

Supplementary Materials: Supplementary materials can be found at http://www.mdpi.com/1422-0067/21/9/3371/s1.

Author Contributions: F.P., T.G.B. and M.P. were all involved in writing-review and editing, F.P. generated Supplementary Table S1. T.G.B. generated the adenoviral actin constructs. All authors have read and agreed to the published version of the manuscript.

Funding: This research was funded by the British Heart Foundation research grants PG/15/2/31208 and PG/07/095/23743.

Acknowledgments: We would like to thank the Human Genome Database (HGMD) for short term access to the full database, to obtain up-to-date mutations for this review.

Conflicts of Interest: The authors declare no conflict of interest.

References

1. Pollard, T.D. Actin and actin-binding proteins. *Cold Spring Harb. Perspect. Biol.* **2016**, *8*. [CrossRef]
2. Varland, S.; Vandekerckhove, J.; Drazic, A. Actin post-translational modifications: The Cinderella of cytoskeletal control. *Trends Biochem. Sci.* **2019**, *44*, 502–516. [CrossRef] [PubMed]
3. Dominguez, R.; Holmes, K.C. Actin structure and function. *Annu. Rev. Biophys.* **2011**, *40*, 169–186. [CrossRef] [PubMed]
4. Behrmann, E.; Muller, M.; Penczek, P.A.; Mannherz, H.G.; Manstein, D.J.; Raunser, S. Structure of the rigor actin-tropomyosin-myosin complex. *Cell* **2012**, *150*, 327–338. [CrossRef] [PubMed]
5. Fujii, T.; Iwane, A.H.; Yanagida, T.; Namba, K. Direct visualization of secondary structures of F-actin by electron cryomicroscopy. *Nature* **2010**, *467*, 724–728. [CrossRef]
6. Gurel, P.S.; Kim, L.Y.; Ruijgrok, P.V.; Omabegho, T.; Bryant, Z.; Alushin, G.M. Cryo-EM structures reveal specialization at the myosin VI-actin interface and a mechanism of force sensitivity. *Elife* **2017**, *6*.
7. Murakami, K.; Yasunaga, T.; Noguchi, T.Q.; Gomibuchi, Y.; Ngo, K.X.; Uyeda, T.Q.; Wakabayashi, T. Structural basis for actin assembly, activation of ATP hydrolysis, and delayed phosphate release. *Cell* **2010**, *143*, 275–287. [CrossRef]
8. von der Ecken, J.; Muller, M.; Lehman, W.; Manstein, D.J.; Penczek, P.A.; Raunser, S. Structure of the F-actin-tropomyosin complex. *Nature* **2015**, *519*, 114–117. [CrossRef]
9. Huehn, A.R.; Bibeau, J.P.; Schramm, A.C.; Cao, W.; De La Cruz, E.M.; Sindelar, C.V. Structures of cofilin-induced structural changes reveal local and asymmetric perturbations of actin filaments. *Proc. Natl. Acad. Sci. USA* **2020**, *117*, 1478–1484. [CrossRef]
10. Rao, J.N.; Madasu, Y.; Dominguez, R. Mechanism of actin filament pointed-end capping by tropomodulin. *Science* **2014**, *345*, 463–467. [CrossRef]
11. Nowak, K.J.; Ravenscroft, G.; Laing, N.G. Skeletal muscle alpha-actin diseases (actinopathies): Pathology and mechanisms. *Acta Neuropathol.* **2013**, *125*, 19–32. [CrossRef] [PubMed]
12. Wallgren-Pettersson, C.; Sewry, C.A.; Nowak, K.J.; Laing, N.G. Nemaline myopathies. *Semin. Pediatr. Neurol.* **2011**, *18*, 230–238. [CrossRef] [PubMed]
13. Guo, D.C.; Pannu, H.; Tran-Fadulu, V.; Papke, C.L.; Yu, R.K.; Avidan, N.; Bourgeois, S.; Estrera, A.L.; Safi, H.J.; Sparks, E.; et al. Mutations in smooth muscle alpha-actin (ACTA2) lead to thoracic aortic aneurysms and dissections. *Nat. Genet.* **2007**, *39*, 1488–1493. [CrossRef] [PubMed]
14. Cuoco, J.A.; Busch, C.M.; Klein, B.J.; Benko, M.J.; Stein, R.; Nicholson, A.D.; Marvin, E.A. ACTA2 cerebral arteriopathy: Not just a puff of smoke. *Cerebrovasc. Dis.* **2018**, *46*, 161–171. [CrossRef] [PubMed]
15. Whiffin, N.; Minikel, E.; Walsh, R.; O'Donnell-Luria, A.H.; Karczewski, K.; Ing, A.Y.; Barton, P.J.R.; Funke, B.; Cook, S.A.; MacArthur, D.; et al. Using high-resolution variant frequencies to empower clinical genome interpretation. *Genet. Med.* **2017**, *19*, 1151–1158. [CrossRef]
16. Mogensen, J.; Klausen, I.C.; Pedersen, A.K.; Egeblad, H.; Bross, P.; Kruse, T.A.; Gregersen, N.; Hansen, P.S.; Baandrup, U.; Borglum, A.D. Alpha-cardiac actin is a novel disease gene in familial hypertrophic cardiomyopathy. *J. Clin. Invest.* **1999**, *103*, R39–43. [CrossRef]
17. Olson, T.M.; Michels, V.V.; Thibodeau, S.N.; Tai, Y.S.; Keating, M.T. Actin mutations in dilated cardiomyopathy, a heritable form of heart failure. *Science* **1998**, *280*, 750–752. [CrossRef]
18. Maron, B.J.; Shirani, J.; Poliac, L.C.; Mathenge, R.; Roberts, W.C.; Mueller, F.O. Sudden death in young competitive athletes. Clinical, demographic, and pathological profiles. *Jama* **1996**, *276*, 199–204. [CrossRef]
19. Maron, B.J. Hypertrophic cardiomyopathy: A systematic review. *Jama* **2002**, *287*, 1308–1320. [CrossRef]
20. Chang, A.N.; Potter, J.D. Sarcomeric protein mutations in dilated cardiomyopathy. *Heart Fail. Rev.* **2005**, *10*, 225–235. [CrossRef]
21. Seidman, J.G.; Seidman, C. The genetic basis for cardiomyopathy: From mutation identification to mechanistic paradigms. *Cell* **2001**, *104*, 557–567. [CrossRef]

22. Riviere, J.B.; van Bon, B.W.; Hoischen, A.; Kholmanskikh, S.S.; O'Roak, B.J.; Gilissen, C.; Gijsen, S.; Sullivan, C.T.; Christian, S.L.; Abdul-Rahman, O.A.; et al. De novo mutations in the actin genes ACTB and ACTG1 cause Baraitser-Winter syndrome. *Nat. Genet.* **2012**, *44*, 440–444. [CrossRef] [PubMed]
23. Verloes, A.; Di Donato, N.; Masliah-Planchon, J.; Jongmans, M.; Abdul-Raman, O.A.; Albrecht, B.; Allanson, J.; Brunner, H.; Bertola, D.; Chassaing, N.; et al. Baraitser-Winter cerebrofrontofacial syndrome: Delineation of the spectrum in 42 cases. *Eur. J. Hum. Genet.* **2015**, *23*, 292–301. [CrossRef] [PubMed]
24. Yates, T.M.; Turner, C.L.; Firth, H.V.; Berg, J.; Pilz, D.T. Baraitser-Winter cerebrofrontofacial syndrome. *Clin. Genet.* **2017**, *92*, 3–9. [CrossRef]
25. Cuvertino, S.; Stuart, H.M.; Chandler, K.E.; Roberts, N.A.; Armstrong, R.; Bernardini, L.; Bhaskar, S.; Callewaert, B.; Clayton-Smith, J.; Davalillo, C.H.; et al. ACTB Loss-of-Function Mutations Result in a Pleiotropic Developmental Disorder. *Am. J. Hum. Genet.* **2017**, *101*, 1021–1033. [CrossRef]
26. Latham, S.L.; Ehmke, N.; Reinke, P.Y.A.; Taft, M.H.; Eicke, D.; Reindl, T.; Stenzel, W.; Lyons, M.J.; Friez, M.J.; Lee, J.A.; et al. Variants in exons 5 and 6 of ACTB cause syndromic thrombocytopenia. *Nat. Commun.* **2018**, *9*, 4250. [CrossRef]
27. Zhu, M.; Yang, T.; Wei, S.; DeWan, A.T.; Morell, R.J.; Elfenbein, J.L.; Fisher, R.A.; Leal, S.M.; Smith, R.J.; Friderici, K.H. Mutations in the gamma-actin gene (ACTG1) are associated with dominant progressive deafness (DFNA20/26). *Am. J. Hum. Genet.* **2003**, *73*, 1082–1091. [CrossRef]
28. Belyantseva, I.A.; Perrin, B.J.; Sonnemann, K.J.; Zhu, M.; Stepanyan, R.; McGee, J.; Frolenkov, G.I.; Walsh, E.J.; Friderici, K.H.; Friedman, T.B.; et al. Gamma-actin is required for cytoskeletal maintenance but not development. *Proc. Natl. Acad. Sci. USA* **2009**, *106*, 9703–9708. [CrossRef]
29. Di Donato, N.; Kuechler, A.; Vergano, S.; Heinritz, W.; Bodurtha, J.; Merchant, S.R.; Breningstall, G.; Ladda, R.; Sell, S.; Altmuller, J.; et al. Update on the ACTG1-associated Baraitser-Winter cerebrofrontofacial syndrome. *Am. J. Med. Genet. A* **2016**, *170*, 2644–2651. [CrossRef]
30. Wangler, M.F.; Gonzaga-Jauregui, C.; Gambin, T.; Penney, S.; Moss, T.; Chopra, A.; Probst, F.J.; Xia, F.; Yang, Y.; Werlin, S.; et al. Heterozygous de novo and inherited mutations in the smooth muscle actin (ACTG2) gene underlie megacystis-microcolon-intestinal hypoperistalsis syndrome. *PLoS Genet.* **2014**, *10*, e1004258. [CrossRef]
31. Ravenscroft, G.; Pannell, S.; O'Grady, G.; Ong, R.; Ee, H.C.; Faiz, F.; Marns, L.; Goel, H.; Kumarasinghe, P.; Sollis, E.; et al. Variants in ACTG2 underlie a substantial number of Australasian patients with primary chronic intestinal pseudo-obstruction. *Neurogastroenterol. Motil.* **2018**, *30*, e13371. [CrossRef] [PubMed]
32. Collins, R.R.J.; Barth, B.; Megison, S.; Pfeifer, C.M.; Rice, L.M.; Harris, S.; Timmons, C.F.; Rakheja, D. ACTG2-associated visceral myopathy with chronic intestinal pseudoobstruction, intestinal malrotation, hypertrophic pyloric stenosis, choledochal cyst, and a novel missense mutation. *Int. J. Surg. Pathol.* **2019**, *27*, 77–83. [CrossRef] [PubMed]
33. Kabsch, W.; Mannherz, H.G.; Suck, D.; Pai, E.F.; Holmes, K.C. Atomic structure of the actin:DNase I complex. *Nature* **1990**, *347*, 37–44. [CrossRef] [PubMed]
34. Huehn, A.; Cao, W.; Elam, W.A.; Liu, X.; De La Cruz, E.M.; Sindelar, C.V. The actin filament twist changes abruptly at boundaries between bare and cofilin-decorated segments. *J. Biol. Chem.* **2018**, *293*, 5377–5383. [CrossRef]
35. von der Ecken, J.; Heissler, S.M.; Pathan-Chhatbar, S.; Manstein, D.J.; Raunser, S. Cryo-EM structure of a human cytoplasmic actomyosin complex at near-atomic resolution. *Nature* **2016**, *534*, 724–728. [CrossRef]
36. Hegyi, G.; Premecz, G.; Sain, B.; Muhlrad, A. Selective carbethoxylation of the histidine residues of actin by diethylpyrocarbonate. *Eur. J. Biochem.* **1974**, *44*, 7–12. [CrossRef]
37. Normand, E.A.; Braxton, A.; Nassef, S.; Ward, P.A.; Vetrini, F.; He, W.; Patel, V.; Qu, C.; Westerfield, L.E.; Stover, S.; et al. Clinical exome sequencing for fetuses with ultrasound abnormalities and a suspected Mendelian disorder. *Genome Med.* **2018**, *10*, 74. [CrossRef]
38. Posey, J.E.; Harel, T.; Liu, P.; Rosenfeld, J.A.; James, R.A.; Coban Akdemir, Z.H.; Walkiewicz, M.; Bi, W.; Xiao, R.; Ding, Y.; et al. Resolution of Disease Phenotypes Resulting from Multilocus Genomic Variation. *N. Engl. J. Med.* **2017**, *376*, 21–31. [CrossRef]
39. Walsh, R.; Thomson, K.L.; Ware, J.S.; Funke, B.H.; Woodley, J.; McGuire, K.J.; Mazzarotto, F.; Blair, E.; Seller, A.; Taylor, J.C.; et al. Reassessment of Mendelian gene pathogenicity using 7,855 cardiomyopathy cases and 60,706 reference samples. *Genet. Med.* **2017**, *19*, 192–203. [CrossRef]

40. Proost, D.; Vandeweyer, G.; Meester, J.A.; Salemink, S.; Kempers, M.; Ingram, C.; Peeters, N.; Saenen, J.; Vrints, C.; Lacro, R.V.; et al. Performant mutation identification using targeted next-generation sequencing of 14 thoracic aortic aneurysm genes. *Hum. Mutat.* **2015**, *36*, 808–814. [CrossRef]
41. Fremont, S.; Romet-Lemonne, G.; Houdusse, A.; Echard, A. Emerging roles of MICAL family proteins - From actin oxidation to membrane trafficking during cytokinesis. *J. Cell Sci.* **2017**, *130*, 1509–1517. [CrossRef] [PubMed]
42. Grintsevich, E.E.; Yesilyurt, H.G.; Rich, S.K.; Hung, R.J.; Terman, J.R.; Reisler, E. F-actin dismantling through a redox-driven synergy between Mical and cofilin. *Nat. Cell Biol.* **2016**, *18*, 876–885. [CrossRef] [PubMed]
43. Alto, L.T.; Terman, J.R. MICALs. *Curr Biol* **2018**, *28*, R538–541. [CrossRef] [PubMed]
44. Hesketh, S.; Srisawat, K.; Sutherland, H.; Jarvis, J.; Burniston, J. On the rate of synthesis of individual proteins within and between different striated muscles of the rat. *Proteomes* **2016**, 4. [CrossRef] [PubMed]
45. Zou, Y.; Wang, J.; Liu, X.; Wang, Y.; Chen, Y.; Sun, K.; Gao, S.; Zhang, C.; Wang, Z.; Zhang, Y.; et al. Multiple gene mutations, not the type of mutation, are the modifier of left ventricle hypertrophy in patients with hypertrophic cardiomyopathy. *Mol. Biol. Rep.* **2013**, *40*, 3969–3976. [CrossRef] [PubMed]
46. Fischer, J.; Weide, T.; Barnekow, A. The MICAL proteins and rab1: A possible link to the cytoskeleton? *Biochem. Biophys. Res. Commun.* **2005**, *328*, 415–423. [CrossRef]
47. Laing, N.G.; Dye, D.E.; Wallgren-Pettersson, C.; Richard, G.; Monnier, N.; Lillis, S.; Winder, T.L.; Lochmuller, H.; Graziano, C.; Mitrani-Rosenbaum, S.; et al. Mutations and polymorphisms of the skeletal muscle alpha-actin gene (ACTA1). *Hum. Mutat.* **2009**, *30*, 1267–1277. [CrossRef]
48. Beuchle, D.; Schwarz, H.; Langegger, M.; Koch, I.; Aberle, H. Drosophila MICAL regulates myofilament organization and synaptic structure. *Mech. Dev.* **2007**, *124*, 390–406. [CrossRef]
49. Hoffjan, S.; Waldmuller, S.; Blankenfeldt, W.; Kotting, J.; Gehle, P.; Binner, P.; Epplen, J.T.; Scheffold, T. Three novel mutations in the ACTA2 gene in German patients with thoracic aortic aneurysms and dissections. *Eur J. Hum. Genet.* **2011**, *19*, 520–524. [CrossRef]
50. Disabella, E.; Grasso, M.; Gambarin, F.I.; Narula, N.; Dore, R.; Favalli, V.; Serio, A.; Antoniazzi, E.; Mosconi, M.; Pasotti, M.; et al. Risk of dissection in thoracic aneurysms associated with mutations of smooth muscle alpha-actin 2 (ACTA2). *Heart* **2011**, *97*, 321–326. [CrossRef]
51. Morin, M.; Bryan, K.E.; Mayo-Merino, F.; Goodyear, R.; Mencia, A.; Modamio-Hoybjor, S.; del Castillo, I.; Cabalka, J.M.; Richardson, G.; Moreno, F.; et al. In vivo and in vitro effects of two novel gamma-actin (ACTG1) mutations that cause DFNA20/26 hearing impairment. *Hum. Mol. Genet.* **2009**, *18*, 3075–3089. [CrossRef] [PubMed]
52. Sparrow, J.C.; Nowak, K.J.; Durling, H.J.; Beggs, A.H.; Wallgren-Pettersson, C.; Romero, N.; Nonaka, I.; Laing, N.G. Muscle disease caused by mutations in the skeletal muscle alpha-actin gene (ACTA1). *NMD* **2003**, *13*, 519–531. [CrossRef]
53. Bhoj, E.J.; Haye, D.; Toutain, A.; Bonneau, D.; Nielsen, I.K.; Lund, I.B.; Bogaard, P.; Leenskjold, S.; Karaer, K.; Wild, K.T.; et al. Phenotypic spectrum associated with SPECC1L pathogenic variants: New families and critical review of the nosology of Teebi, Opitz GBBB, and Baraitser-Winter syndromes. *Eur. J. Med. Genet.* **2019**, *62*, 103588. [CrossRef] [PubMed]
54. Whittington, J.R.; Poole, A.T.; Dutta, E.H.; Munn, M.B. A novel mutation in ACTG2 gene in mother with chronic intestinal pseudoobstruction and fetus with megacystis microcolon intestinal hypoperistalsis syndrome. *Case Rep. Genet.* **2017**, *2017*, 9146507. [CrossRef] [PubMed]
55. Guo, D.C.; Papke, C.L.; Tran-Fadulu, V.; Regalado, E.S.; Avidan, N.; Johnson, R.J.; Kim, D.H.; Pannu, H.; Willing, M.C.; Sparks, E.; et al. Mutations in smooth muscle alpha-actin (ACTA2) cause coronary artery disease, stroke, and Moyamoya disease, along with thoracic aortic disease. *Am. J. Hum. Genet.* **2009**, *84*, 617–627. [CrossRef]
56. Meng, L.; Pammi, M.; Saronwala, A.; Magoulas, P.; Ghazi, A.R.; Vetrini, F.; Zhang, J.; He, W.; Dharmadhikari, A.V.; Qu, C.; et al. Use of exome sequencing for infants in intensive care units: Ascertainment of severe single-gene disorders and effect on medical management. *JAMA Pediatr.* **2017**, *171*, e173438. [CrossRef]
57. Batchelor, M.; Wolny, M.; Dougan, L.; Paci, E.; Knight, P.J.; Peckham, M. Myosin tails and single alpha-helical domains. *Biochem. Soc. Trans.* **2015**, *43*, 58–63. [CrossRef]

58. Iglesias, A.; Anyane-Yeboa, K.; Wynn, J.; Wilson, A.; Truitt Cho, M.; Guzman, E.; Sisson, R.; Egan, C.; Chung, W.K. The usefulness of whole-exome sequencing in routine clinical practice. *Genet. Med.* **2014**, *16*, 922–931. [CrossRef]
59. Weerakkody, R.; Ross, D.; Parry, D.A.; Ziganshin, B.; Vandrovcova, J.; Gampawar, P.; Abdullah, A.; Biggs, J.; Dumfarth, J.; Ibrahim, Y.; et al. Targeted genetic analysis in a large cohort of familial and sporadic cases of aneurysm or dissection of the thoracic aorta. *Genet. Med.* **2018**, *20*, 1414–1422. [CrossRef]
60. Reza, N.; Garg, A.; Merrill, S.L.; Chowns, J.L.; Rao, S.; Owens, A.T. ACTA1 novel likely pathogenic variant in a family with dilated cardiomyopathy. *Circ. Genom. Precis. Med.* **2018**, *11*, e002243. [CrossRef]
61. Rainger, J.; Williamson, K.A.; Soares, D.C.; Truch, J.; Kurian, D.; Gillessen-Kaesbach, G.; Seawright, A.; Prendergast, J.; Halachev, M.; Wheeler, A.; et al. A recurrent de novo mutation in ACTG1 causes isolated ocular coloboma. *Hum. Mutat.* **2017**, *38*, 942–946. [CrossRef] [PubMed]
62. Sandestig, A.; Green, A.; Jonasson, J.; Vogt, H.; Wahlstrom, J.; Pepler, A.; Ellnebo, K.; Biskup, S.; Stefanova, M. Could dissimilar phenotypic effects of ACTB missense mutations reflect the actin conformational change? Two novel mutations and literature review. *Mol. Syndromol.* **2019**, *9*, 259–265. [CrossRef] [PubMed]
63. Ohlsson, M.; Tajsharghi, H.; Darin, N.; Kyllerman, M.; Oldfors, A. Follow-up of nemaline myopathy in two patients with novel mutations in the skeletal muscle alpha-actin gene (ACTA1). *NMD* **2004**, *14*, 471–475. [CrossRef] [PubMed]
64. Regalado, E.S.; Guo, D.C.; Prakash, S.; Bensend, T.A.; Flynn, K.; Estrera, A.; Safi, H.; Liang, D.; Hyland, J.; Child, A.; et al. Aortic disease presentation and outcome associated with ACTA2 mutations. *Circ. Cardiovasc. Genet.* **2015**, *8*, 457–864. [CrossRef] [PubMed]
65. Ilkovski, B.; Cooper, S.T.; Nowak, K.; Ryan, M.M.; Yang, N.; Schnell, C.; Durling, H.J.; Roddick, L.G.; Wilkinson, I.; Kornberg, A.J.; et al. Nemaline myopathy caused by mutations in the muscle alpha-skeletal-actin gene. *Am. J. Hum. Genet.* **2001**, *68*, 1333–1343. [CrossRef] [PubMed]
66. Weitensteiner, V.; Zhang, R.; Bungenberg, J.; Marks, M.; Gehlen, J.; Ralser, D.J.; Hilger, A.C.; Sharma, A.; Schumacher, J.; Gembruch, U.; et al. Exome sequencing in syndromic brain malformations identifies novel mutations in ACTB, and SLC9A6, and suggests BAZ1A as a new candidate gene. *Birth Defects Res.* **2018**, *110*, 587–597. [CrossRef] [PubMed]
67. Lukoyanova, N.; VanLoock, M.S.; Orlova, A.; Galkin, V.E.; Wang, K.; Egelman, E.H. Each actin subunit has three nebulin binding sites: Implications for steric blocking. *Curr. Biol.* **2002**, *12*, 383–388. [CrossRef]
68. Pappas, C.T.; Bliss, K.T.; Zieseniss, A.; Gregorio, C.C. The Nebulin family: An actin support group. *Trends Cell Biol.* **2011**, *21*, 29–37. [CrossRef]
69. Kiss, B.; Lee, E.J.; Ma, W.; Li, F.W.; Tonino, P.; Mijailovich, S.M.; Irving, T.C.; Granzier, H.L. Nebulin stiffens the thin filament and augments cross-bridge interaction in skeletal muscle. *Proc. Natl. Acad. Sci. USA* **2018**, *115*, 10369–10374. [CrossRef]
70. Hernandez, D.A.; Bennett, C.M.; Dunina-Barkovskaya, L.; Wedig, T.; Capetanaki, Y.; Herrmann, H.; Conover, G.M. Nebulette is a powerful cytolinker organizing desmin and actin in mouse hearts. *Mol. Biol. Cell* **2016**, *27*, 3869–3882. [CrossRef]
71. Bang, M.L.; Chen, J. Roles of nebulin family members in the heart. *Circ. J.* **2015**, *79*, 2081–2087. [CrossRef] [PubMed]
72. Meyer, L.C.; Wright, N.T. Structure of giant muscle proteins. *Front. Physiol.* **2013**, *4*, 368. [CrossRef] [PubMed]
73. Zhang, F.; Saha, S.; Shabalina, S.A.; Kashina, A. Differential arginylation of actin isoforms is regulated by coding sequence-dependent degradation. *Science* **2010**, *329*, 1534–1537. [CrossRef] [PubMed]
74. Saha, S.; Mundia, M.M.; Zhang, F.; Demers, R.W.; Korobova, F.; Svitkina, T.; Perieteanu, A.A.; Dawson, J.F.; Kashina, A. Arginylation regulates intracellular actin polymer level by modulating actin properties and binding of capping and severing proteins. *Mol. Biol. Cell* **2010**, *21*, 1350–1361. [CrossRef] [PubMed]
75. Drazic, A.; Aksnes, H.; Marie, M.; Boczkowska, M.; Varland, S.; Timmerman, E.; Foyn, H.; Glomnes, N.; Rebowski, G.; Impens, F.; et al. NAA80 is actin's N-terminal acetyltransferase and regulates cytoskeleton assembly and cell motility. *Proc. Natl. Acad. Sci. USA* **2018**, *115*, 4399–4404. [CrossRef] [PubMed]
76. Yoon, J.; Terman, J.R. MICAL redox enzymes and actin remodeling: New links to classical tumorigenic and cancer pathways. *Mol. Cell Oncol.* **2018**, *5*, e1384881. [CrossRef] [PubMed]
77. Milligan, R.A. Protein-protein interactions in the rigor actomyosin complex. *Proc. Natl. Acad. Sci. USA* **1996**, *93*, 21–26. [CrossRef]

78. Milligan, R.A.; Whittaker, M.; Safer, D. Molecular structure of F-actin and location of surface binding sites. *Nature* **1990**, *348*, 217–221. [CrossRef]
79. Marston, S. The molecular mechanisms of mutations in actin and myosin that cause inherited myopathy. *Int. J. Mol. Sci.* **2018**, 19. [CrossRef]
80. Olson, T.M.; Doan, T.P.; Kishimoto, N.Y.; Whitby, F.G.; Ackerman, M.J.; Fananapazir, L. Inherited and de novo mutations in the cardiac actin gene cause hypertrophic cardiomyopathy. *J. Mol. Cell Cardiol.* **2000**, *32*, 1687–1694. [CrossRef]
81. Vang, S.; Corydon, T.J.; Borglum, A.D.; Scott, M.D.; Frydman, J.; Mogensen, J.; Gregersen, N.; Bross, P. Actin mutations in hypertrophic and dilated cardiomyopathy cause inefficient protein folding and perturbed filament formation. *Febs J.* **2005**, *272*, 2037–2049. [CrossRef] [PubMed]
82. Bookwalter, C.S.; Trybus, K.M. Functional consequences of a mutation in an expressed human alpha-cardiac actin at a site implicated in familial hypertrophic cardiomyopathy. *J. Biol. Chem.* **2006**, *281*, 16777–16784. [CrossRef] [PubMed]
83. Wolny, M.; Colegrave, M.; Colman, L.; White, E.; Knight, P.J.; Peckham, M. Cardiomyopathy mutations in the tail of beta-cardiac myosin modify the coiled-coil structure and affect integration into thick filaments in muscle sarcomeres in adult cardiomyocytes. *J. Biol. Chem.* **2013**, *288*, 31952–31962. [CrossRef] [PubMed]
84. Hayashi, Y.; Ono, K.; Ono, S. Mutations in Caenorhabditis elegans actin, which are equivalent to human cardiomyopathy mutations, cause abnormal actin aggregation in nematode striated muscle. *F1000Res.* **2019**, *8*, 279. [CrossRef] [PubMed]
85. Song, W.; Dyer, E.; Stuckey, D.J.; Copeland, O.; Leung, M.C.; Bayliss, C.; Messer, A.; Wilkinson, R.; Tremoleda, J.L.; Schneider, M.D.; et al. Molecular mechanism of the E99K mutation in cardiac actin (ACTC Gene) that causes apical hypertrophy in man and mouse. *J. Biol. Chem.* **2011**, *286*, 27582–27593. [CrossRef]
86. Rommelaere, H.; Waterschoot, D.; Neirynck, K.; Vandekerckhove, J.; Ampe, C. A method for rapidly screening functionality of actin mutants and tagged actins. *Biol. Proced. Online* **2004**, *6*, 235–249. [CrossRef]
87. Brault, V.; Sauder, U.; Reedy, M.C.; Aebi, U.; Schoenenberger, C.A. Differential epitope tagging of actin in transformed Drosophila produces distinct effects on myofibril assembly and function of the indirect flight muscle. *Mol. Biol. Cell* **1999**, *10*, 135–149. [CrossRef]
88. Debold, E.P.; Saber, W.; Cheema, Y.; Bookwalter, C.S.; Trybus, K.M.; Warshaw, D.M.; Vanburen, P. Human actin mutations associated with hypertrophic and dilated cardiomyopathies demonstrate distinct thin filament regulatory properties in vitro. *J. Mol. Cell Cardiol.* **2010**, *48*, 286–292. [CrossRef]
89. Barua, B.; Fagnant, P.M.; Winkelmann, D.A.; Trybus, K.M.; Hitchcock-DeGregori, S.E. A periodic pattern of evolutionarily conserved basic and acidic residues constitutes the binding interface of actin-tropomyosin. *J. Biol. Chem.* **2013**, *288*, 9602–9609. [CrossRef]
90. Bai, F.; Caster, H.M.; Rubenstein, P.A.; Dawson, J.F.; Kawai, M. Using baculovirus/insect cell expressed recombinant actin to study the molecular pathogenesis of HCM caused by actin mutation A331P. *J. Mol. Cell Cardiol.* **2014**, *74*, 64–75. [CrossRef]
91. Chow, M.L.; Shaffer, J.F.; Harris, S.P.; Dawson, J.F. Altered interactions between cardiac myosin binding protein-C and alpha-cardiac actin variants associated with cardiomyopathies. *Arch. Biochem. Biophys.* **2014**, *550–551*, 28–32. [CrossRef] [PubMed]
92. Song, W.; Dyer, E.; Stuckey, D.; Leung, M.C.; Memo, M.; Mansfield, C.; Ferenczi, M.; Liu, K.; Redwood, C.; Nowak, K.; et al. Investigation of a transgenic mouse model of familial dilated cardiomyopathy. *J. Mol. Cell Cardiol.* **2010**, *49*, 380–389. [CrossRef] [PubMed]
93. Marston, S.; Zamora, J.E. Troponin structure and function: A view of recent progress. *J. Muscle Res. Cell Motil.* **2020**, *41*, 71–89. [CrossRef] [PubMed]
94. Mundia, M.M.; Demers, R.W.; Chow, M.L.; Perieteanu, A.A.; Dawson, J.F. Subdomain location of mutations in cardiac actin correlate with type of functional change. *PLoS ONE* **2012**, *7*, e36821. [CrossRef] [PubMed]
95. Dahari, M.; Dawson, J.F. Do cardiac actin mutations lead to altered actomyosin interactions? *Biochem. Cell Biol.* **2015**, *93*, 330–334. [CrossRef]
96. Muller, M.; Mazur, A.J.; Behrmann, E.; Diensthuber, R.P.; Radke, M.B.; Qu, Z.; Littwitz, C.; Raunser, S.; Schoenenberger, C.A.; Manstein, D.J.; et al. Functional characterization of the human alpha-cardiac actin mutations Y166C and M305L involved in hypertrophic cardiomyopathy. *Cell Mol. Life Sci.* **2012**, *69*, 3457–3479. [CrossRef]

97. Lindqvist, J.; Penisson-Besnier, I.; Iwamoto, H.; Li, M.; Yagi, N.; Ochala, J. A myopathy-related actin mutation increases contractile function. *Acta Neuropathol.* **2012**, *123*, 739–746. [CrossRef]
98. Fan, J.; Chan, C.; McNamara, E.L.; Nowak, K.J.; Iwamoto, H.; Ochala, J. Molecular consequences of the myopathy-related D286G mutation on actin function. *Front. Physiol.* **2018**, *9*, 1756. [CrossRef]
99. Chan, C.; Fan, J.; Messer, A.E.; Marston, S.B.; Iwamoto, H.; Ochala, J. Myopathy-inducing mutation H40Y in ACTA1 hampers actin filament structure and function. *Biochim. Biophys. Acta* **2016**, *1862*, 1453–1458. [CrossRef]
100. Sevdali, M.; Kumar, V.; Peckham, M.; Sparrow, J. Human congenital myopathy actin mutants cause myopathy and alter Z-disc structure in Drosophila flight muscle. *Neuromuscul. Disord.* **2013**, *23*, 243–255. [CrossRef]
101. Anthony Akkari, P.; Nowak, K.J.; Beckman, K.; Walker, K.R.; Schachat, F.; Laing, N.G. Production of human skeletal alpha-actin proteins by the baculovirus expression system. *Biochem. Biophys. Res. Commun.* **2003**, *307*, 74–79. [CrossRef]

© 2020 by the authors. Licensee MDPI, Basel, Switzerland. This article is an open access article distributed under the terms and conditions of the Creative Commons Attribution (CC BY) license (http://creativecommons.org/licenses/by/4.0/).

Review

Dendritic Spines in Alzheimer's Disease: How the Actin Cytoskeleton Contributes to Synaptic Failure

Silvia Pelucchi, Ramona Stringhi and Elena Marcello *

Department of Pharmacological and Biomolecular Sciences, Università degli Studi di Milano, 20133 Milan, Italy; silvia.pelucchi@unimi.it (S.P.); stringhi.ramona@gmail.com (R.S.)
* Correspondence: elena.marcello@unimi.it; Tel.: +39-02-50318314

Received: 24 December 2019; Accepted: 26 January 2020; Published: 30 January 2020

Abstract: Alzheimer's disease (AD) is a neurodegenerative disorder characterized by Aβ-driven synaptic dysfunction in the early phases of pathogenesis. In the synaptic context, the actin cytoskeleton is a crucial element to maintain the dendritic spine architecture and to orchestrate the spine's morphology remodeling driven by synaptic activity. Indeed, spine shape and synaptic strength are strictly correlated and precisely governed during plasticity phenomena in order to convert short-term alterations of synaptic strength into long-lasting changes that are embedded in stable structural modification. These functional and structural modifications are considered the biological basis of learning and memory processes. In this review we discussed the existing evidence regarding the role of the spine actin cytoskeleton in AD synaptic failure. We revised the physiological function of the actin cytoskeleton in the spine shaping and the contribution of actin dynamics in the endocytosis mechanism. The internalization process is implicated in different aspects of AD since it controls both glutamate receptor membrane levels and amyloid generation. The detailed understanding of the mechanisms controlling the actin cytoskeleton in a unique biological context as the dendritic spine could pave the way to the development of innovative synapse-tailored therapeutic interventions and to the identification of novel biomarkers to monitor synaptic loss in AD.

Keywords: synaptopathy; actin cytoskeleton; actin-binding proteins; amyloid; synaptic plasticity

1. Introduction

Alzheimer's disease (AD) is the most common cause of dementia, characterized by decline in memory and thinking and by the impairment of at least two domains of cognition [1]. AD progression is associated with a significant and progressive disability throughout the disease course, with death generally occurring within 5–12 years of symptom onset [2]. Therefore, the burden on caregivers and the public health sector is enormous, leading to high nonmedical cost [3]. Unfortunately, therapies that may prevent or slow the rate of AD progression are not available. In such a scenario, it is fundamental to develop a disease-modifying intervention to produce an enduring change in the clinical progression of AD by interfering with the pathophysiological mechanisms.

So far, AD pathogenesis relies on the amyloid hypothesis [4], according to which the Amyloid-β (Aβ) peptide aggregation plays a critical role at the beginning of the cascade of events leading to dementia. Aβ is a small peptide of 40–42 amino acids that was identified as the main constituent of amyloid neuritic plaques [5]. The concerted action of β-secretase BACE1 and γ-secretase determines the release of Aβ from the β-amyloid precursor protein (APP) [6]. Neuronal activity drives APP into BACE1-containing acidic organelles via clathrin-dependent endocytosis [7], where the vast majority of BACE1 cleavage of APP occurs [8] (Figure 1). Alternatively, APP can undergo a non-amyloidogenic pathway that involves the α-secretase ADAM10, a metalloprotease able to cleave APP within the sequence corresponding to Aβ [9,10]. ADAM10 cleavage not only prevents Aβ generation but also increases the release of the neurotrophic and neuroprotective sAPPα fragment [11]. The non-amyloidogenic ADAM10 processing of APP occurs largely on the plasma membrane [12] and in the trans-Golgi network [13] (Figure 1). APP and the secretases are all transmembrane proteins. Therefore, the trafficking mechanisms can control APP shedding and Aβ generation in neuronal cells. For example, ADAM10 synaptic localization and activity towards APP are finely tuned by its binding partners SAP97 and the clathrin adaptor protein AP2, that regulate ADAM10 forward trafficking and endocytosis respectively [14–17]. Moreover, perturbation of BACE1 post-Golgi trafficking results in an increase in BACE1 cleavage of APP and increased production of Aβ [18].

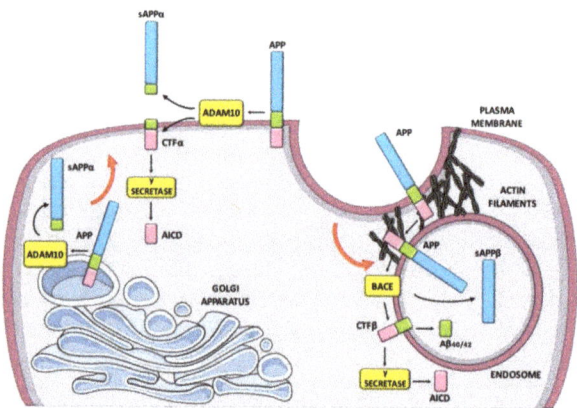

Figure 1. Schematic representation of β-amyloid precursor protein (APP) processing. The vast majority of BACE1 cleavage of APP occurs into BACE1-containing acidic organelles via clathrin-dependent endocytosis. The subsequent γ-secretase activity determines the release of Aβ. Alternatively, APP can undergo a non-amyloidogenic pathway that involves the α-secretase ADAM10 and occurs largely on the plasma membrane and in the trans-Golgi network. Red arrow, exocytosis/endocytosis pathways; black arrows, enzymatic cleavage of APP.

In addition to amyloid plaques, intracellular deposits of hyperphosphorylated tau protein, named neurofibrillary tangles, represent the other main hallmark for AD. Several evidences suggest that the increase in Aβ levels may trigger the progression of tau pathology in AD. For instance, experiments performed in a three-dimensional in vitro human neural cell culture system revealed that elevated Aβ levels alone are sufficient to drive tau pathology in human neurons [19]. Furthermore, the injection of Aβ fibrils into tau transgenic mice and the generation of animal models harboring both pathologies have demonstrated that Aβ accelerates tau pathology and neurodegeneration, whereas Aβ pathology is generally unaffected by concurrent tau pathology [20–24].

However, Aβ leads to cognitive impairment independent of its effects on tau pathology. Soluble Aβ oligomers isolated from the AD brain have been shown to induce synaptic loss [25] and to deeply affect the activity-dependent synaptic plasticity phenomena, such as long-term potentiation (LTP) and long-term depression (LTD) [26]. Moreover, Aβ oligomers cause impairment in cognitive tasks when injected into the lateral ventricles of rodent models [27]. Plasticity is a normal and essential part of cognition and LTP and LTD are considered the biological basis of learning and memory formation processes. Therefore, the small Aβ oligomers impair the main functional units of the brain and negatively affect their capability to store memories. These results have prompted the idea that AD is principally a disorder of synaptic function, i.e., a "synaptopathy" [28]. Indeed, late stage AD involves an incontrovertible and substantial loss of neurons and synapses [29]. Moreover, data obtained in the brains of AD patients and of subjects affected by Mild Cognitive Impairment showed that synapse loss is an early event in the disease process and a structural correlate involved in cognitive decline [30], thus supporting the idea of AD as synaptopathy.

Considering that Aβ oligomer-triggered synaptic dysfunction is causally linked to the early cognitive symptoms detected in AD patients, it is fundamental to decipher which synaptic pathways are affected in early phases of AD. Here we review the role of the actin cytoskeleton in dendritic spines as one of the main biological pathways crucial for both synaptic plasticity and synaptic pathology in early stages of AD. The identification of the synaptic molecules involved in the Aβ-mediated actin cytoskeleton failure in spines will be fundamental to developing synapse-tailored therapies and to assess in vivo biomarkers able to track synaptic integrity over time in patients.

2. The Actin Cytoskeleton as the Architect of Spines

In the mammalian brain, the postsynaptic compartment of glutamatergic excitatory synapses is localized in small protrusions along dendrites named dendritic spines [31,32]. Mature spines are characterized by a mushroom shape consisting of a head connected to the dendrite shaft by a narrower neck, while "stubby" spines lack the neck and filopodia-like are "headless" spines [31]. This distinctive morphology depends on an underlying cytoskeletal structure [33]. Whereas the dendritic shaft cytoplasm is dominated by microtubules, actin is the major cytoskeletal component of dendritic spines [34], where it is organized in a complex network of long and short branching filaments within the spine neck and in the spine head [35]. The actin cytoskeleton is a very dynamic structure able to self-assemble its building block G-actin (globular and monomeric form) in an ATP hydrolysis-dependent manner. This reaction generates a filamentous structure called F-actin (polymeric state) that can disassemble back into the monomer pool. G-actin is arranged head-to-tail to give the filament a molecular polarity, so F-actin is an asymmetric polymer with two ends that are dynamically different, called barbed and pointed ends. The first one is the more dynamic end, in fact, it elongates 10 times faster than the pointed end [36]. These processes occur on short time scales, allowing the cell to rapidly respond to internal or external stimuli [37,38]. Indeed, both the monomeric and filamentous forms of actin are present in the spine and the G-actin/F-actin ratio influences the various aspects of dendritic spine morphology and synaptic function [39].

The spine cytoskeleton is localized in the region just underneath the postsynaptic density (PSD) and is closely associated with a disk-shaped array of proteins attached to the postsynaptic membrane. The PSD provides a structural framework for postsynaptic signaling and plasticity [40]. Indeed, the PSD is fundamental for localizing molecules, including glutamate receptors and signaling molecules, in a functional organized structure of scaffolding proteins [40]. Importantly, scaffolding proteins of the PSD, such as SHANK and PSD-95, are associated with the dendritic actin cytoskeleton through interaction with actin F-binding proteins like cortactin and α-actinin [41,42]. In particular, dendritic spines contain two different pools of F-actin: (i) a very dynamic pool exists below the spine surface and interacts directly or indirectly with AMPA receptors, NMDA receptors, and PSD scaffolding and signaling proteins; (ii) a more internal and stable pool of F-actin serves as the main scaffold that supports the overall spine structure [35,43]. More recently, super-resolution microscopy techniques revealed periodic actin structures in dendrites and in the neck of dendritic spines [44,45] that may provide mechanical support and elasticity to this structure [46].

2.1. Synaptic Actin-Binding Proteins Orchestrating Actin Cytoskeleton Dynamics

The general function of actin is to allow the cell to respond to internal or external stimuli and, therefore, its capability to self-assemble is specifically spatiotemporally controlled [37,38]. To modulate the complex dynamic of the actin filaments there are several different proteins, called actin-binding proteins, able to bind and regulate actin dynamics. Cooperatively, the actin-binding proteins maintain a large pool of G-actin monomers available for polymerization, nucleate assembly of new filaments, promote elongation, cap barbed or pointed ends to terminate elongation, sever filaments, and cross-link filaments (reviewed in [47]). Therefore, the actin-binding proteins regulate the physiology of actin, giving to the actin the characteristic dynamism and stability.

In dendritic spines the most important and investigated actin-binding proteins are:

2.1.1. Actin-Related Proteins-2/3

Actin-Related Proteins-2/3 (Arp2/3) are a complex made of different subunits, among which are Arp2, Arp3, ARPC1, ARPC2, ARPC3, ARPC4, and ARPC. Arp2/3 is the principal actin filaments nucleator [35] and, thereby, the major actin-binding protein with polymerizing and filament branching activities [48]. It can bind the filamentous actin to both sides and allows the insertion and creation of an additional filament [49]. It is enriched in the PSD and its downregulation results in an impairment in the spine head formation [50,51]. Several proteins, such as Cortactin, Abi2, WAVE-1 (WASp-family verprolin homology protein-1), N-WASP (neural Wiskott-Aldrich syndrome protein), and Abp1 activate Arp2/3 and their deletion is associated to memory deficits [52–56].

2.1.2. Profilin

This class of proteins is fundamental for actin polymerization, since they are responsible for the ADP to ATP nucleotide exchange on actin. These proteins catalyze actin polymerization in a concentration-dependent manner: they are catalysts at lower concentrations and inhibitors at higher levels [57]. Profilin 2 is the principal isoform in the mammalian brain [35], even if profilin 1 is also expressed [58].

2.1.3. Rho Family of GTPases

There are different components of this family of Ras proteins, including Ras homolog gene family member A (RhoA), Ras-related C3 botulinum toxin substrate 1 (Rac1), and Cell division control protein 42 homolog (Cdc42). All the members of this family have been studied in neuronal cells since they are involved in the neuronal morphogenesis. In the dendritic spine, the Rho activation is fundamental for the cofilin phosphorylation and therefore for the actin stabilization of the spine [35]. On the other hand, Rac1 and Cdc42 activation lead to an enlargement of the head spine [51], promoting the formation of Arp2/3 complex.

2.1.4. ADF/Cofilin

The cofilin isoforms, i.e., cofilin-1 and cofilin-2, belong to a highly conserved protein family, as the actin depolymerizing factor (ADF) [59]. Cofilin can promote the actin turnover because it exerts a bidirectional effect on F-actin, depending on its relative concentration to actin. At low concentrations, cofilin promotes F-actin disassembly by cutting the actin filaments (severing) or by facilitating the removal of the actin monomers (depolymerization) [60]. F-actin severing can actually result in an increased rate of actin polymerization if there are enough actin monomers available, due to the creation of free barbed filament ends. The role of this complex is fundamental for continuous treadmilling of actin. In fact, since the actin monomers are fundamental for a fast reorganization of the actin cytoskeleton, the complex ADF/cofilin promotes the depolymerization of actin and creates a new pool of G-actin monomers available for the formation of other filaments [61]. The result of this activity is the correct maintenance for the morphology of the spine [51]. Cofilin is inactivated by phosphorylation on Ser3 by LIM kinase 1 (LIMK1) [62] and activated by slingshot homolog 1 (SSH1)-mediated dephosphorylation of Ser3 [63]. The actin interacting protein 1 (Aip1) is a cofilin accessory protein that selectively binds to cofilin-decorated actin filaments to induce the capping [64,65] or the destabilization of the filaments [66].

2.1.5. Cyclase-Associated Proteins

Cyclase-associated proteins (CAP) can control filament turnover by recycling actin monomers and severing actin filaments [67]. In particular, it has been shown that the N-terminal region of the yeast homolog Svr2/CAP is responsible for the role that CAP plays in synergy with cofilin to accelerate actin filament depolymerization [68–71]. In mammals, two CAP homologs are expressed: CAP1 shows a wide tissue distribution, whereas CAP2 is primarily present in brain, heart, and skeletal muscle, skin, and testis [72], suggesting that these proteins have distinct functional roles. The deletion of CAP2 affects the F-actin/G-actin ratio in neurons, leading to an accumulation of F-actin in intracellular structures. The lack of CAP2 modifies neuronal architecture and spine morphology. CAP2 C-terminal domain interacts with cofilin, and such association depends on cofilin phosphorylation on Ser3. The ablation of CAP2 leads to increased levels of dephosphorylated (active) cofilin along with cofilin intracellular aggregation [73].

2.1.6. Epidermal Growth Factor Receptor Pathway Substrate 8

Epidermal Growth Factor Receptor Pathway Substrate 8 (Eps8) is a capping protein. Capping proteins are involved in actin polymerization through the bond with barbed ends of F-actin, in order to block addition and removal of actin subunits [74]. The capping proteins are distributed in the dendritic spine and their function is relevant to inhibiting the filopodia formation [75]. The actin-capping proteins control the organization of filopodia. They can cap the newly branched filaments created by the Arp2/3 complex, thus controlling the elongation of the filopodia and the concentration of the free G-actin monomers.

2.1.7. Myosins V and VI

Myosins V and VI are actin-based motor proteins that hydrolyze ATP to generate the mechanical force required for movement along actin filaments [76,77]. This enables myosins to propel the sliding of actin filaments, to produce tension on actin filaments, and to walk along these filaments. As a result, myosins can regulate the structure and dynamics of the actin cytoskeleton and affect the localization and transport of cellular components [76]. Motor proteins can be classified in either plus- or minus-end-directed motors. Plus-directed motor proteins move towards the barbed (plus) end of actin filaments, directing the cargo to the cell periphery, while minus-end-directed motors move towards the pointed (minus) end of actin filaments and have a major role in the movement of endocytic vesicles away from the plasma membrane [78]. The myosins V and VI are involved in the forward trafficking and internalization of the AMPA receptor [76].

2.1.8. Tropomyosins

Tropomyosins (Tpms) are a family of actin-associated proteins relevant for the regulation of the actin cytoskeleton. Tpms are organized in coiled-coil dimers that form a head-to-tail polymer along the length of actin filaments. In mammals there four Tpm genes that can produce a range of Tpm isoforms by alternative exon splicing [79], but in neuronal cells isoforms deriving from *TPM1*, *TPM3*, and *TPM4* genes are found [80]. Tpms control the interaction of actin filaments with myosin motors and actin-binding proteins in an isoform-specific manner [81]. In particular, it has been shown that Tpm isoforms (i) bind along the sides of filaments and protect them from severing proteins and pointed-end depolymerization in vitro, (ii) affect the actin filament branching and nucleation, influencing the Arp2/3 complex activity [82], (iii) compete with ADF/cofilin family for actin binding in a Tpm isoform-dependent manner [83]. Neuronal-specific Tpm isoforms are differentially expressed, both temporally and spatially. Some Tpm isoforms have higher expression; during development their expression is required for the maintenance of the neuronal phenotype [84–86], whereas the expression of other isoforms increases with maturity [86,87]. For example, the Tpm3.1 isoform promotes axon length, cone size growth, and dendritic branching [88]. Such spatial and temporally controlled Tpm expression throughout the mouse brain, as well as in different sub-cellular compartments of neurons, is relevant for actin filament identity and specificity [85]. Indeed, while Tpm1.12 is found in the presynaptic compartment of cultured hippocampal neurons, *TPM3* and *TPM4* gene products localize to the postsynaptic region in mouse hippocampal neurons [89].

2.1.9. Drebrin

Drebrin has two isoforms, embryonic-type drebrin E and adult-type drebrin A, that change during development from E to A [90]. Drebrin accumulates in dendritic spines where it creates a stable pool of slow turn-over F-actin and bundles filaments by crosslinking them together [91–93]. Drebrin associates with other actin-binding proteins such as myosins (I, II, V) and gelsolin [91,93] and provides a direct interaction with microtubules, since it associates to end-binding proteins that are microtubule plus-end tracking proteins localized to the growing plus-ends of dynamic microtubules [94]. Drebrin localization is regulated by activity-dependent synaptic plasticity and NMDA receptor activation [90].

2.1.10. Ca^{2+}/calmodulin dependent protein kinase II β

The Ca^{2+}/calmodulin dependent protein kinase II (CaMKII) is a ubiquitous serine/threonine protein kinase that plays a key role in postsynaptic LTP and learning/memory [95,96]. Among the four CaMKII isoforms, the α and β subunits are the most abundant in the brain [97]. The β subunit features an F-actin-binding domain that targets CaMKIIβ to actin filaments in cells and particularly in dendritic spines [98]. CaMKIIα isoform weakly binds actin and this interaction is important for its localization to the dendritic spine [99]. On the other hand, CaMKIIβ does not only bind to actin, but also bundles actin filaments together [98] through the formation of hetero-oligomers with the CaMKIIα subunit [100]. Therefore, CaMKIIβ is a protein relevant for the maintenance of the spine structure and for plasticity-driven remodeling. CaMKIIβ appears to be anchored to a protein complex composed of drebrin-binding F-actin during the resting state. NMDA receptor activation releases CaMKIIβ from drebrin, resulting in CaMKIIβ association with PSD [101].

2.1.11. α-actinin

α-actinin assembles in an antiparallel fashion and such dimers crosslink actin filaments [102]. The isoform α-actinin2 localizes to dendritic spines, enriched within the PSD and implicated in actin organization [103]. α-actinin selectively stabilizes CaMKII association with GluN2B-containing glutamate receptors [104]. The isoform α-actinin-4 is an interacting partner of metabotropic glutamate receptors and orchestrates spine dynamics and morphogenesis in neurons through a CaMKIIβ-dependent process [105].

2.2. The Actin Cytoskeleton in Spines: A Key Player of Activity-Dependent Synaptic Plasticity Events

During development and in adulthood, synapse formation, maintenance, and elimination come along with changes in dendritic spine number and morphology to establish and shape the connectivity of neuronal circuits [106]. The actin cytoskeleton is important for the stabilization of postsynaptic proteins in mature spines [107] and modulation of spine morphological adaptation (shape and number) in response to postsynaptic stimuli [108,109]. Indeed, at the cellular level, a synaptic function and spine shape modifications are strictly interdependent, especially during activity-dependent synaptic plasticity events, such as LTP and LTD [110]. In this framework the actin cytoskeleton dynamics are finely tuned since they are a critical element for the remodeling of the spine morphology, as well as for the endocytosis processes that control glutamate receptors levels at the membrane.

2.2.1. Actin Cytoskeleton Remodeling to Change Spines Structure

The equilibrium between F-actin and G-actin is stable under basal neuronal activity, but it can be rapidly modulated, both positively and negatively, by synaptic activity. Indeed, LTP shifts the G-actin/F-actin ratio toward F-actin (rise in spine actin filaments) and results in spine enlargement, while LTD shifts the G-actin/F-actin ratio toward G-actin (decrease in spine actin filaments) and results in spine shrinkage. Actin-binding proteins orchestrate these precise changes in actin cytoskeleton dynamics and have different localizations in spines during the different phases of LTP [109] (Figure 2).

The first phase of LTP is characterized by a transient but profound modification of the overall protein composition of the spine with a rapid increase of the actin levels and of actin polymerization in the spine [108]. During this phase of 1 to 7 min, the actin cytoskeleton is more unstable and susceptible to reorganization. In addition to the increase in F-actin, there is also a change in the composition of the actin-binding proteins in the spine. During these first minutes, the spine is significantly enriched in proteins able to largely modify F-actin through severing (cofilin), branching (Arp2/3), or capping (Aip1). At the same time, the concentration of proteins known to stabilize the suprastructure of the actin cytoskeleton by bundling F-actin or linking F-actin to the PSD (drebrin, CaMKIIβ, and α-actinin) is transiently reduced in the spine [109,111]. During this time window of about 5 min, therefore, actin filaments can lose their supramolecular organization (bundling and cross-linking) and allow the access to other actin-binding factors that can reorganize the actin cytoskeleton. This switch of actin-binding protein type from actin-stabilizers to actin-modifiers in the earliest phase of synapse potentiation creates a time window in which the actin cytoskeleton becomes susceptible to major reorganization (Figure 2). Afterwards, the concentration of drebrin, CaMKIIβ, and α-actinin in the spine progressively returns to basal levels of concentration [109].

One of the major players in such LTP-induced actin cytoskeleton remodeling is cofilin. It has been shown that upon the activation of the NMDA receptors, cofilin is translocated to the spine, where it severs the actin cytoskeleton, leading to the formation of new barbed ends that nucleate new filament growth [112]. The severing allows the formation of new F-actin, which is the preferred site of Arp2/3 nucleating and branching activity [48]. The concerted action of cofilin and Arp2/3 is fundamental for the maintenance of spine expansion and for the control of protein delivery to the synaptic membrane, such as AMPA receptors [47,113]. Indeed, the perturbation of cofilin inactivation by phosphorylation prevents the maintenance, but not the first phase, of spine enlargement, indicating that cofilin is required for the structural consolidation of the spine.

During the second phase (7–60 min after the LTP induction) the actin concentration goes back to basal levels, leading to the stabilization of the spine structure, while cofilin moves to the neck of the spine after its phosphorylation on Ser3 and, thereby, its inactivation. The complex cofilin-actin can stop the actin rearrangement, giving the spine the possibility to enlarge its structure. Indeed, after LTP induction, actin stabilizes the spine apparatus, representing an anchoring for several molecules [114]. In the third phase (corresponding to late LTP) that occurs 60 min after induction, the PSD is structurally remodeled by the increase in PSD proteins and newly synthesized factors [115]. Concerning the structure of the spine, it has been demonstrated that the F-actin nucleation, mediated by WAVE complex, an activator of the Arp2/3 complex, occurs in the central structure of the spine, while the elongation occurs at the tip of finger-like protrusions. For that reason, the proteins involved in the branching of the already assembled filaments are localized in the central part of the PSD, while next to the membrane the filament elongator proteins are confined. The synaptic plasticity modifies the distribution patterns of actin-binding proteins and induces also the redistribution of branched F-actin regulators in spines to create an enlargement also in the distal part of the spine [114].

2.2.2. Actin Cytoskeleton and Endocytosis

Postsynaptic composition is tightly controlled and rapidly modulated during activity-dependent synaptic plasticity events. This synaptic property requires coordinated mechanisms of protein trafficking that are themselves under the control of the actin cytoskeleton.

For instance, the precise regulation of glutamate receptor number and subtype at the synapse, which is crucial to excitatory neurotransmission and synaptic plasticity, can be affected by actin cytoskeleton dynamics modulation. Cultured neurons exposed to latrunculin, a G-actin sequestering drug that blocks polymerization, showed reduced clustering of synaptic NMDA receptors and GluA1-containing AMPA receptors in dendritic spines [116] and specifically decreased AMPA receptor neurotransmission [117]. In addition, the F-actin stabilizing drug Jasplakinolide blocked glutamate-stimulated AMPAR internalization [118]. Indeed, AMPA receptors dynamically cycle between the plasma membrane and intracellular compartments through endo-exocytic events (reviewed in [119]). In addition to endocytosis and exocytosis, lateral diffusion of receptors in the plane of the membrane and exchange between synaptic and extrasynaptic sites emerged as key steps for modifying receptor numbers at synapses (reviewed in [120]). Constitutive endocytosis of AMPA receptors at the postsynaptic membrane is believed to be clathrin-independent [121], even though constitutive clathrin-mediated endocytosis (CME) of the receptor, as well as other cargos, was reported to occur in this subcompartment as well as in dendrites and in the soma [122]. On the other hand, it is widely accepted that the implementation of LTD requires CME of postsynaptic AMPA receptors [122–124] and is relevant for learning in vivo [125].

The CME is a complex process that requires coordination of the molecular events responsible for cargo sorting, membrane invagination, vesicle scission, and vesicle targeting. A vast body of research has revealed an intricate network of numerous protein-protein interactions within the endocytic pathway [126]. Genetic studies in yeast have firmly established a functional connection between actin and endocytosis, and experiments performed with drugs that interfere with actin cytoskeleton dynamics provided significant evidence that, in mammalian cells, CME relies on an active actin cytoskeleton [126]. For instance, the inhibition of actin polymerization by using latrunculin blocked clathrin-coated structure dynamics in neuronal dendrites [127].

In mammalian cells the CME was divided into several distinct stages that included coat assembly on membranes, invagination, fission, movement of vesicles away from the plasma membrane, and finally, uncoating [128,129]. Actin is crucial during these different stages of endocytic internalization and actin networks that form at sites of endocytosis must be tightly regulated for efficient internalization. Indeed, sites of endocytosis contain Arp2/3 to form a branched actin network, capping proteins, which limits the length of filaments and depolymerization factors such as cofilin, which turn over older filaments for recycling of actin subunits [130]. During the first phase, when endocytic coat proteins are recruited and the endocytic structure remains at the cell membrane undergoing relatively minor movements, the regulators of actin assembly are recruited, and near the end of this stage, actin polymerization begins. Therefore, actin is involved in specifying sites of coated-pit formation on the plasma membrane and may provide a scaffold for the assembly and anchoring of the endocytic machinery during clathrin-coated vesicle formation. It was shown that the clathrin-adaptor complex AP2, which mediates attachment of clathrin to the plasma membrane and induces clathrin polymerization into a coat, colocalizes with actin stress fibers [131]. The Huntingtin interacting protein 1 (HIP1) and Hip1R were suggested as potential linkers between the clathrin-coated pit and actin cytoskeleton. HIP1 binds AP2 and clathrin and is present in clathrin-coated vesicles [132], while Hip1R associates to both actin and clathrin [131]. The heterodimerization of HIP1 and Hip1R may be a mechanism that allows HIP1 to indirectly bind to F-actin and that also allows AP2 to be recruited to clathrin nucleation sites defined by the HIP proteins at the plasma membrane. During the second phase, the proteins make a short movement away from the membrane into the cytoplasm. At the end of this stage, endocytic proteins are lost from the vesicle, presumably coinciding with scission of the vesicle from the plasma membrane. At this stage, actin polymerization occurs at endocytic sites, thus providing a force during coated-pit neck constriction, fission, and detachment of clathrin-coated vesicles [133,134]. Using alternating evanescent field and epifluorescence illumination, Merrifield and colleagues showed that clathrin-coated pit invagination and scission are tightly coupled, with scission coinciding with maximal displacement of the clathrin-coated pit from the plasma membrane and with peak recruitment of cortactin, a dynamin and F-actin-binding protein. Indeed, perturbing actin polymerization reduces the efficiency of membrane scission [135]. In addition, a role for Tpm3.1 was hypothesized during the fission of the endosomes from the plasma membrane. Tpm3.1 can recruit non-muscle myosin II to stabilize actin filaments [136]. In light of this consideration, Tpm3.1 could be involved in maintaining the nascent endocytic neck and in stabilizing and recruiting myosin II, allowing the stabilization and constriction of the actin ring surrounding the neck of nascent bulk endosomes in preparation for their fission from the plasma membrane [137].

In dendritic spines, there are endocytosis-specialized regions named endocytic zones near the postsynaptic membrane and lateral to the PSD, where they develop and persist independent of synaptic activity, akin to the PSD itself [127]. In such specialized regions there are actin-binding proteins relevant for glutamate receptor endocytosis, such as the Candidate Plasticity Gene 2 (CPG2) that colocalizes with clathrin at postsynaptic endocytic zones [138,139]. CPG2, through a direct physical interaction, recruits endophilin B2 to F-actin, thus anchoring the endocytic machinery to the spine cytoskeleton and facilitating glutamate receptor internalization [124]. Specific disruption of endophilin B2 or the CPG2-endophilin B2 interaction impairs activity-dependent, but not constitutive, internalization of both NMDA- and AMPA-type glutamate receptors [124] (Figure 2).

As far as concern glutamate AMPA receptors, different linker proteins mediate the association of the receptor subunits to the actin cytoskeleton [113] and modulate AMPA receptor localization, internalization, and forward trafficking. The F-actin-binding proteins 4.1 were shown to stabilize AMPA receptors, providing a link to actin filaments [140], and to be involved in receptor exocytosis [141]. The deletion of the CAP2 protein impairs the LTP-triggered increase in GluA1 surface expression [73]. The protein PICK1, which is involved in LTD-induced internalization of AMPA receptors [142], binds GluA2/3 subunits, F-actin, and the Arp2/3 complex [143]. PICK1 is a negative regulator of Arp2/3-mediated actin polymerization that is critical for a specific form of vesicle trafficking and in particular for NMDA-induced AMPA internalization [143]. The reversion-induced LIM protein (RIL) is another protein linker that associates AMPA receptors with the actin cytoskeleton because it binds both the GluA1 C-terminus and the F-actin cross-linking protein α-actinin. It was proposed that such association plays a role in enhancing surface and synaptic expression of AMPA receptors by regulating endosomal recycling [144]. A protein relevant for AMPA receptor sorting is the actin-regulatory protein cortactin that interacts with the GluA2 subunit. Disrupting GluA2-cortactin binding in neurons causes the targeting of GluA2/A3-containing receptors to lysosomes and their consequent degradation, resulting in a loss of surface and synaptic GluA2 under basal conditions and an occlusion of subsequent LTD expression [145]. In addition, AMPA receptor trafficking is also dependent on the interaction with different myosins isoforms. Myosin Va, which is a plus-end-directed motor, binds GluA1 tail and is required for the LTP-induced forward trafficking of the receptor [146]. On the other hand, myosin VI exists in a complex with the AMPA receptors, AP2 and SAP97 in the brain, suggesting that myosin VI could play a role in the clathrin-mediated endocytosis of AMPA receptors [147] (Figure 2).

Sorting mechanisms are also important for amyloid generation in neuronal cells. Trafficking events play key roles in the convergence of APP and BACE1, putting in close proximity the substrate and the cleaving enzyme, respectively. Considering that BACE1 is optimally active in an acidic environment, the endosomes can be contemplated as the biological locus where the amyloidogenic pathway initiates [148]. Most studies on APP and BACE1 trafficking were done in non-neuronal cells and showed that trans-Golgi network, plasma membrane, and endosomes are the main sorting stations for APP and BACE1 [149]. In cultured hippocampal neurons, APP and BACE1 interact in both endoplasmic reticulum/Golgi and endocytic compartments, particularly in recycling zones such as dendritic spines and presynaptic boutons [150]. Abnormal residence time of APP and BACE1 in endosomes can lead to increased cleavage of APP, and this might contribute to sporadic AD [151]. After the synthesis, APP is conveyed in Golgi-derived vesicles to the plasma membrane where it can be internalized via a clathrin-dependent mechanism [7,152]. After internalization a significant population of APP endocytosed from the plasma membrane enters LAMP-1-positive late endosomes/lysosomes [150], while a portion of internalized APP can be routed into BACE1-positive recycling endosomes [7] Upon endocytosis APP can also be transported back to the trans-Golgi network via a retromer-dependent pathway [153]. A key regulator of the APP retrieval pathway is sorting protein-related receptor with A-type repeats (sorLA) [154], a scaffold protein linking APP to the retromer complex [155]. Overexpression of sorLA in neurons causes redistribution of APP to the Golgi and decreased processing to Aβ, whereas ablation of sorLA expression in knockout mice results in increased levels of Aβ in the brain [154]. Remarkably, genome-wide association studies showed an association of sorLA with sporadic, late-onset AD [156], strengthening the relevance of endocytosis in AD pathogenesis.

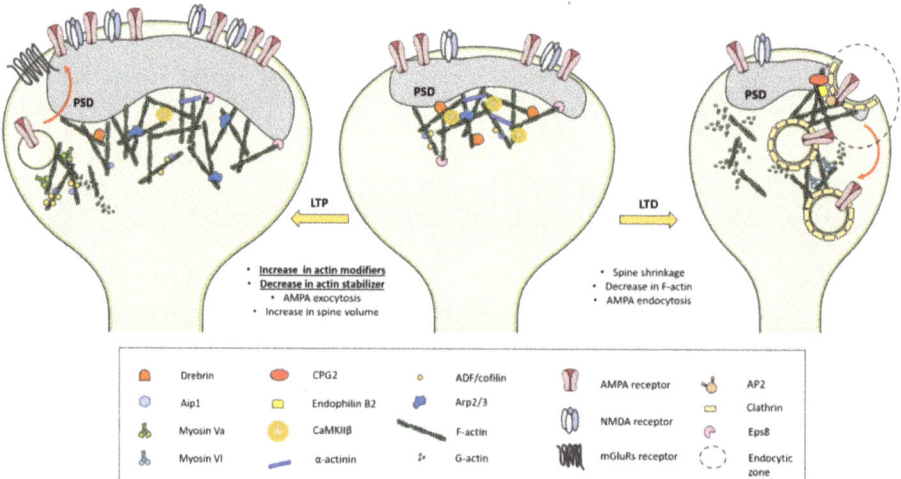

Figure 2. Dendritic spine actin cytoskeleton dynamics in activity-dependent synaptic plasticity phenomena. During the first phase of long-term potentiation (LTP) (1–7 min), there is a rearrangement of the actin cytoskeleton characterized by (i) the increase in the synaptic level of actin-binding proteins that are able to modify F-actin through several process, such as severing (cofilin), branching (Arp2/3), or capping (Aip1); (ii) the decrease in actin-stabilizer proteins (drebrin, CaMKIIβ, and α-actinin) responsible for stabilizing actin filaments and linking them to the postsynaptic density (PSD). After this phase, cofilin is inactivated to enlarge the spine and stabilize changes in structure. During long-term depression (LTD), the spine shrinkage is associated to a decrease in spine actin filaments. The actin cytoskeleton and actin-binding proteins are also involved in AMPA receptor trafficking: LTP triggers AMPA receptor forward trafficking, while LTD promotes clathrin-mediated endocytosis of AMPA receptors.

3. Actin Cytoskeleton Pathology in AD

Given the large contribution of the cytoskeleton to the mechanisms involved in synaptic plasticity, an impairment in actin cytoskeleton dynamics can contribute to AD pathology and underlie the synaptic failure in AD [28]. Several actin-binding proteins are altered in AD brains and AD animal models (Table 1). For example, the expression of drebrin is decreased in the hippocampus of aged AD mice compared with age-matched wild-type and young adult AD mice. In AD mice the overexpression of drebrin ameliorates cognitive ability and attenuates pathological lesions [157]. Moreover, the potential contribution of drebrin in AD synaptic failure is strengthened by studies showing its decrease in the hippocampal dendritic spines of AD brain patients [158] and in cortical areas, including the frontal and temporal cortices [159], in relation to the cognitive impairment associated with normal aging [160].

The involvement of the Rho family of small GTPases (Rho, Rac, and Cdc42) in AD pathology was shown in different AD models. In neuronal cultures exposed to fibrillar Aβ peptide, an increased localization and activity of Rac1/Cdc42 Rho GTPases, promoted by Tiam1, were observed together with a consequent enhancement in actin polymerization [161]. Rac1 is increased in plasma samples of AD patients [162] and, in the hippocampus of AD mice, Rac1 was abnormally activated at the early stages of the pathology, even if the total protein levels decreased at full-blown pathology stage. These data suggest that, in an initial stage, Rac1 deregulation might represent a triggering co-factor due to the direct effect on Aβ and tau [162].

Table 1. Actin-binding proteins: physiological role and involvement in AD.

Protein	Physiological Function	Role in AD
Arp2/3	Actin filaments nucleator: major actin-binding protein with polymerizing and filament branching activities	Required for Hirano body formation in *Dictyostelium*
Profilin	Responsible for the ADP to ATP nucleotide exchange on actin	Required for model Hirano body formation in *Dictyostelium*
Rho family of GTPases	Intermediaries between external signals and internal actin organization	- Increased activity of Rac1/Cdc42 Rho GTPases upon fibrillar Aβ exposure - Rac1 is increased in plasma samples of AD patients and abnormally activated in AD mice at early stages
ADF/cofilin	Promote the actin turnover	- Identified in the Hirano bodies and main component of actin rods - Alterations of its activation in in vitro and in vivo AD models
Myosins V and VI	Actin-based motor proteins	Myosin VI colocalizes with tau protein accumulated in neurons of AD patients
Drebrin	Bundles actin filaments by crosslinking	Decreased in the hippocampal dendritic spines of AD patients and in the hippocampus of AD mice
α-actinin	Form dimers that crosslink actin filaments	Component of the Hirano bodies
Tropomyosins	Form head-to-tail polymers to stabilize the actin filament and recruit myosin	Component of neurofibrillary pathology and Hirano bodies

Evidence of actin involvement in AD comes from studies of the main neuropathological hallmarks of AD, i.e., neurofibrillary tangles and the senile plaque, and of the additional lesions that are detectable in AD brains, such as the Hirano bodies and the actin rods. For example, the screening of 1250 mutant Drosophila lines allowed the identification of 30 specific modifiers of tau-induced neurodegeneration, among which were several components of the actin cytoskeleton, such as Tpm [163] and myosin VI, which colocalize with tau protein accumulated in the neurons of AD patients and of tauopathy animal models [164].

Hirano bodies are bright eosinophilic intracytoplasmic inclusions encountered within the CA1 region of the hippocampus. These lesions were first identified by Asao Hirano and their frequency increases with age and in a number of neurodegenerative diseases including AD [165,166]. Immunocytochemistry and immunoelectron-microscopy studies revealed the diffuse presence of actin, in the F-state, and cytoskeletal proteins throughout the Hirano body [167]. In addition to APP and several microtubule-associated proteins, actin-associated proteins such as Tpm, vinculin, α-actinin, ADF, and cofilin are components of these intracellular inclusions [168]. The association between AD and Tpm dysregulation can be isoform-specific. For example, a proteomics study showed that *TPM1* and *TPM3* gene products are increased in the white matter of AD patients when compared with controls [169]. In addition, a proteomic analysis of hippocampal glycoproteins revealed an increase in *TPM3* gene products in AD brains [170]. The strong presence of actin-binding proteins in the Hirano bodies supports the hypothesis that these lesions are derived from an abnormal organization of the neuronal cytoskeleton. To strengthen this hypothesis, it has been shown that profilin, Arp/2/3, WASH, and de novo actin polymerization are required for model Hirano body formation in *Dictyostelium* [171].

In addition to Hirano bodies, ADF and cofilin were detected in another form of inclusion, named actin rods, that is prominent in hippocampal and cortical neurites of the post-mortem brains of AD patients, especially in neurites contacting amyloid deposits [172,173], and in AD mice models [174]. Actin rods are cytoplasmic rod-shaped bundles of filaments composed of ADF/cofilin-actin in a 1:1 complex. In addition to cofilin and actin, cytoplasmic rods isolated from either cell lines or cultured primary neurons contain other proteins, but only during late stages of rod maturation, indicating that these are not core components of the rods. Vesicles containing APP, BACE1, and presenilin-1, a component of the γ-secretase complex, accumulate at rods, suggesting that actin rod formation blocks the transport of APP and enzymes involved in its processing to Aβ [174]. Rod formation has a local effect since transport deficits of mitochondria and early endosomes, a decline in spine numbers, and glutamate receptor response occur within neurites that form spontaneous rods, while neurites from the same neuron without rods remain unaffected [175].

In order to investigate the primary cause of actin rod formation, several neurodegenerative stimuli were investigated, such as exposure to excitotoxic levels of glutamate, mitochondrial inhibitors, peroxide, Aβ peptides, and proinflammatory cytokines [172,174]. ATP depletion is a major trigger for cofilin-actin rod formation at a stoichiometric proportion of 1:1 [172]. A decrease in ATP shifts the balance of kinase/phosphatase activity toward the active (dephosphorylated) form of cofilin, the augment of the ADP-actin fraction of total actin, and the increase of reactive oxygen species (ROS) [172,176]. In such environments, cofilin is able to saturate regions of F-actin that are readily severed, producing small stable fragments [177]. In the presence of high amounts of ROS, the newly generated F-actin fragments lead to the formation of rods through the direct bundling of these fragments and/or intermolecular disulfide cross-linking of cofilin [172,176]. In summary, the formation of actin rods requires (i) cofilin in the activated (dephosphorylated) form, (ii) cofilin saturation of local regions of F-actin in the presence of abnormally high levels of ADP-actin, which preferentially binds to cofilin, (iii) formation of intermolecular disulfide linkages via oxidation of several key cysteine residues of cofilin [178]. The formation of rods actually may also have a protective function. Considering that in neurons actin dynamics utilize a significant amount of ATP [179], in a short period (approximately 60 min) immediately after initial rod formation cofilin sequestration transiently delays the decline in mitochondrial membrane potential and helps maintain ATP levels by reducing actin filament turnover [180].

Even though the active dephosphorylated form is the main component of actin rods [174], data about the activation state of cofilin in AD are conflicting. Increased cofilin phosphorylation/inactivation with age and AD pathology was reported both in vivo and in vitro [181], while a decrease in cofilin phosphorylation was described in the frontal lobe of younger patients [182]. In APP/PS1 mice (a mouse model of AD) an increased cofilin activation/dephosphorylation was observed [183,184]. The analysis of the PSD-enriched fraction of both APP/PS1 mice and AD brains revealed a significant increase in cofilin phosphorylation/inactivation in the postsynaptic compartment [185]. Another study showed that cofilin phosphorylation state depends on the stage of the pathology, since an increased activation (dephosphorylation) is detected in APP/PS1 mice brains at four months of age while cofilin is strongly phosphorylated (inactivated) at 10 months of age, corresponding to a full-blown pathology stage [181].

Considering the crucial involvement of cofilin in several biological pathways related to the actin cytoskeleton, the alteration of cofilin activation and its sequestration in the actin rods could affect different cellular functions [60]. First, an alteration of cofilin levels and activation profoundly affects the biological pathways implicated in learning and memory processes. Indeed, cofilin is required for LTP, as demonstrated by the generation of cofilin knockout mice, which showed a complete lack of LTP in CA1 neurons of hippocampal slices, aberrant spine morphology, and impaired associative learning [186]. Cofilin controls extrasynaptic excitatory AMPA receptor diffusion [186], which affects the probability of a surface-expressed receptor being incorporated into the synapse [187], and cofilin activation is required for the insertion of the AMPA receptor subunit GluA1 upon LTP induction [188]. Second, the lack of cofilin can affect cellular transport. Indeed, cofilin directly competes with tau for microtubule binding and can be implicated in in tauopathy and destabilization of tau-regulated microtubule dynamics [189]. Third, cofilin is relevant for the regulation of mitochondria morphology and function [190,191]. In oxidative stress conditions, cofilin cysteine residues are oxidized, thus promoting intramolecular disulfide bridging of cofilin [192]. In such a state, cofilin loses affinity for actin and translocates to mitochondria, where it induces swelling, a drop in mitochondrial membrane potential, and cytochrome c release by promoting the opening of the permeability transition pore [192–194]. Mitochondrial targeting of cofilin and subsequent cytochrome c release mediate neuronal apoptosis in response to Aβ oligomers [183,195]. Finally, cofilin is also implicated in APP processing and Aβ metabolism.

The knockdown of cofilin in primary neurons significantly reduces Aβ production by increasing surface APP levels. Expression of active, but not inactive, cofilin reduces membrane APP levels by enhancing APP endocytosis. In APP/PS1 mice, genetic reduction of cofilin reduces Aβ deposition together with significantly increased microglial activation that has a greater ability to uptake and clear Aβ. These results indicate a significant role for cofilin in Aβ accumulation via different neuronal and microglial mechanisms [196].

4. Conclusions

Even though in the last 25 years the pharmaceutical industry has invested in anti-amyloid therapies and in numerous phase 3 clinical trials, no amyloid-targeting therapy has improved or limited the progression of cognitive impairment in symptomatic AD. These data suggest that while amyloid accumulation may be key in beginning the pathological process, other downstream events such as synaptic loss and tau accumulation may be the main drivers of neurodegeneration [197]. In addition, considering that synaptic failure is causally linked to the early cognitive symptoms detected in AD patients, AD can be considered a synaptopathy, such as several other neurological disorders.

Notably, growing evidence demonstrates that the actin cytoskeleton plays a key role in synaptic function and plasticity. Therefore, the disturbance of actin-binding proteins, which orchestrate actin cytoskeleton dynamics, could be a common principle in many synaptopathies [198,199]. Several studies analyzing AD post-mortem tissue, animal and cellular models suggest that AD pathology has a deleterious effect on the pathways governing the actin cytoskeleton.

The actin cytoskeleton is a critical element during activity-dependent synaptic plasticity phenomena, since it takes part in different aspects of the coordinated machinery that modulates synaptic transmission. First, actin filament remodeling shapes the dendritic spine in response to the received stimulus. Secondly, actin polymerization is the driving force of clathrin-mediated endocytosis and actin-binding proteins are involved vesicle sorting. Therefore, the endocytosis controls not only the membrane levels of glutamate receptors but also the colocalization of the amyloid cascade players, thus affecting the production of Aβ peptide.

In light of these considerations, the actin cytoskeleton can be positioned at the crossroad of pathways contributing to AD pathogenesis, i.e., the amyloid cascade and the synaptic dysfunction. However, a great effort is needed to investigate in detail the mechanisms controlling actin cytoskeleton machinery in a specific cellular compartment that is the dendritic spine. Indeed, several studies on actin have been performed in test tubes and in yeast, but the main challenge is to understand the actions of actin and actin-binding proteins in a proper and highly specialized cellular context, i.e., the dendritic spine. The knowledge of the physiological mechanisms controlling synaptic plasticity is fundamental to understand the biological basis of memory formation and, thereby, to shed light on the pathological modifications that drive cognitive impairment in AD.

Author Contributions: S.P., R.S., and E.M. wrote the manuscript; E.M. conceived and revised the manuscript. All authors have read and agreed to the published version of the manuscript.

Funding: This work was supported by the Fondazione Cariplo (Grant n. 2018-0511 to E.M.), AIRAlzh Onlus-COOP Italia (fellowship to S.P.), the Italian Ministry of Education, University and Research (PRIN 2017B9NCSX to E.M.), Fondo per il Finanziamento delle Attività Base di Ricerca (FFABR18_10 to E.M.), MIUR Progetto Eccellenza, PON "Ricerca e Innovazione" PerMedNet project (ARS01_01226), Fondo di sviluppo unimi- linea2-PSR2019.

Conflicts of Interest: The authors declare no conflict of interest. The funders had no role in the design of the study; in the collection, analyses, or interpretation of data; in the writing of the manuscript, or in the decision to publish the results.

References

1. McKhann, G.M.; Knopman, D.S.; Chertkow, H.; Hyman, B.T.; Jack, C.R.; Kawas, C.H.; Klunk, W.E.; Koroshetz, W.J.; Manly, J.J.; Mayeux, R.; et al. The diagnosis of dementia due to Alzheimer's disease: Recommendations from the National Institute on Aging-Alzheimer's Association workgroups on diagnostic guidelines for Alzheimer's disease. *Alzheimers Dement.* **2011**, *7*, 263–269. [CrossRef]
2. Vermunt, L.; Sikkes, S.A.M.; van den Hout, A.; Handels, R.; Bos, I.; van der Flier, W.M.; Kern, S.; Ousset, P.-J.; Maruff, P.; Skoog, I.; et al. Alzheimer Disease Neuroimaging Initiative; AIBL Research Group; ICTUS/DSA study groups Duration of preclinical, prodromal, and dementia stages of Alzheimer's disease in relation to age, sex, and APOE genotype. *Alzheimers Dement.* **2019**, *15*, 888–898. [CrossRef]
3. DiLuca, M.; Olesen, J. The Cost of Brain Diseases: A Burden or a Challenge? *Neuron* **2014**, *82*, 1205–1208. [CrossRef]
4. Hardy, J.A.; Higgins, G.A. Alzheimer's disease: The amyloid cascade hypothesis. *Science* **1992**, *256*, 184–185. [CrossRef] [PubMed]
5. Masters, C.L.; Simms, G.; Weinman, N.A.; Multhaup, G.; McDonald, B.L.; Beyreuther, K. Amyloid plaque core protein in Alzheimer disease and Down syndrome. *Proc. Natl. Acad. Sci. USA* **1985**, *82*, 4245–4249. [CrossRef] [PubMed]
6. Haass, C.; Selkoe, D.J. Cellular processing of beta-amyloid precursor protein and the genesis of amyloid beta-peptide. *Cell* **1993**, *75*, 1039–1042. [CrossRef]
7. Das, U.; Scott, D.A.; Ganguly, A.; Koo, E.H.; Tang, Y.; Roy, S. Activity-induced convergence of APP and BACE-1 in acidic microdomains via an endocytosis-dependent pathway. *Neuron* **2013**, *79*, 447–460. [CrossRef]
8. Vassar, R.; Bennett, B.D.; Babu-Khan, S.; Kahn, S.; Mendiaz, E.A.; Denis, P.; Teplow, D.B.; Ross, S.; Amarante, P.; Loeloff, R.; et al. Beta-secretase cleavage of Alzheimer's amyloid precursor protein by the transmembrane aspartic protease BACE. *Science* **1999**, *286*, 735–741. [CrossRef]
9. Lammich, S.; Buell, D.; Zilow, S.; Ludwig, A.-K.; Nuscher, B.; Lichtenthaler, S.F.; Prinzen, C.; Fahrenholz, F.; Haass, C. Expression of the anti-amyloidogenic secretase ADAM10 is suppressed by its 5'-untranslated region. *J. Biol. Chem.* **2010**, *285*, 15753–15760. [CrossRef]
10. Kuhn, P.-H.; Wang, H.; Dislich, B.; Colombo, A.; Zeitschel, U.; Ellwart, J.W.; Kremmer, E.; Rossner, S.; Lichtenthaler, S.F. ADAM10 is the physiologically relevant, constitutive alpha-secretase of the amyloid precursor protein in primary neurons. *Embo J.* **2010**, *29*, 3020–3032. [CrossRef]
11. Richter, M.C.; Ludewig, S.; Winschel, A.; Abel, T.; Bold, C.; Salzburger, L.R.; Klein, S.; Han, K.; Weyer, S.W.; Fritz, A.-K.; et al. Distinct in vivo roles of secreted APP ectodomain variants APPsα and APPsβ in regulation of spine density, synaptic plasticity, and cognition. *Embo J.* **2018**, *37*, 116. [CrossRef] [PubMed]
12. Lammich, S.; Kojro, E.; Postina, R.; Gilbert, S.; Pfeiffer, R.; Jasionowski, M.; Haass, C.; Fahrenholz, F. Constitutive and regulated alpha-secretase cleavage of Alzheimer's amyloid precursor protein by a disintegrin metalloprotease. *Proc. Natl. Acad. Sci. USA* **1999**, *96*, 3922–3927. [CrossRef] [PubMed]
13. Tan, J.Z.A.; Gleeson, P.A. The trans-Golgi network is a major site for α-secretase processing of amyloid precursor protein in primary neurons. *J. Biol. Chem.* **2019**, *294*, 1618–1631. [CrossRef] [PubMed]
14. Marcello, E.; Gardoni, F.; Mauceri, D.; Romorini, S.; Jeromin, A.; Epis, R.; Borroni, B.; Cattabeni, F.; Sala, C.; Padovani, A.; et al. Synapse-associated protein-97 mediates alpha-secretase ADAM10 trafficking and promotes its activity. *J. Neurosci.* **2007**, *27*, 1682–1691. [CrossRef]
15. Marcello, E.; Saraceno, C.; Musardo, S.; Vara, H.; de la Fuente, A.G.; Pelucchi, S.; Di Marino, D.; Borroni, B.; Tramontano, A.; Pérez-Otaño, I.; et al. Endocytosis of synaptic ADAM10 in neuronal plasticity and Alzheimer's disease. *J. Clin. Invest.* **2013**, *123*, 2523–2538. [CrossRef]
16. Saraceno, C.; Marcello, E.; Di Marino, D.; Borroni, B.; Claeysen, S.; Perroy, J.; Padovani, A.; Tramontano, A.; Gardoni, F.; Di Luca, M. SAP97-mediated ADAM10 trafficking from Golgi outposts depends on PKC phosphorylation. *Cell Death Dis.* **2014**, *5*, 1547. [CrossRef]
17. Marcello, E.; Musardo, S.; Vandermeulen, L.; Pelucchi, S.; Gardoni, F.; Santo, N.; Antonucci, F.; Di Luca, M. Amyloid-β Oligomers Regulate ADAM10 Synaptic Localization Through Aberrant Plasticity Phenomena. *Mol. Neurobiol.* **2019**, *27*, 457. [CrossRef]
18. Tan, J.Z.A.; Fourriere, L.; Wang, J.; Perez, F.; Boncompain, G.; Gleeson, P.A. Distinct anterograde trafficking pathways of BACE1 and amyloid precursor protein from the TGN and the regulation of amyloid-β production. *Mol. Biol. Cell* **2019**, *31*, 27–44. [CrossRef]

19. Choi, S.H.; Kim, Y.H.; Hebisch, M.; Sliwinski, C.; Lee, S.; D'Avanzo, C.; Chen, H.; Hooli, B.; Asselin, C.; Muffat, J.; et al. A three-dimensional human neural cell culture model of Alzheimer's disease. *Nature* **2014**, *515*, 274–278. [CrossRef]
20. Bolmont, T.; Clavaguera, F.; Meyer-Luehmann, M.; Herzig, M.C.; Radde, R.; Staufenbiel, M.; Lewis, J.; Hutton, M.; Tolnay, M.; Jucker, M. Induction of tau pathology by intracerebral infusion of amyloid-beta -containing brain extract and by amyloid-beta deposition in APP x Tau transgenic mice. *Am. J. Pathol.* **2007**, *171*, 2012–2020. [CrossRef]
21. Götz, J.; Chen, F.; van Dorpe, J.; Nitsch, R.M. Formation of neurofibrillary tangles in P301l tau transgenic mice induced by Abeta 42 fibrils. *Science* **2001**, *293*, 1491–1495. [CrossRef] [PubMed]
22. Hurtado, D.E.; Molina-Porcel, L.; Iba, M.; Aboagye, A.K.; Paul, S.M.; Trojanowski, J.Q.; Lee, V.M.-Y. A{beta} accelerates the spatiotemporal progression of tau pathology and augments tau amyloidosis in an Alzheimer mouse model. *Am. J. Pathol.* **2010**, *177*, 1977–1988. [CrossRef] [PubMed]
23. Lewis, J.; Dickson, D.W.; Lin, W.L.; Chisholm, L.; Corral, A.; Jones, G.; Yen, S.H.; Sahara, N.; Skipper, L.; Yager, D.; et al. Enhanced neurofibrillary degeneration in transgenic mice expressing mutant tau and APP. *Science* **2001**, *293*, 1487–1491. [CrossRef] [PubMed]
24. Pooler, A.M.; Polydoro, M.; Maury, E.A.; Nicholls, S.B.; Reddy, S.M.; Wegmann, S.; William, C.; Saqran, L.; Cagsal-Getkin, O.; Pitstick, R.; et al. Amyloid accelerates tau propagation and toxicity in a model of early Alzheimer's disease. *Acta Neuropathol. Commun.* **2015**, *3*, 14. [CrossRef] [PubMed]
25. Wei, W.; Nguyen, L.N.; Kessels, H.W.; Hagiwara, H.; Sisodia, S.; Malinow, R. Amyloid beta from axons and dendrites reduces local spine number and plasticity. *Nat. Neurosci.* **2010**, *13*, 190–196. [CrossRef] [PubMed]
26. Shankar, G.M.; Bloodgood, B.L.; Townsend, M.; Walsh, D.M.; Selkoe, D.J.; Sabatini, B.L. Natural oligomers of the Alzheimer amyloid-beta protein induce reversible synapse loss by modulating an NMDA-type glutamate receptor-dependent signaling pathway. *J. Neurosci.* **2007**, *27*, 2866–2875. [CrossRef]
27. Shankar, G.M.; Li, S.; Mehta, T.H.; Garcia-Munoz, A.; Shepardson, N.E.; Smith, I.; Brett, F.M.; Farrell, M.A.; Rowan, M.J.; Lemere, C.A.; et al. Amyloid-beta protein dimers isolated directly from Alzheimer's brains impair synaptic plasticity and memory. *Nat. Med.* **2008**, *14*, 837–842. [CrossRef]
28. Selkoe, D.J. Alzheimer's disease is a synaptic failure. *Science* **2002**, *298*, 789–791. [CrossRef]
29. Jack, C.R.; Knopman, D.S.; Jagust, W.J.; Shaw, L.M.; Aisen, P.S.; Weiner, M.W.; Petersen, R.C.; Trojanowski, J.Q. Hypothetical model of dynamic biomarkers of the Alzheimer's pathological cascade. *Lancet Neurol.* **2010**, *9*, 119–128. [CrossRef]
30. Scheff, S.W.; Price, D.A.; Schmitt, F.A.; Mufson, E.J. Hippocampal synaptic loss in early Alzheimer's disease and mild cognitive impairment. *Neurobiol. Aging* **2006**, *27*, 1372–1384. [CrossRef]
31. Bourne, J.N.; Harris, K.M. Balancing Structure and Function at Hippocampal Dendritic Spines. *Annu. Rev. Neurosci.* **2008**, *31*, 47–67. [CrossRef] [PubMed]
32. Sala, C.; Segal, M. Dendritic spines: The locus of structural and functional plasticity. *Physiol. Rev.* **2014**, *94*, 141–188. [CrossRef] [PubMed]
33. Matus, A. Actin-based plasticity in dendritic spines. *Science* **2000**, *290*, 754–758. [CrossRef] [PubMed]
34. Landis, D.M.; Reese, T.S. Cytoplasmic organization in cerebellar dendritic spines. *J. Cell Biol.* **1983**, *97*, 1169–1178. [CrossRef] [PubMed]
35. Hotulainen, P.; Hoogenraad, C.C. Actin in dendritic spines: Connecting dynamics to function. *J. Cell Biol.* **2010**, *189*, 619–629. [CrossRef] [PubMed]
36. Pollard, T.D. The cytoskeleton, cellular motility and the reductionist agenda. *Nature* **2003**, *422*, 741–745. [CrossRef] [PubMed]
37. Skruber, K.; Read, T.-A.; Vitriol, E.A. Reconsidering an active role for G-actin in cytoskeletal regulation. *J. Cell. Sci.* **2018**, *131*, jcs203760. [CrossRef]
38. Dugina, V.B.; Shagieva, G.S.; Kopnin, P.B. Biological Role of Actin Isoforms in Mammalian Cells. *Biochem. Mosc.* **2019**, *84*, 583–592. [CrossRef]
39. Hlushchenko, I.; Koskinen, M.; Hotulainen, P. Dendritic spine actin dynamics in neuronal maturation and synaptic plasticity. *Cytoskeleton* **2016**, *73*, 435–441. [CrossRef]
40. Sheng, M.; Hoogenraad, C.C. The Postsynaptic Architecture of Excitatory Synapses: A More Quantitative View. *Annu. Rev. Biochem.* **2007**, *76*, 823–847. [CrossRef]

41. MacGillavry, H.D.; Kerr, J.M.; Kassner, J.; Frost, N.A.; Blanpied, T.A. Shank-cortactin interactions control actin dynamics to maintain flexibility of neuronal spines and synapses. *Eur. J. Neurosci.* **2016**, *43*, 179–193. [CrossRef] [PubMed]
42. Matt, L.; Kim, K.; Hergarden, A.C.; Patriarchi, T.; Malik, Z.A.; Park, D.K.; Chowdhury, D.; Buonarati, O.R.; Henderson, P.B.; Gökçek Saraç, Ç.; et al. α-Actinin Anchors PSD-95 at Postsynaptic Sites. *Neuron* **2018**, *97*, 1094–1109. [CrossRef] [PubMed]
43. Cingolani, L.A.; Goda, Y. Actin in action: The interplay between the actin cytoskeleton and synaptic efficacy. *Nat. Rev. Neurosci.* **2008**, *9*, 344–356. [CrossRef] [PubMed]
44. Xu, K.; Zhong, G.; Zhuang, X. Actin, spectrin, and associated proteins form a periodic cytoskeletal structure in axons. *Science* **2013**, *339*, 452–456. [CrossRef]
45. Bär, J.; Kobler, O.; van Bommel, B.; Mikhaylova, M. Periodic F-actin structures shape the neck of dendritic spines. *Sci. Rep.* **2016**, *6*, 37136. [CrossRef] [PubMed]
46. Bucher, M.; Fanutza, T.; Mikhaylova, M. Cytoskeletal makeup of the synapse: Shaft vs. spine. *Cytoskelet* **2019**, cm.21583. [CrossRef] [PubMed]
47. Pollard, T.D. Actin and Actin-Binding Proteins. *Cold Spring Harb. Perspect. Biol.* **2016**, *8*, a018226. [CrossRef]
48. Ichetovkin, I.; Grant, W.; Condeelis, J. Cofilin produces newly polymerized actin filaments that are preferred for dendritic nucleation by the Arp2/3 complex. *Curr. Biol.* **2002**, *12*, 79–84. [CrossRef]
49. Goley, E.D.; Ohkawa, T.; Mancuso, J.; Woodruff, J.B.; D'Alessio, J.A.; Cande, W.Z.; Volkman, L.E.; Welch, M.D. Dynamic nuclear actin assembly by Arp2/3 complex and a baculovirus WASP-like protein. *Science* **2006**, *314*, 464–467. [CrossRef]
50. Wegner, A. Head to tail polymerization of actin. *J. Mol. Biol.* **1976**, *108*, 139–150. [CrossRef]
51. Hotulainen, P.; Llano, O.; Smirnov, S.; Tanhuanpää, K.; Faix, J.; Rivera, C.; Lappalainen, P. Defining mechanisms of actin polymerization and depolymerization during dendritic spine morphogenesis. *J. Cell Biol.* **2009**, *185*, 323–339. [CrossRef] [PubMed]
52. Haeckel, A.; Ahuja, R.; Gundelfinger, E.D.; Qualmann, B.; Kessels, M.M. The actin-binding protein Abp1 controls dendritic spine morphology and is important for spine head and synapse formation. *J. Neurosci.* **2008**, *28*, 10031–10044. [CrossRef]
53. Hering, H.; Sheng, M. Activity-dependent redistribution and essential role of cortactin in dendritic spine morphogenesis. *J. Neurosci.* **2003**, *23*, 11759–11769. [CrossRef]
54. Soderling, S.H.; Guire, E.S.; Kaech, S.; White, J.; Zhang, F.; Schutz, K.; Langeberg, L.K.; Banker, G.; Raber, J.; Scott, J.D. A WAVE-1 and WRP signaling complex regulates spine density, synaptic plasticity, and memory. *J. Neurosci.* **2007**, *27*, 355–365. [CrossRef] [PubMed]
55. Wegner, A.M.; Nebhan, C.A.; Hu, L.; Majumdar, D.; Meier, K.M.; Weaver, A.M.; Webb, D.J. N-wasp and the arp2/3 complex are critical regulators of actin in the development of dendritic spines and synapses. *J. Biol. Chem.* **2008**, *283*, 15912–15920. [CrossRef] [PubMed]
56. Guo, S.; Sokolova, O.S.; Chung, J.; Padrickm, S.; Gelles, J.; Goode, B.L. Abp1 promotes Arp2/3 complex-dependent actin nucleation and stabilizes branch junctions by antagonizing GMF. *Nat. Commun.* **2018**, *24*, 2895. [CrossRef] [PubMed]
57. Yarmola, E.G.; Bubb, M.R. How depolymerization can promote polymerization: The case of actin and profilin. *Bioessays* **2009**, *31*, 1150–1160. [CrossRef]
58. Witke, W.; Sutherland, J.D.; Sharpe, A.; Arai, M.; Kwiatkowski, D.J. Profilin I is essential for cell survival and cell division in early mouse development. *Proc. Natl. Acad. Sci. USA* **2001**, *98*, 3832–3836. [CrossRef]
59. Bravo-Cordero, J.J.; Magalhaes, M.A.O.; Eddy, R.J.; Hodgson, L.; Condeelis, J. Functions of cofilin in cell locomotion and invasion. *Nat. Rev. Mol. Cell Biol.* **2013**, *14*, 405–415. [CrossRef]
60. Bamburg, J.R.; Bernstein, B.W. Actin dynamics and cofilin-actin rods in alzheimer disease. *Cytoskelet* **2016**, *73*, 477–497. [CrossRef]
61. Hotulainen, P.; Paunola, E.; Vartiainen, M.K.; Lappalainen, P. Actin-depolymerizing factor and cofilin-1 play overlapping roles in promoting rapid F-actin depolymerization in mammalian nonmuscle cells. *Mol. Biol. Cell* **2005**, *16*, 649–664. [CrossRef] [PubMed]
62. Arber, S.; Barbayannis, F.A.; Hanser, H.; Schneider, C.; Stanyon, C.A.; Bernard, O.; Caroni, P. Regulation of actin dynamics through phosphorylation of cofilin by LIM-kinase. *Nature* **1998**, *393*, 805–809. [CrossRef] [PubMed]

63. Niwa, R.; Nagata-Ohashi, K.; Takeichi, M.; Mizuno, K.; Uemura, T. Control of actin reorganization by Slingshot, a family of phosphatases that dephosphorylate ADF/cofilin. *Cell* **2002**, *108*, 233–246. [CrossRef]
64. Okreglak, V.; Drubin, D.G. Loss of Aip1 reveals a role in maintaining the actin monomer pool and an in vivo oligomer assembly pathway. *J. Cell Biol.* **2010**, *188*, 769–777. [CrossRef] [PubMed]
65. Ono, S. Regulation of actin filament dynamics by actin depolymerizing factor/cofilin and actin-interacting protein 1: New blades for twisted filaments. *Biochemistry* **2003**, *42*, 13363–13370. [CrossRef] [PubMed]
66. Nadkarni, A.V.; Brieher, W.M. Aip1 destabilizes cofilin-saturated actin filaments by severing and accelerating monomer dissociation from ends. *Curr. Biol.* **2014**, *24*, 2749–2757. [CrossRef] [PubMed]
67. Ono, S. The role of cyclase-associated protein in regulating actin filament dynamics-more than a monomer-sequestration factor. *J. Cell. Sci.* **2013**, *126*, 3249–3258. [CrossRef] [PubMed]
68. Moriyama, K.; Yahara, I. Human CAP1 is a key factor in the recycling of cofilin and actin for rapid actin turnover. *J. Cell. Sci.* **2002**, *115*, 1591–1601.
69. Quintero-Monzon, O.; Jonasson, E.M.; Bertling, E.; Talarico, L.; Chaudhry, F.; Sihvo, M.; Lappalainen, P.; Goode, B.L. Reconstitution and dissection of the 600-kDa Srv2/CAP complex: Roles for oligomerization and cofilin-actin binding in driving actin turnover. *J. Biol. Chem.* **2009**, *284*, 10923–10934. [CrossRef]
70. Kotila, T.; Wioland, H.; Enkavi, G.; Kogan, K.; Vattulainen, I.; Jégou, A.; Romet-Lemonne, G.; Lappalainen, P. Mechanism of synergistic actin filament pointed end depolymerization by cyclase-associated protein and cofilin. *Nat. Commun.* **2019**, *10*, 5320. [CrossRef]
71. Shekhar, S.; Chung, J.; Kondev, J.; Gelles, J.; Goode, B.L. Synergy between Cyclase-associated protein and Cofilin accelerates actin filament depolymerization by two orders of magnitude. *Nat. Commun.* **2019**, *10*, 5319. [CrossRef] [PubMed]
72. Swiston, J.; Hubberstey, A.; Yu, G.; Young, D. Differential expression of CAP and CAP2 in adult rat tissues. *Gene* **1995**, *165*, 273–277. [CrossRef]
73. Kumar, A.; Paeger, L.; Kosmas, K.; Kloppenburg, P.; Noegel, A.A.; Peche, V.S. Neuronal Actin Dynamics, Spine Density and Neuronal Dendritic Complexity Are Regulated by CAP2. *Front. Cell Neurosci.* **2016**, *10*, 180. [CrossRef] [PubMed]
74. Edwards, T.J.; Hammarlund, M. Syndecan promotes axon regeneration by stabilizing growth cone migration. *Cell Rep.* **2014**, *8*, 272–283. [CrossRef] [PubMed]
75. Menna, E.; Disanza, A.; Cagnoli, C.; Schenk, U.; Gelsomino, G.; Frittoli, E.; Hertzog, M.; Offenhauser, N.; Sawallisch, C.; Kreienkamp, H.-J.; et al. Eps8 regulates axonal filopodia in hippocampal neurons in response to brain-derived neurotrophic factor (BDNF). *PLoS Biol.* **2009**, *7*, e1000138. [CrossRef] [PubMed]
76. Kneussel, M.; Wagner, W. Myosin motors at neuronal synapses: Drivers of membrane transport and actin dynamics. *Nat. Rev. Neurosci.* **2013**, *14*, 233–247. [CrossRef]
77. Soldati, T.; Schliwa, M. Powering membrane traffic in endocytosis and recycling. *Nat. Rev. Mol. Cell Biol.* **2006**, *7*, 897–908. [CrossRef]
78. Hartman, M.A.; Finan, D.; Sivaramakrishnan, S.; Spudich, J.A. Principles of unconventional myosin function and targeting. *Annu. Rev. Cell Dev. Biol.* **2011**, *27*, 133–155. [CrossRef]
79. Gunning, P.; O'Neill, G.; Hardeman, E. Tropomyosin-based regulation of the actin cytoskeleton in time and space. *Physiol. Rev.* **2008**, *88*, 1–35. [CrossRef]
80. Brettle, M.; Patel, S.; Fath, T. Tropomyosins in the healthy and diseased nervous system. *Brain Res. Bull.* **2016**, *126*, 311–323. [CrossRef]
81. Gunning, P.W.; Hardeman, E.C.; Lappalainen, P.; Mulvihill, D.P. Tropomyosin-master regulator of actin filament function in the cytoskeleton. *J. Cell. Sci.* **2015**, *128*, 2965–2974. [CrossRef]
82. Blanchoin, L.; Pollard, T.D.; Hitchcock-DeGregori, S.E. Inhibition of the Arp2/3 complex-nucleated actin polymerization and branch formation by tropomyosin. *Curr. Biol.* **2001**, *11*, 1300–1304. [CrossRef]
83. Ono, S.; Ono, K. Tropomyosin inhibits ADF/cofilin-dependent actin filament dynamics. *J. Cell Biol.* **2002**, *156*, 1065–1076. [CrossRef]
84. Faivre-Sarrailh, C.; Had, L.; Ferraz, C.; Sri Widada, J.S.; Liautard, J.P.; Rabié, A. Expression of tropomyosin genes during the development of the rat cerebellum. *J. Neurochem.* **1990**, *55*, 899–906. [CrossRef]
85. Weinberger, R.; Schevzov, G.; Jeffrey, P.; Gordon, K.; Hill, M.; Gunning, P. The molecular composition of neuronal microfilaments is spatially and temporally regulated. *J. Neurosci.* **1996**, *16*, 238–252. [CrossRef]
86. Dufour, C.; Weinberger, R.P.; Gunning, P. Tropomyosin isoform diversity and neuronal morphogenesis. *Immunol. Cell Biol.* **1998**, *76*, 424–429. [CrossRef] [PubMed]

87. Hook, J.; Lemckert, F.; Qin, H.; Schevzov, G.; Gunning, P. Gamma tropomyosin gene products are required for embryonic development. *Mol. Cell. Biol.* **2004**, *24*, 2318–2323. [CrossRef] [PubMed]
88. Schevzov, G.; Bryce, N.S.; Almonte-Baldonado, R.; Joya, J.; Lin, J.J.-C.; Hardeman, E.; Weinberger, R.; Gunning, P. Specific features of neuronal size and shape are regulated by tropomyosin isoforms. *Mol. Biol. Cell* **2005**, *16*, 3425–3437. [CrossRef]
89. Guven, K.; Gunning, P.; Fath, T. TPM3 and TPM4 gene products segregate to the postsynaptic region of central nervous system synapses. *Bioarchitecture* **2011**, *1*, 284–289. [CrossRef]
90. Koganezawa, N.; Hanamura, K.; Sekino, Y.; Shirao, T. The role of drebrin in dendritic spines. *Mol. Cell. Neurosci.* **2017**, *84*, 85–92. [CrossRef]
91. Hayashi, K.; Ishikawa, R.; Ye, L.H.; He, X.L.; Takata, K.; Kohama, K.; Shirao, T. Modulatory role of drebrin on the cytoskeleton within dendritic spines in the rat cerebral cortex. *J. Neurosci.* **1996**, *16*, 7161–7170. [CrossRef] [PubMed]
92. Mikati, M.A.; Grintsevich, E.E.; Reisler, E. Drebrin-induced stabilization of actin filaments. *J. Biol. Chem.* **2013**, *288*, 19926–19938. [CrossRef] [PubMed]
93. Worth, D.C.; Daly, C.N.; Geraldo, S.; Oozeer, F.; Gordon-Weeks, P.R. Drebrin contains a cryptic F-actin-bundling activity regulated by Cdk5 phosphorylation. *J. Cell Biol.* **2013**, *202*, 793–806. [CrossRef] [PubMed]
94. Gordon-Weeks, P.R. The role of the drebrin/EB3/Cdk5 pathway in dendritic spine plasticity, implications for Alzheimer's disease. *Brain Res. Bull.* **2016**, *126*, 293–299. [CrossRef]
95. Colbran, R.J.; Soderling, T.R. Calcium/calmodulin-dependent protein kinase II. *Curr. Top. Cell. Regul.* **1990**, *31*, 181–221.
96. Lisman, J.; Schulman, H.; Cline, H. The molecular basis of CaMKII function in synaptic and behavioural memory. *Nat. Rev. Neurosci.* **2002**, *3*, 175–190. [CrossRef]
97. Bennett, M.K.; Erondu, N.E.; Kennedy, M.B. Purification and characterization of a calmodulin-dependent protein kinase that is highly concentrated in brain. *J. Biol. Chem.* **1983**, *258*, 12735–12744.
98. Okamoto, K.-I.; Narayanan, R.; Lee, S.H.; Murata, K.; Hayashi, Y. The role of CaMKII as an F-actin-bundling protein crucial for maintenance of dendritic spine structure. *Proc. Natl. Acad. Sci. USA* **2007**, *104*, 6418–6423. [CrossRef]
99. Khan, S.; Conte, I.; Carter, T.; Bayer, K.U.; Molloy, J.E. Multiple CaMKII Binding Modes to the Actin Cytoskeleton Revealed by Single-Molecule Imaging. *Biophys. J.* **2016**, *111*, 395–408. [CrossRef]
100. Brocke, L.; Chiang, L.W.; Wagner, P.D.; Schulman, H. Functional implications of the subunit composition of neuronal CaM kinase II. *J. Biol. Chem.* **1999**, *274*, 22713–22722. [CrossRef]
101. Yamazaki, H.; Sasagawa, Y.; Yamamoto, H.; Bito, H.; Shirao, T. CaMKIIβ is localized in dendritic spines as both drebrin-dependent and drebrin-independent pools. *J. Neurochem.* **2018**, *146*, 145–159. [CrossRef] [PubMed]
102. Djinović-Carugo, K.; Young, P.; Gautel, M.; Saraste, M. Structure of the alpha-actinin rod: Molecular basis for cross-linking of actin filaments. *Cell* **1999**, *98*, 537–546. [CrossRef]
103. Hodges, J.L.; Vilchez, S.M.; Asmussen, H.; Whitmore, L.A.; Horwitz, A.R. α-Actinin-2 mediates spine morphology and assembly of the post-synaptic density in hippocampal neurons. *PLoS ONE* **2014**, *9*, e101770. [CrossRef] [PubMed]
104. Jalan-Sakrikar, N.; Bartlett, R.K.; Baucum, A.J.; Colbran, R.J. Substrate-selective and calcium-independent activation of CaMKII by α-actinin. *J. Biol. Chem.* **2012**, *287*, 15275–15283. [CrossRef] [PubMed]
105. Kalinowska, M.; Chávez, A.E.; Lutzu, S.; Castillo, P.E.; Bukauskas, F.F.; Francesconi, A. Actinin-4 Governs Dendritic Spine Dynamics and Promotes Their Remodeling by Metabotropic Glutamate Receptors. *J. Biol. Chem.* **2015**, *290*, 15909–15920. [CrossRef] [PubMed]
106. Holtmaat, A.; Svoboda, K. Experience-dependent structural synaptic plasticity in the mammalian brain. *Nat. Rev. Neurosci.* **2009**, *10*, 647–658. [CrossRef]
107. Renner, M.; Choquet, D.; Triller, A. Control of the postsynaptic membrane viscosity. *J. Neurosci.* **2009**, *29*, 2926–2937. [CrossRef]
108. Okamoto, K.-I.; Nagai, T.; Miyawaki, A.; Hayashi, Y. Rapid and persistent modulation of actin dynamics regulates postsynaptic reorganization underlying bidirectional plasticity. *Nat. Neurosci.* **2004**, *7*, 1104–1112. [CrossRef]

109. Bosch, M.; Castro, J.; Saneyoshi, T.; Matsuno, H.; Sur, M.; Hayashi, Y. Structural and molecular remodeling of dendritic spine substructures during long-term potentiation. *Neuron* **2014**, *82*, 444–459. [CrossRef]
110. Kasai, H.; Fukuda, M.; Watanabe, S.; Hayashi-Takagi, A.; Noguchi, J. Structural dynamics of dendritic spines in memory and cognition. *Trends Neurosci.* **2010**, *33*, 121–129. [CrossRef]
111. Sekino, Y.; Tanaka, S.; Hanamura, K.; Yamazaki, H.; Sasagawa, Y.; Xue, Y.; Hayashi, K.; Shirao, T. Activation of N-methyl-D-aspartate receptor induces a shift of drebrin distribution: Disappearance from dendritic spines and appearance in dendritic shafts. *Mol. Cell. Neurosci.* **2006**, *31*, 493–504. [CrossRef] [PubMed]
112. Oser, M.; Condeelis, J. The cofilin activity cycle in lamellipodia and invadopodia. *J. Cell. Biochem.* **2009**, *108*, 1252–1262. [CrossRef] [PubMed]
113. Hanley, J.G. Actin-dependent mechanisms in AMPA receptor trafficking. *Front. Cell Neurosci.* **2014**, *8*, 381. [CrossRef] [PubMed]
114. Okamoto, K.; Bosch, M.; Hayashi, Y. The roles of CaMKII and F-actin in the structural plasticity of dendritic spines: A potential molecular identity of a synaptic tag? *Physiol. (Bethesda)* **2009**, *24*, 357–366. [CrossRef]
115. Tada, T.; Sheng, M. Molecular mechanisms of dendritic spine morphogenesis. *Curr. Opin. Neurobiol.* **2006**, *16*, 95–101. [CrossRef]
116. Allison, D.W.; Gelfand, V.I.; Spector, I.; Craig, A.M. Role of actin in anchoring postsynaptic receptors in cultured hippocampal neurons: Differential attachment of NMDA versus AMPA receptors. *J. Neurosci.* **1998**, *18*, 2423–2436. [CrossRef]
117. Kim, C.H.; Lisman, J.E. A role of actin filament in synaptic transmission and long-term potentiation. *J. Neurosci.* **1999**, *19*, 4314–4324. [CrossRef]
118. Zhou, Q.; Xiao, M.; Nicoll, R.A. Contribution of cytoskeleton to the internalization of AMPA receptors. *Proc. Natl. Acad. Sci. USA* **2001**, *98*, 1261–1266. [CrossRef]
119. Malinow, R.; Malenka, R.C. AMPA receptor trafficking and synaptic plasticity. *Annu. Rev. Neurosci.* **2002**, *25*, 103–126. [CrossRef]
120. Choquet, D.; Triller, A. The dynamic synapse. *Neuron* **2013**, *80*, 691–703. [CrossRef]
121. Fujii, S.; Tanaka, H.; Hirano, T. Detection and characterization of individual endocytosis of AMPA-type glutamate receptor around postsynaptic membrane. *Genes Cells* **2017**, *22*, 583–590. [CrossRef]
122. Rosendale, M.; Jullié, D.; Choquet, D.; Perrais, D. Spatial and Temporal Regulation of Receptor Endocytosis in Neuronal Dendrites Revealed by Imaging of Single Vesicle Formation. *Cell Rep.* **2017**, *18*, 1840–1847. [CrossRef]
123. Lee, S.H.; Liu, L.; Wang, Y.T.; Sheng, M. Clathrin adaptor AP2 and NSF interact with overlapping sites of GluR2 and play distinct roles in AMPA receptor trafficking and hippocampal LTD. *Neuron* **2002**, *36*, 661–674. [CrossRef]
124. Loebrich, S.; Benoit, M.R.; Konopka, J.A.; Cottrell, J.R.; Gibson, J.; Nedivi, E. CPG2 Recruits Endophilin B2 to the Cytoskeleton for Activity-Dependent Endocytosis of Synaptic Glutamate Receptors. *Curr. Biol.* **2016**, *26*, 296–308. [CrossRef]
125. Kakegawa, W.; Katoh, A.; Narumi, S.; Miura, E.; Motohashi, J.; Takahashi, A.; Kohda, K.; Fukazawa, Y.; Yuzaki, M.; Matsuda, S. Optogenetic Control of Synaptic AMPA Receptor Endocytosis Reveals Roles of LTD in Motor Learning. *Neuron* **2018**, *99*, 985–998. [CrossRef]
126. Engqvist-Goldstein, A.E.Y.; Drubin, D.G. Actin assembly and endocytosis: From yeast to mammals. *Annu. Rev. Cell Dev. Biol.* **2003**, *19*, 287–332. [CrossRef]
127. Blanpied, T.A.; Scott, D.B.; Ehlers, M.D. Dynamics and regulation of clathrin coats at specialized endocytic zones of dendrites and spines. *Neuron* **2002**, *36*, 435–449. [CrossRef]
128. Brodsky, F.M.; Chen, C.Y.; Knuehl, C.; Towler, M.C.; Wakeham, D.E. Biological basket weaving: Formation and function of clathrin-coated vesicles. *Annu. Rev. Cell Dev. Biol.* **2001**, *17*, 517–568. [CrossRef]
129. Jarousse, N.; Kelly, R.B. Endocytotic mechanisms in synapses. *Curr. Opin. Cell Biol.* **2001**, *13*, 461–469. [CrossRef]
130. Mooren, O.L.; Galletta, B.J.; Cooper, J.A. Roles for actin assembly in endocytosis. *Annu. Rev. Biochem.* **2012**, *81*, 661–686. [CrossRef]
131. Bennett, E.M.; Chen, C.Y.; Engqvist-Goldstein, A.E.; Drubin, D.G.; Brodsky, F.M. Clathrin hub expression dissociates the actin-binding protein Hip1R from coated pits and disrupts their alignment with the actin cytoskeleton. *Traffic* **2001**, *2*, 851–858. [CrossRef] [PubMed]

132. Waelter, S.; Scherzinger, E.; Hasenbank, R.; Nordhoff, E.; Lurz, R.; Goehler, H.; Gauss, C.; Sathasivam, K.; Bates, G.P.; Lehrach, H.; et al. The huntingtin interacting protein HIP1 is a clathrin and alpha-adaptin-binding protein involved in receptor-mediated endocytosis. *Hum. Mol. Genet.* **2001**, *10*, 1807–1817. [CrossRef] [PubMed]
133. Yarar, D.; Waterman-Storer, C.M.; Schmid, S.L. A dynamic actin cytoskeleton functions at multiple stages of clathrin-mediated endocytosis. *Mol. Biol. Cell* **2005**, *16*, 964–975. [CrossRef] [PubMed]
134. Perrais, D.; Merrifield, C.J. Dynamics of endocytic vesicle creation. *Dev. Cell* **2005**, *9*, 581–592. [CrossRef] [PubMed]
135. Merrifield, C.J.; Perrais, D.; Zenisek, D. Coupling between clathrin-coated-pit invagination, cortactin recruitment, and membrane scission observed in live cells. *Cell* **2005**, *121*, 593–606. [CrossRef]
136. Bryce, N.S.; Schevzov, G.; Ferguson, V.; Percival, J.M.; Lin, J.J.-C.; Matsumura, F.; Bamburg, J.R.; Jeffrey, P.L.; Hardeman, E.C.; Gunning, P.; et al. Specification of actin filament function and molecular composition by tropomyosin isoforms. *Mol. Biol. Cell* **2003**, *14*, 1002–1016. [CrossRef]
137. Gormal, R.; Valmas, N.; Fath, T.; Meunier, F. A role for tropomyosins in activity-dependent bulk endocytosis? *Mol. Cell. Neurosci.* **2017**, *84*, 112–118. [CrossRef]
138. Cottrell, J.R.; Borok, E.; Horvath, T.L.; Nedivi, E. CPG2: A brain- and synapse-specific protein that regulates the endocytosis of glutamate receptors. *Neuron* **2004**, *44*, 677–690.
139. Loebrich, S.; Djukic, B.; Tong, Z.J.; Cottrell, J.R.; Turrigiano, G.G.; Nedivi, E. Regulation of glutamate receptor internalization by the spine cytoskeleton is mediated by its PKA-dependent association with CPG2. *Proc. Natl. Acad. Sci. USA* **2013**, *110*, E4548–E4556. [CrossRef]
140. Shen, L.; Liang, F.; Walensky, L.D.; Huganir, R.L. Regulation of AMPA receptor GluR1 subunit surface expression by a 4. 1N-linked actin cytoskeletal association. *J. Neurosci.* **2000**, *20*, 7932–7940. [CrossRef]
141. Lin, D.-T.; Makino, Y.; Sharma, K.; Hayashi, T.; Neve, R.; Takamiya, K.; Huganir, R.L. Regulation of AMPA receptor extrasynaptic insertion by 4.1N, phosphorylation and palmitoylation. *Nat. Neurosci.* **2009**, *12*, 879–887. [CrossRef] [PubMed]
142. Kim, C.H.; Chung, H.J.; Lee, H.K.; Huganir, R.L. Interaction of the AMPA receptor subunit GluR2/3 with PDZ domains regulates hippocampal long-term depression. *Proc. Natl. Acad. Sci. USA* **2001**, *98*, 11725–11730. [CrossRef] [PubMed]
143. Rocca, D.L.; Martin, S.; Jenkins, E.L.; Hanley, J.G. Inhibition of Arp2/3-mediated actin polymerization by PICK1 regulates neuronal morphology and AMPA receptor endocytosis. *Nat. Cell Biol.* **2008**, *10*, 259–271. [CrossRef] [PubMed]
144. Schulz, T.W.; Nakagawa, T.; Licznerski, P.; Pawlak, V.; Kolleker, A.; Rozov, A.; Kim, J.; Dittgen, T.; Köhr, G.; Sheng, M.; et al. Actin/alpha-actinin-dependent transport of AMPA receptors in dendritic spines: Role of the PDZ-LIM protein RIL. *J. Neurosci.* **2004**, *24*, 8584–8594. [CrossRef]
145. Parkinson, G.T.; Chamberlain, S.E.L.; Jaafari, N.; Turvey, M.; Mellor, J.R.; Hanley, J.G. Cortactin regulates endo-lysosomal sorting of AMPARs via direct interaction with GluA2 subunit. *Sci. Rep.* **2018**, *8*, 4155. [CrossRef]
146. Correia, S.S.; Bassani, S.; Brown, T.C.; Lisé, M.-F.; Backos, D.S.; El-Husseini, A.; Passafaro, M.; Esteban, J.A. Motor protein-dependent transport of AMPA receptors into spines during long-term potentiation. *Nat. Neurosci.* **2008**, *11*, 457–466. [CrossRef]
147. Osterweil, E.; Wells, D.G.; Mooseker, M.S. A role for myosin VI in postsynaptic structure and glutamate receptor endocytosis. *J. Cell Biol.* **2005**, *168*, 329–338. [CrossRef]
148. Sun, J.; Roy, S. The physical approximation of APP and BACE-1: A key event in alzheimer's disease pathogenesis. *Dev. Neurobiol.* **2018**, *78*, 340–347. [CrossRef]
149. Greenfield, J.P.; Tsai, J.; Gouras, G.K.; Hai, B.; Thinakaran, G.; Checler, F.; Sisodia, S.S.; Greengard, P.; Xu, H. Endoplasmic reticulum and trans-Golgi network generate distinct populations of Alzheimer beta-amyloid peptides. *Proc. Natl. Acad. Sci. USA* **1999**, *96*, 742–747. [CrossRef]
150. Das, U.; Wang, L.; Ganguly, A.; Saikia, J.M.; Wagner, S.L.; Koo, E.H.; Roy, S. Visualizing APP and BACE-1 approximation in neurons yields insight into the amyloidogenic pathway. *Nat. Neurosci.* **2016**, *19*, 55–64. [CrossRef]
151. Gowrishankar, S.; Wu, Y.; Ferguson, S.M. Impaired JIP3-dependent axonal lysosome transport promotes amyloid plaque pathology. *J. Cell Biol.* **2017**, *216*, 3291–3305. [CrossRef] [PubMed]

152. Lai, A.; Sisodia, S.S.; Trowbridge, I.S. Characterization of sorting signals in the beta-amyloid precursor protein cytoplasmic domain. *J. Biol. Chem.* **1995**, *270*, 3565–3573. [CrossRef] [PubMed]
153. Vieira, S.I.; Rebelo, S.; Esselmann, H.; Wiltfang, J.; Lah, J.; Lane, R.; Small, S.A.; Gandy, S.; da Cruz, E.; Silva, E.F.; et al. Retrieval of the Alzheimer's amyloid precursor protein from the endosome to the TGN is S655 phosphorylation state-dependent and retromer-mediated. *Mol. Neurodegener.* **2010**, *5*, 40. [CrossRef] [PubMed]
154. Andersen, O.M.; Reiche, J.; Schmidt, V.; Gotthardt, M.; Spoelgen, R.; Behlke, J.; von Arnim, C.A.F.; Breiderhoff, T.; Jansen, P.; Wu, X.; et al. Neuronal sorting protein-related receptor sorLA/LR11 regulates processing of the amyloid precursor protein. *Proc. Natl. Acad. Sci. USA* **2005**, *102*, 13461–13466. [CrossRef]
155. Fjorback, A.W.; Seaman, M.; Gustafsen, C.; Mehmedbasic, A.; Gokool, S.; Wu, C.; Militz, D.; Schmidt, V.; Madsen, P.; Nyengaard, J.R.; et al. Retromer binds the FANSHY sorting motif in SorLA to regulate amyloid precursor protein sorting and processing. *J. Neurosci.* **2012**, *32*, 1467–1480. [CrossRef]
156. Lambert, J.C.; Ibrahim-Verbaas, C.A.; Harold, D.; Naj, A.C.; Sims, R.; Bellenguez, C.; DeStafano, A.L.; Bis, J.C.; Beecham, G.W.; Grenier-Boley, B.; et al. Meta-analysis of 74,046 individuals identifies 11 new susceptibility loci for Alzheimer's disease. *Nat. Genet.* **2013**, *45*, 1452–1458. [CrossRef]
157. Liu, Y.; Xu, Y.-F.; Zhang, L.; Huang, L.; Yu, P.; Zhu, H.; Deng, W.; Qin, C. Effective expression of Drebrin in hippocampus improves cognitive function and alleviates lesions of Alzheimer's disease in APP (swe)/PS1 (ΔE9) mice. *CNS Neurosci.* **2017**, *23*, 590–604. [CrossRef]
158. Harigaya, Y.; Shoji, M.; Shirao, T.; Hirai, S. Disappearance of actin-binding protein, drebrin, from hippocampal synapses in Alzheimer's disease. *J. Neurosci. Res.* **1996**, *43*, 87–92. [CrossRef]
159. Counts, S.E.; Nadeem, M.; Lad, S.P.; Wuu, J.; Mufson, E.J. Differential expression of synaptic proteins in the frontal and temporal cortex of elderly subjects with mild cognitive impairment. *J. Neuropathol. Exp. Neurol.* **2006**, *65*, 592–601. [CrossRef]
160. Hatanpää, K.; Isaacs, K.R.; Shirao, T.; Brady, D.R.; Rapoport, S.I. Loss of proteins regulating synaptic plasticity in normal aging of the human brain and in Alzheimer disease. *J. Neuropathol. Exp. Neurol.* **1999**, *58*, 637–643. [CrossRef]
161. Mendoza-Naranjo, A.; Gonzalez-Billault, C.; Maccioni, R.B. Abeta1-42 stimulates actin polymerization in hippocampal neurons through Rac1 and Cdc42 Rho GTPases. *J. Cell. Sci.* **2007**, *120*, 279–288. [CrossRef] [PubMed]
162. Borin, M.; Saraceno, C.; Catania, M.; Lorenzetto, E.; Pontelli, V.; Paterlini, A.; Fostinelli, S.; Avesani, A.; Di Fede, G.; Zanusso, G.; et al. Rac1 activation links tau hyperphosphorylation and Aβ dysmetabolism in Alzheimer's disease. *Acta Neuropathol. Commun.* **2018**, *6*, 61. [CrossRef] [PubMed]
163. Galloway, P.G.; Mulvihill, P.; Siedlak, S.; Mijares, M.; Kawai, M.; Padget, H.; Kim, R.; Perry, G. Immunochemical demonstration of tropomyosin in the neurofibrillary pathology of Alzheimer's disease. *Am. J. Pathol.* **1990**, *137*, 291–300. [PubMed]
164. Feuillette, S.; Deramecourt, V.; Laquerriere, A.; Duyckaerts, C.; Delisle, M.-B.; Maurage, C.-A.; Blum, D.; Buée, L.; Frébourg, T.; Campion, D.; et al. Filamin-A and Myosin VI colocalize with fibrillary Tau protein in Alzheimer's disease and FTDP-17 brains. *Brain Res.* **2010**, *1345*, 182–189. [CrossRef] [PubMed]
165. Hirano, A. Hirano bodies and related neuronal inclusions. *Neuropathol. Appl. Neurobiol.* **1994**, *20*, 3–11. [CrossRef]
166. Mitake, S.; Ojika, K.; Hirano, A. Hirano bodies and Alzheimer's disease. *Kaohsiung J. Med. Sci.* **1997**, *13*, 10–18.
167. Galloway, P.G.; Perry, G.; Gambetti, P. Hirano body filaments contain actin and actin-associated proteins. *J. Neuropathol. Exp. Neurol.* **1987**, *46*, 185–199. [CrossRef]
168. Maciver, S.K.; Harrington, C.R. Two actin binding proteins, actin depolymerizing factor and cofilin, are associated with Hirano bodies. *Neuroreport* **1995**, *6*, 1985–1988. [CrossRef]
169. Castaño, E.M.; Maarouf, C.L.; Wu, T.; Leal, M.C.; Whiteside, C.M.; Lue, L.-F.; Kokjohn, T.A.; Sabbagh, M.N.; Beach, T.G.; Roher, A.E. Alzheimer disease periventricular white matter lesions exhibit specific proteomic profile alterations. *Neurochem. Int.* **2013**, *62*, 145–156. [CrossRef]
170. Owen, J.B.; Di Domenico, F.; Sultana, R.; Perluigi, M.; Cini, C.; Pierce, W.M.; Butterfield, D.A. Proteomics-determined differences in the concanavalin-A-fractionated proteome of hippocampus and inferior parietal lobule in subjects with Alzheimer's disease and mild cognitive impairment: Implications for progression of AD. *J. Proteome Res.* **2009**, *8*, 471–482. [CrossRef]

171. Dong, Y.; Shahid-Salles, S.; Sherling, D.; Fechheimer, N.; Iyer, N.; Wells, L.; Fechheimer, M.; Furukawa, R. De novo actin polymerization is required for model Hirano body formation in Dictyostelium. *Biol. Open* **2016**, *5*, 807–818. [CrossRef] [PubMed]

172. Minamide, L.S.; Striegl, A.M.; Boyle, J.A.; Meberg, P.J.; Bamburg, J.R. Neurodegenerative stimuli induce persistent ADF/cofilin-actin rods that disrupt distal neurite function. *Nat. Cell Biol.* **2000**, *2*, 628–636. [CrossRef] [PubMed]

173. Rahman, T.; Davies, D.S.; Tannenberg, R.K.; Fok, S.; Shepherd, C.; Dodd, P.R.; Cullen, K.M.; Goldsbury, C. Cofilin rods and aggregates concur with tau pathology and the development of Alzheimer's disease. *J. Alzheimers Dis.* **2014**, *42*, 1443–1460. [CrossRef] [PubMed]

174. Maloney, M.T.; Minamide, L.S.; Kinley, A.W.; Boyle, J.A.; Bamburg, J.R. Beta-secretase-cleaved amyloid precursor protein accumulates at actin inclusions induced in neurons by stress or amyloid beta: A feedforward mechanism for Alzheimer's disease. *J. Neurosci.* **2005**, *25*, 11313–11321. [CrossRef]

175. Cichon, J.; Sun, C.; Chen, B.; Jiang, M.; Chen, X.A.; Sun, Y.; Wang, Y.; Chen, G. Cofilin aggregation blocks intracellular trafficking and induces synaptic loss in hippocampal neurons. *J. Biol. Chem.* **2012**, *287*, 3919–3929. [CrossRef]

176. Bernstein, B.W.; Shaw, A.E.; Minamide, L.S.; Pak, C.W.; Bamburg, J.R. Incorporation of cofilin into rods depends on disulfide intermolecular bonds: Implications for actin regulation and neurodegenerative disease. *J. Neurosci.* **2012**, *32*, 6670–6681. [CrossRef]

177. Chen, B.; Wang, Y. Cofilin rod formation in neurons impairs neuronal structure and function. *Cns Neurol. Disord. Drug Targets* **2015**, *14*, 554–560. [CrossRef]

178. Kang, D.E.; Woo, J.A. Cofilin, a Master Node Regulating Cytoskeletal Pathogenesis in Alzheimer's Disease. *J. Alzheimers Dis.* **2019**, *278*, 1–14. [CrossRef]

179. Bernstein, B.W.; Bamburg, J.R. Actin-ATP hydrolysis is a major energy drain for neurons. *J. Neurosci.* **2003**, *23*, 1–6. [CrossRef]

180. Bernstein, B.W.; Chen, H.; Boyle, J.A.; Bamburg, J.R. Formation of actin-ADF/cofilin rods transiently retards decline of mitochondrial potential and ATP in stressed neurons. *Am. J. Physiol. Cell Physiol.* **2006**, *291*, C828–C839. [CrossRef]

181. Barone, E.; Mosser, S.; Fraering, P.C. Inactivation of brain Cofilin-1 by age, Alzheimer's disease and γ-secretase. *Biochim. Biophys. Acta* **2014**, *1842*, 2500–2509. [CrossRef]

182. Kim, T.; Vidal, G.S.; Djurisic, M.; William, C.M.; Birnbaum, M.E.; Garcia, K.C.; Hyman, B.T.; Shatz, C.J. Human LilrB2 is a β-amyloid receptor and its murine homolog PirB regulates synaptic plasticity in an Alzheimer's model. *Science* **2013**, *341*, 1399–1404. [CrossRef] [PubMed]

183. Woo, J.A.; Jung, A.R.; Lakshmana, M.K.; Bedrossian, A.; Lim, Y.; Bu, J.H.; Park, S.A.; Koo, E.H.; Mook-Jung, I.; Kang, D.E. Pivotal role of the RanBP9-cofilin pathway in Aβ-induced apoptosis and neurodegeneration. *Cell Death Differ.* **2012**, *19*, 1413–1423. [CrossRef] [PubMed]

184. Lakshmana, M.K.; Chung, J.Y.; Wickramarachchi, S.; Tak, E.; Bianchi, E.; Koo, E.H.; Kang, D.E. A fragment of the scaffolding protein RanBP9 is increased in Alzheimer's disease brains and strongly potentiates amyloid-beta peptide generation. *Faseb J.* **2010**, *24*, 119–127. [CrossRef] [PubMed]

185. Rush, T.; Martinez-Hernandez, J.; Dollmeyer, M.; Frandemiche, M.L.; Borel, E.; Boisseau, S.; Jacquier-Sarlin, M.; Buisson, A. Synaptotoxicity in Alzheimer's Disease Involved a Dysregulation of Actin Cytoskeleton Dynamics through Cofilin 1 Phosphorylation. *J. Neurosci.* **2018**, *38*, 10349–10361. [CrossRef]

186. Rust, M.B.; Gurniak, C.B.; Renner, M.; Vara, H.; Morando, L.; Görlich, A.; Sassoè-Pognetto, M.; Banchaabouchi, M.A.; Giustetto, M.; Triller, A.; et al. Learning, AMPA receptor mobility and synaptic plasticity depend on n-cofilin-mediated actin dynamics. *Embo J.* **2010**, *29*, 1889–1902. [CrossRef] [PubMed]

187. Opazo, P.; Choquet, D. A three-step model for the synaptic recruitment of AMPA receptors. *Mol. Cell. Neurosci.* **2011**, *46*, 1–8. [CrossRef] [PubMed]

188. Gu, J.; Lee, C.W.; Fan, Y.; Komlos, D.; Tang, X.; Sun, C.; Yu, K.; Hartzell, H.C.; Chen, G.; Bamburg, J.R.; et al. ADF/cofilin-mediated actin dynamics regulate AMPA receptor trafficking during synaptic plasticity. *Nat. Neurosci.* **2010**, *13*, 1208–1215. [CrossRef] [PubMed]

189. Woo, J.-A.A.; Liu, T.; Fang, C.C.; Cazzaro, S.; Kee, T.; LePochat, P.; Yrigoin, K.; Penn, C.; Zhao, X.; Wang, X.; et al. Activated cofilin exacerbates tau pathology by impairing tau-mediated microtubule dynamics. *Commun. Biol.* **2019**, *2*, 112. [CrossRef]

190. Hoffmann, L.; Rust, M.B.; Culmsee, C. Actin(g) on mitochondria—a role for cofilin1 in neuronal cell death pathways. *Biol. Chem.* **2019**, *400*, 1089–1097. [CrossRef]

191. Rehklau, K.; Hoffmann, L.; Gurniak, C.B.; Ott, M.; Witke, W.; Scorrano, L.; Culmsee, C.; Rust, M.B. Cofilin1-dependent actin dynamics control DRP1-mediated mitochondrial fission. *Cell Death Dis.* **2017**, *8*, e3063. [CrossRef] [PubMed]
192. Klamt, F.; Zdanov, S.; Levine, R.L.; Pariser, A.; Zhang, Y.; Zhang, B.; Yu, L.-R.; Veenstra, T.D.; Shacter, E. Oxidant-induced apoptosis is mediated by oxidation of the actin-regulatory protein cofilin. *Nat. Cell Biol.* **2009**, *11*, 1241–1246. [CrossRef] [PubMed]
193. Wang, C.; Zhou, G.-L.; Vedantam, S.; Li, P.; Field, J. Mitochondrial shuttling of CAP1 promotes actin- and cofilin-dependent apoptosis. *J. Cell. Sci.* **2008**, *121*, 2913–2920. [CrossRef]
194. Chua, B.T.; Volbracht, C.; Tan, K.O.; Li, R.; Yu, V.C.; Li, P. Mitochondrial translocation of cofilin is an early step in apoptosis induction. *Nat. Cell Biol.* **2003**, *5*, 1083–1089. [CrossRef] [PubMed]
195. Roh, S.-E.; Woo, J.A.; Lakshmana, M.K.; Uhlar, C.; Ankala, V.; Boggess, T.; Liu, T.; Hong, Y.-H.; Mook-Jung, I.; Kim, S.J.; et al. Mitochondrial dysfunction and calcium deregulation by the RanBP9-cofilin pathway. *Faseb J.* **2013**, *27*, 4776–4789. [CrossRef] [PubMed]
196. Liu, T.; Woo, J.-A.A.; Yan, Y.; LePochat, P.; Bukhari, M.Z.; Kang, D.E. Dual role of cofilin in APP trafficking and amyloid-β clearance. *Faseb J.* **2019**, *33*, 14234–14245. [CrossRef] [PubMed]
197. Long, J.M.; Holtzman, D.M. Alzheimer Disease: An Update on Pathobiology and Treatment Strategies. *Cell* **2019**, *179*, 312–339. [CrossRef]
198. Penzes, P.; Cahill, M.E.; Jones, K.A.; Vanleeuwen, J.-E.; Woolfrey, K.M. Dendritic spine pathology in neuropsychiatric disorders. *Nat. Neurosci.* **2011**, *14*, 285–293. [CrossRef]
199. Penzes, P.; Vanleeuwen, J.-E. Impaired regulation of synaptic actin cytoskeleton in Alzheimer's disease. *Brain Res. Rev.* **2011**, *67*, 184–192. [CrossRef]

© 2020 by the authors. Licensee MDPI, Basel, Switzerland. This article is an open access article distributed under the terms and conditions of the Creative Commons Attribution (CC BY) license (http://creativecommons.org/licenses/by/4.0/).

Article

Defects in G-Actin Incorporation into Filaments in Myoblasts Derived from Dysferlinopathy Patients Are Restored by Dysferlin C2 Domains

Ximena Báez-Matus [1], Cindel Figueroa-Cares [1], Arlek M. Gónzalez-Jamett [1], Hugo Almarza-Salazar [1], Christian Arriagada [2], María Constanza Maldifassi [1], María José Guerra [1], Vincent Mouly [3], Anne Bigot [3], Pablo Caviedes [4,5] and Ana M. Cárdenas [1,*]

1. Centro Interdisciplinario de Neurociencia de Valparaíso, Facultad de Ciencias, Universidad de Valparaíso, Valparaíso 2360102, Chile; ximena.baez@cinv.cl (X.B.-M.); cindel.figueroa@postgrado.uv.cl (C.F.-C.); arlek.gonzjam@gmail.com (A.M.G.-J.); hugo.almarza@cinv.cl (H.A.-S.); constanza.maldifassi@cinv.cl (M.C.M.); mjguerraf@gmail.com (M.J.G.)
2. Departamento de Anatomía y Medicina Legal, Facultad de Medicina, Universidad de Chile, Santiago 8389100, Chile; carriagada@med.uchile.cl
3. Sorbonne Université, Inserm, Institut de Myologie, UMRS 974, Center for Research in Myology, 75013 Paris, France; vincent.mouly@upmc.fr (V.M.); anne.bigot@upmc.fr (A.B.)
4. Programa de Farmacología Molecular y Clínica, ICBM, Facultad de Medicina, Universidad de Chile, Santiago 8389100, Chile; pablo.caviedes@cicef.cl
5. Centro de Biotecnología y Bioingeniería (CeBiB), Departamento de Ingeniería Química, Biotecnología y Materiales, Facultad de Ciencias Físicas y Matemáticas, Universidad de Chile, Santiago 8370456, Chile
* Correspondence: ana.cardenas@uv.cl; Tel.: +56-322-508-052

Received: 12 September 2019; Accepted: 16 December 2019; Published: 19 December 2019

Abstract: Dysferlin is a transmembrane C-2 domain-containing protein involved in vesicle trafficking and membrane remodeling in skeletal muscle cells. However, the mechanism by which dysferlin regulates these cellular processes remains unclear. Since actin dynamics is critical for vesicle trafficking and membrane remodeling, we studied the role of dysferlin in Ca^{2+}-induced G-actin incorporation into filaments in four different immortalized myoblast cell lines (DYSF2, DYSF3, AB320, and ER) derived from patients harboring mutations in the *dysferlin* gene. As compared with immortalized myoblasts obtained from a control subject, dysferlin expression and G-actin incorporation were significantly decreased in myoblasts from dysferlinopathy patients. Stable knockdown of dysferlin with specific shRNA in control myoblasts also significantly reduced G-actin incorporation. The impaired G-actin incorporation was restored by the expression of full-length dysferlin as well as dysferlin N-terminal or C-terminal regions, both of which contain three C2 domains. DYSF3 myoblasts also exhibited altered distribution of annexin A2, a dysferlin partner involved in actin remodeling. However, dysferlin N-terminal and C-terminal regions appeared to not fully restore such annexin A2 mislocation. Then, our results suggest that dysferlin regulates actin remodeling by a mechanism that does to not involve annexin A2.

Keywords: dysferlin; actin; C2 domains; annexin A2; dysferlinopathy

1. Introduction

Dysferlin is a transmembrane protein containing seven cytosolic C2 domains, which bind Ca^{2+} and acidic phospholipids with different affinities [1,2]. Its most well-known function is to facilitate Ca^{2+}-dependent aggregation and fusion of vesicles at wounded plasmalemma [3–5] by a mechanism that has not been completely elucidated. Dysferlin plays a role in other cellular processes in the skeletal muscle tissue including cytokine secretion and membrane receptor recycling in myoblasts [6,7] as well as biogenesis, remodeling, and maintenance of the T-tubule system [8,9]. Among the proteins involved in vesicle trafficking and/or membrane remodeling that interact with dysferlin are the SNAREs (acronym for soluble N-ethylmaleimide-sensitive factor activating protein receptors) syntaxin-4 and SNAP23 [10], annexins A1 and A2 [4], α-tubulin [11], and Mitsugumin 53 [12,13].

Mutations in the *dysferlin* gene cause a group of autosomal recessive muscular dystrophies known as dysferlinopathies [14]. The most common forms of dysferlinopathy are Miyoshi myopathy, limb-girdle muscular dystrophy type 2B, and distal anterior compartment myopathy [15,16]. Dysferlinopathy phenotypes include progressive atrophy of limb muscles, elevated serum creatine kinase levels, reduced expression of plasmalemmal dysferlin, and prevalence of immature muscle fibers [17,18]. As expected, skeletal muscle cells derived from dysferlinopathy patients [19] and dysferlin-deficient mice [3] display a defective Ca^{2+}-dependent plasmalemma repair. Furthermore, when dysferlin expression is reduced, vesicles accumulate beneath the plasmalemma [20–22], suggesting a role of dysferlin in vesicle trafficking. A critical element for vesicle trafficking and membrane repair is the actin cytoskeleton [23–27]. Skeletal muscle cells express two cytoskeletal actin isoforms, β-actin and γ-actin, that localize in sub-plasmalemmal regions [28,29]. Skeletal muscle-specific ablation of β-actin or γ-actin causes a progressive myopathy, characterized by myofiber degeneration/regeneration and muscle weakness [29,30], thus emphasizing the critical role of the cytoskeletal actin network in the function of skeletal muscle cells.

Interplay between dysferlin and the actin cytoskeleton has been observed during plasmalemma repair [25,26]. Moreover, dysferlin interacts with proteins important for actin organization and remodeling such as annexin A2 [4], suggesting the implication of dysferlin in actin dynamics. Alterations in dysferlin expression, such as those occurring in dysferlinopathies, could then potentially affect actin dynamics in muscle cells. With this in mind, we studied whether the dynamics of the cytoskeletal actin is affected in myoblasts derived from skeletal muscle of dysferlinopathy patients. Our data show that the expression of dysferlin is dramatically reduced in dysferlinopathy-derived myoblasts compared to myoblasts from a healthy subject. Moreover, dysferlinopathy myoblasts exhibit a reduced capability to incorporate new actin monomers to the pre-existing actin filament (F-actin) network compared to control myoblasts, suggesting defects in actin cytoskeleton remodeling. Finally, the expression of a construct harboring the full-length dysferlin, as well as the expression of its N-terminal or its C-terminal regions, successfully restores actin dynamics in dysferlin-deficient myoblasts. These results support a role of dysferlin in actin cytoskeleton dynamics in muscle cells and suggest that this mechanism could be deregulated in dysferlinopathy.

2. Results

2.1. Dysferlin Expression in the Dysferlinopathy Cell Lines

Four different cell lines of immortalized myoblasts were derived from skeletal muscle biopsies from dysferlinopathy patients. These cell lines named DYSF2 (also called 107), DYSF3 (also called 379), AB320, and ER myoblasts were previously characterized [31,32]. Table 1 describes the origin of each cell line, including the mutations carried by donors. All of them are heterozygous with the exception of ER cells. As a control, we used the cell line C25, which was derived from a biopsy of semitendinosus muscle of a 25 year old male who did not suffer from any skeletal muscle disease [33]. All these cell lines were obtained from the platform for the immortalization of human cells from the Institut de Myologie (Paris, France), and their characterizations were previously reported [33–35]. All analyses were performed on undifferentiated myoblasts.

We first analyzed the relative expression of dysferlin in the cell lines DYSF2, DYSF3, AB320, and ER and compared it with the expression of the protein in C25 control myoblasts. Figure 1a shows a typical immunoblot stained with antibodies against dysferlin (upper bands) and β-tubulin (lower, loading control). Figure 1b shows dysferlin/β-tubulin ratios from five independent experiments. As compared with C25 myoblasts, the expression of dysferlin was reduced by 68%, 87%, 88%, and 83%

in DYSF2, DYSF3, AB320, and ER myoblasts, respectively. These results agree with the expected dysferlin expression in these myoblasts [31,32,36,37] and validate these cell lines as in vitro models of dysferlinopathy.

Table 1. Description of the immortalized human skeletal myoblasts used in this study.

Cell Line	Muscle Biopsy	Patient	Mutations
DYSF2	Vastus lateralis	37 year old male	Exon 8:c.855 + 1delG, mRNAdecay Exon 9: c.895G > A, r.895G > A, p.G299R
DYSF3	Vastus lateralis	36 year old female	Exon 16: c.1448C > A, p.S483X Exon 55:c.*107T > A, 3′UTR
AB320	Quadriceps	29 year old female	Intron 4: c.342-1G > A Exon 32: c.3516–3517delTT, p.S1173X
ER	Quadriceps	17 year old male	Homozygous Exon 44: c.4882G > A, p.G1628R

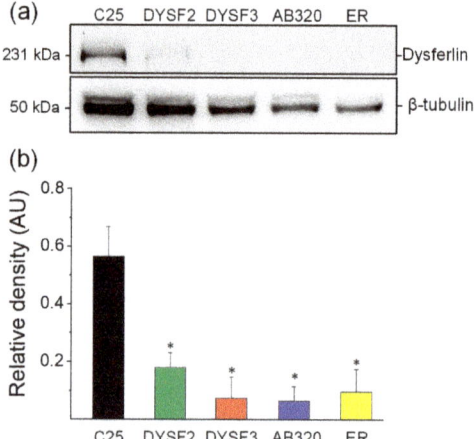

Figure 1. Reduced expression of dysferlin in immortalized myoblasts obtained from dysferlinopathy patients. Dysferlin expression was determined by immunoblotting of total protein extracts from C25 myoblasts obtained from an unaffected donor, or DYSF2, DYSF3, AB320, or ER myoblasts obtained from dysferlinopathy patients. (**a**) Example of immunoblot detection of dysferlin (upper bands) and β-tubulin (loading control; bottom bands). (**b**) Relative density (dysferlin/β-tubulin ratio). Data are means ± SEM from 5 independent immunoblots. * $p < 0.05$ compared to C25 myoblasts (t-test).

2.2. Cytoskeleton Actin Dynamics in the Dysferlinopathy Myoblasts

The cytoskeletal actin network is a highly dynamic structure that is rearranged in skeletal muscle cells during vesicle trafficking and membrane repair [23,25–27,38,39], and such actin remodeling in skeletal myoblasts, evaluated as G-actin incorporation, seems to depend on high cytosolic Ca^{2+} concentrations [27]. Therefore, we measured the incorporation of actin monomers (G-actin) into pre-existing actin filaments in permeabilized myoblasts by using a previously reported assay [27,40]. In this assay, myoblasts were permeabilized with digitonin in a solution containing green-fluorescent Alexa-Fluor-488 G-actin and 2 mM ATP-Mg^{2+}, in the absence or presence of Ca^{2+}. As shown in the supplementary Figure S1, G-actin incorporation was significantly higher in the presence of 10 µM free Ca^{2+}. Therefore, experiments were performed in this latter condition. Figure 2a shows examples of myoblasts with fluorescent actin filaments, indicative of incorporation of Alexa Fluor-labelled actin monomers into the F-actin network. Figure 2b shows the quantification of fluorescent G-actin incorporation. As compared with control C25 myoblasts, all dysferlinopathy myoblasts (DYSF2,

DYSF3, AB320, and ER) displayed significant reduction in G-actin incorporation to the pre-existent cytoskeletal actin network ($p < 0.05$). In DYSF3 myoblasts, G-actin incorporation was reduced by 50%, whereas in the DYSF2, AB320, and ER myoblasts it was reduced by 36%, 35%, and 42%, respectively. Supplementary Figure S2 shows 1D intensity profiles of the incorporated fluorescently tagged G-actin. Higher-intensity fluorescence was observed in the cell periphery of C25 myoblasts, whereas an overall reduction of cell fluorescence was observed in the dysferlinopathy myoblasts (Figure S2). The peripheral distribution of the fluorescently-tagged G-actin probably corresponds to submembrane cortical cytoskeleton actin [28,29,40].

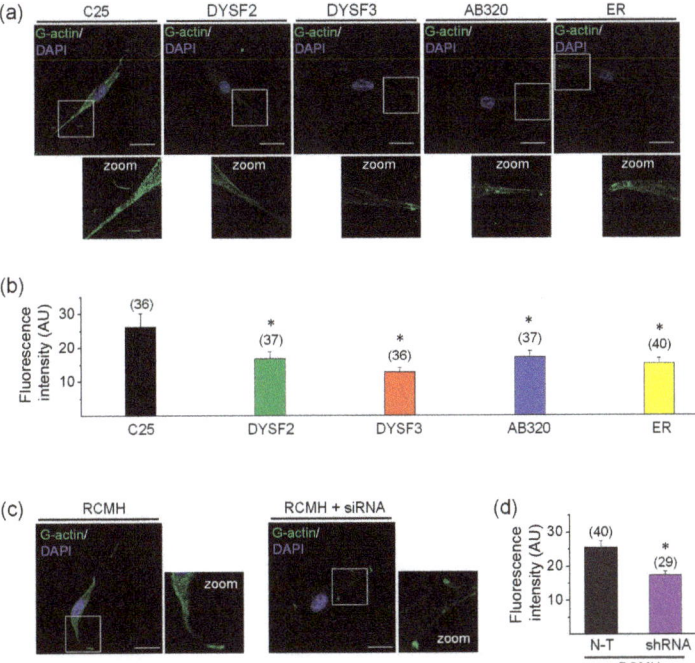

Figure 2. G-actin incorporation into filaments in dysferlin-deficient myoblasts. Fluorescently tagged G-actin incorporation into filaments was assayed in myoblasts permeabilized with 20 µM digitonin in the presence of 300 nM Alexa-Fluor-488 actin, 2 mM ATP-Mg^{2+}, and 10 µM free Ca^{2+} during 6 min at 37 °C. After permeabilization, cells were fixed and nuclei were stained with DAPI. Confocal images were acquired at the equatorial plane of the cells using identical exposure settings between compared samples. (**a–c**) Fluorescent actin filaments in control C25 and dysferlinopathy DYSF2, DYSF3, AB320, and ER myoblasts (**a**) and RCMH myoblasts non-transfected (N-T) or stably transfected with shRNA for dysferlin (**c**). Scale bar = 20 µm. Insets show digital zooms of the boxed areas; brightness and contrast were increased to appreciate better the pattern of fluorescent actin. (**b–d**) Data are means ± SEM. Actin fluorescence intensity was measured in a single focal plane at the equator of cells and normalized by the cell area. The number of analyzed cells from five different cultures is indicated in parentheses. * $p < 0.05$ compared to C25 (b) or N-T RCMH myoblasts (c) (*t*-test).

To rule out that the differences in G-actin incorporation in the control and dysferlinopathy myoblasts was a consequence of the size of pores formed by digitonin, we evaluated the incorporation of a protein with a size similar to G-actin into the cells. Therefore, we incubated digitonin-permeabilized C25 and DYSF3 myoblasts for 6 min at 37 °C in K$^+$-glutamate /EGTA/Pipes (KGEP) buffer containing the GST-amphiphysin-1 SH3 domain fusion protein. Incorporation of GST-amphiphysin-1 SH3 was determined in fixed cells using a monoclonal anti-GST and secondary CY2-conjugated antibodies.

As shown in Table S1, no significant differences were encountered in the GST relative fluorescence intensity in C25 and DYSF3 cells.

We also performed experiments in RCMH myoblasts stably transfected with shRNA for dysferlin, a strategy to knockdown dysferlin expression previously reported by our group [32]. The RCMH cell line was established from a quadriceps biopsy of a healthy human [41], and it has been previously characterized [41,42]. The supplementary Figure S3 shows the reduced expression of dysferlin in stably transfected RCMH myoblasts. As compared with non-transfected RCMH myoblasts, the stable knockdown of dysferlin significantly reduced G-actin incorporation into filaments (Figure 2c–d).

Together, these data strongly suggest that Ca^{2+}-dependent G-actin incorporation is impaired when dysferlin is reduced, such as in the dysferlinopathy context.

2.3. Expression of Full-Length Dysferlin or Its C or N-Terminal Regions Restores G-Actin Incorporation in Dysferlinopathy Myoblasts

Next, we decided to evaluate whether the expression of dysferlin constructs can restore the impaired actin dynamics in DYSF3 myoblasts. We chose this myoblast cell line, and not the others, for its lower expression of dysferlin and reduced G-actin incorporation (Figures 1 and 2). We first analyzed the incorporation of Alexa Fluor-labelled actin monomers in myoblasts expressing full-length dysferlin-HA. Representative images of these experiments are shown in Figure 3a. Quantification of fluorescent actin filaments in these cells shows that dysferlin-HA expression restored G-actin incorporation in DYSF3 myoblasts to levels comparable to C25 cells also expressing dysferlin-HA (Figure 3b).

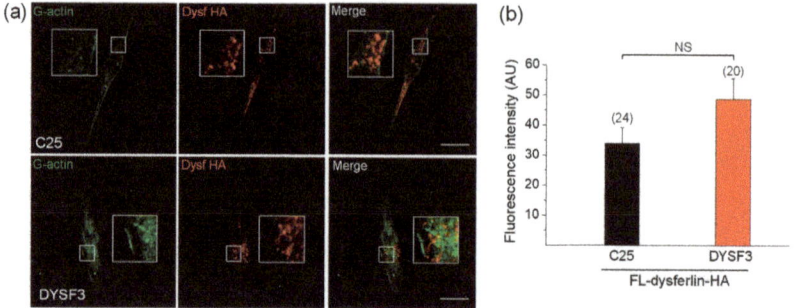

Figure 3. Full-length dysferlin-HA restores G-actin incorporation in dysferlinopathy myoblasts. Control C25 or dysferlinopathy DYSF3 myoblasts were transfected with full-length dysferlin-HA (FL-dysferlin-HA), and 24 h fluorescently tagged G-actin incorporation was assayed as previously described. FL-dysferlin-HA expression was assayed by immunofluorescence using a monoclonal antibody against dysferlin and a Cy3-conjugated anti-rabbit secondary antibody. Confocal images were acquired at the equatorial plane of the cells using identical exposure settings between compared samples. (**a**) C25 and DYSF3 myoblasts with fluorescent actin filaments (green) and FL-dysferlin-HA immunostaining (red). Scale bar = 10 μm. Insets show digital magnification of the boxed areas. (**b**) Bars represent means ± SEM. Actin fluorescence intensity was measured in a single focal plane at the equator of cells and normalized by the cell area. The number of cells analyzed from four different cultures is indicated in parentheses. No significant differences (NS) were found (*t*-test).

We then evaluated how different dysferlin regions contributed to restore the impaired actin dynamics in the DYSF3 myoblasts. We specifically evaluated the N-terminal dysferlin region that comprises the Ca^{2+} and lipid-binding C2A, C2B, and C2C domains and FerI domain, and in parallel we examined the contribution of the C-terminal region containing a transmembrane region plus C2E, C2F, and C2G Ca^{2+}-binding domains. Both dysferlin regions were fused to a mCherry reporter protein and transfected in C25 and DYSF3 myoblasts. Myoblasts expressing the empty mCherry vector were used as controls. Transfection efficiency was around 25%, 10%, and 35% in cells transfected with mCherry, dysferlin N-terminal, or C-terminal constructs. Fluorescence intensities of expression of these constructs are provided in the supplementary Table S2.

A scheme of dysferlin domains is shown in Figure 4a. Images of fluorescent actin filaments and expression of the constructs are shown in Figure 4b. Figure 4c shows quantification of fluorescent actin filaments. The analyses show that either N-terminal or C-terminal parts of dysferlin increased actin fluorescence threefold in DYSF3 myoblasts, as compared to that observed in myoblasts expressing empty mCherry (Figure 4c). Furthermore, both dysferlin regions also increased G-actin incorporation by 1.5-fold to 1.8-fold in C25 myoblasts. This strongly suggests the involvement of both dysferlin regions in actin dynamics in skeletal myoblasts.

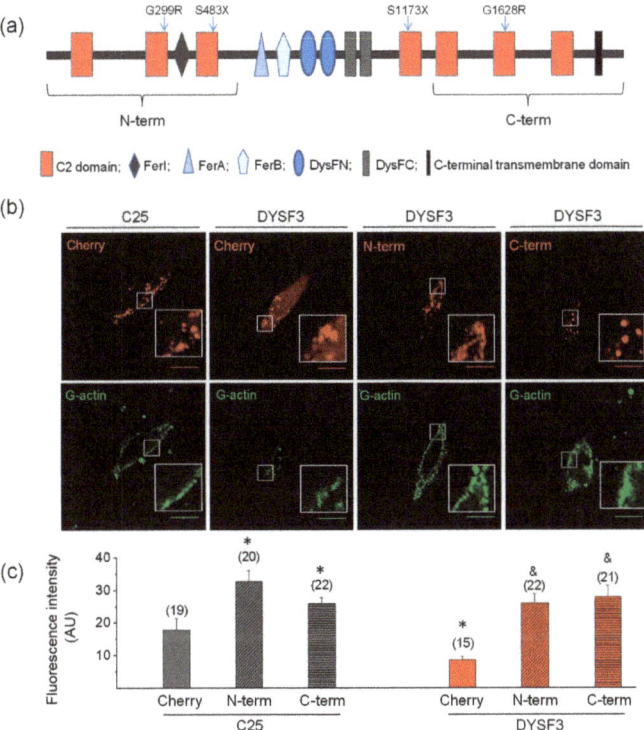

Figure 4. The N- or C- terminal dysferlin regions are efficient in restoring G-actin incorporation in dysferlinopathy myoblasts. (**a**) Schematic representation of dysferlin regions included in the N-terminal and C-terminal constructs. Blue arrows indicate positions of dysferlin mutations carried by the cell lines (see Table 1). (**b,c**) Control C25 or dysferlinopathy DYSF3 myoblasts were transfected with mCherry alone (Cherry), or dysferlin N-terminal (N-term) or C-terminal (C-term) fused to mCherry. Twenty-four hours later, fluorescently tagged G-actin incorporation was assayed as described in Methods, and confocal images were acquired at the equatorial plane of the cells using identical exposure settings between compared samples. (**b**) Fluorescent actin filaments in C25 or DYSF3 myoblasts expressing Cherry, and DYSF3 myoblasts expressing N-term or C-term. Scale bar = 10 μm. Insets show digital magnification of the boxed areas. (**c**) Actin fluorescence intensity was measured in a single focal plane and normalized by the cell area. Bars represent means ± SEM. Actin fluorescence intensity was measured in a single focal plane at the equator of cells and normalized by the cell area. The number of analyzed cells from four different cultures is indicated in parentheses. * $p < 0.05$ compared with C25 myoblasts expressing mCherry, & $p < 0.05$ compared with DYSF3 expressing mCherry (*t*-test).

2.4. Annexin A2 Distribution in Dysferlin-Deficient Myoblasts

Annexin A2 is a Ca^{2+}-binding protein that is recruited to membranes upon increments in cytosolic Ca^{2+}, where it facilitates actin assembly [43,44]. Annexin A2 also associates to dysferlin in a Ca^{+2}-dependent manner [4], and it is suggested that both are mutually required for their recruitment to injury sites [25,45]. Therefore, we analyzed the distribution of annexin A2 and its colocalization with newly incorporated G-actin in control C25 and dysferlinopathy DYSF3 myoblasts. These experiments were carried out in myoblasts permeabilized in the presence of 10 μM Ca^{2+} and Alexa Fluor-labelled G-actin, thus allowing the analysis of the colocalization of annexin A2 with the incorporated fluorescent G-actin. Figure 5a shows images of C25 and DYSF3 myoblasts immunostained with annexin A2 (red) and fluorescent actin (green). Annexin A2 localizes throughout the cytosol in C25 myoblasts, whereas it is mainly restricted to nuclear and perinuclear areas in DYSF3 myoblasts. Indeed, the ratio of the mean fluorescence intensity of the cytosol/whole cell of annexin A2 was 0.80 ± 0.02 in C25 myoblasts and 0.6 ± 0.03 in DYSF3 myoblasts (Figure 5b). Supplementary Figure S4 shows 1D intensity profiles of annexin A2 and fluorescently tagged G-actin. Both stains showed a similar distribution pattern in C25 myoblasts, whereas their distribution pattern was different in DYSF3 myoblasts, wherein annexin A2 showed a higher nuclear localization (Figure S4).

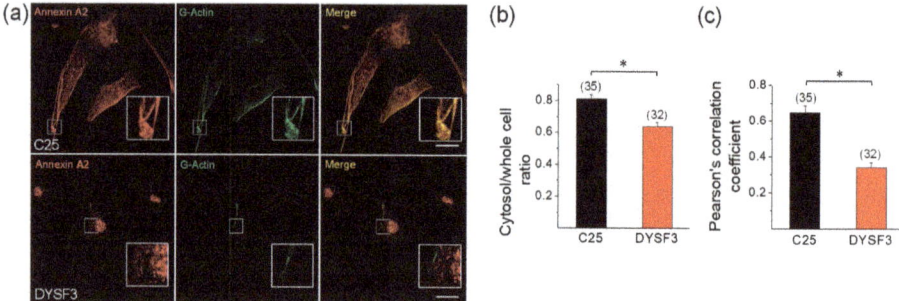

Figure 5. Annexin A2 distribution in dysferlinopathy myoblasts. Distribution of annexin A2 was analyzed by immunofluorescence using an anti-annexin A2 antibody and Cy3-conjugated anti-rabbit secondary antibody in C25 and DYSF3 myoblasts permeabilized with 20 μM digitonin in the presence of 300 nM Alexa-Fluor-488 actin, 2 mM ATP-Mg^{2+}, and 10 μM free Ca^{2+}. Confocal images were captured at the equatorial plane of the cells using identical exposure settings between compared samples; therefore, annexin A2 distribution was measured in a single focal plane. (**a**) C25 and DYSF3 myoblasts immunostained with annexin A2 (red) and fluorescent actin incorporated into filaments (green). Scale bar = 20 μm. Insets show digital magnification of the boxed areas. (**b,c**) Bars represent means ± SEM of the ratio of the mean fluorescence intensity of the cytosol/whole cell of annexin A2 immunostaining (**b**) and Pearson correlation coefficient for colocalization of annexin A2 with fluorescent actin filaments (**c**). The number of analyzed cells from four different cultures is indicated in parentheses. * $p < 0.05$ (*t*-test).

Pearson correlation coefficient analysis indicated a significant colocalization of annexin A2 with newly incorporated G-actin (0.65 ± 0.04; $n = 35$) in control C25 myoblasts, whereas this colocalization was lost in DYSF3 myoblasts (Pearson correlation coefficient of 0.34 ± 0.03; $n = 32$; Figure 5c).

2.5. Annexin A2 Distribution in Dysferlin-Deficient Myoblasts Expressing Dysferlin Constructs

Next, we determined whether the expression of full-length dysferlin-HA restores annexin A2 distribution in DYSF3 myoblasts. Figure 6A shows images of annexin A2 immunostaining (green) in C25 and DYSF3 myoblasts expressing full-length dysferlin-HA (red). Analyses of annexin A2 distribution are shown in Figure 6B. In DYSF3 myoblasts expressing full-length dysferlin-HA, annexin A2 cytosol/whole cell ratio was 0.76 ± 0.03 ($n = 37$), a value comparable with that of C25 myoblasts without transfection (0.80 ± 0.02). However, in C25 myoblasts expressing dysferlin-HA, annexin A2 cytosol/whole cell ratio was 0.46 ± 0.03. The change in annexin A2 distribution in the latter condition might be caused by the excess of dysferlin, since in mock condition this ratio was 0.86 ± 0.03 (Figure 6B). No colocalization of dysferlin with annexin A2 was observed in C25 or DYSF3 myoblasts expressing full-length dysferlin-HA (Pearson correlation coefficient of 0.4 ± 0.02 and 0.2 ± 0.03, respectively).

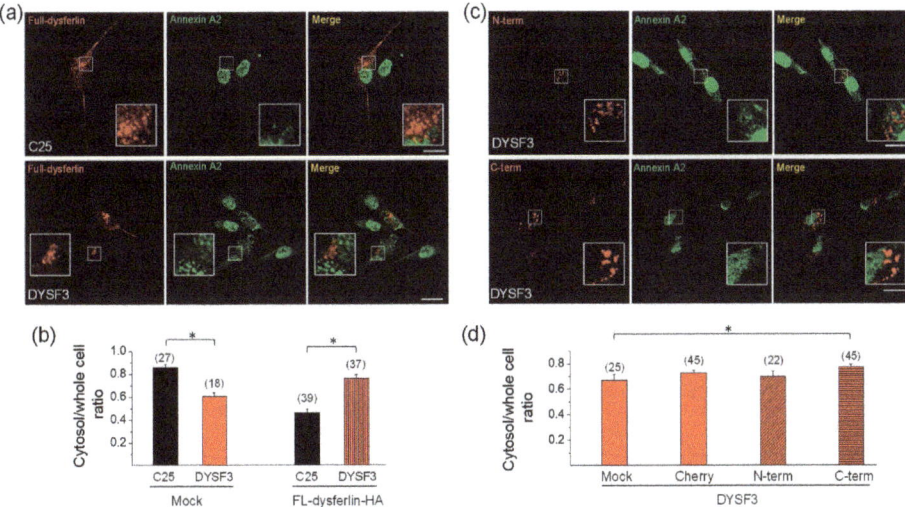

Figure 6. Annexin A2 distribution in DYSF3 myoblasts expressing dysferlin constructs. Distribution of annexin A2 was analyzed by immunofluorescence using an anti-annexin A2 antibody and Cy2-conjugated anti-rabbit secondary antibody in C25 and DYSF3 myoblasts expressing full-length dysferlin-HA (FL-dysferlin-HA) (**a**,**b**) and mCherry (Cherry), or dysferlin N-terminal (N-term) or C-terminal (C-term) fused to mCherry (**c**,**d**). Experiments were performed in digitonin-permeabilized cells in the presence of 10 µM free Ca^{2+}. Confocal images were captured at the equatorial plane of the cells using identical exposure settings between compared samples; therefore, annexin A2 distribution was measured in a single focal plane. (**a**,**c**) Annexin A2 stained (green) in C25 or DYSF3 myoblasts in a mock condition or expressing FL-dysferlin-HA (**a**) and N-term or C-term (**c**). Scale bar = 20 µm. Insets show digital magnification of the boxed areas. (**b**,**d**) Data show means ± SEM of the ratio of the mean fluorescence intensity of the cytosol/whole cell of annexin A2 immunostaining. The number of analyzed cells from four different cultures is indicated in parentheses. * $p < 0.05$ (*t*-test).

To determine whether dysferlin N- or C-terminal regions were capable of restoring annexin A2 mislocation in dysferlin-deficient myoblasts, we evaluated annexin A2 distribution in DYSF3 cells transfected with the mCherry constructs described above. Here, we included a control consisting of DYSF3 cells treated with the transfection reagents but without addition of the plasmid (mock). Figure 6C shows images of annexin A2 immunostaining (green) in DYSF3 myoblasts expressing the dysferlin N- or C-terminal constructs. Annexin A2 cytosol/whole cell ratios were 0.73 ± 0.02 (n = 45), 0.70 ± 0.04 (n = 22), and 0.77 ± 0.02 (n = 45) for DYSF3 myoblasts expressing mCherry or the N- or C-terminal constructs, respectively. For the mock experiment, the annexin A2 cytosol/whole cell ratio was 0.67 ± 0.04 (n = 25). Then, the annexin A2 cytosol/whole cell ratio in DYSF3 myoblasts expressing the C-terminal construct was comparable with that of C25 myoblasts without transfection (0.80 ± 0.02), and significantly different from the mock control, but not from the mCherry control (Figure 6d). No colocalization of annexin A2 with dysferlin N- or C-terminal was observed (Pearson correlation coefficient of 0.3 ± 0.03 and 0.2 ± 0.02, respectively).

Together, these data show that expression of full-length dysferlin restores annexin A2 distribution in dysferlinopathy myoblasts. However, the contribution of its C-terminal is not clear.

3. Discussion

Dysferlin and cytoskeletal actin play critical roles at various differentiation stages of skeletal muscle physiology; they are required for vesicle trafficking and exocytosis in myoblasts [6,7,27,46,47] and plasmalemma repair in myofibers [3–5]. Regarding the plasmalemma repair in the skeletal muscle, two mechanisms have been proposed. One of them includes Ca^{+2}-induced recruitment of vesicles and other membranous organelles to repair sites, where they fuse together to form a membrane patch that seals the damage membrane [5,18]. The second mechanism involves Ca^{+2}-induced exocytosis of lysosomes with the consequent release of acid sphingomyelinase, which in turn promotes the endocytosis of lesion areas [46,48]. It has been proposed that dysferlin is involved in both types of mechanisms, favoring vesicle recruitment and fusion in the first [5,18] and promoting lysosome exocytosis in the second [46]. The presence of dysferlin in vesicles or other membranous organelles seems to be important for such functions [5]. Annexin A2 is also critical for plasmalemma repair [4,49], and its ablation causes progressive muscle weakening [49]. Interplay between dysferlin, annexin A2, and actin remodeling has been observed during plasmalemmal repair. In this regard, actin filaments accumulate at membrane injury sites [26], and their disruption with cytochalasin D inhibits recruitment of dysferlin [25]. Likewise, Latrunculin A, a G-actin sequestering agent, delays the formation of the repair complex composed by annexins A1, A2, A5, and A6 [26]. In turn, annexin A2 facilitates actin remodeling [43] and contributes to the resealing of the plasma membrane by promoting actin polymerization [50]. Here, we found that dysferlin may also contribute to actin remodeling, as observed by a reduced G-actin incorporation into filaments in dysferlin-lacking myoblasts (Figure 2). Noteworthy, the poor G-actin incorporation in dysferlinopathy myoblasts was restored by expression of both N- and C-terminal regions of dysferlin (Figure 3); however, the contribution of these dysferlin regions to annexin A2 distribution remains unclear (Figure 6). Then, it is not possible to establish a causal relation between annexin A2 mislocation and impaired actin dynamics in dysferlin-deficient myoblasts.

Another critical element for actin remodeling is cytosolic Ca^{2+} level [51]. Indeed, high cytosolic Ca^{2+} concentrations promote both severing and formation of actin filaments [40]. The mechanism involves the participation of Ca^{2+}-sensitive actin severing proteins, such as scinderin [52], as well as Src kinases and Rho GTPases, which lead to the activation of actin nucleation promoting factors such as N-WASP and cortactin [40,53,54]. Here, we observed that a high Ca^{2+} concentration (10 µM) favors G-actin incorporation in skeletal myoblasts (Figure S1), and such G-actin incorporation appears to be regulated by dysferlin Ca^{+2}-sensitive C2 domains.

The fact that both N- and C-terminal regions of dysferlin restore actin dynamics suggests that dysferlin comprises functionally redundant modules. Both N- and C-terminal dysferlin constructs contain three C2 domains (Figure 4a), which bind Ca^{2+} and phospholipids with different affinity and Ca^{2+} dependency [2]. Synaptotagmin-1, another protein with tandem C2 domains, also promotes actin remodeling by binding phosphatidylinositol 4,5-bisphosphate $PI(4,5)P_2$ via a polybasic region [55]. Reportedly, $PI(4,5)P_2$ promotes actin remodeling by recruiting actin binding proteins, such as formins [56,57], a family of proteins involved in actin nucleation and assembly [58]. Dysferlin also binds and recruits $PI(4,5)P_2$ to membranes and promotes membrane remodeling and T-tubule biogenesis in a $PI(4,5)P_2$-dependent manner [59]. Then, the interaction of dysferlin C2 domains with $PI(4,5)P_2$ might explain its ability to remodel the actin cytoskeleton. However, only the isolated C2A domain of dysferlin binds $PI(4,5)P_2$ in vitro in a Ca^{2+}-dependent way, whereas C2F binds weakly to phosphatidylinositols. The other isolated C2 domains exhibited no detectable associations [1]. Since the study of Therrien et al. (2009) [1] was carried out with isolated C2 domains, it did not consider potential cooperative effects between tandem C2 domains, as reported for synaptotagmin-1 C2 domains for its association to $PI(4,5)P_2$ [60]. Then, further studies are necessary to achieve a better comprehension of the mechanism by which dysferlin promotes actin dynamics.

Full-length dysferlin-HA, as well as the dysferlin N-terminal and C-terminal constructs, exhibit a granular cytoplasmic pattern. This granular pattern might reflect the distribution pattern of dysferlin, but it might also be determined by the distribution pattern of the constructs, since mCherry seems to form aggregates (Figure 4B). A granular cytoplasmic expression of dysferlin has been observed in myoblasts [46], differentiated L6 myotubes [5], and skeletal myofibers [61]. PI(4,5)P$_2$ also associates to intracellular compartments, such as vesicles, lysosomes, endosomes, and the nucleus, among others [62,63]. Then, dysferlin might favor PI(4,5)P$_2$-dependent actin remodeling in intracellular compartments, contributing to the described dysferlin-dependent processes in myoblasts, such as vesicle trafficking and exocytosis [7,46]. However, additional studies are necessary to demonstrate the role of dysferlin in actin remodeling during plasmalemmal repair in differentiated skeletal muscle cells, since our study was performed in digitonin-permeabilized myoblasts, and dysferlin seems to be unable to repair membrane damage induced by nonionic detergents [64].

This work might contribute to a better understanding of the mechanisms involved in the pathogenesis of muscular dystrophies caused by mutations in dysferlin.

4. Materials and Methods

4.1. Reagents

Actin, from rabbit muscle, Alexa Fluor™ 488 conjugate (Thermo Fisher Scientific, Waltham, MA, USA); ATP (Sigma-Aldrich, St. Louis, MO, USA); bovine serum albumin (Sigma-Aldrich, St. Louis, MO, USA); dexamethasone (Sigma-Aldrich, St. Louis, MO, USA); 40,6-diamidino-2-phenylindole (Sigma-Aldrich, St. Louis, MO, USA); digitonin (Sigma-Aldrich St. Louis, MO, USA); dimethyl sulfoxide (Merck Company, Darmstadt, Germany); DMEM/F-12 medium (Gibco, BRL, Gaithersburg, MD, USA); Dulbecco modified Eagle's minimal essential medium (Gibco BRL, Gaithersburg, MD, USA); EDTA (Calbiochem, La Jolla, CA); EGTA (Sigma-Aldrich, St. Louis, MO, USA); fetal bovine serum (Gibco BRL, Gaithersburg, MD, USA); fetuin (Sigma-Aldrich, St. Louis, MO, USA); gentamicin (Gibco/Life Technology, China); glutamic acid (Sigma-Aldrich, St. Louis, MO, USA); HEPES (Calbiochem, La Jolla, CA); human insulin (Eli Lilly and company, Indianapolis, USA); Lipofectamine 2000 (Invitrogen, Carlsbad, CA, USA); 199 medium (Sigma-Aldrich St. Louis, MO, USA); NaF (Merck Company, Darmstadt, Germany); Na$_3$VO$_4$ (Merck Company, Darmstadt, Germany); Opti-Mem (Gibco BRL, Gaithersburg, MD, USA); penicillin (OPKO, Santiago, Chile); p-formaldehyde (Sigma-Aldrich St. Louis, MO, USA); phenylmethyl sulfonylflouride (Sigma-Aldrich, St. Louis, MO, USA); PIPES (Sigma-Aldrich, St. Louis, MO, USA); poly-l-lysine (Sigma-Aldrich, St. Louis, MO, USA); recombinant human basic fibroblast growth factor (Gibco BRL, Gaithersburg, MD, USA); protease inhibitor cocktail (Sigma-Aldrich, St. Louis, MO, USA); recombinant human epidermal growth factor (Gibco BRL, Gaithersburg, MD, USA); triton X-100 (Merck Company, Darmstadt, Germany); trypsin- EDTA 0.25% (Sigma-Aldrich, St. Louis, MO, USA); and Tween-20 (Merck Company, Darmstadt, Germany) were purchased.

4.2. Antibodies

Anti-annexin A2 (Santa Cruz Biotechnology, Inc, Dallas, Texas, USA); anti-β-tubulin (Cytoskeleton, St. Denver, CO, USA); anti-dysferlin-HAMLET (Novocastra TM Lyophilized Leica; Newcastle, United Kingdom); CY2-conjugated goat anti-rabbit IgG (H+L) (Jackson Immunoresearch, West Grove, PA, USA); Cy3-conjugated goat anti-rabbit IgG (H+L) (Jackson Immunoresearch, West Grove, PA, USA); horseradish peroxidase (HRP)-conjugated donkey anti-sheep (R & D Systems, Minneapolis, USA); HRP-conjugated goat anti-mouse IgG (H+L) (Thermo Fisher Scientific, Waltham, MA, USA); and peroxidase affiniPure F(ab')$_2$ fragment donkey anti-rabbit IgG (H+L) (Jackson ImmunoResearch, West Grove, PA, USA) were purchased.

4.3. cDNA Constructs and Plasmids

Plasmid dysferlin-HA (Addgene plasmid 29767) was provided by The Jain Foundation (www.jain-foundation.org). The mCherry-tagged plasmids containing dysferlin N-terminal (amino acids 1–572) or C-terminal (amino acids 1169–2119) were constructed by GenScript Corporation (Nanjing, China) by cloning the appropriate DNA sequences (Homo sapiens dysferlin isoform 1, NCBI; reference Sequence: NP_001124459.1) into a mCherry_pcDNA3.1(+) vector. mCherry was fused to the C-terminal end in dysferlin N-terminal and fused to the N-terminal end in dysferlin C-terminal. The plasmid encoding for the amphiphysin SH3 domain fused to a glutathione-S-transferase (GST)-tag (amphiphysin-SH3) was provided by Dr. Patricia Hidalgo (Institut für Neurophysiologie, Medizinische Hochschule Hannover, Germany). shRNA plasmids against dysferlin with puromycin resistance were obtained from Santa Cruz Biotech (Dallas, Tx).

4.4. Culture of Cell Lines and Transfection

The C25 cell line was established from human biopsies of semitendinosus of unaffected individuals [33,35]. The dysferlinopathy cell lines DYSF2, DYSF3, AB320, and ER were established from human biopsies of patients bearing different dysferlin mutations (c.855 + 1delG c.895G > A; c.1448C > A c.107 T > A; c.342-1G > A HTZ c.3516_3517delTT (p.Ser1173X) HTZ; G1628R (c.4882G > A) HMZ, respectively). All these cell lines (C25, DYSF2, DYSF3, AB320, and ER myoblasts) were obtained from the platform for the immortalization of human cells from the Institut de Myologie (Paris, France), with agreement of the subjects through signature of an informed consent and anonymization before immortalization, according to the EU GDPR regulation. Their characterization has been previously reported [31,33,35]. They were cultivated in a mix of 199 medium/Dulbecco modified Eagles minimal essential medium (1:4 ratio) supplemented with 20% fetal bovine serum (FBS), 25 µg/mL fetuin, 0.5 ng/mL basic fibroblast growth factor, 5 ng/mL epidermal growth factor, 0.2 µg/mL dexamethasone, 5 µg/mL insulin, 50 U/mL penicillin, and 100 µg/mL gentamicin at a density of 3×10^5 cells/mL in 25 mm glass coverslips and incubated at 37 °C in a 5% CO_2 atmosphere until experimentation. For transfections, cells were incubated in 50 µL of Opti-Mem containing 1 µg DNA and 1.5 µL of Lipofectamine 2000 for 20 min. Then, 250 µL of Opti-Mem was added, and cells were kept at 37 °C in a 5% CO_2 atmosphere for 5 h. After that period, 1 mL of the culture medium described above was added, and cells were kept at 37 °C in a 5% CO_2 atmosphere for 24 h.

RCMH myoblasts were cultured in DMEM/F-12 medium supplemented with 10% fetal bovine serum and incubated at 37 °C in a 5% CO_2 atmosphere. Transfection of shRNA plasmid against dysferlin with puromycin resistance was performed with Lipofectamine 2000 as described above. A stable transfected pool was selected with 10 µg/mL puromycin in the cultured medium.

4.5. Immunoblotting Analyses

Cells were lysed in a non-denaturing lysis buffer composed of 300 mM NaCl, 5 mM EDTA, 50 mM TRIS HCl, 1% Triton X-100, and supplemented with 0.1% (v/v) protease inhibitor cocktail, 50 mM NaF, and 0.2 mM Na_3VO_4. Total protein content was determined using the Quant-it Protein Assay Kit (Invitrogen, Carlsbad, CA, USA). Total proteins (100 µg) were separated by sodium dodecyl sulfate-polyacrylamide gel electrophoresis (10% polyacrylamide gels) and electrophoretically transferred to polyvinylidene difluoride membranes (GE Healthcare Life Sciences, Piscataway, NJ, USA). Blots were first incubated with Tris-buffered saline containing 5% bovine serum albumin and 1% Tween-20 for 1 h at room temperature. Then, the membranes were cut at approximately 40 kDa for parallel incubation with the antibody against β-tubulin (control loading) or the tested antibody. Afterwards, membranes were incubated overnight at 4 °C with a specific antibody against dysferlin (1:500), or β-tubulin (1:1000). Later, membranes were washed and incubated with a secondary antibody (anti-rabbit HRP, 1:5000 or anti-sheep HRP, 1:2500) for 1 h. Immunoreactive bands were detected using ECL Select Western Blotting Detection Reagent (GE Healthcare Bio-Sciences Corp., Piscataway, NJ, USA) and an image acquisition system Epichemi3 Darkroom (Cambridge Scientific, Watertown, MA, USA). ImageJ 1.43 m (NIH, Bethesda, MD, USA) was used for quantification. To measure the relative expression levels of dysferlin in the cell lines, the signal intensity of dysferlin bands obtained after background subtractions were normalized with respect to the signal intensity of the loading control (β-tubulin) detected on the same blot.

4.6. G-actin Incorporation into Filaments and Immunofluorescence

Non-transfected or transfected myoblasts were incubated 6 min at 37 °C in KGEP buffer (139 mM K$^+$ glutamate, 20 mM PIPES, 5 mM EGTA, 2 mM ATP-Mg^{2+}, 0.3 µM Alexa Fluor 488-G-actin conjugate, and 20 µM digitonin, pH 6.9) in the absence or presence of 10 µM free Ca^{2+} [27,40]. The online software Ca-EGTA Calculator v1.2 (University of California, Davis, CA, USA; https://somapp.ucdmc.ucdavis.edu/pharmacology/bers/maxchelator/CaEGTA-NIST.htm) was used to estimate the Ca^{+2} and EGTA concentrations to achieve 10 µM free Ca^{2+} with parameters of pH of 6.9 and temperature equal to 25 °C. Next, samples were fixed with 4% p-formaldehyde (PFA), stained with 5 mg/mL 4,6-diamidino-2-phenylindole (DAPI), and visualized by confocal microscopy. In cells transfected with dysferlin-HA, after PFA fixation, samples were incubated with the anti-dysferlin antibody (1:50), washed three times, and incubated with a Cy3-conjugated anti-rabbit secondary antibody (1:100). For immunodetection of annexin A2, after PFA fixation, samples were incubated with the anti-annexin A2 antibody, washed three times, and incubated with Cy2-conjugated anti-rabbit secondary antibody (1: 250). Images were captured at the equatorial plane of the cells using identical exposure settings between compared samples. To quantify G-actin incorporation, cell fluorescence of each individual cell was divided by the cell area, which was determined by manually drawing the cell outline using the differential interference contrast. For annexin A2 distribution, annexin A2 immunostaining was analyzed also in a single focal plane, and data are shown as the ratio of the mean fluorescence intensity of the cytosol/whole cell of annexin A2. All images were analyzed and processed using ImageJ software (NIH, Bethesda, USA).

4.7. Statistics

Results were expressed as means ± SEM. Normality of data was checked using the Kolmogorov–Smirnov test. Statistical comparisons were performed using a *t*-test. All statistical analyses were performed using InStat3 software (GraphPad Software Inc, La Jolla, CA, USA).

4.8. Ethics Statement

The investigators declare to know the Manual of Biosafety Regulations stipulated by CONICYT (Chile), version 2008; CDC (USA) Biosafety Manual 4th Edition; and Laboratory Biosafety, WHO, Geneva, 2005 (mainly in reference to experiments with recombinant DNA and RNA and the manipulation of cell lines). This research was approved by the Biosafety and Bioethics committees of Universidad de Valparaíso (Chile), approval identification numbers BS002/2016 and BEA080-216, respectively.

Supplementary Materials: The following are available online at http://www.mdpi.com/1422-0067/21/1/37/s1; Figure S1: High Ca^{2+} concentrations increase G actin incorporation into filaments; Figure S2: Intensity profiles of G-actin in C25 and dysferlin-deficient myoblasts; Table S1: Incorporation GST- amphiphysin-1 SH3 domain fusion protein in permeabilized C25 and DYSF3 myoblasts; Figure S3: Dysferlin expression in RCMH cells with stable dysferlin knockdown; Table S2: mCherry fluorescence intensity in C25 and DYSF3 myoblasts; Figure S4: Intensity profiles of annexin A2 (red) and G-actin (green) in C25 and DYSF3 myoblasts.

Author Contributions: X.B.-M., C.F.-C., H.A-S., and C.A. performed experiments and analysis. M.J.G. performed experiments and contributed to sample preparation. A.M.G.-J. performed experiments and critically revised the manuscript. A.B. and V.M. provided the cell lines used in this study as well as technical help and critical reading of the manuscript. M.C.M. and P.C. critically revised the manuscript and contributed to writing the manuscript. A.M.C. conceived the study, designed experiments, and drafted the manuscript. All authors contributed to interpretation of data and revised the final version of the manuscript. All authors have read and agreed to the published version of the manuscript.

Funding: This work has been supported by the FONDECYT (CONICYT, Chile) grant 1160495, P09-022-F from ICM-ECONOMIA, Chile, and CONICYT for funding of Basal Centre, CeBiB, FB0001 and P09-022-F from ICM-ECONOMIA, Chile. The Centro Interdisciplinario de Neurociencia de Valparaíso (CINV) is a Millennium Institute supported by the Millennium Scientific Initiative of the Ministerio de Economía, Fomento y Turismo.

Acknowledgments: We thank the platform for the immortalization of human cells from the Institut de Myologie (Paris, France) for providing the cell lines C25, DYSF2, DYSF3, AB320, and ER, and Dr. E. Gallardo, N. Levy, and S. Spuler for providing the original material. We also thank The Jain Foundation for providing the plasmid dysferlin-HA.

Conflicts of Interest: The authors declare no conflicts of interest.

Abbreviations

DAPI	4,6-diamidino-2-phenylindole
F-actin	Actin filament
HRP	Horseradish peroxidase
PI(4,5)P$_2$	Phosphatidylinositol 4,5-bisphosphate
PFA	P-formaldehyde

References

1. Therrien, C.; Di Fulvio, S.; Pickles, S.; Sinnreich, M. Characterization of lipid binding specificities of dysferlin C2 domains reveals novel interactions with phosphoinositides. *Biochemistry* **2009**, *48*, 2377–2384. [CrossRef]
2. Abdullah, N.; Padmanarayana, M.; Marty, N.J.; Johnson, C.P. Quantitation of the calcium and membrane binding properties of the C2 domains of dysferlin. *Biophys. J.* **2014**, *106*, 382–389. [CrossRef]
3. Bansal, D.; Miyake, K.; Vogel, S.S.; Groh, S.; Chen, C.C.; Williamson, R.; McNeil, P.L.; Campbell, K.P. Defective membrane repair in dysferlin-deficient muscular dystrophy. *Nature* **2003**, *423*, 168–172. [CrossRef]
4. Lennon, N.J.; Kho, A.; Bacskai, B.J.; Perlmutter, S.L.; Hyman, B.T.; Brown, R.H., Jr. Dysferlin interacts with annexins A1 and A2 and mediates sarcolemmal wound-healing. *J. Biol. Chem.* **2003**, *278*, 50466–50473. [CrossRef]
5. McDade, J.R.; Michele, D.E. Membrane damage-induced vesicle-vesicle fusion of dysferlin-containing vesicles in muscle cells requires microtubules and kinesin. *Hum. Mol. Genet.* **2014**, *23*, 1677–1686. [CrossRef]
6. Chiu, Y.H.; Hornsey, M.A.; Klinge, L.; Jørgensen, L.H.; Laval, S.H.; Charlton, R.; Barresi, R.; Straub, V.; Lochmüller, H.; Bushby, K. Attenuated muscle regeneration is a key factor in dysferlin-deficient muscular dystrophy. *Hum. Mol. Genet.* **2009**, *18*, 1976–1989. [CrossRef]
7. Demonbreun, A.R.; Fahrenbach, J.P.; Deveaux, K.; Earley, J.U.; Pytel, P.; McNally, E.M. Impaired muscle growth and response to insulin like growth factor 1 in dysferlin mediated muscular dystrophy. *Hum. Mol. Genet.* **2011**, *20*, 779–789. [CrossRef]
8. Kerr, J.P.; Ziman, A.P.; Mueller, A.L.; Muriel, J.M.; Kleinhans-Welte, E.; Gumerson, J.D.; Vogel, S.S.; Ward, C.W.; Roche, J.A.; Bloch, R.J. Dysferlin stabilizes stress-induced Ca2+ signaling in the transverse tubule membrane. *Proc. Natl. Acad. Sci. USA* **2013**, *110*, 20831–20836. [CrossRef]
9. Demonbreun, A.R.; Rossi, A.E.; Alvarez, M.G.; Swanson, K.E.; Deveaux, H.K.; Earley, J.U.; Hadhazy, M.; Vohra, R.; Walter, G.A.; Pytel, P.; et al. Dysferlin and myoferlin regulate transverse tubule formation and glycerol sensitivity. *Am. J. Pathol.* **2014**, *184*, 248–259. [CrossRef]
10. Codding, S.; Marty, M.; Abdullah, N.; Johnson, C. Dysferlin binds snares and stimulates membrane fusion in a calcium sensitive manner. *J. Biol. Chem.* **2016**, *291*, 14575–14584. [CrossRef]
11. Azakir, B.A.; Di Fulvio, S.; Therrien, C.; Sinnreich, M. Dysferlin interacts with tubulin and microtubules in mouse skeletal muscle. *PLoS ONE* **2010**, *5*, e10122. [CrossRef]
12. Cai, C.; Weisleder, N.; Ko, J.K.; Komazaki, S.; Sunada, Y.; Nishi, M.; Takeshima, H.; Ma, J. Membrane repair defects in muscular dystrophy are linked to altered interaction between MG53, caveolin-3, and dysferlin. *J. Biol. Chem.* **2009**, *284*, 15894–15902. [CrossRef]
13. Matsuda, C.; Miyake, K.; Kameyama, K.; Keduka, E.; Takeshima, H.; Imamura, T.; Araki, N.; Nishino, I.; Hayashi, Y. The C2A domain in dysferlin is important for association with MG53 (TRIM72). *PLoS Curr.* **2012**, *4*, e5035add8caff4. [CrossRef]
14. Liu, J.; Aoki, M.; Illa, I.; Wu, C.; Fardeau, M.; Angelini, C.; Serrano, C.; Urtizberea, J.A.; Hentati, F.; Hamida, M.B.; et al. Dysferlin, a novel skeletal muscle gene, is mutated in Miyoshi myopathy and limb girdle muscular dystrophy. *Nat. Genet.* **1998**, *20*, 31–36. [CrossRef]

15. Nguyen, K.; Bassez, G.; Bernard, R.; Krahn, M.; Labelle, V.; Figarella-Branger, D.; Pouget, J.; Hammouda el, H.; Béroud, C.; Urtizberea, A.; et al. Dysferlin mutations in LGMD2B, Miyoshi myopathy, and atypical dysferlinopathies. *Hum. Mutat.* **2005**, *26*, 165. [CrossRef]
16. Nguyen, K.; Bassez, G.; Krahn, M.; Bernard, R.; Laforêt, P.; Labelle, V.; Urtizberea, J.A.; Figarella-Branger, D.; Romero, N.; Attarian, S.; et al. Phenotypic study in 40 patients with dysferlin gene mutations: High frequency of atypical phenotypes. *Arch. Neurol.* **2007**, *64*, 1176–1182. [CrossRef]
17. Amato, A.A.; Brown, R.H., Jr. Dysferlinopathies. *Handb. Clin. Neurol.* **2011**, *101*, 111–118.
18. Cárdenas, A.M.; González-Jamett, A.M.; Cea, L.A.; Bevilacqua, J.A.; Caviedes, P. Dysferlin function in skeletal muscle: Possible pathological mechanisms and therapeutical targets in dysferlinopathies. *Exp. Neurol.* **2016**, *283*, 246–254. [CrossRef]
19. Lostal, W.; Bartoli, M.; Roudaut, C.; Bourg, N.; Krahn, M.; Pryadkina, M.; Borel, P.; Suel, L.; Roche, J.A.; Stockholm, D.; et al. Lack of correlation between outcomes of membrane repair assay and correction of dystrophic changes in experimental therapeutic strategy in dysferlinopathy. *PLoS ONE* **2012**, *7*, e38036. [CrossRef]
20. Selcen, D.; Stilling, G.; Engel, A.G. The earliest pathologic alterations in dysferlinopathy. *Neurology* **2001**, *56*, 1472–1481. [CrossRef]
21. Ho, M.; Post, C.M.; Donahue, L.R.; Lidov, H.G.; Bronson, R.T.; Goolsby, H.; Watkins, S.C.; Cox, G.A.; Brown, R.H., Jr. Disruption of muscle membrane and phenotype divergence in two novel mouse models of dysferlin deficiency. *Hum. Mol. Genet.* **2004**, *13*, 1999–2010. [CrossRef] [PubMed]
22. Cenacchi, G.; Fanin, M.; De Giorgi, L.B.; Angelini, C. Ultrastructural changes in dysferlinopathy support defective membrane repair mechanism. *J. Clin. Pathol.* **2005**, *58*, 190–195. [CrossRef] [PubMed]
23. Kim, S.; Shilagardi, K.; Zhang, S.; Hong, S.N.; Sens, K.L.; Bo, J.; Gonzalez, G.A.; Chen, E.H. A critical function for the actin cytoskeleton in targeted exocytosis of prefusion vesicles during myoblast fusion. *Dev. Cell.* **2007**, *12*, 571–586. [CrossRef] [PubMed]
24. Chiu, T.T.; Patel, N.; Shaw, A.E.; Bamburg, J.R.; Klip, A. Arp2/3- and cofilin-coordinated actin dynamics is required for insulin-mediated GLUT4 translocation to the surface of muscle cells. *Mol. Biol. Cell.* **2010**, *20*, 3529–3539. [CrossRef]
25. McDade, J.R.; Archambeau, A.; Michele, D.E. Rapid actin-cytoskeleton-dependent recruitment of plasma membrane-derived dysferlin at wounds is critical for muscle membrane repair. *FASEB J.* **2014**, *28*, 3660–3670. [CrossRef]
26. Demonbreun, A.R.; Quattrocelli, M.; Barefield, D.Y.; Allen, M.V.; Swanson, K.E.; McNally, E.M. An actin-dependent annexin complex mediates plasma membrane repair in muscle. *J. Cell Biol.* **2016**, *213*, 705–718. [CrossRef]
27. González-Jamett, A.M.; Baez-Matus, X.; Olivares, M.J.; Hinostroza, F.; Guerra-Fernández, M.J.; Vasquez-Navarrete, J.; Bui, M.T.; Guicheney, P.; Romero, N.B.; Bevilacqua, J.A.; et al. Dynamin-2 mutations linked to Centronuclear Myopathy impair actin-dependent trafficking in muscle cells. *Sci. Rep.* **2017**, *7*, 4580. [CrossRef]
28. Hanft, L.M.; Rybakova, I.N.; Patel, J.R.; Rafael-Fortney, J.A.; Ervasti, J.M. Cytoplasmic gamma-actin contributes to a compensatory remodeling response in dystrophin-deficient muscle. *Proc. Natl. Acad. Sci. USA* **2006**, *103*, 5385–5390. [CrossRef]
29. Prins, K.W.; Call, J.A.; Lowe, D.A.; Ervasti, J.M. Quadriceps myopathy caused by skeletal muscle-specific ablation of β(cyto)-actin. *J. Cell Sci.* **2011**, *124*, 951–957. [CrossRef]
30. Sonnemann, K.J.; Fitzsimons, D.P.; Patel, J.R.; Liu, Y.; Schneider, M.F.; Moss, R.L.; Ervasti, J.M. Cytoplasmic gamma-actin is not required for skeletal muscle development but its absence leads to a progressive myopathy. *Dev. Cell.* **2006**, *11*, 387–397. [CrossRef]
31. Philippi, S.; Bigot, A.; Marg, A.; Mouly, V.; Spuler, S.; Zacharias, U. Dysferlin-deficient immortalized human myoblasts and myotubes as a useful tool to study dysferlinopathy. *PLoS Curr.* **2012**, *4*, RRN1298. [CrossRef] [PubMed]
32. Cea, L.A.; Bevilacqua, J.A.; Arriagada, C.; Cárdenas, A.M.; Bigot, A.; Mouly, V.; Sáez, J.C.; Caviedes, P. The absence of dysferlin induces the expression of functional connexin-based hemichannels in human myotubes. *BMC Cell Biol.* **2016**, *17* (Suppl. 1), 15. [CrossRef] [PubMed]

33. Thorley, M.; Duguez, S.; Mazza, E.M.C.; Valsoni, S.; Bigot, A.; Mamchaoui, K.; Harmon, B.; Voit, T.; Mouly, V.; Duddy, W. Skeletal muscle characteristics are preserved in hTERT/cdk4 human myogenic cell lines. *Skelet Muscle.* **2016**, *6*, 43. [CrossRef] [PubMed]
34. Mamchaoui, K.; Trollet, C.; Bigot, A.; Negroni, E.; Chaouch, S.; Wolff, A.; Kandalla, P.K.; Marie, S.; Di Santo, J.; St Guily, J.L.; et al. Immortalized pathological human myoblasts: Towards a universal tool for the study of neuromuscular disorders. *Skelet. Muscle* **2011**, *1*, 34. [CrossRef] [PubMed]
35. Arandel, L.; Polay Espinoza, M.; Matloka, M.; Bazinet, A.; De Dea Diniz, D.; Naouar, N.; Rau, F.; Jollet, A.; Edom-Vovard, F.; Mamchaoui, K.; et al. Immortalized human myotonic dystrophy muscle cell lines to assess therapeutic compounds. *Dis. Model Mech.* **2017**, *10*, 487–497. [CrossRef] [PubMed]
36. Krahn, M.; Béroud, C.; Labelle, V.; Nguyen, K.; Bernard, R.; Bassez, G.; Figarella-Branger, D.; Fernandez, C.; Bouvenot, J.; Richard, I.; et al. Analysis of the DYSF mutational spectrum in a large cohort of patients. *Hum. Mutat.* **2009**, *30*, E345–E375. [CrossRef]
37. Gallardo, E.; de Luna, N.; Diaz-Manera, J.; Rojas-García, R.; Gonzalez-Quereda, L.; Flix, B.; de Morrée, A.; van der Maarel, S.; Illa, I. Comparison of dysferlin expression in human skeletal muscle with that in monocytes for the diagnosis of dysferlin myopathy. *PLoS ONE* **2011**, *6*, e29061. [CrossRef]
38. Cheng, X.; Zhang, X.; Yu, L.; Xu, H. Calcium signaling in membrane repair. *Semin. Cell Dev. Biol.* **2015**, *45*, 24–31. [CrossRef]
39. Barthélémy, F.; Defour, A.; Lévy, N.; Krahn, M.; Bartoli, M. Muscle Cells Fix Breaches by Orchestrating a Membrane Repair Ballet. *J. Neuromuscul. Dis.* **2018**, *5*, 21–28. [CrossRef]
40. Olivares, M.J.; González-Jamett, A.M.; Guerra, M.J.; Baez-Matus, X.; Haro-Acuña, V.; Martínez-Quiles, N.; Cárdenas, A.M. Src kinases regulate de novo actin polymerization during exocytosis in neuroendocrine chromaffin cells. *PLoS ONE* **2014**, *9*, e99001. [CrossRef]
41. Caviedes, R.; Liberona, J.L.; Hidalgo, J.; Tascon, S.; Salas, K.; Jaimovich, E. A human skeletal muscle cell line obtained from an adult donor. *Biochim. Biophys. Acta* **1992**, *1134*, 247–255. [CrossRef]
42. Liberona, J.L.; Caviedes, P.; Tascón, S.; Hidalgo, J.; Giglio, J.R.; Sampaio, S.V.; Caviedes, R.; Jaimovich, E. Expression of ion channels during differentiation of a human skeletal muscle cell line. *J. Muscle Res. Cell Motil.* **1997**, *18*, 587–598. [CrossRef] [PubMed]
43. Hayes, M.J.; Shao, D.; Bailly, M.; Moss, S.E. Regulation of actin dynamics by annexin 2. *EMBO J.* **2006**, *25*, 1816–1826. [CrossRef] [PubMed]
44. Rescher, U.; Gerke, V. Annexins—Unique membrane binding proteins with diverse functions. *J. Cell Sci.* **2004**, *117*, 2631–2639. [CrossRef] [PubMed]
45. Roostalu, U.; Strähle, U. In vivo imaging of molecular interactions at damaged sarcolemma. *Dev. Cell.* **2012**, *22*, 515–529. [CrossRef] [PubMed]
46. Defour, A.; Van der Meulen, J.H.; Bhat, R.; Bigot, A.; Bashir, R.; Nagaraju, K.; Jaiswal, J.K. Dysferlin regulates cell membrane repair by facilitating injury-triggered acid sphingomyelinase secretion. *Cell Death Dis.* **2014**, *5*, e1306. [CrossRef]
47. Tunduguru, R.; Chiu, T.T.; Ramalingam, L.; Elmendorf, J.S.; Klip, A.; Thurmond, D.C. Signaling of the p21-activated kinase (PAK1) coordinates insulin-stimulated actin remodeling and glucose uptake in skeletal muscle cells. *Biochem. Pharmacol.* **2014**, *92*, 380–388. [CrossRef]
48. Andrews, N.W.; Corrotte, M.; Castro-Gomes, T. Above the fray: Surface remodeling by secreted lysosomal enzymes leads to endocytosis-mediated plasma membrane repair. *Semin. Cell. Dev. Biol.* **2015**, *45*, 10–17. [CrossRef]
49. Defour, A.; Medikayala, S.; Van der Meulen, J.H.; Hogarth, M.W.; Holdreith, N.; Malatras, A.; Duddy, W.; Boehler, J.; Nagaraju, K.; Jaiswal, J.K. Annexin A2 links poor myofiber repair with inflammation and adipogenic replacement of the injured muscle. *Hum. Mol. Genet.* **2017**, *26*, 1979–1991. [CrossRef]
50. Jaiswal, J.K.; Lauritzen, S.P.; Scheffer, L.; Sakaguchi, M.; Bunkenborg, J.; Simon, S.M.; Kallunki, T.; Jäättelä, M.; Nylandsted, J. S100A11 is required for efficient plasma membrane repair and survival of invasive cancer cells. *Nat. Commun.* **2014**, *5*, 3795. [CrossRef]
51. Izadi, M.; Hou, W.; Qualmann, B.; Kessels, M.M. Direct effects of Ca2+/calmodulin on actin filament formation. *Biochem. Biophys. Res. Commun.* **2018**, *506*, 355–360. [CrossRef] [PubMed]
52. Trifaró, J.M.; Rosé, S.D.; Marcu, M.G. Scinderin, a Ca2+-dependent actin filament severing protein that controls cortical actin network dynamics during secretion. *Neurochem. Res.* **2000**, *25*, 133–144. [CrossRef] [PubMed]

53. Weed, S.A.; Du, Y.; Parsons, J.T. Translocation of cortactin to the cell periphery is mediated by the small GTPase Rac1. *J. Cell. Sci.* **1998**, *111*, 2433–2443. [PubMed]
54. González-Jamett, A.M.; Guerra, M.J.; Olivares, M.J.; Haro-Acuña, V.; Baéz-Matus, X.; Vásquez-Navarrete, J.; Momboisse, F.; Martinez-Quiles, N.; Cárdenas, A.M. The F-Actin Binding Protein Cortactin Regulates the Dynamics of the Exocytotic Fusion Pore through its SH3 Domain. *Front. Cell. Neurosci.* **2017**, *11*, 130. [CrossRef]
55. Johnsson, A.K.; Karlsson, R. Synaptotagmin 1 causes phosphatidyl inositol lipid-dependent actin remodeling in cultured non-neuronal and neuronal cells. *Exp. Cell Res.* **2012**, *318*, 114–126. [CrossRef]
56. van Gisbergen, P.A.; Li, M.; Wu, S.Z.; Bezanilla, M. Class II formin targeting to the cell cortex by binding PI(3,5)P(2) is essential for polarized growth. *J. Cell. Biol.* **2012**, *198*, 235–250. [CrossRef]
57. Bucki, R.; Wang, Y.H.; Yang, C.; Kandy, S.K.; Fatunmbi, O.; Bradley, R.; Pogoda, K.; Svitkina, T.; Radhakrishnan, R.; Janmey, P.A. Lateral distribution of phosphatidylinositol 4,5-bisphosphate in membranes regulates formin- and ARP2/3-mediated actin nucleation. *J. Biol. Chem.* **2019**, *294*, 4704–4722. [CrossRef]
58. Vizcarra, C.L.; Bor, B.; Quinlan, M.E. The role of formin tails in actin nucleation, processive elongation, and filament bundling. *J. Biol. Chem.* **2014**, *289*, 30602–30613. [CrossRef]
59. Hofhuis, J.; Bersch, K.; Büssenschütt, R.; Drzymalski, M.; Liebetanz, D.; Nikolaev, V.O.; Wagner, S.; Maier, L.S.; Gärtner, J.; Klinge, L.; et al. Dysferlin mediates membrane tubulation and links T-tubule biogenesis to muscular dystrophy. *J. Cell Sci.* **2017**, *130*, 841–852. [CrossRef]
60. Tran, H.T.; Anderson, L.H.; Knight, J.D. Membrane-Binding Cooperativity and Coinsertion by C2AB Tandem Domains of Synaptotagmins 1 and 7. *Biophys. J.* **2019**, *116*, 1025–1036. [CrossRef]
61. Ampong, B.N.; Imamura, M.; Matsumiya, T.; Yoshida, M.; Takeda, S. Intracellular localization of dysferlin and its association with the dihydropyridine receptor. *Acta Myol.* **2005**, *24*, 134–144. [PubMed]
62. Tans, X.; Thapa, N.; Choi, S.; Anderson, R.A. Emerging roles of PtdIns(4,5)P2–beyond the plasma membrane. *J. Cell Sci.* **2015**, *128*, 4047–4056.
63. Martin, T.F. PI(4,5)P$_2$-binding effector proteins for vesicle exocytosis. *Biochim. Biophys. Acta* **2015**, *1851*, 785–793. [CrossRef] [PubMed]
64. Zhao, P.; Xu, L.; Ait-Mou, Y.; de Tombe, P.P.; Han, R. Equal force recovery in dysferlin-deficient and wild-type muscles following saponin exposure. *J. Biomed. Biotechnol.* **2011**, *2011*, 235216. [CrossRef]

© 2019 by the authors. Licensee MDPI, Basel, Switzerland. This article is an open access article distributed under the terms and conditions of the Creative Commons Attribution (CC BY) license (http://creativecommons.org/licenses/by/4.0/).

Review

Pathogenic Puppetry: Manipulation of the Host Actin Cytoskeleton by *Chlamydia trachomatis*

Liam Caven [1,2] and Rey A. Carabeo [2,*]

1. School of Molecular Biosciences, Washington State University, Pullman, WA 99164, USA; liam.caven@wsu.edu
2. Department of Pathology and Microbiology, University of Nebraska Medical Center, Omaha, NE 68198-5900, USA
* Correspondence: rey.carabeo@unmc.edu; Tel.: +1-402-836-9778

Received: 4 December 2019; Accepted: 19 December 2019; Published: 21 December 2019

Abstract: The actin cytoskeleton is crucially important to maintenance of the cellular structure, cell motility, and endocytosis. Accordingly, bacterial pathogens often co-opt the actin-restructuring machinery of host cells to access or create a favorable environment for their own replication. The obligate intracellular organism *Chlamydia trachomatis* and related species exemplify this dynamic: by inducing actin polymerization at the site of pathogen-host attachment, *Chlamydiae* induce their own uptake by the typically non-phagocytic epithelium they infect. The interaction of chlamydial adhesins with host surface receptors has been implicated in this effect, as has the activity of the chlamydial effector TarP (translocated actin recruitment protein). Following invasion, *C. trachomatis* dynamically assembles and maintains an actin-rich cage around the pathogen's membrane-bound replicative niche, known as the chlamydial inclusion. Through further induction of actin polymerization and modulation of the actin-crosslinking protein myosin II, *C. trachomatis* promotes egress from the host via extrusion of the inclusion. In this review, we present the experimental findings that can inform our understanding of actin-dependent chlamydial pathogenesis, discuss lingering questions, and identify potential avenues of future study.

Keywords: actin cytoskeleton; chlamydia; bacterial pathogenesis

1. Introduction

The cytoskeleton is a highly dynamic structural framework composed of actin, microtubules, intermediate filaments, and septins. Restructuring of the cytoskeleton's actin component is critical for a variety of cellular processes, including endocytosis, motility, nutrient acquisition, and mitosis. It should therefore be unsurprising that modulation of the actin structure and function is a common theme amongst intracellular and extracellular pathogenic bacteria: by manipulating the cytoskeleton, these organisms can induce their own uptake by host cells, scavenge nutrients from host organelles, and ultimately establish a niche that facilitates their own replication.

The structure of the actin cytoskeleton is dynamically regulated by the recruitment of actin-polymerizing (or depolymerizing) factors, which consequently alter the balance of monomeric/globular actin (G-actin) and filamentous actin (F-actin) in the cytosol. Nucleation of a new actin filament requires the formation of a thermodynamically unfavorable actin trimer; to bypass this requirement, the Arp2/3 (actin-related protein) complex recruits a single G-actin monomer alongside two structural analogs to form a site of nucleation [1–3]. Subsequent F-actin branching by Arp2/3 is regulated by nucleation promotion factors (NPFs), such as N-WASP, SCAR/WAVE, and WASH [4,5]. The binding of NPFs to G-actin (Type I) or F-actin (Type II) facilitates conformational changes in Arp2/3 that enhance the complex's F-actin branching activity. NPF activation is in turn regulated by the Rho family GTPases RhoA-C, Rac1, and Cdc42—all of which are targets for modulation by pathogens seeking to restructure actin and thereby facilitate pathogenesis [6,7].

The manipulation of host actin can promote a wide variety of beneficial outcomes for the pathogen. *Salmonella* spp. translocate the effectors SopE and SopE2 into host cells—these guanine exchange factor (GEF) mimics enhance the activity of Rac1 and Cdc42, creating localized concentrations of F-actin at the apical surface of mucosal epithelia [8–11]. The result is extensive ruffling of the plasma membrane at the site of *Salmonella* attachment, leading to internalization of the pathogen via micropinocytosis [8,12–14]. Upon internalization and escape into the host cytosol, the Gram-positive intracellular pathogen *Listeria monocytogenes* induces the polymerization of actin on the bacterial surface through the activity of ActA, a surface protein functionally analogous to the nucleation promotion factor WASP [15]. ActA recruits an Arp2/3 complex to the bacterial pole, resulting in branched actin polymerization producing a comet-shaped structure that propels *Listeria* across the cytosol and into adjacent uninfected cells [16–20].

Indeed, this dynamic can be observed even in non-invasive bacterial pathogens. Enteropathogenic and enterohemorrhagic *E. coli* (EPEC/EHEC) induce the formation of distinctive, actin-rich pedestals that facilitate their attachment to gastric epithelia. The virulence factor Tir is responsible for this effect: upon delivery into host cells by the *E. coli* type III secretion system (T3SS), Tir is incorporated into the plasma membrane, promoting EPEC/EHEC attachment via binding to the bacterial adhesin intimin [21,22]. This clusters Tir at the site of attachment, inducing the phosphorylation of Tir's cytosolic domain by host kinases and the subsequent recruitment of Nck [23,24]. Nck is an adaptor protein that binds and activates N-WASP—consequently, the downstream effect of the Tir/Nck interaction is the recruitment of N-WASP and Arp2/3 complexes at sites of EPEC/EHEC attachment [25]. The resulting polymerization of branched actin produces pedestal formation, effacing the microvillar structure of the gastric mucosa and facilitating EPEC/EHEC colonization of the gastrointestinal tract [26,27].

The Gram-negative *Chlamydia* spp. constitute a valuable model for the study of actin modulation by bacterial pathogens. As obligate intracellular parasites, *Chlamydia trachomatis* and related species restructure actin in a variety of ways, to facilitate host invasion, maintain their replicative niche, and egress from host epithelial cells. Multiple *C. trachomatis* serovars have been isolated with distinct tissue tropism in the host: serovars A–C infect the conjunctival epithelium (producing the species' eponymous fibrotic trachoma), whereas serovars D–K and L1–L3 colonize the urogenital and anogenital tracts, respectively [28,29]. This extensive tissue tropism demonstrates a capability to modulate actin in multiple epithelial cell types, further borne out by the observation of pathogen-directed actin rearrangement by the respiratory pathogen *C. pneumoniae* [30,31], as well as the mouse- and guinea pig-infecting *C. muridarum* and *C. caviae* [32–34]. The study of chlamydial pathogenesis thus has the potential to reveal striking insight into both the pathogenic and steady-state regulation of actin in the host. In this review, we will summarize the field's current understanding of actin modulation by *Chlamydiae* both during and after host invasion, as well as discuss potential avenues of further research.

2. A Multilayered Assault: *Chlamydia* Redistributes the Actin Cytoskeleton to Invade Host Cells

The initial study of chlamydial invasion emphasized the importance of actin recruitment at sites where the infectious form of *Chlamydiae* (the elementary body, or EB) adheres to the host cell surface [35,36]. This early observation of in vitro infections occurred concomitant with the formation of microvillar structures that surround (and presumably internalize) invading *Chlamydia* [36,37]. The pharmacological disruption of F-actin (via cytochalasin D) or sequestration of G-actin (via latrunculin B) substantially inhibits chlamydial invasion and microvillar formation, suggesting that actin polymerization (not simply recruitment) is critical to fostering entry of the pathogen [35–37]. Furthermore, live-cell imaging of invasion events after cytochalasin D washout revealed the selective reestablishment of microvilli sites of EB attachment, indicating that this phenomenon is highly specific and pathogen-directed [36]. Clearly, the attachment of *Chlamydia* to the host cell surface is a critical first step to invasion, but how does EB attachment lead to actin rearrangement in the host? Extensive study of chlamydial invasion over the past two decades has suggested two, likely complementary, mechanisms: first, the engagement of host receptors by chlamydial adhesins facilitates the induction of actin polymerization indirectly, and second, bacterial effectors delivered by *Chlamydia* into the host remodel the actin cytoskeleton directly. The findings underlying these hypotheses are reviewed in the following sections.

2.1. Actin Modulation during Transient Chlamydial Adhesion

An early observation of *C. trachomatis* infections in vitro was that host attachment appeared to occur in two distinct stages: a reversible and temperature-insensitive interaction, followed by irreversible adhesion that requires physiological temperature [38–40]. Initial, transient EB attachment has since been shown to be an electrostatic interaction mediated by the glycosaminoglycan (GAG) heparan sulfate [41,42]. Heparan sulfate is demonstrably required for invasion by chlamydial serovars L1–L3, with other serovars exhibiting varying levels of GAG requirement [38,43]. Produced by a variety of cell types, heparan sulfate exhibits a strong negative charge that permits its binding to bacterial surface proteins; indeed, an early proposed mechanism for EB attachment posited the formation of a tripartite "molecular bridge" between heparan sulfate and host/bacterial GAG-binding proteins [44–47]. The bacterial surface protein OmcB is considered the primary GAG-binding protein in serovars whose adhesion is dependent on heparan sulfate [48–50]. OmcB has been shown to bind the alternative GAG heparin in vitro, and exhibits a variant amino acid sequence in the GAG-independent serovar E [50]. The chlamydial major outer membrane protein (MOMP) of *C. muridarum* has also been shown to bind GAGs to mediate invasion: treating host cells with recombinant MOMP or OmcB markedly reduces EB-host binding, as does treatment with monoclonal antibodies against *C. muridarum* MOMP [51].

The host component of this GAG-mediated bridge between EB and the host has remained elusive; however, recent work suggests an alternative, GAG-independent OmcB/MOMP attachment mechanism. Mounting evidence indicates that both proteins are post-translationally modified via glycosylation, as OmcB/MOMP recovered from EBs shows evidence of modification by N-linked high-mannose oligosaccharides [52–54]. Furthermore, both proteins exhibit reactivity to *Erythrina crista-galli* lectin, which binds to sites of N-acetyllactosamine glycosylation [54]. It has been further shown that glycosylated OmcB and MOMP are recognized by the protein galectin-1 (Gal1), which is both secreted by the host and bound on the plasma membrane to surface receptors [54]. Given that galectin-1 can demonstrably bridge EBs to host Gal1 receptors [54], both its secreted and membrane-associated state could conceivably facilitate EB-host binding. However, the latter state presents an intriguing avenue for the chlamydial induction of actin recruitment and invasion: in neurons, surface-associated Gal1 has been shown to initiate clathrin-independent endocytosis upon ligand binding, and Gal1 internalization is associated with F-actin polymerization during axonal growth [55].

A clathrin-independent mechanism of chlamydial invasion is of particular interest, given that clathrin's role in chlamydial invasion is somewhat controversial. In 1999, Boleti et al. assessed the importance of clathrin-mediated endocytosis to invasion via the expression of a dominant negative mutant of Eps15 [56]. Eps15 interacts with the clathrin adaptor protein AP-2 at clathrin-coated pits [57,58]; ablating this interaction thus inhibits clathrin-mediated endocytosis [59,60]. Importantly, the invasion of both *C. trachomatis* serovar L2 and *C. caviae* was unaffected by the expression of dominant negative Eps15, suggesting that clathrin is largely dispensable for chlamydial invasion [56]. However, a more recent study involving RNAi-mediated clathrin knockdown during serovar L2 infection presented a small (but statistically significant) invasion defect [61]. A likely explanation for these conflicting results is that *Chlamydia* employs multiple, functionally redundant mechanisms of invasion, each of varying clathrin dependence. The interaction of OmcB and MOMP with surface-associated Gal1 may enhance clathrin-independent invasion, either by other adhesin-receptor binding or the action of chlamydial effectors. Experiments studying invasion in galectin-knockout cells are therefore warranted, in order to assess the relative importance of this interaction to chlamydial invasion generally, and its potential role in actin recruitment specifically.

2.2. Actin Modulation during Irreversible Chlamydial Attachment

Irreversible, temperature-sensitive attachment of *Chlamydia* to the host is the result of bacterial adhesin binding to specific host receptors. Indeed, the study of chlamydial invasion has identified an extensive portfolio of host and bacterial proteins that promote attachment (reviewed in more detail by Romero et al. [62]). Of these, several host receptors bound by *Chlamydiae* promote actin recruitment and endocytosis elsewhere, suggesting their possible role in the pathogen's modulation of the actin cytoskeleton (Figure 1). For example, *C. trachomatis* serovar E and *C. muridarum* have been shown to directly bind the fibroblast growth factor (FGF), enabling molecular bridge formation between EBs and the host FGF receptor (FGFR) [63]. Intriguingly, the engagement of FGF-bound EBs with FGFR activated downstream signaling: phosphorylation of the FGFR substrate and docking protein FRS2α was increased by infection, and phosphorylated FRS2α was recruited to sites of EB attachment [63]. Typical binding of FGFR to its ligand results in internalization of the ligand–receptor complex, via Src/Eps8-dependent, clathrin-mediated endocytosis [64]. The involvement of Src is particularly significant, given that the kinase has been implicated in the recruitment of Rac1, WAVE, and Arp2/3 by the chlamydial invasion effector TarP (described below). Taken together, these observations suggest that *Chlamydia*-mediated activation of FGFR may aid in the recruitment of Src to the site of invasion.

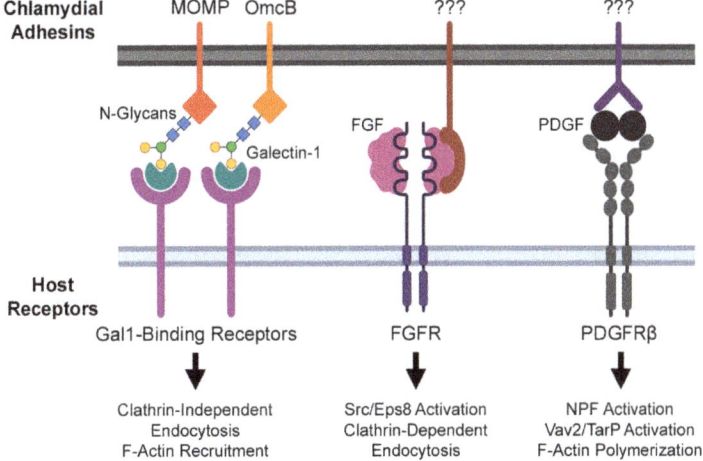

Figure 1. Summary of known chlamydial adhesin/host receptor interactions implicated in *Chlamydia*-directed modulation of actin.

An RNA interference screen of host factors contributing to invasion by *C. muridarum* identified the beta-isoform of the platelet-derived growth factor receptor (PDGFRβ) as another target of chlamydial attachment [33]. Much like FGFR, PDGFRβ is activated by EB binding and thereby promotes the tyrosine phosphorylation of multiple actin-modulatory proteins previously implicated in chlamydial invasion, including the nucleation promoting factors (NPFs) WAVE2 and cortactin, the Rac1 activator Vav2, and the chlamydial invasion effector TarP [33]. However, PDGFRβ activation was found to be dispensable for invasion, due to a compensatory mechanism dependent on activation of the Abelson (Abl) family of kinases [33]. This result suggests that PDGFRβ and Abl are elements of complementary invasion pathways, likely acting through a TarP-dependent mechanism given previous data linking TarP function to Vav2 and Abl (see below). Ultimately, a more thorough understanding of underlying mechanisms of each pathway is required, of which an essential first step is the identification of chlamydial factors that bind PDGFRβ or activate Abl kinases. Recent advances in *Chlamydia* transposon mutant strains may enable further study in this area, as would the use of Abl/PDGFRβ-deleted host cell lines generated via CRISPR/Cas9. Regardless, the substantial variety of host receptors

2.3. TarP, a Multifunctional Actin-Recruiting Effector

As noted previously, the invasion of host cells by EBs occurs concomitantly with extensive tyrosine phosphorylation of proteins at the site of EB attachment [65–67]. Pulldown of these species by tyrosine phosphoantibodies and subsequent characterization revealed the presence of a bacterial effector, now known as TarP (translocated actin recruitment protein) [67]. TarP has since been shown to be translocated into the host cytosol by the chlamydial T3SS, where it has a multifunctional role in facilitating actin polymerization and subsequent internalization of the pathogen. TarP has been repeatedly implicated in the invasion of C. trachomatis and is highly conserved amongst all Chlamydia species with sequenced genomes [34,68], emphasizing a critical role for this effector in chlamydial pathogenesis.

The structure and functional domains of TarP can be divided into C- and N-terminal halves (Figure 2). The C-terminal domains are highly conserved amongst C. trachomatis serovars and other Chlamydiae [34], containing a proline-rich domain (PRD), an actin-binding domain (ABD), a leucine-aspartate (LD) motif that binds focal adhesion kinase (FAK), and a vinculin-binding domain (VBD). At sites of EB attachment and engagement of the chlamydial T3SS, the TarP PRD allows translocated TarP molecules to oligomerize [69]. This is presumed to create a high local concentration of ABD-bound actin at the invasion site, likely facilitating the formation of a trimeric actin nucleus for subsequent F-actin polymerization (Figure 3A) [69]. By contrast, the LD and VBD domains of TarP stimulate actin polymerization indirectly. Recruitment of the signaling kinase FAK by the TarP LD induces actin remodeling in an Arp2/3-dependent fashion (Figure 3B) [70]. Binding of the VBD to the actin adaptor protein vinculin promotes F-actin recruitment, which leads to further actin polymerization via an uncharacterized mechanism (Figure 3C) [71].

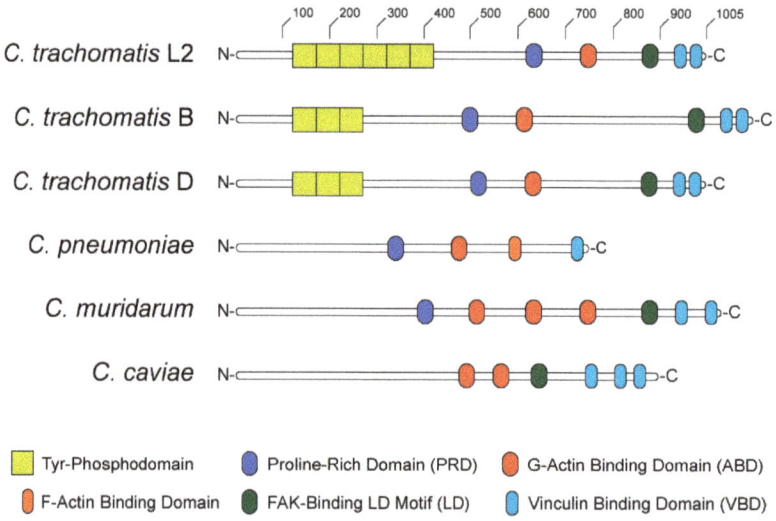

Figure 2. Diagram of TarP homologs in anogenital (L2), ocular (B), and genital (D) serovars of Chlamydia trachomatis, as well as C. pneumoniae, C. muridarum, and C. caviae.

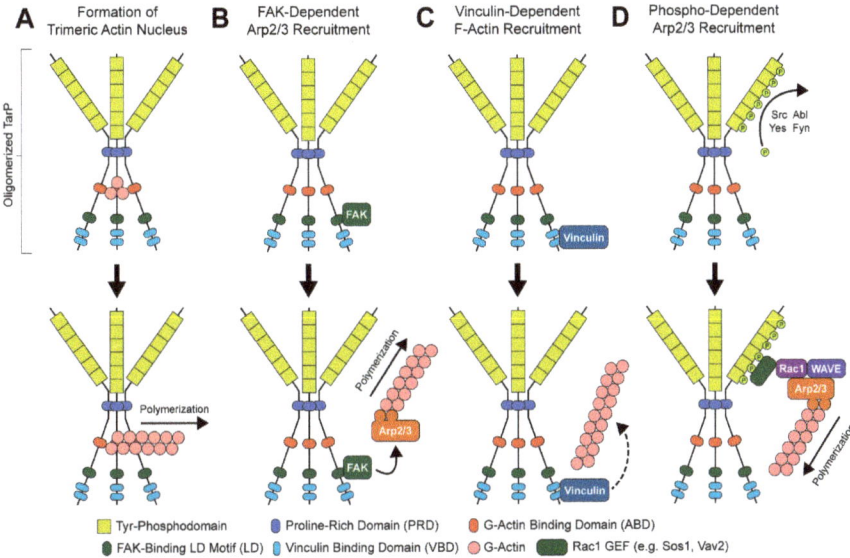

Figure 3. Mechanisms of TarP-directed actin modulation: (**A**) oligomerization-dependent formation of trimeric actin nucleus; (**B**) focal adhesion kinase (FAK)-dependent recruitment of the Arp2/3 complex; (**C**) vinculin-dependent recruitment of F-actin; (**D**) phosphorylation-dependent recruitment of an Arp2/3-activating protein complex.

Intriguingly, the complex of host and chlamydial proteins recruited at the invasion site bears substantial similarity to a focal adhesion—A membrane-associated protein complex that couples the actin cytoskeleton to the extracellular matrix (ECM). Focal adhesions are assembled in an FAK- and Src-dependent fashion around the cytosolic domain of ECM-bound integrins [72–76], subsequently recruiting F-actin stress fibers to play an important role in host cell motility and signaling [77]. The resemblance of the chlamydial invasion complex to the focal adhesion is supported by a number of additional observations: the focal adhesion component paxillin is also recruited at sites of EB invasion, the invasion efficiency is markedly decreased during *C. caviae* infection of FAK- or vinculin-knockout MEFs, and an interaction between the chlamydial surface protein Ctad1 and host integrin-β1 supports the adhesion of *C. trachomatis* serovar E to host cells in vitro [71,78]. Given the predominantly basolateral localization of integrins in the polarized epithelial cells *Chlamydiae* typically infect, access to integrin-β1 receptors would require the disruption of intercellular junctions via micro-abrasions in the epithelium or loss of apicobasolateral polarity, as occurs during documented the epithelial-to-mesenchymal transition (EMT) of *Chlamydia*-infected epithelial cells [79]. The functional consequences of the recruitment of a focal adhesion-like structure to the site of chlamydial invasion of polarized cells are largely unexplored. Additionally, given that TarP expression is detectable between 8 and 18 h post-infection (hpi)—Long after invasion, but before the differentiation of infectious EBs—The possibility that this effector may have a post-invasion role modulating FAK- or vinculin-containing structures in the host is difficult to ignore [80].

The N-terminal half of *C. trachomatis* TarP contains between one and twelve copies of a tyrosine phosphodomain, with the copy number varying by serovar (Figure 2) [34,67,68,81,82]. This domain exhibits a high sequence similarity to targets of Src- and Abl-family kinases; accordingly, p60-Src, Yes, Fyn, and Abl are capable of phosphorylating TarP in vitro [83]. A variety of host proteins have been shown to interact with the TarP N-terminus upon its phosphorylation, including two guanidine exchange factors (Sos1 and Vav2) that enhance the activity of the GTPase Rac1 [84]. Rac1 activation by TarP subsequently initiates actin branching and polymerization, via the Rac1- and WAVE-dependent recruitment of an Arp2/3 complex (Figure 3D) [84,85]. Additionally, it has been demonstrated that the recruitment of Arp2/3 may enhance the actin-nucleating function of TarP's C-terminal domains: the same in vitro pyrene-actin assay that established this function of TarP exhibited enhanced actin polymerization when Arp2/3 complexes were added [69]. The prevailing interpretation of this result is that the PRD/ABD domains and phosphodomains of TarP act in concert: the PRD and ABD domains promote the formation and extension of an initial (mother) actin filament, while the phosphodomains recruit Arp2/3 to facilitate branching from the mother filament.

The study of TarP has generally focused upon the *C. trachomatis* serovar homologs of the protein; perhaps unsurprisingly, characterization of the TarP orthologs has substantially complicated our understanding of the effector's various functions. *C. muridarum* and *C. caviae* possess TarP orthologs absent of any phosphodomains [34], suggesting that TarP phosphorylation is largely dispensable for the invasion of mouse and guinea pig epithelial cells. Additionally, recent functional characterization of the *C. pneumoniae* TarP homolog CPn0572 has revealed an alternative actin-modulating function: CPn0572 expressed in HEK293T cells exhibited an alternative localization pattern to that of TarP, associating with F-actin filaments projecting from the actin-rich aggregates where TarP is typically observed [31]. When ectopically expressed in yeast, CPn0572 promotes the increased incidence of short F-actin bundles [30,31], which may be reminiscent of the F-actin bundles formed by *C. trachomatis* TarP [86]. This effect is dependent on the CPn0572 VBD, as well as a novel domain that binds F-actin (in contrast to the G-actin-binding TarP/CPn0572 ABD) [30]. Furthermore, CPn0572 binding displaces the actin depolymerizing factor cofilin, suggesting that CPn0572 promotes the formation of actin bundles indirectly by inhibiting their disassembly [31]. The purpose of these structures to *C. pneumoniae* invasion is unclear, though cofilin and cofilin-displacing effectors have been implicated in the invasion of other intracellular pathogens [87,88].

The TarP orthologs of *C. pneumoniae*, *C. muridarum*, and *C. caviae* have the potential to provide critical insight into how requirements for the induction of phagocytosis differ, depending on the host tissue type (genital/ocular vs. pulmonary epithelium) or host organism (human vs. murine). Heterologous complementation of virulence factors has been used to interrogate the mechanism of host tropism for other pathogens. However, the relative intractability of *Chlamydia* to genetic manipulation generally and TarP knockout specifically has hitherto precluded this approach [89]. The recent development of *Chlamydia* conditional expression/knockout strains may enable the use of this method [90,91], thereby illustrating the relative contribution of TarP's many actin-modulating functions to invasion.

While the field's understanding of TarP's functional domains is extensive, the mechanism by which TarP translocation is regulated by pathogen adhesion to host cells remains somewhat unclear, beyond a presumed dependence on the activity of the chlamydial type III secretion system. In the Gram-negative pathogen *Yersinia pestis*, activation of the type III secretion system is cell contact-dependent, and possibly mediated by the YopN-TyeA-YscB-SycN complex—termed the calcium plug due to the complex's calcium-dependent restriction of effector secretion in a cell-free context [92]. Only one complex member has been identified in *Chlamydia*: CopN (homologous to YopN) [93]. However, the chlamydial T3SS exhibits a similar sensitivity to calcium-mediated inhibition—indeed, Jamison and Hackstadt observed that Ca^{2+} chelation by EGTA was required for cell-free chlamydial effector translocation to occur [94]. Cell-free TarP translocation by chlamydial EBs is demonstrably induced by cholesterol and sphingomyelin-enriched liposomes [94], which suggests a contact-dependent model of T3SS activation dependent on specific lipids. We refer the reader to reviews from Betts-Hampikian, Ferrell, and Fields for a comprehensive discussion of the chlamydial T3SS [95,96].

The field's current understanding of TarP's various functions indicates that this effector serves as a signaling scaffold, targeting host actin-remodeling machinery to the site of chlamydial invasion. The potential of TarP to nucleate actin likely serves to enhance the actin-dependent invasion of non-phagocytic cell types. It is possible that invasion is mechanistically linked with the nascent *Chlamydia*-containing vacuole's subsequent evasion of fusion with bactericidal host lysosomes, as has been observed with other intracellular pathogens. For example, recruitment of the small GTPase Rab5 to the vesicular surface during infection by *Brucella abortus* has been implicated in trafficking of the *Brucella* to a hospitable intracellular compartment; critically, Rab5 recruitment has been observed concomitant with invasion, during invagination of the clathrin-coated *Brucella*-containing vesicle [97]. The action of TarP and other actin-modulating effectors may similarly promote trafficking to a pathogen-favorable environment—elucidating the mechanistic connection between the invasion and chlamydial modulation of endocytic machinery is thus an intriguing topic for further study.

2.4. Actin-Depolymerizing Chlamydial Effectors

It is important to note that the induction of actin polymerization and bundling is not a terminal state for chlamydial invasion. After restructuring the cytoskeleton to initiate endocytosis, it is then necessary to restore steady-state actin dynamics in the host for invasion to proceed. Multiple candidates for this function have been identified in *Chlamydiae*—one such example is the *C. trachomatis* protein TmeA, a T3SS effector shown to associate with host AHNAK by a yeast two-hybrid assay [98,99]. AHNAK is a ubiquitous phosphoprotein implicated in cortical actin maintenance [100]; specifically, a C-terminal fragment of the protein has been shown to induce F-actin bundling [101]. Accordingly, the ectopic expression of TmeA in HeLa cells produced striking morphological changes consistent with the redistribution of actin [98]. This implies that TmeA may relieve the actin-bundling effects of TarP after invasion has begun; however, further study has complicated our understanding of TmeA's true role in pathogenesis.

A recent report by McKuen et al. adds more nuance to the relationship between TmeA and AHNAK [102]. While TmeA can recruit AHNAK to the site of invasion and inhibit AHNAK-mediated actin bundling in vitro, *C. trachomatis* invasion of AHNAK-knockout mouse embryonic fibroblasts (MEFs) had no invasion defect relative to that of wild-type cells [102]. Combined with the reported AHNAK-independent invasion defect in a TmeA-knockout *C. trachomatis* strain, these results suggest that TmeA facilitates invasion via an unknown, AHNAK-independent mechanism [102]. That being the case, how does TmeA-mediated inhibition AHNAK affect the host and benefit the pathogen? One possibility is that TmeA possesses an AHNAK-dependent, post-invasion function. However, the lack of an observed defect in replication or EB recovery in AHNAK-knockout MEFs would seem to contradict this hypothesis [102]. It is possible that any effect of AHNAK deficiency on pathogenesis was obscured by the choice of host; a murine, mesenchymal model for *C. trachomatis* infection likely has differing requirements for invasion relative to the polarized genital/ocular epithelium the pathogen infects in vivo. Unfortunately, the total knockdown of AHNAK has proven technically challenging in many cell lines [100,103]; continuing advances in CRISPR/Cas9 and related gene-editing techniques may thus be required to address this lingering question in a more physiologically relevant model.

The chlamydial toxin CT166 has also been proposed to indirectly mediate actin depolymerization, acting as an inhibitor of host actin-polymerizing factors. CT166 contains a DXD amino acid motif with homology to the *Clostridium difficile* toxin TcdB, a glucosylator of the Rho-family GTPases Rac1 and Cdc42 (both well-known regulators of actin polymerization) [104,105]. Glucosylation of Rac1/Cdc42 by TcdB is inhibitory and irreversible, occurring at a catalytic threonine residue essential for GTP binding [105]. The downstream effect of TcdB activity is a dramatic redistribution of filamentous actin, resulting in cell shrinking, the loss of stress fibers, and host cell death [105,106].

CT166 is highly variable between *C. trachomatis* serovars: serovar D possesses a severely truncated CT166 that nevertheless retains the TcdB-homologous glucosylating domain, whereas serovar L2 lacks the gene entirely [104]. The inoculation of HeLa cells with *C. trachomatis* serovar D or *C. muridarum* (which possesses a full-length copy of CT166) at a high multiplicity of infection resulted in morphological defects and cytotoxicity, reminiscent of the effects of TcdB [104]. Importantly, cytotoxicity was not observed in a comparable infection using the CT166-deficient *C. trachomatis* serovar L2, suggesting that this phenomenon was CT166-specific [104]. It has since been shown that CT166 expressed in HeLa cells inactivates the Rac/Cdc42 relative Ras, and that this effect could be ablated via mutation of the DXD domain [107,108].

CT166's role during invasion is somewhat unclear, given that the CT166-deficient *C. trachomatis* serovars can invade host cells in vitro with a comparable efficiency to those that possess the effector [104]. As with TmeA, this may be a product of the choice of host model; it remains to be seen whether CT166's glucosylating activity is equally dispensable in three-dimensional or organismal models of chlamydial infection. Given that the proposed function of CT166 runs counter to chlamydial invasion's established dependence on actin polymerization, it follows that the timing of CT166 delivery and activation is tightly regulated, in order to ensure that the toxin's activity is beneficial to the pathogen. Further study of CT166 may therefore inform the timing and kinetics of chlamydial effector delivery, as well as provide insight into the regulation of actin dynamics by the host.

3. Life after Invasion: *Chlamydia* Modulates Actin for Inclusion Stability and Host Egress

Given the robust and multifactorial nature of cytoskeletal restructuring during chlamydial invasion, it should not be surprising that the pathogen continues to manipulate the cytoskeleton once internalized by the host. Initial study of inclusion development has shown that *Chlamydiae* restructure the microtubule network to foster the development of their replicative niche. Another early observation of infections in vitro was that the treatment of infected cells with the microtubule polymerization inhibitor colchicine produces a marked increase in inclusion size [109]. Furthermore, rapid and inclusion-proximate microtubule assembly has been observed after washout of the microtubule-disrupting agent nocodazole [110,111]. It has been shown that *Chlamydia*-directed microtubule restructuring traffics the nascent *C. trachomatis* inclusion to the microtubule-organizing center (MTOC)—an outcome that is presumed to enable the chlamydial scavenging of nutrients from the host [111,112]. Indeed, the inclusion demonstrably interacts with a variety of nutrient-rich, MTOC-associated organelles, including the ER and multivesicular bodies (MVBs) [113–117]; pharmacological disruption of inclusion-ER/MVB association impairs chlamydial growth [113,115], consistent with a model of the chlamydial theft of nutrients from these organelles.

The role of actin in post-invasion chlamydial pathogenesis is more poorly characterized. In 1989, Campbell et al. discovered that mature inclusions are surrounded by a cage of actin and intermediate filaments [112], but it was only in the past decade that a mechanistic understanding of this structure's assembly and importance became clear. Additionally, recent study has shown that actin and actin-related structures play a role in pathogen egress, by facilitating the extrusion of intact chlamydial inclusions from the host.

3.1. Actin-Mediated Reinforcement of the Inclusion

As noted previously, early live-cell imaging experiments established the rapid recruitment of actin at the site of chlamydial invasion of the host [40]. Critically, the actin-rich structures surrounding invasion sites are transient, meaning that the actin-rich cage observed surrounding nascent inclusions is a distinct structure assembled later, likely with a separate function [112,118,119]. It has since been demonstrated that actin affords considerable stability to the inclusion: strikingly, treatment with 1% Triton X-100 does not significantly alter the morphology of the inclusion, despite the near-total solubilization of its membrane [118]. This apparent resistance to nonionic detergents was subsequently shown to depend on an F-actin ring surrounding the inclusion—actin disruption via latrunculin-A/B compromised the inclusion membrane integrity, leading to the detection of bacterial lipopolysaccharide (LPS) and antibody-labeled *Chlamydia* in the cytosol of infected cells [118]. Importantly, microtubule disruption with nocodazole did not produce similar changes in inclusion morphology. Taken together, these results imply a specific role for actin in structurally reinforcing the maturing inclusion [118].

The mechanism of actin recruitment to the inclusion remains somewhat unclear. Actin-rich structures in the host are largely indistinguishable from uninfected cells, in stark contrast to the pathogen's extensive remodeling of the microtubule network [110,112]. Furthermore, F-actin rings associated with maturing inclusions (30 hpi) are unaffected by the pharmacological disruption of stress fibers and cortical actin, including the inhibition of Rac1, Cdc42, ROCK, and myosin II [118]. Intriguingly, the inhibition of RhoA (either via siRNA knockdown or treatment with the *Clostridium botulinum* toxin C3-transferase) reduced the incidence of F-actin rings, and ablated the inclusion resistance to TX-100 [118]. Collectively, these results imply that the cage's actin component is assembled independently of host actin-rich structures.

More recent study of actin recruitment to the inclusion has complicated the model of cage assembly. Live-cell imaging of actin recruitment to mature inclusions (44 and 68 hpi) demonstrates sensitivity to formin inhibition (but not Arp2/3) [119], suggesting that cage formation may indeed depend upon the de novo, unbranched polymerization of actin at inclusions. However, actin recruitment at these time points occurred independently of RhoA and was sensitive to myosin II inhibition—seemingly in direct contradiction to ring assembly dynamics at earlier stages of infection [118,119]. How might data from these two stages of infection be reconciled? One explanation is that initial assembly and subsequent maintenance of the cage's actin component occurs via two distinct mechanisms—the former requiring RhoA, and the latter requiring formins and myosin II. A more longitudinal study of chlamydial actin recruitment is required to test this hypothesis, and may reveal how the mechanism of cage assembly changes with maturation of the inclusion.

Intermediate filaments have also been shown to contribute to the stability and function of the inclusion cage. While the infection of vimentin-knockout MEFs (a cell line also deficient in cytokeratins) produced F-actin rings comparable to the infection of wild-type cells, these rings lacked the highly compact and ordered morphology of their wild-type counterparts [118]. Inclusions in vimentin-knockout cells also lacked the previously observed resistance to TX-100. Taken together, these results suggest that the recruitment or assembly of F-actin rings surrounding the inclusion provides a scaffold for further inclusion reinforcement by intermediate filaments [118]. The chlamydial protease CPAF has been shown to cleave vimentin (as well as cytokeratin-8/18) within the protein's head domain, partially inhibiting its ability to form filamentous structures [118,120,121]. It is postulated that this interaction permits highly dynamic maintenance of the actin/filament cage, allowing the structure to accommodate an ever-expanding inclusion.

Collectively, these findings suggest that *Chlamydiae* dynamically reinforce the developing inclusion with F-actin and intermediate filaments (Figure 4). While the precise mechanism of F-actin synthesis and recruitment to the inclusion is somewhat unclear, the interconnected nature of the cytoskeleton would suggest that pathogen-directed actin restructuring may affect other cytoskeletal components, like the microtubule network. Given the established importance of the MTOC and vesicular trafficking to chlamydial growth (reviewed in detail by Nogueira et al. [122]), disruption of the microtubule network by any means has significant implications for the pathogen. Therefore, further study of the actin cage seems warranted, in order to evaluate a possible role for this structure in the chlamydial modulation of microtubule dynamics.

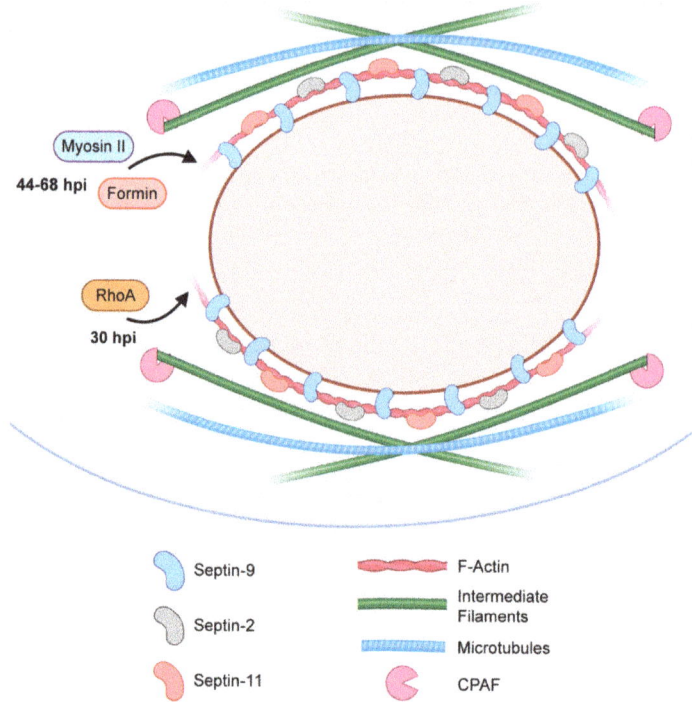

Figure 4. Diagram of the filamentous cage surrounding the maturing chlamydial inclusion.

3.2. The Role of Actin in Chlamydial Extrusion

Like other intracellular pathogens, *Chlamydiae* can exit the host via the induction of host cell lysis. However, another, less destructive, pathway exists as well. Early indication of an alternative mechanism of egress came in the observation of "scarred" host cells that—while intact—lacked a substantial portion of their plasma membranes [123]. Hybiske and Stephens proved the existence of non-lytic egress by taking advantage of the tendency of the chlamydial inclusion to exclude host cytosolic proteins [61]. Upon infecting GFP-expressing HeLa cells with *C. trachomatis*, they determined that while half of the observed inclusions became GFP-permeable during host cell lysis, the remaining half fused with the plasma membrane, excluding GFP throughout egress from the host, suggesting that an exocytotic event had occurred [61]. The incidence of this event (termed extrusion) was sensitive to disruption of the actin cytoskeleton by latrunculin B; intriguingly, the same did not hold true for nocodazole treatment, indicating that extrusion occurs independent of the microtubule network and conventional vesicular trafficking [61]. Further inhibitor treatments revealed a dependence on the nucleation promoting factor N-WASP, Rho-family GTPases, and myosin II. Taken together, these data imply an extrusion mechanism reliant on F-actin polymerization and bundling [61].

In 2013, Lutter et al. observed the recruitment of a complex of myosin-related host proteins to the inclusion surface, including myosin IIa; myosin IIb; and the phosphorylated, active forms of myosin light chain 2 (MLC2) and myosin light chain kinase (MLCK) [124,125]. The depletion of these myosin-activating factors via siRNA knockdown reduced the incidence of extrusion events, suggesting that *Chlamydia*-directed activation of myosin II promotes non-lytic egress of the pathogen [124]. Accordingly, the inclusion surface protein CT228 has been shown to directly interact with MYPT1, a negative regulator of myosin II activity. MYPT1 acts as a regulatory subunit of the phosphatase PP2—upon binding, MYPT1 alters the phosphatase's binding specificity, resulting in the PP2-mediated dephosphorylation (and, thereby, inactivation) of MLC2 [126]. The binding of MYPT1 to PP2 can be inhibited by the former's phosphorylation—given that CT228 preferentially recruits phosphorylated MYPT1 to the inclusion surface, it was initially proposed that CT228 might facilitate extrusion via MYPT1 inhibition [124].

However, the recent finding that the TargeTron-mediated knockout of CT228 increases the extrusion incidence (consistent with the siRNA-mediated knockdown of MYPT1) indicates an alternative, extrusion-inhibitory role for CT228 [127]. Combined with the earlier observation of the robust dephosphorylation of MYPT1 late in infection (42 hpi), this result suggests that CT228-mediated MYPT1 recruitment may result in MYPT1 activation instead [127], perhaps serving as a mechanism to regulate myosin II activity at the inclusion (and thereby regulating extrusion-mediated egress of the pathogen). Extrusion is further regulated by the interaction of the chlamydial inclusion membrane protein MrcA with IPTR3, an inositol 1,4,5-triphosphate receptor that acts as a Ca^{2+} channel [128]. Accordingly, the regulatory action of this complex further required the host cell Ca2+ sensor STIM1 [128]. It is important to note that premature host egress (i.e. prior to the differentiation of *Chlamydia* into infectious EBs) is a disastrous outcome for the pathogen that precludes subsequent infection of the surrounding tissue. The temporal regulation of MYPT1 activity via the expression of CT228 may therefore constitute a means by which *Chlamydiae* inhibit extrusion until differentiation is complete, thereby ensuring the release of invasion-competent organisms.

Further study has put forward a connection between the inclusion's actin cage and extrusion, specifically involving the septin family of GTP-binding proteins [129]. The function of septins is primarily structural—upon binding to GTP, septins oligomerize to form filamentous structures that associate with F-actin and other cytoskeletal components [130]. Accordingly, septin-2, 9, and 11 were recently observed to colocalize with the actin-rich structures surrounding the maturing inclusion [129]. Septin-9 knockdown both ablated actin recruitment to the inclusion and resulted in a 2 to 3-fold reduction of extrusion events [129], suggesting that these processes are functionally interrelated. This exciting result prompts several questions: does the structural content of the inclusion cage bias chlamydial egress via one pathway over another? How might the various structural components of the cage contribute to the activity of CT228 and myosin II? Further research in this area may indicate how actin, intermediate filaments, and septins contribute to vesicular trafficking and exocytosis, as well as provide insight into their specific role in chlamydial egress.

4. Conclusions and Summary

While mechanistic study of the actin modulation by *Chlamydiae* is both robust and ongoing, it is important to acknowledge that actin remodeling does not occur in a vacuum, and that the effect of *Chlamydia*-restructured actin on the host cell has been largely unassessed. The actin cytoskeleton is much more than a structural network for the host—in addition to driving endocytosis and cellular motility, mounting evidence implicates actin dynamics in transcriptional regulation.

The tensile force transmitted across actin filaments has been demonstrated to induce changes in gene expression, via the activity of actin- and/or tension-responsive signaling pathways. For example, the serum response factor (SRF) is demonstrably activated by regulators of actin dynamics, such as the cofilin-inhibiting LIM kinase-1 [131]. The application of tensile force to cardiac fibroblasts in vitro has shown the tension-dependent expression of α-smooth muscle actin, an SRF regulatory target [132]. Dysregulation of SRF activity has been associated with induction of the epithelial-to-mesenchymal transition (EMT)—a process through which epithelial cells transdifferentiate into scar-forming fibroblasts [133,134]. Intriguingly, *Chlamydia* infection has recently been shown to induce EMT, which likely promotes the scar-forming pathology observed in chronic *C. trachomatis* infections [79,135–137]. Could chlamydial restructuring of actin contribute to this effect? Further characterization of actin-regulated gene expression in *Chlamydia*-infected cells appears warranted.

Force transduction across the actin cytoskeleton also has demonstrable effects on the nuclear architecture. The inner nuclear membrane (INM) is reinforced by a network of intermediate filaments and associated proteins—including lamin-A/C and emerin—that are structurally integrated with the cytoskeleton via the LINC (linker of nucleoskeleton and cytoskeleton) complex [138]. Tension-mediated phosphorylation of emerin is associated with redistribution of the LINC complex and nuclear actin, which in turn are respectively associated with chromatin remodeling and myocardin-related transcription factor (MRTF) activity [139–142]. Given the multifaceted way in which *Chlamydiae* restructure actin and modulate the activity of myosin II—a contractile, actin-crosslinking protein—it is plausible that infection has the secondary effect of modulating tension- and actin-responsive gene expression, and that characterization of this effect may provide insight into chlamydial pathogenesis.

The polarized epithelial mucosa that *Chlamydiae* typically infect often react to bacterial insult via the concerted exfoliation and apoptosis of infected cells—a process that requires the degradation of focal adhesions and related cell-ECM adherent structures, as well as extensive remodeling of the actin cytoskeleton [143]. Exfoliation and apoptosis of the host cell is clearly an unfavorable outcome for any intracellular pathogen. Accordingly, there is evidence indicating that other intracellular bacteria inhibit this host response to infection. The *Shigella flexneri* effector OspE has been shown to reinforce sites of cell-ECM contact through an interaction with integrin-linked kinase (ILK), which results in an increased incidence of integrin-β1 on the host cell surface and the consequent assembly of focal adhesions [144]. A secondary effect of the OspE modulation of ILK is the stabilization of focal adhesions, via the inhibition of focal adhesion kinase (FAK). Given that the chlamydial effector TarP is known to recruit FAK (albeit in the context of invasion) [70], it is tempting to speculate that *Chlamydia* might inhibit exfoliation in a similar fashion. Ultimately, further characterization of host cell adherence and exfoliation during chlamydial infection is necessary to address this hypothesis.

In summary, ongoing study of actin modulation by *Chlamydiae* has revealed the multifaceted way in which these pathogens restructure the host actin cytoskeleton. During the two-stage attachment of chlamydial EBs to the host cell surface, the pathogen engages with a variety of host surface proteins—including galectin-1, FGFR, and PDGFRβ—that in turn facilitate the recruitment of signaling factors promoting actin polymerization at the attachment site [33,52–54,63]. The subsequent delivery of chlamydial effectors by the pathogen's type III secretion system then induces actin-dependent invasion of the host cell. The chlamydial effector TarP contributes to this effect in multiple ways: via the recruitment of F-actin and Arp2/3 actin-polymerizing complexes, via direct nucleation of actin polymerization, and by providing a scaffold for actin-modulatory signaling [69–71,84,85]. While the mechanism by which actin regulation at the invasion site returns to a steady state remains somewhat unclear, the chlamydial effectors TmeA and CT166 have been demonstrated to inhibit F-actin bundling and polymerization, respectively, and may contribute to this process [98,99,102,104,107,108]. Once invasion is complete, actin is recruited in either an RhoA- or ROCK-dependent fashion to the maturing inclusion alongside intermediate filaments and septins, providing dynamic structural reinforcement to *Chlamydia*'s replicative niche [118,119]. Finally, non-lytic chlamydial egress from the host depends on actin polymerization, as well as the modulation of myosin II by the inclusion membrane protein CT228 [61,124,127]. While the mechanistic underpinnings of actin restructuring by *Chlamydiae* are coming into focus, the effects of this restructuring on the host are relatively uncharacterized, presenting intriguing opportunities for future study.

Author Contributions: Conceptualization, L.C. and R.A.C.; writing—original draft preparation, L.C.; writing—review and editing, R.A.C.; visualization, L.C.; supervision, R.A.C.; project administration, R.A.C.; funding acquisition, R.A.C. All authors have read and agreed to the published version of the manuscript.

Funding: This research was funded by the National Institutes of Health, grant number AI065545.

Acknowledgments: The authors acknowledge Amanda Brinkworth for a critical review of the manuscript. Figure 1; Figure 4 were created with BioRender.

Conflicts of Interest: The authors declare no conflicts of interest.

Abbreviations

G-actin	Globular actin
F-actin	Filamentous actin
Arp2/3	Actin-related protein 2/3
NPF	Nucleation promotion factor
N-WASP	Neuronal Wiskott–Aldrich Syndrome protein
WAVE	WASP-family verprolin-homologous family protein
WASH	Wiskott–Aldrich syndrome and SCAR homolog
GTP	Guanosine-5′-triphosphate
EPEC	Enteropathogenic *Escherichia coli*
EHEC	Enterohemorrhagic *Escherichia coli*
T3SS	Type III Secretion System
EB	Elementary body
GAG	Glycosaminoglycan
MOMP	Major outer membrane protein
FGF	Fibroblast growth factor
FGFR	Fibroblast growth factor receptor
PDGFRβ	Platelet-derived growth factor receptor beta
Abl	Abelson-family kinase
TarP	Translocated actin recruitment protein
EGTA	Ethylene glycol-bis(β-aminoethyl ether)-N,N,N′,N′-tetraacetic acid
PRD	Proline-rich domain
ABD	Actin binding domain
FAK	Focal adhesion kinase
LD	Leucine-aspartatic acid repeat domain
VBD	Vinculin binding domain
ECM	Extracellular matrix
EMT	Epithelial-mesenchymal transition
hpi	Hours post-infection
MEF	Mouse embryonic fibroblast
MTOC	Microtubule organizing center
ER	Endoplasmic reticulum
MVB	Multivesicular body
ROCK	Rho-associated protein kinase
CPAF	Chlamydial protease/proteasome-like activity factor
MLC2	Myosin light chain 2
MLCK	Myosin light chain kinase
MYPT1	Myosin phosphatase target subunit 1
PP2	Protein phosphatase 2
IPTR3	Inositol-1,4,5-triphosphate receptor, type 3
SRF	Serum response factor
MRTF	Myocardin-related transcription factor
INM	Inner nuclear membrane
LINC	Linker of nucleoskeleton and cytoskeleton
ILK	Integrin-linked kinase

References

1. Goley, E.D.; Welch, M.D. The ARP2/3 complex: An actin nucleator comes of age. *Nat. Rev. Mol. Cell Biol.* **2006**, *7*, 713–726. [CrossRef] [PubMed]
2. Pollard, T.D. Regulation of actin filament assembly by Arp2/3 complex and formins. *Annu. Rev. Biophys. Biomol. Struct.* **2007**, *36*, 451–477. [CrossRef] [PubMed]
3. Rotty, J.D.; Wu, C.; Bear, J.E. New insights into the regulation and cellular functions of the ARP2/3 complex. *Nat. Rev. Mol. Cell Biol.* **2013**, *14*, 7–12. [CrossRef] [PubMed]

4. Machesky, L.M.; Mullins, R.D.; Higgs, H.N.; Kaiser, D.A.; Blanchoin, L.; May, R.C.; Hall, M.E.; Pollard, T.D. Scar, a WASp-related protein, activates nucleation of actin filaments by the Arp2/3 complex. *Proc. Natl. Acad. Sci. USA* **1999**, *96*, 3739–3744. [CrossRef] [PubMed]
5. Linardopoulou, E.V.; Parghi, S.S.; Friedman, C.; Osborn, G.E.; Parkhurst, S.M.; Trask, B.J. Human Subtelomeric WASH Genes Encode a New Subclass of the WASP Family. *PLoS Genet.* **2007**, *3*, e237. [CrossRef] [PubMed]
6. Hodge, R.G.; Ridley, A.J. Regulating Rho GTPases and their regulators. *Nat. Rev. Mol. Cell Biol.* **2016**, *17*, 496–510. [CrossRef] [PubMed]
7. Popoff, M.R. Bacterial factors exploit eukaryotic Rho GTPase signaling cascades to promote invasion and proliferation within their host. *Small GTPases* **2014**, *5*, e983863. [CrossRef]
8. Zhou, D.; Chen, L.M.; Hernandez, L.; Shears, S.B.; Galán, J.E. A Salmonella inositol polyphosphatase acts in conjunction with other bacterial effectors to promote host cell actin cytoskeleton rearrangements and bacterial internalization. *Mol. Microbiol.* **2001**, *39*, 248–259. [CrossRef]
9. Stender, S.; Friebel, A.; Linder, S.; Rohde, M.; Mirold, S.; Hardt, W.-D. Identification of SopE2 from Salmonella typhimurium, a conserved guanine nucleotide exchange factor for Cdc42 of the host cell. *Mol. Microbiol.* **2000**, *36*, 1206–1221. [CrossRef]
10. Hardt, W.D.; Chen, L.M.; Schuebel, K.E.; Bustelo, X.R.; Galán, J.E.S. typhimurium encodes an activator of Rho GTPases that induces membrane ruffling and nuclear responses in host cells. *Cell* **1998**, *93*, 815–826. [CrossRef]
11. Friebel, A.; Ilchmann, H.; Aepfelbacher, M.; Ehrbar, K.; Machleidt, W.; Hardt, W.-D. SopE and SopE2 from Salmonella typhimurium Activate Different Sets of RhoGTPases of the Host Cell. *J. Biol. Chem.* **2001**, *276*, 34035–34040. [CrossRef] [PubMed]
12. Hardt, W.D.; Urlaub, H.; Galán, J.E. A substrate of the centisome 63 type III protein secretion system of Salmonella typhimurium is encoded by a cryptic bacteriophage. *Proc. Natl. Acad. Sci. USA* **1998**, *95*, 2574–2579. [CrossRef] [PubMed]
13. Garcia-del Portillo, F.; Finlay, B.B. Salmonella invasion of nonphagocytic cells induces formation of macropinosomes in the host cell. *Infect. Immun.* **1994**, *62*, 4641–4645. [PubMed]
14. Humphreys, D.; Davidson, A.; Hume, P.J.; Koronakis, V. Salmonella virulence effector SopE and Host GEF ARNO cooperate to recruit and activate WAVE to trigger bacterial invasion. *Cell Host Microbe* **2012**, *11*, 129–139. [CrossRef]
15. Boujemaa-Paterski, R.; Gouin, E.; Hansen, G.; Samarin, S.; Le Clainche, C.; Didry, D.; Dehoux, P.; Cossart, P.; Kocks, C.; Carlier, M.F.; et al. Listeria protein ActA mimics WASp family proteins: It activates filament barbed end branching by Arp2/3 complex. *Biochemistry* **2001**, *40*, 11390–11404. [CrossRef]
16. Welch, M.D.; Rosenblatt, J.; Skoble, J.; Portnoy, D.A.; Mitchison, T.J. Interaction of human Arp2/3 complex and the Listeria monocytogenes ActA protein in actin filament nucleation. *Science* **1998**, *281*, 105–108. [CrossRef]
17. Skoble, J.; Portnoy, D.A.; Welch, M.D. Three regions within ActA promote Arp2/3 complex-mediated actin nucleation and Listeria monocytogenes motility. *J. Cell Biol.* **2000**, *150*, 527–538. [CrossRef]
18. Ireton, K. Molecular mechanisms of cell-cell spread of intracellular bacterial pathogens. *Open Biol.* **2013**, *3*, 130079. [CrossRef]
19. Welch, M.D.; Way, M. Arp2/3-mediated actin-based motility: A tail of pathogen abuse. *Cell Host Microbe* **2013**, *14*, 242–255. [CrossRef]
20. Lamason, R.L.; Welch, M.D. Actin-based motility and cell-to-cell spread of bacterial pathogens. *Curr. Opin. Microbiol.* **2017**, *35*, 48–57. [CrossRef]
21. Kenny, B.; DeVinney, R.; Stein, M.; Reinscheid, D.J.; Frey, E.A.; Finlay, B.B. Enteropathogenic E. coli (EPEC) Transfers Its Receptor for Intimate Adherence into Mammalian Cells. *Cell* **1997**, *91*, 511–520. [CrossRef]
22. Deibel, C.; Krämer, S.; Chakraborty, T.; Ebel, F. EspE, a novel secreted protein of attaching and effacing bacteria, is directly translocated into infected host cells, where it appears as a tyrosine-phosphorylated 90 kDa protein. *Mol. Microbiol.* **1998**, *28*, 463–474. [CrossRef] [PubMed]
23. Warawa, J.; Kenny, B. Phosphoserine modification of the enteropathogenic Escherichia coli Tir molecule is required to trigger conformational changes in Tir and efficient pedestal elongation. *Mol. Microbiol.* **2001**, *42*, 1269–1280. [CrossRef] [PubMed]
24. Campellone, K.G.; Rankin, S.; Pawson, T.; Kirschner, M.W.; Tipper, D.J.; Leong, J.M. Clustering of Nck by a 12-residue Tir phosphopeptide is sufficient to trigger localized actin assembly. *J. Cell Biol.* **2004**, *164*, 407–416. [CrossRef]

25. Gruenheid, S.; DeVinney, R.; Bladt, F.; Goosney, D.; Gelkop, S.; Gish, G.D.; Pawson, T.; Finlay, B.B. Enteropathogenic E. coli Tir binds Nck to initiate actin pedestal formation in host cells. *Nat. Cell Biol.* **2001**, *3*, 856–859. [CrossRef]
26. Frankel, G.; Phillips, A.D. Attaching effacing Escherichia coli and paradigms of Tir-triggered actin polymerization: Getting off the pedestal. *Cell. Microbiol.* **2008**, *10*, 549–556. [CrossRef]
27. Garmendia, J.; Frankel, G.; Crepin, V.F. Enteropathogenic and Enterohemorrhagic Escherichia coli Infections: Translocation, Translocation, Translocation. *Infect. Immun.* **2005**, *73*, 2573–2585. [CrossRef]
28. Elwell, C.; Mirrashidi, K.; Engel, J. Chlamydia cell biology and pathogenesis. *Nat. Rev. Microbiol.* **2016**, *14*, 385–400. [CrossRef]
29. Malhotra, M.; Sood, S.; Mukherjee, A.; Muralidhar, S.; Bala, M. Genital Chlamydia trachomatis: An update. *Indian J. Med. Res.* **2013**, *138*, 303–316.
30. Braun, C.; Alcázar-Román, A.R.; Laska, A.; Mölleken, K.; Fleig, U.; Hegemann, J.H. CPn0572, the C. pneumoniae ortholog of TarP, reorganizes the actin cytoskeleton via a newly identified F-actin binding domain and recruitment of vinculin. *PLoS ONE* **2019**, *14*, e0210403. [CrossRef]
31. Zrieq, R.; Braun, C.; Hegemann, J.H. The Chlamydia pneumoniae Tarp Ortholog CPn0572 Stabilizes Host F-Actin by Displacement of Cofilin. *Front. Cell. Infect. Microbiol.* **2017**, *7*, 511. [CrossRef] [PubMed]
32. Jiwani, S.; Alvarado, S.; Ohr, R.J.; Romero, A.; Nguyen, B.; Jewett, T.J. Chlamydia trachomatis Tarp Harbors Distinct G and F Actin Binding Domains That Bundle Actin Filaments. *J. Bacteriol.* **2013**, *195*, 708–716. [CrossRef] [PubMed]
33. Elwell, C.A.; Ceesay, A.; Kim, J.H.; Kalman, D.; Engel, J.N. RNA Interference Screen Identifies Abl Kinase and PDGFR Signaling in Chlamydia trachomatis Entry. *PLoS Pathog.* **2008**, *4*, e1000021. [CrossRef] [PubMed]
34. Clifton, D.R.; Dooley, C.A.; Grieshaber, S.S.; Carabeo, R.A.; Fields, K.A.; Hackstadt, T. Tyrosine Phosphorylation of the Chlamydial Effector Protein Tarp Is Species Specific and Not Required for Recruitment of Actin. *Infect. Immun.* **2005**, *73*, 3860–3868. [CrossRef] [PubMed]
35. Carabeo, R. Bacterial subversion of host actin dynamics at the plasma membrane. *Cell. Microbiol.* **2011**, *13*, 1460–1469. [CrossRef] [PubMed]
36. Carabeo, R.A.; Grieshaber, S.S.; Fischer, E.; Hackstadt, T. Chlamydia trachomatis Induces Remodeling of the Actin Cytoskeleton during Attachment and Entry into HeLa Cells. *Infect. Immun.* **2002**, *70*, 3793–3803. [CrossRef]
37. Ford, C.; Nans, A.; Boucrot, E.; Hayward, R.D. Chlamydia exploits filopodial capture and a macropinocytosis-like pathway for host cell entry. *PLoS Pathog.* **2018**, *14*, e1007051. [CrossRef]
38. Kuo, C.C.; Grayston, T. Interaction of Chlamydia trachomatis organisms and HeLa 229 cells. *Infect. Immun.* **1976**, *13*, 1103–1109.
39. Byrne, G.I. Kinetics of phagocytosis of Chlamydia psittaci by mouse fibroblasts (L cells): Separation of the attachment and ingestion stages. *Infect. Immun.* **1978**, *19*, 607–612.
40. Carabeo, R.A.; Hackstadt, T. Isolation and Characterization of a Mutant Chinese Hamster Ovary Cell Line That Is Resistant to Chlamydia trachomatis Infection at a Novel Step in the Attachment Process. *Infect. Immun.* **2001**, *69*, 5899–5904. [CrossRef]
41. Moulder, J.W. Interaction of chlamydiae and host cells in vitro. *Microbiol. Mol. Biol. Rev.* **1991**, *55*, 143–190.
42. Tiwari, V.; Maus, E.; Sigar, I.M.; Ramsey, K.H.; Shukla, D. Role of heparan sulfate in sexually transmitted infections. *Glycobiology* **2012**, *22*, 1402–1412. [CrossRef] [PubMed]
43. Taraktchoglou, M.; Pacey, A.A.; Turnbull, J.E.; Eley, A. Infectivity of Chlamydia trachomatis serovar LGV but not E is dependent on host cell heparan sulfate. *Infect. Immun.* **2001**, *69*, 968–976. [CrossRef] [PubMed]
44. Sugahara, K.; Kitagawa, H. Heparin and heparan sulfate biosynthesis. *IUBMB Life* **2002**, *54*, 163–175. [CrossRef] [PubMed]
45. Rasmussen-Lathrop, S.J.; Koshiyama, K.; Phillips, N.; Stephens, R.S. Chlamydia-dependent biosynthesis of a heparan sulphate-like compound in eukaryotic cells. *Cell. Microbiol.* **2000**, *2*, 137–144. [CrossRef]
46. Stephens, R.S. Molecular mimicry and Chlamydia trachomatis infection of eukaryotic cells. *Trends Microbiol.* **1994**, *2*, 99–101. [CrossRef]
47. Zhang, J.P.; Stephens, R.S. Mechanism of C. trachomatis attachment to eukaryotic host cells. *Cell* **1992**, *69*, 861–869. [CrossRef]
48. Fadel, S.; Eley, A. Chlamydia trachomatis OmcB protein is a surface-exposed glycosaminoglycan-dependent adhesin. *J. Med. Microbiol.* **2007**, *56*, 15–22. [CrossRef]

49. Moelleken, K.; Hegemann, J.H. The Chlamydia outer membrane protein OmcB is required for adhesion and exhibits biovar-specific differences in glycosaminoglycan binding. *Mol. Microbiol.* **2008**, *67*, 403–419. [CrossRef]
50. Fechtner, T.; Stallmann, S.; Moelleken, K.; Meyer, K.L.; Hegemann, J.H. Characterization of the interaction between the chlamydial adhesin OmcB and the human host cell. *J. Bacteriol.* **2013**, *195*, 5323–5333. [CrossRef]
51. Su, H.; Raymond, L.; Rockey, D.D.; Fischer, E.; Hackstadt, T.; Caldwell, H.D. A recombinant Chlamydia trachomatis major outer membrane protein binds to heparan sulfate receptors on epithelial cells. *Proc. Natl. Acad. Sci. USA* **1996**, *93*, 11143–11148. [CrossRef] [PubMed]
52. Campbell, L.A.; Lee, A.; Kuo, C. Cleavage of the N-linked oligosaccharide from the surfaces of Chlamydia species affects infectivity in the mouse model of lung infection. *Infect. Immun.* **2006**, *74*, 3027–3029. [CrossRef] [PubMed]
53. Swanson, A.F.; Kuo, C.C. Evidence that the major outer membrane protein of Chlamydia trachomatis is glycosylated. *Infect. Immun.* **1991**, *59*, 2120–2125. [PubMed]
54. Lujan, A.L.; Croci, D.O.; Gambarte Tudela, J.A.; Losinno, A.D.; Cagnoni, A.J.; Mariño, K.V.; Damiani, M.T.; Rabinovich, G.A. Glycosylation-dependent galectin-receptor interactions promote Chlamydia trachomatis infection. *Proc. Natl. Acad. Sci. USA* **2018**, *115*, E6000–E6009. [CrossRef]
55. Quintá, H.R.; Wilson, C.; Blidner, A.G.; González-Billault, C.; Pasquini, L.A.; Rabinovich, G.A.; Pasquini, J.M. Ligand-mediated Galectin-1 endocytosis prevents intraneural H_2O_2 production promoting F-actin dynamics reactivation and axonal re-growth. *Exp. Neurol.* **2016**, *283*, 165–178. [CrossRef]
56. Boleti, H.; Benmerah, A.; Ojcius, D.M.; Cerf-Bensussan, N.; Dautry-Varsat, A. Chlamydia infection of epithelial cells expressing dynamin and Eps15 mutants: Clathrin-independent entry into cells and dynamin-dependent productive growth. *J. Cell Sci.* **1999**, *112*, 1487–1496.
57. Benmerah, A.; Gagnon, J.; Bègue, B.; Mégarbané, B.; Dautry-Varsat, A.; Cerf-Bensussan, N. The tyrosine kinase substrate eps15 is constitutively associated with the plasma membrane adaptor AP-2. *J. Cell Biol.* **1995**, *131*, 1831–1838. [CrossRef]
58. Tebar, F.; Sorkina, T.; Sorkin, A.; Ericsson, M.; Kirchhausen, T. Eps15 is a component of clathrin-coated pits and vesicles and is located at the rim of coated pits. *J. Biol. Chem.* **1996**, *271*, 28727–28730. [CrossRef]
59. Benmerah, A.; Lamaze, C.; Bègue, B.; Schmid, S.L.; Dautry-Varsat, A.; Cerf-Bensussan, N. AP-2/Eps15 interaction is required for receptor-mediated endocytosis. *J. Cell Biol.* **1998**, *140*, 1055–1062. [CrossRef]
60. Benmerah, A.; Bayrou, M.; Cerf-Bensussan, N.; Dautry-Varsat, A. Inhibition of clathrin-coated pit assembly by an Eps15 mutant. *J. Cell Sci.* **1999**, *112*, 1303–1311.
61. Hybiske, K.; Stephens, R.S. Mechanisms of host cell exit by the intracellular bacterium Chlamydia. *Proc. Natl. Acad. Sci. USA* **2007**, *104*, 11430–11435. [CrossRef] [PubMed]
62. Romero, M.D.; Mölleken, K.; Hegemann, J.H.; Carabeo, R.A. Chlamydia Adhesion and Invasion. In *Chlamydia Biology: From Genome to Disease*; Caister Academic Press: Poole, UK, 2020; ISBN 978-1-912530-28-1.
63. Kim, J.H.; Jiang, S.; Elwell, C.A.; Engel, J.N. Chlamydia trachomatis co-opts the FGF2 signaling pathway to enhance infection. *PLoS Pathog.* **2011**, *7*, e1002285. [CrossRef] [PubMed]
64. Gotoh, N. Regulation of growth factor signaling by FRS2 family docking/scaffold adaptor proteins. *Cancer Sci.* **2008**, *99*, 1319–1325. [CrossRef]
65. Birkelund, S.; Johnsen, H.; Christiansen, G. Chlamydia trachomatis serovar L2 induces protein tyrosine phosphorylation during uptake by HeLa cells. *Infect. Immun.* **1994**, *62*, 4900–4908. [PubMed]
66. Fawaz, F.S.; van Ooij, C.H.; Homola, E.L.; Mutka, S.C.; Engel, J.N. Infection with Chlamydia trachomatis alters the tyrosine phosphorylation and/or localization of several host cell proteins including cortactin. *Infect. Immun.* **1997**, *65*, 5301–5308.
67. Clifton, D.R.; Fields, K.A.; Grieshaber, S.S.; Dooley, C.A.; Fischer, E.R.; Mead, D.J.; Carabeo, R.A.; Hackstadt, T. A chlamydial type III translocated protein is tyrosine-phosphorylated at the site of entry and associated with recruitment of actin. *Proc. Natl. Acad. Sci. USA* **2004**, *101*, 10166–10171. [CrossRef]
68. Lutter, E.I.; Bonner, C.; Holland, M.J.; Suchland, R.J.; Stamm, W.E.; Jewett, T.J.; McClarty, G.; Hackstadt, T. Phylogenetic Analysis of Chlamydia trachomatis Tarp and Correlation with Clinical Phenotype. *Infect. Immun.* **2010**, *78*, 3678–3688. [CrossRef]
69. Jewett, T.J.; Miller, N.J.; Dooley, C.A.; Hackstadt, T. The Conserved Tarp Actin Binding Domain Is Important for Chlamydial Invasion. *PLoS Pathog.* **2010**, *6*, e1000997. [CrossRef]

70. Thwaites, T.; Nogueira, A.T.; Campeotto, I.; Silva, A.P.; Grieshaber, S.S.; Carabeo, R.A. The Chlamydia Effector TarP Mimics the Mammalian Leucine-Aspartic Acid Motif of Paxillin to Subvert the Focal Adhesion Kinase during Invasion. *J. Biol. Chem.* **2014**, *289*, 30426–30442. [CrossRef]
71. Thwaites, T.R.; Pedrosa, A.T.; Peacock, T.P.; Carabeo, R.A. Vinculin Interacts with the Chlamydia Effector TarP Via a Tripartite Vinculin Binding Domain to Mediate Actin Recruitment and Assembly at the Plasma Membrane. *Front. Cell. Infect. Microbiol.* **2015**, *5*, 88. [CrossRef]
72. Hynes, R.O. Integrins: Bidirectional, allosteric signaling machines. *Cell* **2002**, *110*, 673–687. [CrossRef]
73. Schaller, M.D.; Hildebrand, J.D.; Shannon, J.D.; Fox, J.W.; Vines, R.R.; Parsons, J.T. Autophosphorylation of the focal adhesion kinase, pp125FAK, directs SH2-dependent binding of pp60src. *Mol. Cell. Biol.* **1994**, *14*, 1680–1688. [CrossRef] [PubMed]
74. Parsons, J.T. Focal adhesion kinase: The first ten years. *J. Cell Sci.* **2003**, *116*, 1409–1416. [CrossRef] [PubMed]
75. Calalb, M.B.; Polte, T.R.; Hanks, S.K. Tyrosine phosphorylation of focal adhesion kinase at sites in the catalytic domain regulates kinase activity: A role for Src family kinases. *Mol. Cell. Biol.* **1995**, *15*, 954–963. [CrossRef]
76. Schlaepfer, D.D.; Hunter, T. Signal transduction from the extracellular matrix–a role for the focal adhesion protein-tyrosine kinase FAK. *Cell Struct. Funct.* **1996**, *21*, 445–450. [CrossRef]
77. Huveneers, S.; Danen, E.H.J. Adhesion signaling–crosstalk between integrins, Src and Rho. *J. Cell Sci.* **2009**, *122*, 1059–1069. [CrossRef]
78. Stallmann, S.; Hegemann, J.H. The Chlamydia trachomatis Ctad1 invasin exploits the human integrin β1 receptor for host cell entry. *Cell. Microbiol.* **2016**, *18*, 761–775. [CrossRef]
79. Igietseme, J.U.; Omosun, Y.; Stuchlik, O.; Reed, M.S.; Partin, J.; He, Q.; Joseph, K.; Ellerson, D.; Bollweg, B.; George, Z.; et al. Role of Epithelial-Mesenchyme Transition in Chlamydia Pathogenesis. *PLoS ONE* **2015**, *10*, e0145198. [CrossRef]
80. Belland, R.J.; Nelson, D.E.; Virok, D.; Crane, D.D.; Hogan, D.; Sturdevant, D.; Beatty, W.L.; Caldwell, H.D. Transcriptome analysis of chlamydial growth during IFN-γ-mediated persistence and reactivation. *Proc. Natl. Acad. Sci. USA* **2003**, *100*, 15971–15976. [CrossRef]
81. Carlson, J.H.; Porcella, S.F.; McClarty, G.; Caldwell, H.D. Comparative Genomic Analysis of Chlamydia trachomatis Oculotropic and Genitotropic Strains. *Infect. Immun.* **2005**, *73*, 6407–6418. [CrossRef]
82. Thomson, N.R.; Holden, M.T.G.; Carder, C.; Lennard, N.; Lockey, S.J.; Marsh, P.; Skipp, P.; O'Connor, C.D.; Goodhead, I.; Norbertczak, H.; et al. Chlamydia trachomatis: Genome sequence analysis of lymphogranuloma venereum isolates. *Genome Res.* **2008**, *18*, 161–171. [CrossRef] [PubMed]
83. Jewett, T.J.; Dooley, C.A.; Mead, D.J.; Hackstadt, T. Chlamydia trachomatis TarP is phosphorylated by src family tyrosine kinases. *Biochem. Biophys. Res. Commun.* **2008**, *371*, 339–344. [CrossRef] [PubMed]
84. Lane, B.J.; Mutchler, C.; Al Khodor, S.; Grieshaber, S.S.; Carabeo, R.A. Chlamydial Entry Involves TARP Binding of Guanine Nucleotide Exchange Factors. *PLoS Pathog.* **2008**, *4*, e1000014. [CrossRef] [PubMed]
85. Carabeo, R.A.; Dooley, C.A.; Grieshaber, S.S.; Hackstadt, T. Rac interacts with Abi-1 and WAVE2 to promote an Arp2/3-dependent actin recruitment during chlamydial invasion. *Cell. Microbiol.* **2007**, *9*, 2278–2288. [CrossRef] [PubMed]
86. Ghosh, S.; Park, J.; Thomas, M.; Cruz, E.; Cardona, O.; Kang, H.; Jewett, T. Biophysical characterization of actin bundles generated by the Chlamydia trachomatis Tarp effector. *Biochem. Biophys. Res. Commun.* **2018**, *500*, 423–428. [CrossRef] [PubMed]
87. McGhie, E.J.; Hayward, R.D.; Koronakis, V. Control of actin turnover by a salmonella invasion protein. *Mol. Cell* **2004**, *13*, 497–510. [CrossRef]
88. Bierne, H.; Gouin, E.; Roux, P.; Caroni, P.; Yin, H.L.; Cossart, P. A role for cofilin and LIM kinase in Listeria-induced phagocytosis. *J. Cell Biol.* **2001**, *155*, 101–112. [CrossRef]
89. Bastidas, R.J.; Valdivia, R.H. Emancipating Chlamydia: Advances in the Genetic Manipulation of a Recalcitrant Intracellular Pathogen. *Microbiol. Mol. Biol. Rev.* **2016**, *80*, 411–427. [CrossRef]
90. Wickstrum, J.; Sammons, L.R.; Restivo, K.N.; Hefty, P.S. Conditional Gene Expression in Chlamydia trachomatis using the Tet System. *PLoS ONE* **2013**, *8*, e76743. [CrossRef]
91. Ouellette, S.P. Feasibility of a Conditional Knockout System for Chlamydia Based on CRISPR Interference. *Front. Cell. Infect. Microbiol.* **2018**, *8*, 59. [CrossRef]
92. Yother, J.; Goguen, J.D. Isolation and characterization of Ca^{2+}-blind mutants of Yersinia pestis. *J. Bacteriol.* **1985**, *164*, 704–711. [PubMed]

93. Fields, K.A.; Hackstadt, T. Evidence for the secretion of Chlamydia trachomatis CopN by a type III secretion mechanism. *Mol. Microbiol.* **2000**, *38*, 1048–1060. [CrossRef] [PubMed]
94. Jamison, W.P.; Hackstadt, T. Induction of type III secretion by cell-free Chlamydia trachomatis elementary bodies. *Microb. Pathog.* **2008**, *45*, 435–440. [CrossRef] [PubMed]
95. Betts-Hampikian, H.J.; Fields, K.A. The Chlamydial Type III Secretion Mechanism: Revealing Cracks in a Tough Nut. *Front. Microbiol.* **2010**, *1*, 114. [CrossRef]
96. Ferrell, J.C.; Fields, K.A. A working model for the type III secretion mechanism in Chlamydia. *Microbes Infect.* **2016**, *18*, 84–92. [CrossRef]
97. Lee, J.J.; Kim, D.G.; Kim, D.H.; Simborio, H.L.; Min, W.; Lee, H.J.; Her, M.; Jung, S.C.; Watarai, M.; Kim, S. Interplay between Clathrin and Rab5 Controls the Early Phagocytic Trafficking and Intracellular Survival of Brucella abortus within HeLa cells. *J. Biol. Chem.* **2013**, *288*, 28049–28057. [CrossRef]
98. Hower, S.; Wolf, K.; Fields, K.A. Evidence that CT694 is a novel Chlamydia trachomatis T3S substrate capable of functioning during invasion or early cycle development. *Mol. Microbiol.* **2009**, *72*, 1423–1437. [CrossRef]
99. Chen, Y.-S.; Bastidas, R.J.; Saka, H.A.; Carpenter, V.K.; Richards, K.L.; Plano, G.V.; Valdivia, R.H. The Chlamydia trachomatis Type III Secretion Chaperone Slc1 Engages Multiple Early Effectors, Including TepP, a Tyrosine-phosphorylated Protein required for the Recruitment of CrkI-II to Nascent Inclusions and Innate Immune Signaling. *PLoS Pathog.* **2014**, *10*, e1003954. [CrossRef]
100. Benaud, C.; Gentil, B.J.; Assard, N.; Court, M.; Garin, J.; Delphin, C.; Baudier, J. AHNAK interaction with the annexin 2/S100A10 complex regulates cell membrane cytoarchitecture. *J. Cell Biol.* **2004**, *164*, 133–144. [CrossRef]
101. Haase, H.; Pagel, I.; Khalina, Y.; Zacharzowsky, U.; Person, V.; Lutsch, G.; Petzhold, D.; Kott, M.; Schaper, J.; Morano, I. The carboxyl-terminal ahnak domain induces actin bundling and stabilizes muscle contraction. *FASEB J.* **2004**, *18*, 839–841. [CrossRef]
102. McKuen, M.J.; Mueller, K.E.; Bae, Y.S.; Fields, K.A. Fluorescence-Reported Allelic Exchange Mutagenesis Reveals a Role for Chlamydia trachomatis TmeA in Invasion That Is Independent of Host AHNAK. *Infect. Immun.* **2017**, *85*, e00640-17. [CrossRef] [PubMed]
103. Lee, I.H.; Lim, H.J.; Yoon, S.; Seong, J.K.; Bae, D.S.; Rhee, S.G.; Bae, Y.S. Ahnak Protein Activates Protein Kinase C (PKC) through Dissociation of the PKC-Protein Phosphatase 2A Complex. *J. Biol. Chem.* **2008**, *283*, 6312–6320. [CrossRef] [PubMed]
104. Belland, R.J.; Scidmore, M.A.; Crane, D.D.; Hogan, D.M.; Whitmire, W.; McClarty, G.; Caldwell, H.D. Chlamydia trachomatis cytotoxicity associated with complete and partial cytotoxin genes. *Proc. Natl. Acad. Sci. USA* **2001**, *98*, 13984–13989. [CrossRef] [PubMed]
105. Aktories, K.; Schwan, C.; Jank, T. Clostridium difficile Toxin Biology. *Annu. Rev. Microbiol.* **2017**, *71*, 281–307. [CrossRef] [PubMed]
106. Chandrasekaran, R.; Lacy, D.B. The role of toxins in Clostridium difficile infection. *FEMS Microbiol. Rev.* **2017**, *41*, 723–750. [CrossRef]
107. Thalmann, J.; Janik, K.; May, M.; Sommer, K.; Ebeling, J.; Hofmann, F.; Genth, H.; Klos, A. Actin Re-Organization Induced by Chlamydia trachomatis Serovar D—Evidence for a Critical Role of the Effector Protein CT166 Targeting Rac. *PLoS ONE* **2010**, *5*, e9887. [CrossRef]
108. Bothe, M.; Dutow, P.; Pich, A.; Genth, H.; Klos, A. DXD Motif-Dependent and -Independent Effects of the Chlamydia trachomatis Cytotoxin CT166. *Toxins* **2015**, *7*, 621–637. [CrossRef]
109. Dennis, M.W.; Storz, J. Infectivity of Chlamydia psittaci of bovine and ovine origins for cultured cells. *Am. J. Vet. Res.* **1982**, *43*, 1897–1902.
110. Campbell, S.; Richmond, S.J.; Yates, P.S. The effect of Chlamydia trachomatis infection on the host cell cytoskeleton and membrane compartments. *J. Gen. Microbiol.* **1989**, *135*, 2379–2386. [CrossRef]
111. Grieshaber, S.S.; Grieshaber, N.A.; Hackstadt, T. Chlamydia trachomatis uses host cell dynein to traffic to the microtubule-organizing center in a p50 dynamitin-independent process. *J. Cell Sci.* **2003**, *116*, 3793–3802. [CrossRef]
112. Campbell, S.; Richmond, S.J.; Yates, P. The development of Chlamydia trachomatis inclusions within the host eukaryotic cell during interphase and mitosis. *J. Gen. Microbiol.* **1989**, *135*, 1153–1165. [CrossRef] [PubMed]
113. Derré, I.; Swiss, R.; Agaisse, H. The lipid transfer protein CERT interacts with the Chlamydia inclusion protein IncD and participates to ER-Chlamydia inclusion membrane contact sites. *PLoS Pathog.* **2011**, *7*, e1002092. [CrossRef] [PubMed]

114. Dumoux, M.; Clare, D.K.; Saibil, H.R.; Hayward, R.D. Chlamydiae assemble a pathogen synapse to hijack the host endoplasmic reticulum. *Traffic* **2012**, *13*, 1612–1627. [CrossRef] [PubMed]
115. Beatty, W.L. Trafficking from CD63-positive late endocytic multivesicular bodies is essential for intracellular development of Chlamydia trachomatis. *J. Cell Sci.* **2006**, *119*, 350–359. [CrossRef]
116. Beatty, W.L. Late endocytic multivesicular bodies intersect the chlamydial inclusion in the absence of CD63. *Infect. Immun.* **2008**, *76*, 2872–2881. [CrossRef]
117. Robertson, D.K.; Gu, L.; Rowe, R.K.; Beatty, W.L. Inclusion biogenesis and reactivation of persistent Chlamydia trachomatis requires host cell sphingolipid biosynthesis. *PLoS Pathog.* **2009**, *5*, e1000664. [CrossRef]
118. Kumar, Y.; Valdivia, R.H. Actin and Intermediate Filaments Stabilize the Chlamydia trachomatis Vacuole by Forming Dynamic Structural Scaffolds. *Cell Host Microbe* **2008**, *4*, 159–169. [CrossRef]
119. Chin, E.; Kirker, K.; Zuck, M.; James, G.; Hybiske, K. Actin Recruitment to the Chlamydia Inclusion Is Spatiotemporally Regulated by a Mechanism That Requires Host and Bacterial Factors. *PLoS ONE* **2012**, *7*, e46949. [CrossRef]
120. Dong, F.; Su, H.; Huang, Y.; Zhong, Y.; Zhong, G. Cleavage of host keratin 8 by a Chlamydia-secreted protease. *Infect. Immun.* **2004**, *72*, 3863–3868. [CrossRef]
121. Savijoki, K.; Alvesalo, J.; Vuorela, P.; Leinonen, M.; Kalkkinen, N. Proteomic analysis of Chlamydia pneumoniae-infected HL cells reveals extensive degradation of cytoskeletal proteins. *FEMS Immunol. Med. Microbiol.* **2008**, *54*, 375–384. [CrossRef]
122. Nogueira, A.T.; Pedrosa, A.T.; Carabeo, R.A. Manipulation of the Host Cell Cytoskeleton by Chlamydia. *Curr. Top. Microbiol. Immunol.* **2018**, *412*, 59–80. [PubMed]
123. Todd, W.J.; Caldwell, H.D. The interaction of Chlamydia trachomatis with host cells: Ultrastructural studies of the mechanism of release of a biovar II strain from HeLa 229 cells. *J. Infect. Dis.* **1985**, *151*, 1037–1044. [CrossRef] [PubMed]
124. Lutter, E.I.; Barger, A.C.; Nair, V.; Hackstadt, T. Chlamydia trachomatis inclusion membrane protein CT228 recruits elements of the myosin phosphatase pathway to regulate release mechanisms. *Cell Rep.* **2013**, *3*, 1921–1931. [CrossRef] [PubMed]
125. Mital, J.; Miller, N.J.; Fischer, E.R.; Hackstadt, T. Specific chlamydial inclusion membrane proteins associate with active Src family kinases in microdomains that interact with the host microtubule network. *Cell. Microbiol.* **2010**, *12*, 1235–1249. [CrossRef] [PubMed]
126. Matsumura, F.; Hartshorne, D.J. Myosin phosphatase target subunit: Many roles in cell function. *Biochem. Biophys. Res. Commun.* **2008**, *369*, 149–156. [CrossRef] [PubMed]
127. Shaw, J.H.; Key, C.E.; Snider, T.A.; Sah, P.; Shaw, E.I.; Fisher, D.J.; Lutter, E.I. Genetic Inactivation of Chlamydia trachomatis Inclusion Membrane Protein CT228 Alters MYPT1 Recruitment, Extrusion Production, and Longevity of Infection. *Front. Cell. Infect. Microbiol.* **2018**, *8*, 415. [CrossRef]
128. Nguyen, P.H.; Lutter, E.I.; Hackstadt, T. Chlamydia trachomatis inclusion membrane protein MrcA interacts with the inositol 1,4,5-trisphosphate receptor type 3 (ITPR3) to regulate extrusion formation. *PLoS Pathog.* **2018**, *14*, e1006911. [CrossRef]
129. Volceanov, L.; Herbst, K.; Biniossek, M.; Schilling, O.; Haller, D.; Nölke, T.; Subbarayal, P.; Rudel, T.; Zieger, B.; Häcker, G. Septins Arrange F-Actin-Containing Fibers on the Chlamydia trachomatis Inclusion and Are Required for Normal Release of the Inclusion by Extrusion. *MBio* **2014**, *5*, e01802-14. [CrossRef]
130. Mostowy, S.; Cossart, P. Septins: The fourth component of the cytoskeleton. *Nat. Rev. Mol. Cell Biol.* **2012**, *13*, 183–194. [CrossRef]
131. Geneste, O.; Copeland, J.W.; Treisman, R. LIM kinase and Diaphanous cooperate to regulate serum response factor and actin dynamics. *J. Cell Biol.* **2002**, *157*, 831–838. [CrossRef]
132. Zhao, X.-H.; Laschinger, C.; Arora, P.; Szászi, K.; Kapus, A.; McCulloch, C.A. Force activates smooth muscle alpha-actin promoter activity through the Rho signaling pathway. *J. Cell Sci.* **2007**, *120*, 1801–1809. [CrossRef] [PubMed]
133. Zhao, L.; Chi, L.; Zhao, J.; Wang, X.; Chen, Z.; Meng, L.; Liu, G.; Guan, G.; Wang, F. Serum response factor provokes epithelial-mesenchymal transition in renal tubular epithelial cells of diabetic nephropathy. *Physiol. Genom.* **2016**, *48*, 580–588. [CrossRef] [PubMed]
134. Zhao, X.; He, L.; Li, T.; Lu, Y.; Miao, Y.; Liang, S.; Guo, H.; Bai, M.; Xie, H.; Luo, G.; et al. SRF expedites metastasis and modulates the epithelial to mesenchymal transition by regulating miR-199a-5p expression in human gastric cancer. *Cell Death Differ.* **2014**, *21*, 1900–1913. [CrossRef] [PubMed]

135. Zadora, P.K.; Chumduri, C.; Imami, K.; Berger, H.; Mi, Y.; Selbach, M.; Meyer, T.F.; Gurumurthy, R.K. Integrated Phosphoproteome and Transcriptome Analysis Reveals Chlamydia-Induced Epithelial-to-Mesenchymal Transition in Host Cells. *Cell Rep.* **2019**, *26*, 1286–1302. [CrossRef] [PubMed]
136. Igietseme, J.U.; Omosun, Y.; Nagy, T.; Stuchlik, O.; Reed, M.S.; He, Q.; Partin, J.; Joseph, K.; Ellerson, D.; George, Z.; et al. Molecular Pathogenesis of Chlamydia Disease Complications: Epithelial-Mesenchymal Transition and Fibrosis. *Infect. Immun.* **2018**, *86*, e00585-17. [CrossRef] [PubMed]
137. Rajić, J.; Inic-Kanada, A.; Stein, E.; Dinić, S.; Schuerer, N.; Uskoković, A.; Ghasemian, E.; Mihailović, M.; Vidaković, M.; Grdović, N.; et al. Chlamydia trachomatis Infection Is Associated with E-Cadherin Promoter Methylation, Downregulation of E-Cadherin Expression, and Increased Expression of Fibronectin and α-SMA—Implications for Epithelial-Mesenchymal Transition. *Front. Cell. Infect. Microbiol.* **2017**, *7*, 253. [CrossRef]
138. Thorpe, S.D.; Lee, D.A. Dynamic regulation of nuclear architecture and mechanics—A rheostatic role for the nucleus in tailoring cellular mechanosensitivity. *Nucleus* **2017**, *8*, 287–300. [CrossRef]
139. Lee, K.K.; Haraguchi, T.; Lee, R.S.; Koujin, T.; Hiraoka, Y.; Wilson, K.L. Distinct functional domains in emerin bind lamin A and DNA-bridging protein BAF. *J. Cell Sci.* **2001**, *114*, 4567–4573.
140. Guilluy, C.; Osborne, L.D.; Van Landeghem, L.; Sharek, L.; Superfine, R.; Garcia-Mata, R.; Burridge, K. Isolated nuclei adapt to force and reveal a mechanotransduction pathway in the nucleus. *Nat. Cell Biol.* **2014**, *16*, 376–381. [CrossRef]
141. Ho, C.Y.; Jaalouk, D.E.; Vartiainen, M.K.; Lammerding, J. Lamin A/C and emerin regulate MKL1-SRF activity by modulating actin dynamics. *Nature* **2013**, *497*, 507–511. [CrossRef]
142. Plessner, M.; Melak, M.; Chinchilla, P.; Baarlink, C.; Grosse, R. Nuclear F-actin formation and reorganization upon cell spreading. *J. Biol. Chem.* **2015**, *290*, 11209–11216. [CrossRef] [PubMed]
143. Kim, M.; Ogawa, M.; Mimuro, H.; Sasakawa, C. Reinforcement of epithelial cell adhesion to basement membrane by a bacterial pathogen as a new infectious stratagem. *Virulence* **2010**, *1*, 52–55. [CrossRef] [PubMed]
144. Kim, M.; Ogawa, M.; Fujita, Y.; Yoshikawa, Y.; Nagai, T.; Koyama, T.; Nagai, S.; Lange, A.; Fässler, R.; Sasakawa, C. Bacteria hijack integrin-linked kinase to stabilize focal adhesions and block cell detachment. *Nature* **2009**, *459*, 578–582. [CrossRef] [PubMed]

© 2019 by the authors. Licensee MDPI, Basel, Switzerland. This article is an open access article distributed under the terms and conditions of the Creative Commons Attribution (CC BY) license (http://creativecommons.org/licenses/by/4.0/).

MDPI
St. Alban-Anlage 66
4052 Basel
Switzerland
Tel. +41 61 683 77 34
Fax +41 61 302 89 18
www.mdpi.com

International Journal of Molecular Sciences Editorial Office
E-mail: ijms@mdpi.com
www.mdpi.com/journal/ijms

www.ingramcontent.com/pod-product-compliance
Lightning Source LLC
LaVergne TN
LVHW070507100526
838202LV00014B/1803